Fundamentals of Global Air Transport Geography

The commercial air transport industry can be broadly split into three component parts: airlines, airports and aircraft. Each of these components is shaped by geography, insofar as each is influenced by places, landscapes, environments, people and their various interactions. Conversely, air transport plays a large role in shaping the various themes of geography and the position of our physical, human and environmental world. It connects people, cultures and businesses across every continent and generates economic growth, allows international trade to occur and develops tourism. It can also be involved in creating negative outputs, such as emissions, noise and loss of biodiversity, which can have a large impact on the planet and quality of life. A lack of air transport links can also have a significantly negative impact on world regions in terms of economic and cultural development. In short, air transport and geography are inextricably linked.

Fundamentals of Global Air Transport Geography details the geography of the global commercial air transport industry. The book aims to provide an understanding of these key areas at an introductory level, in order to be accessible to students and non-technical airport/airline management. A key theme throughout the book will not only be how geographical issues have influenced air transport, but also how air transport continues to influence geography. Each chapter boasts a range of features aimed at enhancing the reader's understanding, including learning objectives, discussion questions and case studies, and lecturers can find supporting resources including PowerPoint slides and teaching notes online.

George Arbuckle is a Senior Lecturer and Course Leader at the University of West London. He has been teaching on Air Transport Management and Pilot Training programmes since 2010 and has worked in a range of management roles within the airline industry, including Inflight Product Manager with Monarch Airlines. He has a degree in Geography from the University of Strathclyde in Glasgow, an MSc in Air Transport from Cranfield University, and is a Fellow of the Higher Education Academy, a Fellow of the Royal Geographical Society and a Member of the American Association of Geographers. His first airline responsibility was in 1997, checking in a flight from London Luton to Glasgow for a tiny airline called easyJet – at the time operating five aircraft!

Aviation Fundamentals
Series Editor: Suzanne K. Kearns

Aviation Fundamentals is a series of air transport textbooks that incorporate instructional design principles to present content in a manner that is engaging to the learner, at an accessible level for young adults, allowing for practical application of the content to real-world problems via cases, reflection questions and examples. Each textbook will be supported by a companion website of supplementary materials and a test bank. The series is designed to help facilitate the recruitment and education of the next generation of aviation professionals (NGAP), a task which has been named a 'Global Priority' by the ICAO Assembly. It will also support education for new air transport sectors that are expected to rapidly evolve in future years, such as commercial space and the civil use of remotely piloted aircraft. The objective of *Aviation Fundamentals* is to become the leading source of textbooks for the variety of subject areas that make up aviation college/university degree programmes, evolving in parallel with these curricula.

Fundamentals of Airline Marketing
Scott Ambrose and Blaise Waguespack

Fundamentals of Statistics for Aviation Research
Michael A. Gallo, Brooke E. Wheeler and Isaac M. Silver

Fundamentals of Airport Planning
Theory and Practice
Ravi Lakshmanan

Fundamentals of Sustainable Aviation
Eva Maleviti

Fundamentals of Aviation Crisis and Emergency Management
Gail A. Rowntree

Fundamentals of International Aviation Law and Policy
Second Edition
Benjamyn I. Scott and Andrea Trimarchi

Fundamentals of Global Air Transport Geography
George Arbuckle

For more information about this series, please visit: www.routledge.com/Aviation-Fundamentals/book-series/AVFUND

Fundamentals of Global Air Transport Geography

George Arbuckle

Routledge
Taylor & Francis Group

LONDON AND NEW YORK

Designed cover image: Getty Images / Mongkol Chuewong

First published 2025
by Routledge
4 Park Square, Milton Park, Abingdon, Oxon OX14 4RN

and by Routledge
605 Third Avenue, New York, NY 10158

Routledge is an imprint of the Taylor & Francis Group, an informa business

© 2025 George Arbuckle

The right of George Arbuckle to be identified as author of this work has been asserted in accordance with sections 77 and 78 of the Copyright, Designs and Patents Act 1988.

All rights reserved. No part of this book may be reprinted or reproduced or utilised in any form or by any electronic, mechanical, or other means, now known or hereafter invented, including photocopying and recording, or in any information storage or retrieval system, without permission in writing from the publishers.

Trademark notice: Product or corporate names may be trademarks or registered trademarks, and are used only for identification and explanation without intent to infringe.

British Library Cataloguing-in-Publication Data
A catalogue record for this book is available from the British Library

ISBN: 978-1-032-52126-8 (hbk)
ISBN: 978-1-032-52125-1 (pbk)
ISBN: 978-1-003-40535-1 (ebk)

DOI: 10.4324/9781003405351

Typeset in Times New Roman
by codeMantra

Access the Support Material at www.routledge.com/9781032521251

For Ina, George Sr. and Claire
Taken too soon.

Contents

List of figures xix
List of tables xxv
List of abbreviations xxvii
List of ATG case studies xxxiii
List of ATG trivia xxxvii
Foreword xxxix
Acknowledgements xli

PART I
Introduction to geography and air transport 1

1 Introduction to geography and transport 3
   Chapter outcomes 3
   1.1 Introduction 3
   1.2 Physical geography 5
      1.2.1 Meteorology 6
      1.2.2 Climatology 7
      1.2.3 Geomorphology and landforms 10
      1.2.4 Geophysical hazards 11
      1.2.5 Biogeography 11
      1.2.6 Hydrology 12
   1.3 Human geography 12
      1.3.1 Political geography 13
      1.3.2 Economic geography 14
      1.3.3 Urban geography 15
      1.3.4 Rural geography 17
      1.3.5 Population geography 18
   1.4 Environmental geography 18
      1.4.1 Geomorphic change 19
      1.4.2 Climate change 20
      1.4.3 Air transport noise 22
   1.5 Geography of transport 22

viii   *Contents*

   1.5.1 Concept of space 22
   1.5.2 The elements of transport 23
   1.5.3 Core principles of transport geography 23
   Chapter review questions 25
   ATG trivia: airport codes 26
   References 27

2   The air transport industry 30
   Chapter outcomes 30
   2.1 Introduction 30
   2.2 The origins of air transport 32
      2.2.1 Paris Convention (1919) 32
      2.2.2 Chicago Convention (1944) 33
      2.2.3 ICAO 34
   2.3 The aviation value chain 36
   2.4 The airline industry 36
      2.4.1 Airline industry performance 38
      2.4.2 Airline business models 39
   2.5 Aircraft 48
      2.5.1 Aircraft and engine historical developments 49
      2.5.2 Types of airliner aircraft 53
      2.5.3 Aircraft manufacturers 54
      2.5.4 Aircraft performance and payload range 57
      2.5.5 Aircraft and the future 58
   2.6 Airports 59
      2.6.1 Types of airports 59
      2.6.2 An evolving industry 62
      2.6.3 Airport constraints 63
   2.7 Airspace 65
      2.7.1 Airspace horizontal boundary 65
      2.7.2 Airspace vertical boundary 65
      2.7.3 Flight information regions 66
      2.7.4 Types of airspace 67
      2.7.5 Conclusion 69
   Chapter review questions 69
   ATG trivia: great circle route 70
   References 72

PART II
Physical geography and air transport   75

3   Meteorology and air transport: the atmosphere   77
   Chapter outcomes 77
   3.1 Introduction 77
   3.2 Principles of flight 78
     3.2.1 Newton's laws of motion 78
     3.2.2 The four forces of flight 79
     3.2.3 Aeroplane flight controls 80
     3.2.4 Centre of gravity (CG) 82
   3.3 Components of the atmosphere 82
   3.4 Layers of the atmosphere 83
     3.4.1 Troposphere 83
     3.4.2 Stratosphere 85
     3.4.3 Mesosphere 86
     3.4.4 Thermosphere 86
     3.4.5 Exosphere 86
   3.5 International Standard Atmosphere (ISA) 86
   3.6 Temperature 88
     3.6.1 Defining temperature 88
     3.6.2 Temperature measurement and units 88
     3.6.3 Solar and terrestrial radiation 88
     3.6.4 Conduction 91
     3.6.5 Convection 91
     3.6.6 Advection 91
     3.6.7 Surface temperature variations 92
   3.7 Atmospheric pressure and density 92
     3.7.1 Measurement and units 93
     3.7.2 High and low pressure 93
     3.7.3 Density's effects on pressure 94
     3.7.4 The effects of temperature and water vapour on density 94
   3.8 Altimetry 95
     3.8.1 Types of altitude 95
     3.8.2 QNH 96
     3.8.3 QFE 97
     3.8.4 Standard pressure 97
     3.8.5 Altimeter errors and corrections 97
   Chapter review questions 98
   ATG trivia: hot and cold 98
   References 99

4 Meteorology and air transport: water vapour 101
Chapter outcomes 101
4.1 Introduction 101
4.2 Hydrologic (water) cycle 102
    4.2.1 Evaporation 102
    4.2.2 Transpiration 103
    4.2.3 Condensation 103
    4.2.4 Precipitation 103
    4.2.5 Runoff 103
    4.2.6 Infiltration 103
    4.2.7 Groundwater flow 103
    4.2.8 Plant uptake 103
    4.2.9 Sublimation 104
    4.2.10 Deposition 104
4.3 Humidity 104
    4.3.1 Absolute humidity 104
    4.3.2 Relative humidity 105
    4.3.3 Dewpoint 106
    4.3.4 Latent heat 107
4.4 Atmospheric stability 108
    4.4.1 Adiabatic lapse rate 108
    4.4.2 Types of atmospheric stability 109
4.5 Clouds 110
    4.5.1 Cloud classification 110
    4.5.2 Cloud types – high 113
    4.5.3 Cloud types – middle 114
    4.5.4 Cloud types – low 114
    4.5.5 Cloud formation 116
4.6 Precipitation 118
    4.6.1 Precipitation growth 118
    4.6.2 Types of precipitation 118
4.7 Hazards – visibility 119
    4.7.1 Obstructions to visibility 120
    4.7.2 Precipitation 120
    4.7.3 Dust storm and sandstorm 120
    4.7.4 Fog 120
4.8 Hazards – icing 122
    4.8.1 Airframe icing 123
    4.8.2 Engine icing 123
    4.8.3 Aviation hazards associated with icing 124
    4.8.4 Aircraft ground de-/anti-icing 124
Chapter review questions 126
ATG trivia: cabin humidity 126
References 127

5 Meteorology and air transport: wind — 129

Chapter outcomes 129
5.1 Introduction 129
5.2 Wind direction and speed 130
   5.2.1 Defining and measuring wind direction 130
   5.2.2 Defining and measuring wind speed 130
   5.2.3 Headwinds 133
   5.2.4 Tailwinds 133
   5.2.5 Crosswinds 134
   5.2.6 Wind direction and runway orientation 134
5.3 Forces affecting wind 137
   5.3.1 Pressure gradient force 137
   5.3.2 Coriolis force 138
   5.3.3 Centripetal acceleration 139
   5.3.4 Friction force 139
5.4 Wind types 140
   5.4.1 Geostrophic wind 140
   5.4.2 Cyclostrophic wind 141
   5.4.3 Gradient wind 141
5.5 Local winds 141
   5.5.1 Sea and land breezes 141
   5.5.2 Valley and mountain winds 142
   5.5.3 Foehn winds 144
5.6 Hazards: turbulence 145
   5.6.1 Convective (thermal) turbulence 146
   5.6.2 Mechanical turbulence 146
   5.6.3 Wind shear 148
   5.6.4 Clear air turbulence 149
   5.6.5 Wake vortex turbulence 149
5.7 Hazards: thunderstorms 150
   5.7.1 Thunderstorm life cycle 150
   5.7.2 Thunderstorm hazards to aviation 151
Chapter review questions 153
ATG trivia: how hot do aircraft get? 154
References 154

6 Climatology and air transport — 157

Chapter outcomes 157
6.1 Introduction 157
6.2 Global atmospheric circulation 158
   6.2.1 The three-cell circulation model 159
   6.2.2 The effects of the three-cell circulation 160
6.3 Air masses 163

6.4 Fronts 164
   6.4.1 Cold fronts 164
   6.4.2 Warm fronts 165
   6.4.3 Stationary fronts 165
   6.4.4 Occluded fronts 165
6.5 Jet streams 166
   6.5.1 Jet stream impacts on air transport 167
6.6 Tropical cyclones 168
   6.6.1 Conditions required for tropical cyclone formation 168
   6.6.2 Location of tropical cyclones 168
   6.6.3 Classification of tropical cyclones 170
   6.6.4 Tropical cyclones and their related hazards 170
6.7 Climate change 172
   6.7.1 Climate change and greenhouse gases 172
   6.7.2 Climate change history 172
   6.7.3 Contemporary climate change 173
   6.7.4 IPCC Sixth Assessment Report (AR6) on Climate Change 175
6.8 The impacts of climate change on the air transport industry 175
   6.8.1 How climate change impacts air transport 176
   6.8.2 Impact on airports 177
   6.8.3 Disruption in the air 180
   6.8.4 Disruption to passenger demand 181
Chapter review questions 182
ATG trivia: hurricane hunters 183
References 183

7 Landforms, airports and geophysical hazards 187

Chapter outcomes 187
7.1 Introduction 187
7.2 An overview of geomorphology and landforms 188
   7.2.1 Spatial scales 188
   7.2.2 Landform classification 189
7.3 Physical geography of airports 189
   7.3.1 The early airports 190
7.4 International airport regulations 191
   7.4.1 ICAO Annex 14 – aerodromes 191
   7.4.2 Aerodrome reference codes 192
7.5 Runways 193
   7.5.1 Runway surfaces 195
   7.5.2 Unpaved runway surfaces 195
   7.5.3 Paved runways 199
   7.5.4 Tabletop/mountain runways 201
   7.5.5 Waterways 203
   7.5.6 Runway slopes 207

7.6 Natural hazards 208
7.7 Geophysical hazards – earthquakes 209
   7.7.1 Earthquake hazards and airports 209
   7.7.2 Airport earthquake mitigation 210
7.8 Geophysical hazards – tsunamis 211
   7.8.1 Tsunami hazards and airports 211
   7.8.2 Airport tsunami mitigation 213
7.9 Geophysical hazards – volcanoes 214
   7.9.1 Volcanic hazards and air transport 215
   7.9.2 Volcanic mitigation 217
Chapter review questions 218
ATG trivia: the Galunggung Glider 218
References 219

8 Biogeography, hydrology and air transport   221
Chapter outcomes 221
8.1 Introduction 221
8.2 Biogeography and air transport 222
   8.2.1 Habitat modification and fragmentation 222
8.3 Wildlife management at airports 223
   8.3.1 Non-avian wildlife hazards 223
   8.3.2 Bird strikes 224
   8.3.3 Wildlife hazard mitigation 227
8.4 Biodiversity at airports 228
8.5 Biological hazards 230
   8.5.1 Disease transmission 230
   8.5.2 Air transport and pandemics 231
   8.5.3 Severe Acute Respiratory Syndrome (SARS) 231
   8.5.4 COVID-19 232
8.6 Hydrology 234
   8.6.1 Regulations 235
   8.6.2 Hydrology and air transport 236
8.7 Water supply at airports 236
8.8 Water handling capacity at airports 237
   8.8.1 Airport drainage systems 237
   8.8.2 Hydrological hazards – coastal flooding 238
   8.8.3 Hydrological hazards – river flooding 239
   8.8.4 Landslides 240
8.9 Water disposal at airports 241
   8.9.1 De-icing/anti-icing 241
   8.9.2 Disposal management 242
Chapter review questions 234
ATG trivia: the "Rain Vortex" at the Jewel Changi Airport, Singapore 244
References 244

PART III
Human geography and air transport — 249

9 Political geography and air transport — 251
Chapter outcomes 251
9.1 Introduction 251
9.2 Nationality and sovereignty 252
9.3 Multinational regulations: Chicago Convention 254
  9.3.1 ICAO 254
  9.3.2 Freedoms of the Air 255
9.4 Bilateral air service agreements 258
  9.4.1 Bilateral provisions 259
9.5 USA – airline deregulation 260
  9.5.1 US open skies 261
9.6 Europe 262
  9.6.1 The three packages 262
  9.6.2 European Common Aviation Area (ECAA) 263
  9.6.3 EU/non-EU Open Skies 263
9.7 Latin America and the Caribbean 264
9.8 Middle East 266
9.9 Asia Pacific 268
9.10 Africa 269
  9.10.1 Yamoussoukro Decision (YD) 269
  9.10.2 Single African Air Transport Market (SAATM) 270
Chapter review questions 271
ATG trivia: time travel: GMT, UTC and local times 272
References 272

10 Economic geography and air transport — 275
Chapter outcomes 275
10.1 Introduction 275
10.2 Factors influencing air transport growth 276
  10.2.1 Economic growth 276
  10.2.2 Income growth 278
  10.2.3 Globalisation 279
  10.2.4 Liberalisation 279
  10.2.5 Reduction in airfares 279
  10.2.6 Political stability and security 280
10.3 Aviation value chain 280
  10.3.1 Airlines 282
  10.3.2 Aviation fuel production 282
  10.3.3 Manufacturers 283
10.4 Economic performance of the airline industry 284
  10.4.1 Characteristics of airline operations 284

Figures

0.1	Map illustrating the global locations of the case studies in the book	xlii
1.1	The branches of geography	4
1.2	Subdisciplines of physical geography	5
1.3a	El Niño and La Niña	8
1.3b	El Niño impacts on the US in winter	9
1.4	Denver International Airport layout	11
1.5	Subdisciplines of human geography	13
1.6a	Global rural population as % of total population, 1960–2022	17
1.6b	Rural population as % of total – selected countries, 2022	17
1.7a	Location of Chek Lap Kok Airport	20
1.7b	Chek Lap Kok Airport three-runway system	20
1.8	Point-to-point and hub-and-spoke networks	24
1.9	ICAO four-letter codes – first letter principle	26
2.1	The Five Freedoms of the Air	34
2.2	The commercial aviation value chain	37
2.3	Revenues in the aviation value chain	37
2.4	Airline business models	40
2.5	Qantas route developments between London and Sydney from 1947	55
2.6	Airbus and Boeing aircraft on passengers versus range	58
2.7	Boeing commercial aircraft forecast, 2023–2042	59
2.8a	Total annual airline passengers carried, 2000–2019: China and USA	64
2.8b	Number of civil airports in China	65
2.9	Flight information regions	67
2.10	Great Circle Route between New York and Madrid	71
2.11	Great Circle Route between Tokyo and Louisville, via Anchorage	71
3.1	The four forces of flight	79
3.2	Aircraft wing and lift	79
3.3	Aeroplane parts, flight control surfaces and functions	81
3.4	Flight control surfaces and the three axes of rotation	81
3.5	The basic layers of the atmosphere	84
3.6	Average temperature profile for the lower layers of the atmosphere	85
3.7	Stevenson screen	89
3.8	Solar radiation at the high and low latitudes	90
3.9	Radiation, conduction and convection	91
3.10	Pressure's effects on density in the atmosphere	94
3.11	Pressure, temperature and aircraft performance	95

xx *Figures*

3.12	Altimeter pressure settings	96
4.1	The hydrologic cycle	102
4.2	Temperature effects on relative humidity	106
4.3	Example METAR at London Heathrow and Phoenix Sky Harbor	107
4.4	Latent heat transactions	108
4.5	Cloud types	111
4.6	High-level clouds	113
4.7	Mid-level clouds	114
4.8	Low-level clouds	115
4.9	Cloud trigger mechanisms	117
4.10	Orographic cloud on Athos, Greece	117
4.11	Radiation and advection fog	121
4.12	Orographic, frontal and steam fog	121
4.13	Hazards to the four forces of flight due to icing	124
4.14	Deicing of aircraft	125
5.1	Wind directions in air transport	131
5.2	Windsock and anemometer at Old Warden Aerodrome, Shuttleworth, Bedfordshire	132
5.3	Headwinds and tailwinds	133
5.4a	Wind Rose for Mariscal Sucre International Airport, Quito, between 30/6/1977 and 15/11/2015	136
5.4b	Mariscal Sucre runway orientation	136
5.5a	Wind Rose for Newark Liberty International Airport, between 01/01/1970 and 24/07/2023	137
5.5b	Newark Liberty runway orientation	137
5.6	Isobars, pressure gradient and wind strength	138
5.7	Coriolis effect	139
5.8a	Geostrophic wind formation	140
5.8b	Real winds due to surface friction	140
5.9a	Sea breeze	142
5.9b	Land breeze	142
5.10a	Anabatic winds	143
5.10b	Katabatic winds	143
5.11	Foehn winds	145
5.12	Thermal/convective turbulence	147
5.13	Mechanical turbulence	147
5.14	The old Kai Tak Airport and Checkerboard Hill location	148
5.15	Wake vortex turbulence	149
5.16	The three stages in a thunderstorm life cycle	151
6.1	Global atmospheric circulation	160
6.2a	Location of Lagos and Kano, Nigeria	162
6.2b	Monthly normal rainfall for Lagos and Kano	162
6.3	Air masses affecting the UK and mainland Europe	163
6.4	Weather fronts	165
6.5	Location of jet streams	167
6.6	Global tropical cyclone formation basins	169
6.7	Hurricane Maria, 24 September, 2017	170
6.8	Atmospheric carbon dioxide levels	174
6.9	Temperature anomalies in 2022	175

6.10	Global case study locations for airport climate change impacts	177
7.1	Spatial scale landforms	189
7.2	Dallas Fort Worth International Airport	194
7.3	Gravel kit example	196
7.4	Barra Airport, Scotland	198
7.5a	Antarctic runways	199
7.5b	First Boeing 787 lands in Antarctica	199
7.6	Pavement materials used for the new runway/taxiways at Brisbane Airport	201
7.7a	Talcha Airport, Nepal	202
7.7b	Sedona Airport, Arizona	202
7.8	Juancho E. Yrausquin Airport, Saba	203
7.9	Benoist first airline take-off	204
7.10a	Map of the Maldives	205
7.10b	Seaplanes in the Maldives	206
7.11	Courchevel Altiport, France	208
7.12	Tsunami-vulnerable airports	212
7.13	Sendai Airport, Japan, March 13, 2011	214
7.14a	Damage from volcanic ashfall at Clark Air Force Base, Philippines	216
7.14b	World Airways DC10 sitting on its tail due to the weight of wet volcanic ash	216
7.15a	Iceland location	217
7.15b	Location of ash cloud on April 16, 2010	217
8.1	Kimberley Airport, South Africa	224
8.2	Locations of bird-strike damage	225
8.3a	US Airways Flight 1549 on Hudson River	226
8.3b	Canada geese	226
8.4	Total global passenger numbers, 2010–2022	233
8.5	Global airline passenger and cargo revenues, 2016–2022	233
8.6	Bangkok Don Mueang Airport, 31 October, 2011	240
8.7	De-icing area at Oslo Gardermoen Airport	243
8.8	The Rain Vortex at Jewel Changi	244
9.1	Approximate rerouting of Qatar Airways flight from Doha to Khartoum, June 2017	254
9.2	The Nine Freedoms of the Air	255
9.3	Normal route between Tokyo and London for Japan Airlines Flight 43 versus route avoiding Russian airspace, 4 March, 2022	257
10.1	The 20 countries with the largest GDP in 2022	276
10.2	Annual world % GDP growth, 1982–2022	277
10.3	Annual GDP growth per country, 2022	278
10.4	Annual US domestic average itinerary fare in constant 2023 US dollars	280
10.5	Aviation value chain revenue by subsector, 2022	281
10.6	US Gulf Coast kerosene-type jet fuel spot price FOB, 1990–2024	282
10.7	Airline RPKs, 2004–2023	286
10.8	% annual change in airline yield, 2005–2023	287
10.9	Total revenue of airlines worldwide, 2004–2023	288
10.10	CarTrawler Worldwide estimate of ancillary revenue, 2014–2023	288
10.11	Top 20 airlines for revenue and comparative RASK, 2023	291
10.12	Average passenger load factor of commercial airlines, 2012–2023, with Ryanair as a comparison	292
10.13	Evolution of real airline yield and unit cost, 1970–2023	292

10.14	Labour costs as % of total costs, 2023	293
10.15	RASK and CASK for the top 20 airlines by revenue, 2023	294
10.16	CASK for select LCCs, 2023	295
10.17	Operating profit and operating profit margin for airlines worldwide, 2010–2023	296
10.18	Operating margins for the top 20 airlines for revenue, 2023	297
10.19	Circular and symbiotic relationship between economic growth and air connectivity	301
10.20	Air connectivity and growth rates by region, 2019 versus 2014	303
10.21	Top 20 most connected countries in the world and growth rates, 2019 versus 2014	304
10.22	Top 20 most connected countries for air connectivity relative to population size and growth rates, 2019 versus 2014	304
10.23	Top three cities for connectivity per world region, 2019	305
10.24	Air transport liberalisation benefits	307
10.25	New aircraft demand to 2042	308
11.1	The earth at night	314
11.2	The airports of London	317
11.3	Location of the old and new Mariscal Sucre Airports, Quito	319
11.4	Don Mueang, Suvarnabhumi and U-Tapao Rayong-Pattaya International Airports, Thailand	324
11.5a	Number of people living in urban and rural areas	325
11.5b	Share of people living in urban and rural areas	325
11.6	Location of Bogotá and Medellín, Colombia	326
11.7	Darién National Park, Panama	328
11.8	Small Island Developing States and Associate Members of United Nations Regional Commissions	330
11.9	East Malaysia	332
11.10	Urban air mobility concepts	335
12.1	Map of Indonesia	342
12.2	International migrants at mid-year 2020	348
12.3	Location of Kiribati and Baker and Howland Islands	354
13.1a	Palm Jumeirah, Dubai	363
13.1b	Pearl Island, Doha	363
13.2a	Location of Kansai International and Kobe Airports	364
13.2b	Detailed location of Kansai International Airport	364
13.3a	Hong Kong International Airport, Chek Lap Kok	366
13.3b	The location of Chek Lap Kok and Kowloon	366
13.4	Toncontin Airport and surrounding terrain	368
13.5a	Runway at Madeira Airport with extension	369
13.5b	Madeira runway extension bridge and pillars	369
13.6a	The location of Sangster International Airport on Jamaica	370
13.6b	Runway extension project area	370
13.7a	The runway layout at AMS	374
13.7b	Noise-deflecting ridges at Buitenschot Land Art Park	374
13.8	Aerial view of the aircraft boneyard at Davis-Monthan Air Force Base, in Tucson, Arizona	377
14.1	Global carbon dioxide emissions from aviation, 1940–2018	382
14.2	World air passenger traffic evolution, 1980–2020	383

14.3	HKIA Airport carbon footprint breakdown, 2021	386
14.4	Change in effective radiative forcing from 1750 to 2019	387
14.5	Aircraft contrails	388
14.6	Aircraft contrails from space	389
14.7	High bypass-ratio turbofan engine	390
14.8	Composites as a % of aircraft structural weight	391
14.9	Aircraft wingtip devices	392
14.10	CDO and CCO	395
14.11	IATA strategy towards net zero 2050	398
14.12	Futuristic blended wing body aircraft	399
14.13	The role of hydrogen in reaching net zero in aviation	400
15.1	Most common health effects of aircraft noise exposure	411
15.2a	Jet engine	414
15.2b	Jet engine components	415
15.3	Low-bypass and high-bypass ratio engines	415
15.4	Sources of aircraft noise	416
15.5	Comparative dBA noise sources	419
15.6	ICAO Annex 16 timeline	422
15.7	ICAO Balanced Approach to Aircraft Noise Management	425
15.8a	An early chevron test article with symmetrical notches	427
15.8b	An early chevron test article inside an engine nozzle	427
15.8c	Jet engine nozzle	428

Tables

1.1	Landform changes due to airport building	19
2.1	Annexes 1–19 of the Chicago Convention	34
2.2	Example oldest airlines still operating	37
2.3	Top ten airline groups by passenger traffic (RPKs), 2022	39
2.4	Top ten cargo airlines by CTKs, 2022	39
2.5	FSNC and LCC characteristics	41
2.6	Global share of LCC ASKs, 12 months to August 2023	42
2.7	Airline differences from the "traditional" FSNC or LCC business models	44
2.8	Key developments in commercial aircraft development, 1783–1947	50
2.9	Key developments in commercial aircraft development, 1949–2023	51
2.10	Boeing aircraft currently in production	55
2.11	Airbus aircraft currently in production	56
2.12	Top ten airports for passenger numbers, 2023	61
2.13	Top ten airports for aircraft movements, 2023	61
2.14	Top ten airports for cargo, 2023	62
3.1	Newton's laws of motion	78
3.2	Chemical makeup of the atmosphere excluding water vapour	82
3.3	Chemical makeup of the atmosphere including water vapour	83
3.4	Aeroplane service ceilings	87
3.5	ISA values at sea level	87
3.6	ISA values at a range of altitudes	87
4.1	Characteristics of atmospheric stability	109
4.2	Approximate height of cloud bases above the surface	112
5.1	The Beaufort wind force scale	132
5.2	Turbulence classifications	146
5.3	A380–800 radar wake turbulence separation minima	150
6.1	Classification of tropical cyclones	171
6.2	Hazards related to occurrence of tropical cyclones	171
6.3	Airports in world's top 100 by passenger traffic, currently exposed to high take-off weight-restriction risk	182
7.1	Example earliest airports	190
7.2	Aerodrome reference codes	192
7.3	Top five airports in the world by land area	193
7.4	Example longest commercial runways in the world	194
8.1	UN Sustainable Development Goals	235
9.1	The 22 member states of LACAC	265

9.2	The 15 member and five associate member states of CARICOM	265
9.3	Air transport liberalisation arrangements within ASEAN countries	269
10.1	Key economic terms	285
10.2	Airline ancillary revenue categories and examples	289
10.3	Net profit, operating margin and RPK growth by world region, 2014–2023	298
11.1	Travel and tourism relative contribution to employment in 2019	329
11.2	Travel time for selected destinations in East Malaysia	332
12.1	Top ten global cities by overall score	340
12.2	Total passengers at commercial New York Airports, 2023	341
12.3	Busiest domestic flight routes in the world by seat capacity, 2023	346
12.4	Busiest international flight routes in the world by seat capacity, 2023	347
12.5	Top five countries of migrant origin and destination in 2020, by number and proportion of total population	348
12.6	Most routes operated by Wizz Air as of 31 March, 2024 – selected countries	352
13.1	Anthropogenic processes	361
14.1	Fuel burn percentage by phase of flight for various routes and aircraft type	384
15.1	Boeing 737 variants	424
15.2	Categorisation of types of land-use planning and management measures	429
15.3	Zurich Airport noise charges	430
15.4	Proportion of flights achieving CDO from the cruising altitude at the top 25 European airports	431

Abbreviations

AACO	Arab Air Carriers Organisation
ACA	Airport Carbon Accreditation
ACI	Airports Council International
ACS	Association of Caribbean States
AGL	above ground level
AHU	air handling unit
AI	artificial intelligence
AMSL	above mean sea level
ANSP	Air Navigation Service Provider
AOC	air operators certificate
AOR	advanced open rotor
APU	auxiliary power unit
ASA	air service agreement
ASK	available seat kilometre
ASAM	ASEAN Single Aviation Market
ASEAN	Association of Southeast Asian Nations
ASL	above sea level
ASNP	air service navigation provider
ATC	air traffic control
ATCOs	air traffic controllers
ATK	available tonne kilometre
ATM	air traffic management
AU	African Union
BAP	Biodiversity Action Plan
bbl	barrel
BPR	bypass ratio
BWB	blended wing body
°C	Degrees Celsius
CAA	Civil Aviation Authority
CAAC	Civil Aviation Administration of China
CAAF	Conference on Aviation and Alternative Fuels
CAEE	Committee on Aircraft Engine Emissions
CAEP	Committee on Aviation Environmental Protection
CAGR	compound annual growth rate
CAN	Committee on Aircraft Noise
CARICOM	Caribbean Community

CASK	cost per available seat kilometre
CAT	clear air turbulence
CBD	central business district
CBR	California Bearing Ratio
CCO	continuous climb operation
CDA	continual descent approach
CDO	continuous descent operation
CFC	chlorofluorocarbon
CG	centre of gravity
CO_2	carbon dioxide
CORSIA	carbon offsetting and reduction scheme for international aviation
CTK	cargo tonne kilometre
CWC	carrier-within-a-carrier
DALR	dry adiabatic lapse rate
dB	decibel
dBA	A-weighted decibel
DNL	day–night average sound level
DPD	design peak day
EAS	essential air service
EASA	European Union Aviation Space Agency
EBIT	earnings before interest and tax
ECAA	European Common Aviation Area
ECAC	European Civil Aviation Conference
ECJ	European Court of Justice
ELR	environmental lapse rate
ENSO	El Niño-southern oscillation
EPNdB	effective perceived noise in decibels
EPNL	effective perceived noise level
ERF	effective radiative forcing
ETOPS	extended-range twin-engine operations performance standards
ETS	emissions trading scheme
EU	European Union
°F	Degrees Fahrenheit
FAA	Federal Aviation Administration
FDI	foreign direct investment
FIFO	fly-in, fly-out
FIR	flight information region
FFP	frequent flyer programme
FL	flight level
FMS	flight management system
FSNC	full-service network carrier
GDP	gross domestic product
GDS	global distribution system
GGC	Gulf Cooperation Council
GHG	greenhouse gas
GMT	Greenwich Mean Time
GPU	ground power unit
H2	hydrogen

HEPA	high-efficiency particulate air
hPa	hectopascals
HEFA	hydroprocessed esters and fatty acids
Hz	Hertz
IATA	International Air Transport Association
IBAC	International Business Aviation Council
ICAN	International Commission for Air Navigation
ICAO	International Civil Aviation Organisation
IEA	International Energy Agency
IFE	in-flight entertainment
IFR	instrument flight rules
ILS	instrument landing system
IMF	International Monetary Fund
inHg	inches of mercury
IPCC	Intergovernmental Panel on Climate Change
ISA	International Standard Atmosphere
ITCZ	inter-tropical convergence zone
km	kilometre
kts	knots
LAC	Latin America and the Caribbean
LACAC	Latin American Civil Aviation Commission
LCC	low-cost carrier
LDCs	least developed countries
LF	load factor
LLDCs	landlocked developing countries
LLWAS	low-level wind shear alert system
Lden	day–night average sound level
Leq	equivalent sound level
Lmax	maximum A-weighted noise level
LRAD	long-range acoustic device
LTAG	long-term aspirational goal
mb	millibars
MBM	market-based measures
METAR	Meteorological Terminal Aviation Routine Weather Report
MICE	meetings, incentives, conferences and exhibitions
MoU	Memorandum of Understanding
MRO	maintenance, repair and operations
MSL	mean sea level
mt	megatonne
MTOM	maximum take-off mass
MTOW	maximum take-off weight
NASA	National Aeronautics and Space Administration
NATS	National Air Traffic Services (UK)
NIS	noise insulation scheme
NM	nautical mile
NOAA	National Oceanic and Atmospheric Administration
NOx	nitrogen oxides
NPM	net profit margin

NTSB	National Transportation Safety Board
NWS	National Weather Service (USA)
OAT	outside air temperature
O–D	origin–destination
OPM	operating profit margin
PHEIC	Public Health Emergency of International Concern
PICAO	Provisional International Civil Aviation Organisation
PNL	perceived noise level
PPE	protective personal equipment
PSO	public service obligation
RAS	rural air services
RASK	revenue per available seat kilometre
RASS	remote air services subsidy
RESA	runway end safety area
RF	radiative forcing
RFDS	Royal Flying Doctor Service
RH	relative humidity
RIATS	Roadmap for Integration of Air Travel Sector
RNAV	area navigation
ROIC	return on invested capital
RPK	revenue passenger kilometre
RTK	revenue tonne kilometre
SAATM	Single African Air Transport Market
SAF	sustainable aviation fuel
SAK	synthetic aromatic kerosene
SALR	saturated adiabatic lapse rate
SARPs	standards and recommended practices
SARS	severe acute respiratory syndrome
SDA	step-down approach
SDGs	Sustainable Development Goals (UN)
SEL	sound exposure level
SES	Single European Sky
SESAR	Single European Sky ATM Research
SET	single-engine taxiing
SID	standard instrument departure
SIDS	small island developing states
SOR	start-of-roll
STAR	standard terminal arrival
TCI	Tourism Climate Index
TOA	top of atmosphere
UAE	United Arab Emirates
UAM	urban air mobility
UHBR	ultra-high bypass ratio
ULCC	ultra low-cost carrier
UN	United Nations
UNCLOS	United Nations Convention on the Law of the Sea
UNEP	United Nations Environment Programme
UNHAS	United Nations Humanitarian Air Service

UTC	Coordinated Universal Time
UV	ultraviolet
VAAC	volcanic ash advisory centres
VFR	visual flight rules
VMC	visual meteorological conditions
WACC	weighted average cost of capital
WFP	World Food Programme
WHO	World Health Organization
WMO	World Meteorological Organization

ATG case studies

1.1	El Niño and La Niña – Pacific Ocean	7
1.2	Denver International Airport, Colorado, USA (DEN)	10
1.3	Chek Lap Kok Airport, Hong Kong (HKG)	19
1.4	Transport mode choice – London to Paris	24
1.5	Transport mode choice – France	25
2.1	FSNC – KLM Royal Dutch Airlines (Netherlands)	40
2.2	LCC – GOL Linhas Aéreas Inteligentes S.A (Brazil)	42
2.3	ULCC – Ryanair (Ireland)	43
2.4	Hybrid – JetBlue (USA)	43
2.5	Charter – TUI Group (Germany)	45
2.6	Regional – SkyWest Airlines (USA)	45
2.7	CWC – Go (UK)	46
2.8	Specialist – "The Mail Plane": remote air services subsidy (RASS) scheme – Australia	46
2.9	Business aviation – NetJets	47
2.10	Cargo – Cathay Cargo (Hong Kong)	48
2.11	The "Kangaroo Route" – Qantas, Australia–London	54
2.12	Ethiopian Airlines and Addis Ababa Bole International Airport	62
2.13	Airports in China	64
4.1	The hazard of fog: crash of KLM 4805 and Pan Am 1736 at Tenerife Los Rodeos Airport, 27 March, 1977	122
4.2	The hazard of ice at the airport: serious incident involving ATR72–212A on departure from Manchester Airport, UK on 4 March, 2016	125
4.3	The hazard of ice inflight: accident of British Airways Boeing 777-236ER at London Heathrow Airport, UK on 17 January, 2008	125
5.1	The hazard of tailwinds: crash of Raytheon Hawker 800XP at Aspen, Colorado, 21 February, 2022	134
5.2	Single runway orientation: Mariscal Sucre International Airport (UIO) – Quito, Ecuador	135
5.3	Multiple/crosswind runway orientation: Newark Liberty International Airport (EWR), New Jersey, USA	136
5.4	Tenzing-Hillary Airport, Lukla, Nepal	144
5.5	Kai Tak Airport, Hong Kong (closed 1998)	147
5.6	Aircraft accident involving Dornier DO 228-202 at Bodo Airport, Norway, 4 December, 2003	153
6.1	Nigeria and the ITCZ	161

xxxiv ATG case studies

6.2	U.S. National Weather Service (NWS) StormReady® Programme: Tampa, USA	171
6.3	The impacts of climate change on Greek airports	178
6.4	ECAC region and flood risk	179
6.5	Airports within the top 100 exposed to high take-off weight-restriction risk	181
7.1	Boeing 777–8 and Boeing 777–9 (B777X) – folding wingtips	192
7.2	King Fahd International Airport, Saudi Arabia (DMM)	193
7.3	Air Inuit, Northen Canada	197
7.4	Barra Airport, Outer Hebrides, Scotland	197
7.5	Ice runways, Antarctica	198
7.6	"Soil like toothpaste": the new runway at Brisbane Airport, Australia	201
7.7	Juancho E. Yrausquin Airport, Island of Saba, Caribbean Netherlands Antilles: the shortest commercial runway in the world	203
7.8	Seaplanes in the Maldives	205
7.9	Courchevel Altiport, France	207
7.10	Earthquake in Nepal, 2015	210
7.11	Sabiha Gökçen Airport, Istanbul: seismic engineering	211
7.12	Tsunami near Sendai, Japan, 2011	213
7.13	The 1991 Mount Pinatubo (Philippines) eruptions and their effects on aircraft operations	215
7.14	Eruptions of Eyjafjallajökull Volcano, Iceland, April 2010	217
8.1	Aircraft accident at Kimberley, South Africa, 2010	223
8.2	"Miracle on the Hudson River", New York City, US Airways 1549	226
8.3	Wildlife management at Singapore Changi Airport (SIN)	228
8.4	Vancouver International Airport, Canada (YVR): a salmon-safe airport	229
8.5	Maun International Airport, Botswana (MUB): restoration and regeneration	229
8.6	Salvador Bahia Airport, Brazil: fauna management	230
8.7	London Gatwick Airport, UK (LGW): Biodiversity Action Plan (BAP)	230
8.8	Rajiv Gandhi International Airport, Hyderabad, India: water sustainability through efficient devices, recycling and replenishment	236
8.9	Aircraft "drywashing": Emirates Airline/Dubai International Airport (DXB)	237
8.10	Flood protection at Amsterdam Schiphol Airport, Netherlands (AMS)	238
8.11	Flooding at Don Mueang International Airport, Bangkok, Thailand (DMK)	239
8.12	Flooding at La Vanguardia Airport, Colombia (VVC)	240
8.13	De-icing operations at Oslo Gardermoen Airport (OSL), Norway	242
9.1	Gulf airspace blockade: Qatar and Saudi Arabia	253
9.2	Russian airspace	256
9.3	Bermuda I Agreement	259
9.4	EU–US Open Skies	264
9.5	Liberalisation in Argentina	266
9.6	Air service liberalisation in Nigeria	271
10.1	India	279
10.2	Wizz Air, Hungary	290
10.3	Emirates and Dubai	299
10.4	Air connectivity in Africa	305
11.1	Mariscal Sucre Quito International (UIO) Airport, Ecuador	319
11.2	Incheon Airport, Seoul, South Korea (ICN)	322
11.3	Eastern Airport City Project and U-Tapao Rayong-Pattaya International Airport, Thailand	323

11.4	Colombia – Andean geography	326
11.5	Darién National Park, Panama	327
11.6	Small Island Developing States	329
11.7	Wick John O'Groats Airport (WIC) to Aberdeen International Airport (ABZ), Scotland – PSO	331
11.8	Rural air services in East Malaysia	332
11.9	Royal Flying Doctor Service – Australia	333
12.1	New York airport system	340
12.2	Air transport in Indonesia	341
12.3	Growth in China	343
12.4	Overtourism in Venice	345
12.5	EU enlargement in 2004	350
12.6	Poland and LCCs – post-2004	352
13.1	Kansai International Airport (KIX), Osaka Bay, Japan	364
13.2	Hong Kong International Airport, Chek Lap Kok (HKG)	365
13.3	Toncontin Airport, Tegucigalpa, Honduras (TGU)	367
13.4	Madeira (Funchal) Airport (FNC)	369
13.5	Sangster International Airport, Montego Bay, Jamaica (MBJ)	370
13.6	Stornoway Airport, Isle of Lewis, Scotland (SYY)	371
13.7	Amsterdam Schiphol Airport, Netherlands (AMS)	373
14.1	Fuel burn and % phase of flight	384
14.2	Hong Kong International Airport carbon footprint	386
14.3	American Airlines, Dallas, USA: the legend of the olive	393
14.4	All Nippon Airlines (ANA), Japan: reducing CO_2 emissions by reducing the weight of aeroplanes	395
14.5	Cochin International Airport, India – world's first solar-powered airport	397
14.6	Airbus ZEROe aircraft concepts	401
14.7	Virgin Atlantic and SAF	403
14.8	Galapagos Ecological Airport – one of the world's "greenest" airports	405
14.9	British Airways – "Perfect Flight"	405
15.1	The discussion on the health effects of aircraft noise (DEBATS) study, France	413
15.2	Noise monitoring at London Heathrow, UK (LHR)	420
15.3	Boeing 737 series, Washington, USA	423
15.4	Noise chevrons	426
15.5	Noise insulation scheme, Glasgow Airport, UK (GLA)	429
15.6	Noise-related airport charges: Zurich Airport, Switzerland (ZRH)	430
15.7	Noise reduction initiatives at London City Airport, UK (LCY)	432

ATG trivia

Chapter 1: Airport codes — 26
Chapter 2: Great circle route — 70
Chapter 3: Hot and cold — 98
Chapter 4: Cabin humidity — 126
Chapter 5: How hot do aircraft get? — 154
Chapter 6: Hurricane hunters — 183
Chapter 7: The Galunggung glider — 218
Chapter 8: The "Rain Vortex" at the Jewel Changi Airport, Singapore — 244
Chapter 9: Time travel: GMT, UTC and local times — 272
Chapter 10: The missing minute — 310
Chapter 11: Flying vehicles — 334
Chapter 12: Extreme time zones — 353
Chapter 13: The second lives of aircraft — 376
Chapter 14: Plant a tree? Restore peat swamps? Install "friendly" stoves? The world of carbon offsetting — 406
Chapter 15: Supersonic flight and the sonic boom — 434

Foreword

I write this foreword as I fly at 32,000 feet over the Caribbean Sea, crossing from the Curaçao to the Barranquilla FIR on Copa's flight CM133 from Sint Maarten (TNCM/SXM) to Panama City (MPTO/PTY) on a Boeing 737-700. Next to me is a catering company executive who has been inspecting restaurants on the island and, across the aisle, a Colombian couple is returning home from a short vacation. Among the other passengers are hotel staff from the Dominican Republic; tourists returning to various cities in the United States; students going to college in Florida; and all sorts of people who will be arriving at their destination via Copa's hub in Panama – "the Hub of the Americas". In fact, more than 90% of the passengers on this flight will make connections in Panama.

SXM is fortunate to have Copa. This airline is a coveted prize for all Caribbean islands, as it opens up new markets in Latin America while strengthening connectivity options with North America. Panama's geographic location, combined with Copa's skillful management, has made this airline-airport pair an essential component of aviation in this part of the world, successfully linking South America, North America and the Caribbean. It should therefore come as no surprise that Turkish Airlines has the same number of weekly frequencies (11) to Panama, Latin America's smallest capital, as to Mexico City and Sao Paulo, the two largest cities.

Today's world could not be conceived without air travel. And while this may seem like a truism, it is likely that many of those living in more developed countries are unaware of the importance of aviation to the daily lives of hundreds of millions of citizens around the globe. In fact, some only think of air travel as part of the "necessary evil" of enjoying a weekend skiing trip or the annual summer family vacation. But for many, air travel is an absolute necessity, like getting an air conditioning technician, going to college or seeing the doctor.

Indeed, in many countries, air travel is the only realistic means of transportation available. There are no alternatives if you want anything resembling a "modern lifestyle". Many countries lack rail transport and the geographical and topographical features of some parts of the planet make road transport long and treacherous. For example, the road linking two of Ecuador's most populous cities, Guayaquil and Cuenca, crosses a 4,140-meter mountain pass. In Colombia, the flight time between Bogota and Medellin, the two most populated cities, is 50 minutes, but it takes ten hours by car. And there is no road connecting Panama and Colombia, even though they share a land border.

Air transport will continue to be increasingly important for the social and economic development of nations, regions and cities. Tourism, for example, is the main economic activity in the Caribbean and all the islands want to increase their air connectivity. And the growing demand for perishable exports, such as seasonal fruit and fresh fish, is what keeps the economies of Peru and Chile going.

But air transport also faces numerous challenges, many of them related to the environment and climate. On the one hand, environmental pressures are forcing an urgent search for alternative fuels and propulsion systems to ensure that aviation is socially acceptable. On the other hand, climate change is putting pressure on existing infrastructure: many airports are built at sea level and are already suffering from sea level rise and storm surges.

However, despite the challenges, there are also great opportunities ahead. For example, technological advances in photovoltaic cells and lower prices make it possible to power entire airports with solar panels, thereby significantly reducing carbon emissions and operating costs (energy is a key cost item at island airports). New aeronautical developments, such as electric, VTOL and ground effect aircraft, will open up many opportunities for regional thin routes, which will improve connectivity at affordable prices for many remote communities around the world.

For all these reasons, this book provides a comprehensive and timely analysis of the dynamic relationship between air transport and geography. It provides readers with valuable insights into the spatial organization of air networks, the role of airports and the socioeconomic implications of air transport systems in different regions. This book serves not only as a textbook for students, but as a valuable reference for both practitioners and policy makers. Enjoy the flight!

Rafael Echevarne
Director General, Airports Council International –
Latin America and the Caribbean (ACI-LAC)

Rafael Echevarne is the Director General of Airports Council International – Latin America and the Caribbean. Previously, he was CEO of Montego Bay Airport, Jamaica, and has worked for international airport organizations in Canada, Europe, the Middle East and Australasia. He holds a PhD in Air Transport from Cranfield University, UK.

Acknowledgements

As someone who did not fly on an aircraft until they were 23, nor had parents who ever experienced this pleasure, a career in air transport seemed an unlikely choice. However, watching the lines following aircraft high in the sky (which I later discovered were contrails) where I grew up, 25 miles from Glasgow Airport in Scotland, sowed the seed. It is thanks to the following people, and more, who helped to further develop my twin passions of both air transport and geography, many of whom also inspired me before and during the writing of this book.

To my Geography lecturers at the University of Strathclyde, including Mark Boyle, Graham Hollier and Robert Rogerson. It was completing my dissertation on 'The Growth and Development of Glasgow Airport' in 1996, which provided my first 'geography/air transport' link.

To my Air Transport lecturers at Cranfield University, including Professor Rigas Doganis, Romano Pagliari, Peter Morrell, Ralph Anker and George Williams.

At Monarch Airlines – Eddie Rumsey, Dave Summers, Linda Swan, William Barton, Jane Fuller, Chris Deans, Amanda Robertson and Angelina Ashby.

At the University of West London – James Edmunds, Elitza Iordanova, Ash Mistry, James Goodman, Kanchana Gamage, Cristina Maxim, Peter Smith, Francesco Ragni, Richard Dale, Anil Padhra and Ivan Sikora.

At Bucks New University – George Georgiou, Jenny Tilbury, Nigel Griffiths, Maurice Gledhill, Peter Duquemin and John Furley.

At Coventry University – Jenni Fernando, Ivan Stevenson and James Richardson.

To all my students over the last 15 years – thank you for your inspiration, you are the future!

Thank you also to everyone who has helped with the book – for the advice, chats and ideas. To John Irish for his cloud photos and Jon Sinclair for reviewing the meteorology chapters. To University of West London Air Transport Management Student Representatives Ruby Wood and Zane Santamaria for choosing the front cover image and layout. I would also like to acknowledge Canva.com, which I have used for the design of many of the images which appear in the book.

Finally, to the most important people in my life – Gergina, James and Cameron – for your love, patience and support.

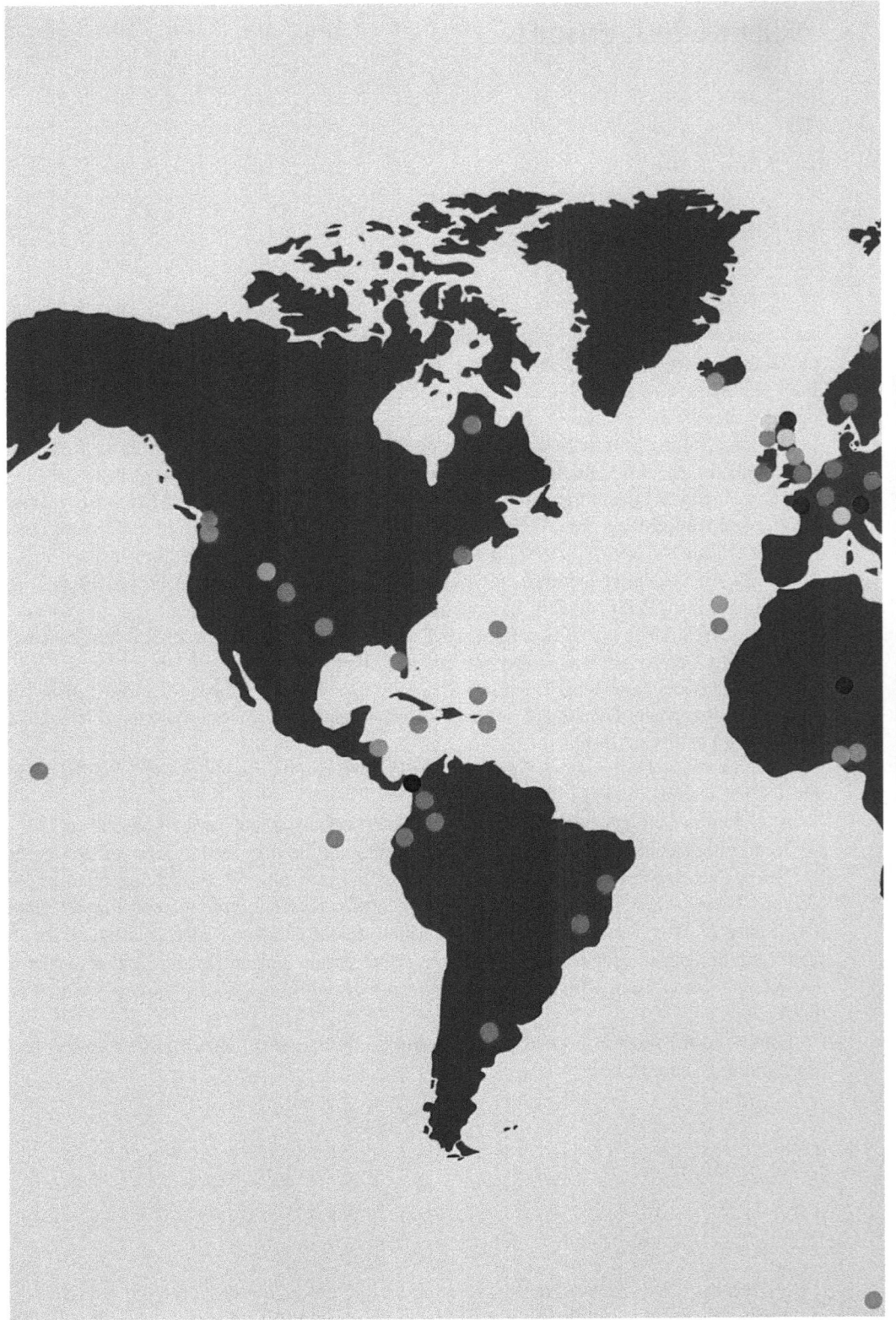

Figure 0.1 Map illustrating the global locations of the case studies in the book

Acknowledgements xliii

Part I
Introduction to geography and air transport

1 Introduction to geography and transport

Chapter outcomes

At the end of this chapter, you will be able to:

- Understand the importance of the role of geography in air transport and the key relationships.
- Explain the key sub-disciplines of physical geography and their relationship with air transport.
- Describe the main aspects of human geography and how these link with the air transport industry.
- Explain the key impacts air transport has on environmental geography and conversely environmental geography has on air transport.
- Understand the importance of the concept of space in transport and the transport modes and elements which are underpinned by this concept.
- Analyse the core principles of transport geography.

1.1 Introduction

The phrase "the only constant is change" could have been written for geography. In the moment it takes you to read this sentence, geographical changes – some large, some extremely small – will have occurred somewhere in our world. The phrase could also have been written for the air transport industry – a global, extremely dynamic and ever evolving industry. The combination and overlap between geography and air transport is more relevant now than it has ever been.

Geography is the study of the physical features of the earth and its atmosphere and how human activity affects, and is affected by, these features. It is the study of places and relationships between people and their environments. Geography looks to understand where things are found, why they happen to be where they are and how they change and develop over time.

4 *Fundamentals of Global Air Transport Geography*

> **ATG Did you know?**
>
> **Geography**
>
> The term "geography" may have been coined by Eratosthenes – a Greek Scholar – around the third century BC. In Greek, geo means "earth" and -graphy means "to write". The concerns of Eratosthenes and other Greeks regarding location, the features of places and the distribution of people and environments, have been a core part of geography since these early days (National Geographic, n.d.).

There are various schools of thought as to the number of branches of geography. This book takes the assumption that there are three main branches: physical geography, human geography and environmental geography (Figure 1.1).

- **Physical geography** is the study of the interactions between the earth's climate system, landscapes, oceans, plants, animals and people (Holden, 2011). It is concerned with the processes that shape the earth's surface, their spatial patterns and the animals and plants which inhabit it. It is the study of the natural features of the earth – its surface, the oceans below and the atmosphere above.
- **Human geography** is a branch of knowledge which seeks to venture descriptions of and explanations for the uneven distribution of human activity across the surface of the earth (Boyle, 2015). It studies the inter-relationships between people, place and environment and their spatial and temporal variations across and between locations.
- **Environmental geography** is centred on the interactions and relations of the biogeophysical environment with human societies (Zimmerer and McSweeney, 2020). The increasing importance of human impacts on the physical world, via global warming, land use changes, etc., has necessitated the extension of a dual branch philosophy of human and physical geography, to include environmental geography as a separate and very much linked stream.

The purpose of air transport is the movement of people and freight from one location to another. The medium is the air, hence the importance of the atmosphere, however aircraft must take off and land. Infrastructure needs to be developed to facilitate the transfer of people and goods, hence the physical features of the earth will have an important part to play in location,

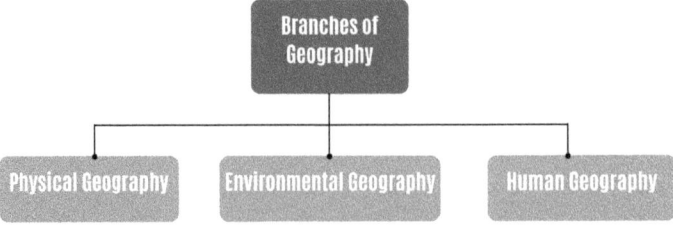

Figure 1.1 The branches of geography
Source: author

via aerodromes and airports. The spatial distribution of geopolitics, economic activity, urban and rural development and population movement and mobilities are influenced by, and have an influence on, the air transport industry. Environmental changes, both being influenced by and having an impact on air transport, are of increasing importance as we move well into the 21st century.

Marshall (2021, p.xiv), in his book *The Power of Geography*, stated that "Geography is a key factor shaping what humanity can and cannot do … the choices people make, now and in the future, are never separate from their physical context". This was stated in the geopolitical context; however, it can be related to the context of air transport. Indeed, it can be argued that the existence of geographical constraints is one of the primary drivers of the need for air transport across the globe.

Part I of this book introduces the branches of geography and their relevance and importance to the air transport industry, introducing the key component of transport geography (Chapter 1) whilst Chapter 2 will explain the key components of the air transport industry relevant to the book. Part II (Chapters 3–8) focuses on physical geography, analysing the sub-fields relevant in the context of the air transport industry – meteorology, climatology, geomorphology, natural hazards, biogeography and hydrology. Part III (Chapters 9–12) will focus on human geography and the key sub-fields of political, economic, urban, rural and population geography. Part IV will examine the interaction of physical and human geography in the context of environmental geography and three of the key areas affecting people and the planet – climate change, geomorphic change and air transport noise.

1.2 Physical geography

Geography asks spatial questions such as how and why things are distributed around our planet. Physical geography focuses on understanding the natural features and processes on earth's surface, in the atmosphere and in our oceans. Spatially, this involves analysis of physical elements and processes such as landforms, climate, vegetation, soils, water and natural resources at both small and larger scales and over different time periods. It is interdisciplinary, drawing on aspects such as chemistry, biology, environmental science, meteorology, pedology and geology. This book considers the impact of physical geography on air transport (and indeed air transport on the physical environment) in several subdisciplines (Figure 1.2).

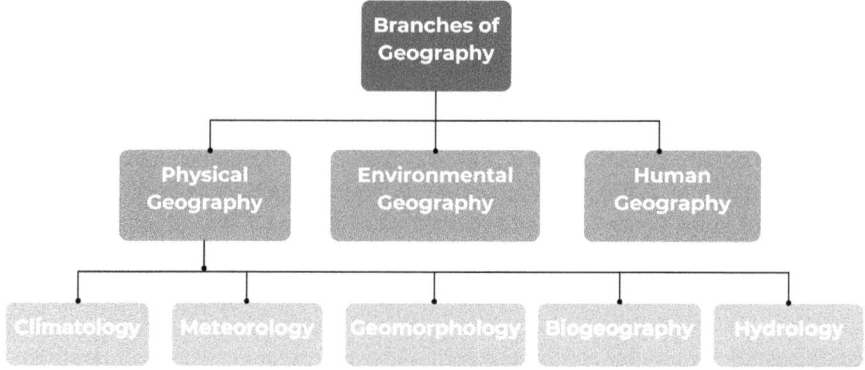

Figure 1.2 Subdisciplines of physical geography
Source: author

1.2.1 Meteorology

Meteorology is the study of the earth's atmosphere and its constituent parts and processes, such as temperature, precipitation, air pressure, humidity and movement of air, over short time scales (minutes to weeks) and enables weather forecasts to be compiled.

> **ATG Did you know?**
>
> **Meteorology**
>
> The word "meteorology" originates from the Ancient Greek – μετέωρος – *metéōros* (meaning "lofty" or "in the air") and -λογία – *logia* – (meaning "to study, explain"). Aristotle (the Greek philosopher) is often cited as being one of the founding fathers of meteorology with his treatise *Meteorologica* (written in 350 BC), which was one of the earliest attempts to understand the atmosphere and the earth's water cycle.
>
> Some sayings and proverbs in use today, such as "red sky at night is a shepherd's delight", were anecdotal observations that a red sky in the evening was more often than not followed by good weather the next day, providing evidence of people throughout history observing and recording the weather above (Met Office, n.d.).

In the context of aviation, a knowledge of meteorology is vital for pilots, who encounter the vagaries of meteorological factors before they have even left the runway and throughout the duration of the flight, as well as airline flight and route planners. Safety critical weather conditions, such as thunderstorms, right through to passenger comfort during flight, necessitates a thorough analysis of meteorology. Airport operations are also hampered by factors such as rain, wind and snow, which can all have impacts from both a safety perspective and the ability to adequately, and in a timely manner, service aeroplanes during turnarounds and whilst aeroplanes are taxiing on the ground.

Chapters 3–5 will explain and assess meteorological factors and their impact on air transport. Chapter 3 will assess the atmosphere and atmospheric principles. It will begin by introducing the principles of flight, which provide the foundation for subsequent sections, assess atmospheric components and layers and focus on the concept of the International Standard Atmosphere (ISA) and its importance. The impacts of temperature, air pressure and air density on air transport will be explained.

Chapter 4 will explain the concept of water vapour, which is the foundation for much of our weather. We may not always see it, but it is always there – as aircraft are continually aware! The importance and stages of the hydrologic cycle will be introduced and the concept of humidity and its importance in the context of flight operations explained. Atmospheric stability (or instability) will be explained before leading on to explaining the more visible components of water vapour – clouds – their phenomena, as well as associated visibility hazards and their impact on flight. Precipitation in its various forms, such as rain, snow and sleet, as well as icing hazards, will also be assessed in the context of air transport operations.

Chapter 5 will assess the concept of wind and its impact on air transport. It usually takes less time flying from west to east (USA to Europe for example) than the opposite direction and this is due to wind direction, wind speed and aspects such as the jet stream. Wind direction is a crucial factor for airport planners to consider when designing airports, as runway direction should be orientated towards taking off and landing into the wind, for safety and general lift requirements. Turbulence can pose risks to air transport, from mild discomfort to passengers during small turbulent events, to major, aircraft safety risk events. Thunderstorms can be some of the most serious risks to air transport and their formation and life cycle will be explained, as well as the hazards they pose to air transport.

1.2.2   Climatology

Climate is the long-term weather pattern in a particular area. What do we wear today? Do we take an umbrella? These are some of the short-term decisions we will take based on the *weather*. However, the question of what we wear will be influenced by the time of year and where we are. Visitors to Tenerife or Phoenix in July are much more likely to consider shorts and t-shirts and these decisions are generally based on *climatic* understanding – it will be hot in Tenerife and Phoenix in July. Next year and the year after that.

Climatology focuses on three main aspects of climate:

1. Weather patterns which govern normal conditions in different world regions.
2. The relationship between different aspects of weather such as temperature and sunlight.
3. The way weather changes over time. Results from these studies have demonstrated that human activities are affecting earth's climate.

ATG Case study 1.1

El Niño and La Niña – Pacific Ocean

Climatologists also research natural changes in air and ocean currents, such as El Niño and La Niña, which are phases in a fluctuating cycle of air and ocean temperature over the Pacific Ocean. During normal conditions, trade winds blow west along the equator, taking warm water from South America towards Asia. To replace that warm water, cold water rises from the depths – a process called upwelling. El Niño and La Niña are two opposing climate patterns that break these normal conditions. Scientists call these phenomena the El Niño-southern oscillation (ENSO) cycle (Figure 1.3a). El Niño and La Niña can both have global impacts on climate, ecosystems and economies (NOAA, 2023).

8 *Fundamentals of Global Air Transport Geography*

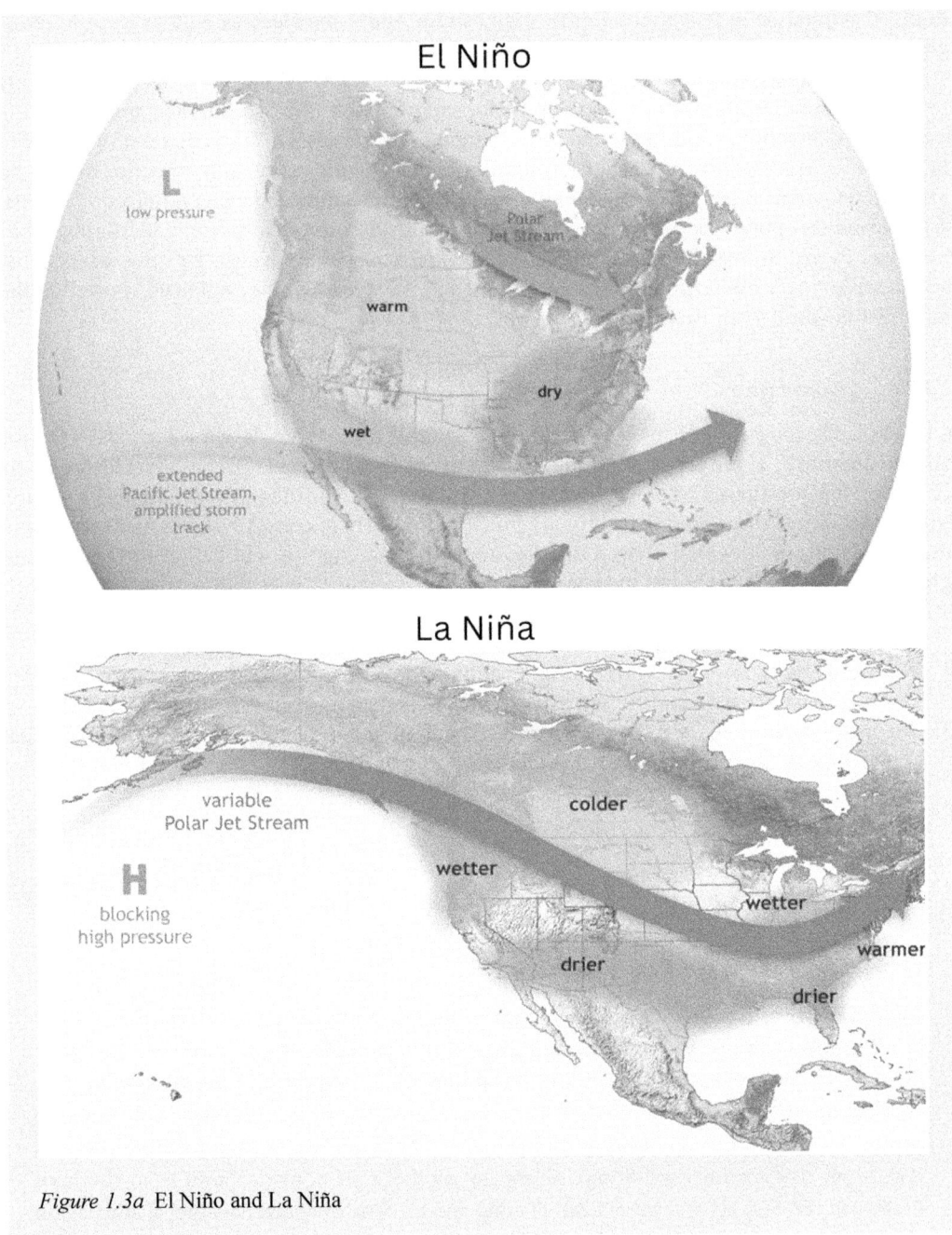

Figure 1.3a El Niño and La Niña

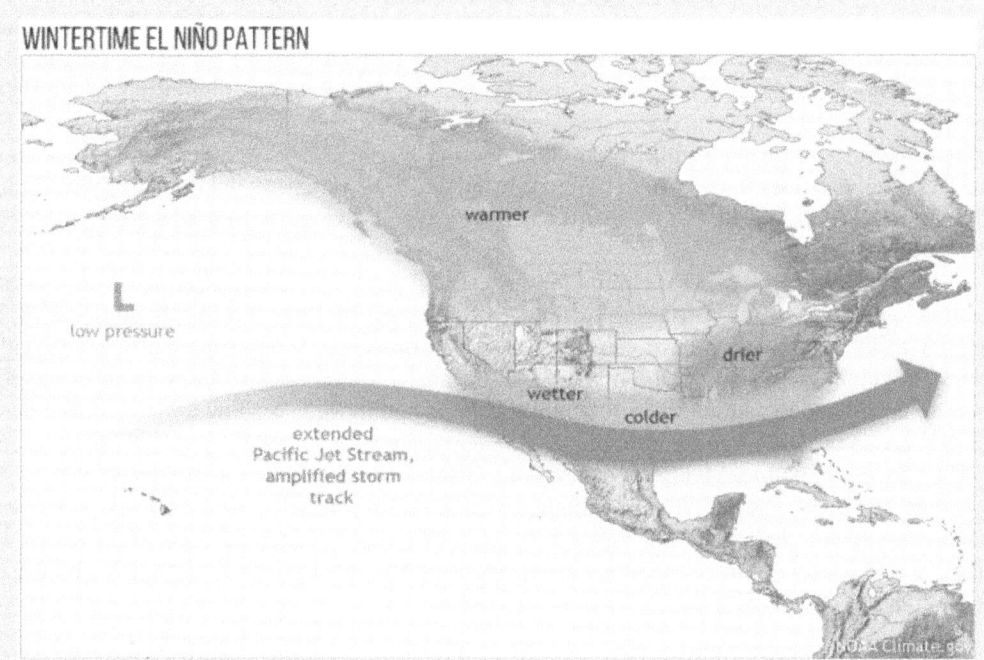

Figure 1.3b El Niño impacts on the US in winter
Source: Climate.gov (2023)

ATG Did you know?

El Niño and La Niña

El Niño means "Little Boy" in Spanish. South American fishermen first noticed periods of unusually warm water in the Pacific Ocean in the 1600s. Its full name is El Niño de Navidad, as El Niño usually peaks around December.

La Niña means "Little Girl" in Spanish and has the opposite effect to El Niño.

The pattern shifts back and forth irregularly every two to seven years, bringing predictable shifts in ocean surface temperature and disrupting the wind and rainfall patterns across the tropics (Climate.gov, 2023).

These changes have a cascade of global side effects. For example, in North America, El Niño and La Niña have their strongest influence on seasonal climates in winter (Figure 1.3b). The warmer waters of El Niño cause the Pacific jet stream to move south of its neutral position. With this shift, areas in the northern USA and Canada are dryer and warmer than usual. But in the USA Gulf Coast and Southeast, these periods are wetter than usual and may have increased flooding (NOAA, 2023).

10 *Fundamentals of Global Air Transport Geography*

Chapter 6 will analyse climatology from an air transport perspective. Climatic factors are key in terms of our decision to travel – especially on holiday – where to travel to and when to travel. Whilst short-term travel decisions may be based on weather forecasts, longer-term travel decisions are generally based on climatic factors – inhabitants of countries in Northern Europe, such as the UK, Netherlands and Germany, often visit Spain, Greece and Portugal in the northern hemisphere summer due to the likelihood of warmer, drier weather. Airlines and tour operators will consider these climatic factors when deciding which destinations to fly to, which originating bases would like to fly there, when they would like to fly and how many flights to offer (frequency).

There are broad atmospheric circulation processes which result in global weather patterns, creating specific climates. Different air masses also bring differing weather conditions, as do weather fronts, and whilst these can vary, their locations are often climatically predictable. Tropical cyclones can bring some of the most hazardous weather conditions for air transport operations and understanding their potential locations and hazards are key to try to mitigate, or at least reduce the risks – to aircraft, passengers and airport personnel.

The concept of climate change is one of the critical issues facing the planet today, with rising global temperatures and sea levels, changing precipitation, changing wind patterns, reductions in biodiversity and desertification. Each of these is extremely important to the air transport industry and the second part of the chapter will focus on the impacts of our changing climate on the air transport industry.

1.2.3 Geomorphology and landforms

Whilst the medium of air transport is the air, landforms are key to facilitating safe and efficient aircraft landing, take-off and processing of passengers and freight.

Geomorphology is the study of earth processes and landforms, whilst landforms themselves are topographic features of the earth's surface. The amount of land area required for airports can be immense, for example at airports such as Chicago O'Hare (ORD), which has more runways than any other in the world (8), as well as numerous terminal buildings and associated facilities. The largest airport in the world, in terms of land use, is King Fahd International Airport in Saudi Arabia – 776 sq.km – with a land area greater than Bahrain!

ATG Case study 1.2

Denver International Airport, Colorado, USA (DEN)

DEN has the largest airport land area in the USA and the second largest in the world, at 137.8sq.km. It has six runways, an overall area twice the size of Manhattan and is larger than the city boundaries of Boston, Miami or San Francisco. It has approximately 300 lane miles of road and more than 31,000 parking spaces. In 2023, it had an average of 213,254 daily passengers (flydenver, n.d.).

Figure 1.4 illustrates the land area of DEN prior to the construction of the sixth runway, which is now the longest commercial runway in the USA at 16,000ft/4,887m, and the land required for a runway of this size.

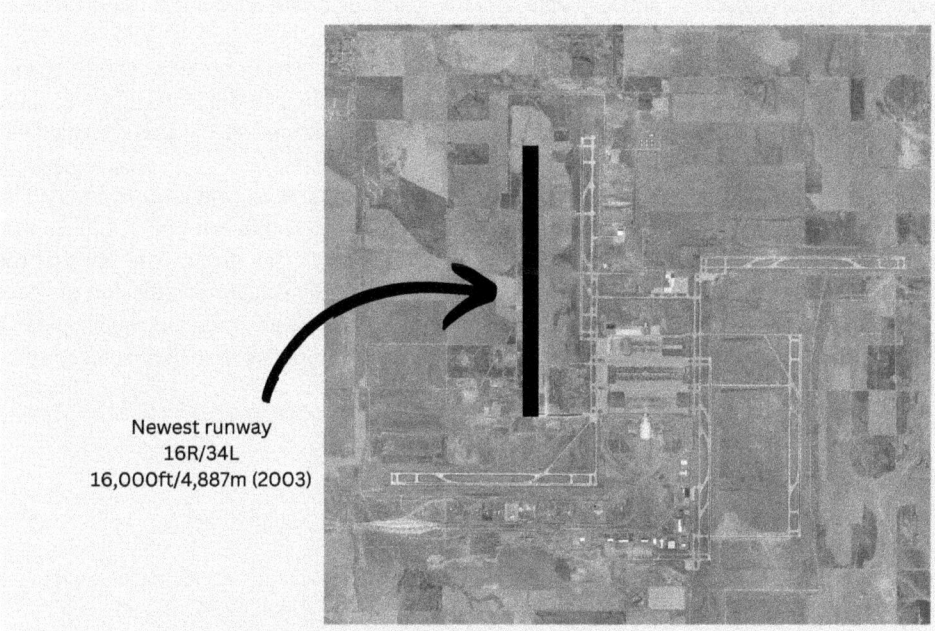

Figure 1.4 Denver International Airport layout
Source: adapted from USGS (2002). Public Domain

Airports require essentially flat land for aircraft to take off and land safely, and trying to find this amount of flat land can be challenging. If it is not available, then land altering may need to occur – the extent of which depends on the landforms and topography (Chapter 13).

An understanding of the factors which influence landforms and the characteristics of landforms themselves is fundamental to the air transport industry, especially for airport design, planning and development, and this will be assessed in Chapter 7. The spatial scale of landforms and surface and slope requirements for runways will also be analysed.

1.2.4 Geophysical hazards

The occurrence of a threatening condition from a natural phenomenon in a defined space or time can be considered a natural hazard, especially where there is a human presence. Geophysical (or geological) hazards originate from internal earth processes, such as earthquakes, volcanic activity and tsunamis – although tsunamis essentially become a hydrological hazard in their manifestation (UNDRR, 2007). These have massive implications for airport resilience and design principles. Chapter 7 will assess the impact to the air transport industry of earthquake, volcanic and tsunami geophysical hazards, as well as possible risk mitigation strategies.

1.2.5 Biogeography

Although landforms are crucial to air transport industry operations, it is not just the form of the land which impacts the air transport industry – it is the plants, animals and other organisms which live there. This, and their geographical location, plays a major role in where airports can

12 *Fundamentals of Global Air Transport Geography*

be built and in any development and expansion plans. Chapter 8 will assess the importance of biogeography in the context of air transport.

Wildlife and habitat management is a crucial aspect of airport environmental planning and is often influenced by international, national and local policies. Their implementation will have impacts on species which inhabit those localities, to airport operations and to the safety of aircraft, because of possible bird and other wildlife strikes.

Some of the smallest organisms can have the deadliest consequences (think mosquitos). The COVID-19 virus decimated the air transport industry in 2020 and it was not until 2023/24 that passenger levels reached 2019 levels (Chapter 2) in most parts of the world. It is not just the economic cost, it is the potential for the air transport industry to facilitate transmission of these organisms and due to its global nature, this can happen quickly. The air transport industry has a responsibility to reduce the risk of biological hazard transmission between different locations.

1.2.6 Hydrology

Hydrology is concerned with:

- the distribution of water on the surface of the earth
- water movement over and beneath the surface of the earth
- water movement through the atmosphere.

> **ATG Did you know?**
>
> **Hydrology**
>
> Hydrology comes from the Greek word for ὕδωρ (*húdōr*) "water", and -λογία (*-logía*) and is the *science or study of the movement, distribution and management of water on Earth*, including the water cycle, water resources and drainage basin sustainability.
>
> There is the same amount of water on earth as when it was formed. The water in your sink could contain molecules that dinosaurs drank! A person can live for around a few weeks without food but only a few days without water. Seventy-five percent of the human brain is water (US EPA, 2016).

In the context of air transport, water in the atmosphere will be analysed in Chapter 4, whereas Chapter 8 will focus on global and local water regulations and the three key areas in the context of airports – water supply, handling capacity and disposal. Flooding is also a major risk to the air transport industry in many parts of the world, both in terms of coastal and river flooding and these will both be assessed.

1.3 Human geography

Human geography is "a branch of knowledge that seeks to venture descriptions of and explanations for the uneven distribution of human activity across the surface of the earth" (Boyle, 2015, p.4). It focuses on how we organise ourselves and our activities in space; our connection to each other and our environment; how we make places and how these shape us and how we think about and organise ourselves locally and globally (Fouberg and Murphy, 2020).

On 15 November, 2022, the world's population reached 8 billion people, growing from an estimated 2.5 billion people in 1950 and adding 1 billion since 2010 (United Nations, 2024). People in much of the world can now cross the globe in less than a day, and access to cars, trains, ships and aeroplanes is widespread. The growth in the air transport industry has revolutionised – and globalised – the world and the concept of globalisation is key in contemporary human geography. This has happened at different rates, in different ways and in different regions, and the impacts are felt differently across individual local areas, countries and continents.

The National Council for Geographic Education and the American Association of Geographers (in Boyle, 2021) state that there are five key themes in human geography, which Part III of the book will have at its core:

- **Location.** How are human activities distributed across the face of the earth?
- **Place.** What is it like in a particular location?
- **Human/environment interaction.** What is the relationship between humans and their environment?
- **Movement.** How and why are places connected with one another?
- **Regions.** How and why is one area like another?

Fouberg and Murphy (2020) state that the subdisciplines of human geography include: political geography, economic geography, urban geography and population geography (Figure 1.5) and Part III (Chapters 9–12) will be structured around these.

1.3.1 *Political geography*

According to Marshall (2015, p.6), "broadly speaking, geopolitics looks at the ways in which international affairs can be understood through geographical factors". Squire and Jackman (2024) define geopolitics as, in simple terms, the relationship between power and space. Dastrup (2020) states that political geography is the study of how humans have divided up the earth's surface for management and control purposes. In the air transport industry, this can be extended further and beyond earth's surface – the notion of sovereignty and territoriality is critical when it comes to

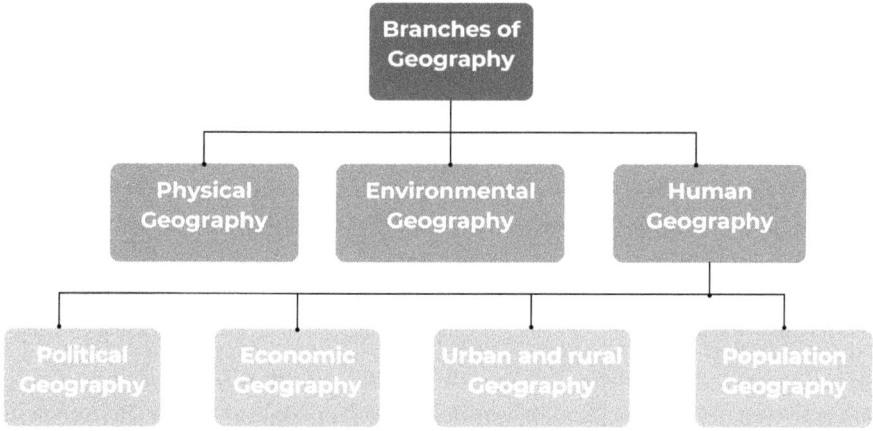

Figure 1.5 Subdisciplines of human geography

Source: author

airspace – who can fly within a country's airspace essentially shapes the air transport industry. This sovereign right to airspace above a nation's territory can be traced back to the Paris Convention of 1919 (Debbage, 2014). An understanding of political geography is therefore paramount to understanding the structure and performance (safety and economic) of the air transport industry.

Imagine if every country had its own standards for aircraft, navigation, communications, licensing, maintenance, airport procedures, runway markings, etc. Every time pilots crossed national boundaries, a new set of complicated regulations may be encountered; crossing Europe, for example, could potentially happen seven or eight times in the space of a couple of hours, with all the safety implications this would involve. For the air transport industry to function safely and efficiently, international standards and recommended practices were developed by the International Civil Aviation Organisation (ICAO). ICAO is a United Nations agency which was established in 1944 following the Chicago Convention, which was convened to develop an acceptable legal framework for the international operation of airlines. ICAO has as its aim "to promote the safe and orderly development of civil aviation around the world" (ICAO, 2018).

One of the key outcomes of the Chicago Convention was the five "Freedoms of the Air" agreements, which underpin air travel agreements, and along with additional freedoms since, are still applicable in the present day. For example, the First Freedom is "the right of an airline of one country to fly over the territory of another country". These Freedoms govern regulations as to which countries and airlines can fly where, how often and for which purpose. As such, they underpin the geography of air transport.

As a result, bilateral air service agreements were developed and traffic rights granted (or not) between individual countries (or Blocs such as the EU). Some parts of the world have adopted progressively more liberal policies more akin to "open skies" agreements, whilst others have remained more protectionist, often to protect their own flag carriers. This competitive landscape also has a huge impact on the geography of air transport.

Ownership restrictions placed on aviation organisations (especially airlines) by countries and Blocs, has limited ownership to foreign nationals in much of the world, again significantly influencing the geography of air transport. Chapter 9 will analyse the importance of political geography to air transport.

ATG Did you know?

Airline ownership restrictions

In the USA, Federal laws limit foreign ownership of US airlines to 25% and require US-citizen control of airlines (US GAO, 2019). In the EU, foreign investment cannot exceed 49% ownership of an EU airline – that is, EU member states or nationals must own more than 50% of the undertaking and control must remain in EU hands (European Parliament, 2019).

1.3.2   Economic geography

Economic geography can be defined as focusing on the spatial distribution and organisation of economic activities, as well as the relationships between people and their environments in the context of economic processes. For the purposes of this book, it can be characterised by the economic performance of the air transport industry, in terms of key metrics such as revenue, cost, profitability and value, as well as economic development to the regions it serves.

ATG Did you know?

The airline industry

In 2023, almost 40 million flights departed, with over 8 billion passenger journeys and more than $8 trillion in trade value carried by commercial air transport.

Air transport is not distributed evenly across continents, countries, regions or even individual locations. This spatial organisation of economic activities is key and the link with the economic performance of the air transport industry on a regional, national and global scale is revealing. Links between air connectivity and economic development can be huge and is one of the main reasons why there are such spatial disparities. Air transport is an economic enabler and those with more air transport links – regionally and globally – are generally those areas which have a better economic performance.

Chapter 10 will not only analyse the trends in global and regional economic development created by air connectivity, but also the trends in air transport economic performance in the context of the broader industry. Globally, the airline industry is characterised over time by marginal profitability – even in the "good" years. Some regions of the world generally struggle to even achieve marginal airline profitability and understanding why is crucial in attempting to improve performance.

It is impossible to disassociate the political and regulatory factors discussed in Chapter 9 with economic performance. The trend to liberalisation in many parts of the world has created a proliferation of new airlines and also in business models. The late 1970s in the USA, the 1990s in Europe and the early 2000s in Asia, for example, witnessed the rise of a new type of airline – the low-cost carrier (LCC) – in addition to the traditional full-service network carrier (FSNC). This has massively altered the geography of air transport, making air travel much more accessible and affordable to many more people, at least in some parts of the world. Where competition is present, fares have generally dropped, and the industry's often ruthless pursuit of cost-cutting to achieve profitability has increased. The growth of airline alliances and other consolidation strategies (often because of ownership rules) has also changed the geography of air transport, and alongside the growth in LCCs, the hub and spoke and point-to-point route structures are keys to understanding the spatial distribution of the economics of air transport.

Doganis (2019) discusses the paradox of the marginal profitability of the airline industry, where the underlying trend has been one of consistently good growth in demand over time and that industries faced with steady long-term growth in demand should be reaping substantial profits. Not so the airline industry, even in the early years when the industry was highly regulated and protected from internal competition. The causes and effects of this, and their spatial variations, largely underpin Chapter 10.

1.3.3 *Urban geography*

Urban geography focuses on the study of cities and urban areas, examining the spatial organisation, development and dynamics of urban spaces. It seeks to understand how cities evolve, how people interact with their urban environments, and the social, economic, cultural and political processes that shape urban landscapes.

ATG Did you know?

Cities

According to the World Bank (2024a), 56% of the world's population lives in cities – 4.4 billion inhabitants. This trend is expected to continue, with the urban population more than doubling its current size by 2050, at which point nearly seven in ten people will live in cities.

More than 80% of global gross domestic product (GDP) is generated in cities and if managed well, urbanisation can contribute to sustainable growth through increased productivity and innovation. The speed and scale bring challenges, however, via affordable housing, viable infrastructure, health, education and transport services.

According to Kaplan and Holloway (2014), urban geographers have centred their attention on the study of cities in two ways and scales – intermetropolitan and intrametropolitan – and this will be the core component of the first half of Chapter 11.

1. *Inter*metropolitan (or urban systems) approach. This stresses relationships *amongst* a system or groups of cities at the regional, national or global level.

From an air transport perspective, this can be researched by focusing firstly on connectivity. So, does the urban area have an airport? If so, how many? For example, London has Heathrow, Gatwick, City, Stansted, Luton and Southend within its (albeit loose!) boundaries. An assessment of route connectivity and destinations needs a focus on all. These routes can be direct, point-to-point services or transfer "hub" flights. Including hub flights may increase connectivity by multiples. For example, flying from London Heathrow to New York with British Airways is one route. Connecting on to partner airlines such as American Airlines at New York means potentially dozens more US domestic and international flights are available for that one ticket from London Heathrow. This, along with the inbound flight connection possibilities, can create a substantially positive economic impact for the urban area, via business, tourism and associated economic activity.

2. *Intra*metropolitan approach. This highlights the internal locational arrangements of humans, activities and institutions *within* metropolitan areas.

The location of the airport and its size has a significant bearing on its impact on and within the city. This can be positive but potentially negative – especially regarding environmental issues. Land use and zoning around airports impact the urban areas as does the transportation infrastructure to and from the airport, in terms of both public transport and availability of an efficient road network. Any expansion plans often involve consultations with local councils and populations, as witnessed with the proposed third runway at London Heathrow, which has met staunch resistance locally.

A concept which is gaining in importance and prominence is the *aerotropolis/airport city* concept proposed by John Kasarda. Airports, like cities, constantly evolve in form and function. Historically, airports have been understood as places where aircraft operate, which facilitates passenger and freight transfer. This traditional understanding is giving way to a much broader concept known as the Airport City. The Airport City model is grounded in the fact that in addition to their core aeronautical infrastructure and services, major airports have developed significant nonaeronautical facilities, services and revenue streams. At the same time, they are extending their commercial reach and economic impact well beyond airport boundaries (Kasarda, 2008).

The aerotropolis consists of a multimodal airport-based commercial core (Airport City) and outlying corridors and clusters of aviation-linked businesses and associated mixed-use commercial/residential developments that feed off each other and their accessibility to the airport. Kasarda has claimed that "The 20th century was about cities building airports. The 21st century will be about airports building cities" (Aerotropolis, 2024). Airports such as Amsterdam Schiphol and Seoul Incheon can be classed under this concept, and alongside the symbiotic relationship between air transport and *global cities*, this will be the final concepts analysed in the first part of Chapter 11.

1.3.4 Rural geography

Whilst air transport is often associated with urban areas and international travel, the importance of air transport in the context of rural geography cannot be under-estimated. Routes to remote or rural communities can often be seen as "niche", however they are certainly not niche to the communities that depend on them. Figure 1.6a shows that the proportion of people living in rural communities has steadily declined over time, from a figure of 66% in 1960 to 43% in 2022 (World Bank, 2024b). However, this still means well over 3 billion people are living in rural areas and some countries are more dominated by rural populations than others (Figure 1.6b).

Since many air services to rural, remote or island destinations may not be commercially viable due to low passenger or cargo volumes, they may not be provided by the market in the

b)

Nation	% rural population (2022)
Niger	83%
Rwanda	82%
Sri Lanka	81%
Ethiopia	77%
India	64%
Philippines	52%
China	36%
Germany	22%
USA	17%
UK	16%
Japan	8%
Singapore	0%

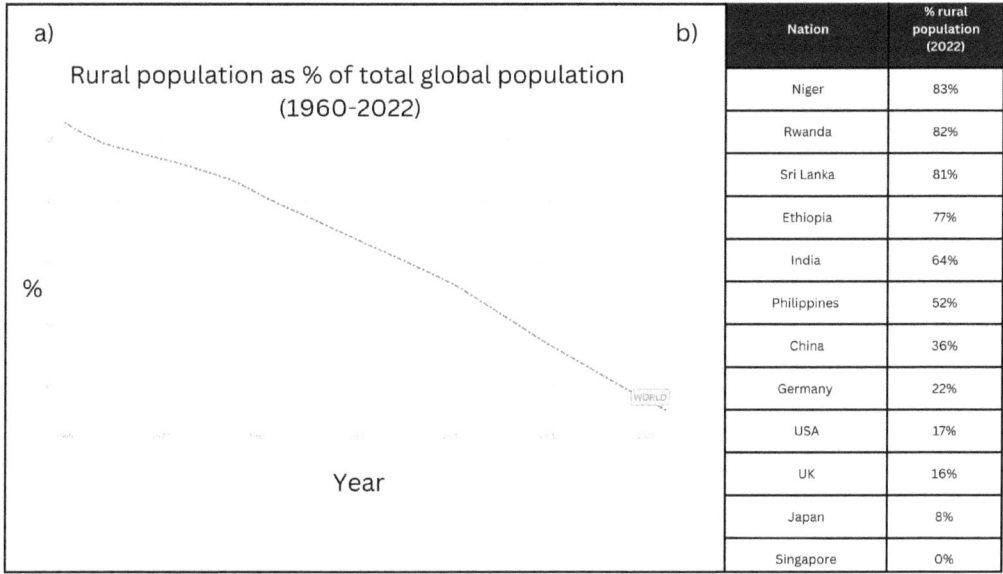

Figure 1.6a Global rural population as % of total population, 1960–2022

Source: adapted from World Bank (2024b)

Figure 1.6b Rural population as % of total – selected countries, 2022

Source: adapted from World Bank (2024b)

absence of government support. The result is that choice may be limited or non-existent. These services may drive economic development, primarily through stimulating inward tourism and investments, and also provide a lifeline for the local communities. Tourism is increasingly being recognised by the international community and its institutions as a focal instrument for development, with special emphasis on the capacity of the sector to help alleviate poverty. For many rural and remote regions in particular, tourism is often, or has the potential to be, their major economic driver (Industry High Level Group, 2019).

The second part of Chapter 11 will focus on rural, remote and island regions and the importance of air transport, analysing accessibility and connectivity, economic development, tourism and access to other facilities and services, as well as operational considerations.

1.3.5 Population geography

Pacione (2011) stated that there was no one definition of population geography. Geographies of population, spatial perspectives in population studies and population geography as a separate entity are three areas of this multi-disciplinary specialism. For the purpose of this book – which is applied in the context of air transport – population geography can be defined as a division of human geography that focuses on the study of people, their spatial distributions, their characteristics and their density. Population geographers seek to understand the society around them, the structure of a population and how populations change through population movements and processes (Newbold, 2017).

The concept of "mobilities" is key in this context, and Hannam, Sheller and Urry (2006) state that this encompasses the large-scale movements of people, objects, capital and information across the world. Air transport is a critical component, and an understanding of population geography and the influence air transport has had on mobility can help to explain larger-scale spatial patterns.

As an example, the growth in LCC travel within Europe in the early 2000s, combined with the enlargement of the EU to incorporate many countries from Eastern Europe, such as Poland in 2004 and Bulgaria and Romania in 2007 (European Union, n.d.) led to large-scale population movements from some of these countries into countries such as the UK. EU expansion enabled free movement of workers from more countries, and rapidly increasing and more affordable air transport links facilitated this movement. Chapter 12 will focus on population, mobilities and air transport in more detail.

1.4 Environmental geography

Environmental geography is at the core of geography. It explains the complex interactions between people and the environment and has an impact in both our own neighbourhoods and the entire planet (Duram, 2018).

> **ATG Did you know?**
>
> **Environmental geography**
>
> "Geo" means "earth" and "graphy" "to write", whilst "environ" means "surroundings" and "ment" is "the result of". Thus, environmental geography is centred on the interactions and relations of the biogeophysical environment with human societies (Zimmerer and McSweeney, 2020). Or more specifically in our case, with air transport.

Introduction to geography and transport 19

Environmental issues are at the forefront of life in the 21st century. Indeed, the future of our planet is in danger if this focus does not stay that way and if our actions in stabilising and reducing our impacts are not successful. In the context of air transport, there are a few areas which impact the physical environment – from those at the local level, such as aircraft noise and air pollution, to biogeographical/hydrological impacts, geomorphic change via airport development and global climate change – and these will be analysed in Part IV.

1.4.1 Geomorphic change

Geomorphology is the study of landforms and the processes that shape them. With the large land requirements for airports and the requirement to construct airports in specific locations, such as near coastlines, in hilly and mountainous areas and even on reclaimed land where enough suitable land is not available, geomorphic change can occur. Local relief, site hydrology and physical obstacles may necessitate these changes, especially from a safety perspective, and it is incumbent on the air transport industry to consider its environmental implications.

As Table 1.1 illustrates, these geomorphic changes can range from major changes, such as hill removal and artificial islands, to second-order effects such as ground erosion and enhanced weathering of exposed rock slopes. The causes, effects and attempts to mitigate geomorphic change due to airport developments will be assessed in Chapter 13.

Table 1.1 Landform changes due to airport building.

Major	*Minor*	*Second-order effects*
Land reclamation	Earth embankments	Ground subsidence
Coastline alterations	Drainage ditches	Enhanced weathering of exposed rock slopes
Hill removal	Sea-bottom dredging	Rill erosion
Artificial islands	Tunnels	

Source: adapted from Pijet-Migon and Migon (2018)

ATG Case study 1.3

Chek Lap Kok Airport, Hong Kong (HKG)

On 6 July 1998, in a mammoth overnight operation, airport operations in Hong Kong moved from Kai Tak Airport, in the congested area of Kowloon, to a small island 30 kilometres to the west (Figure 1.7a). Chek Lap Kok became an enormous infrastructure programme, including bridges, expressways, the Airport Express railway, two runways and vast land reclamation (Cathay Pacific, n.d.).

Originally, the island's area was only 302ha, but 938ha were then reclaimed from the sea and the shoreline was moved 5km further west. A 13-km-long seawall surrounded the levelled airport island. The marine component of the works required the removal of 69 million m^3 of marine mud from the reclamation area and the deposition of 76 million m^3 of sand excavated from offshore borrow pits. The earthworks operation on land entailed the excavation of 122 million m^3 of rock. The rock excavation for the anchorages of the Tsing Ma Suspension Bridge, carrying the road and railway to the airport, required the removal of nearly 1,000,000m^3 of material (Douglas and Lawson, 2003). The new third runway system (Figure 1.7b) added a further 650ha of reclaimed land (Cathay Pacific, n.d.).

20 *Fundamentals of Global Air Transport Geography*

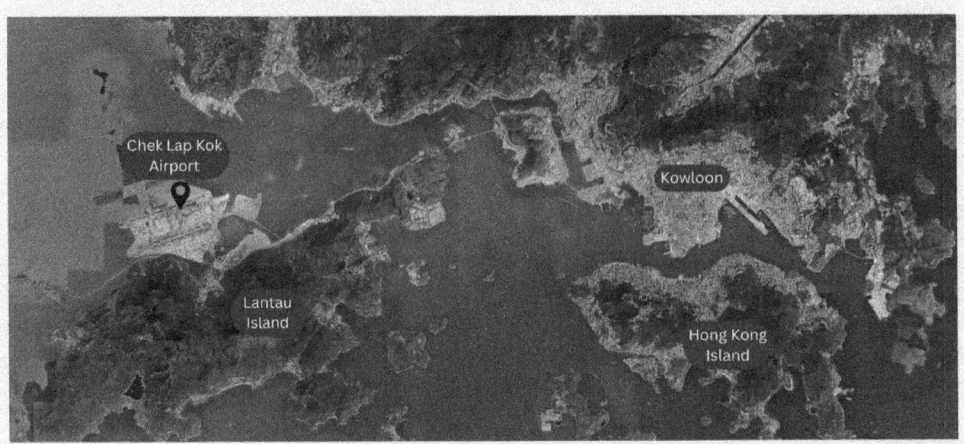

Figure 1.7a Location of Chek Lap Kok Airport
Source: adapted from Google Maps (n.d.)

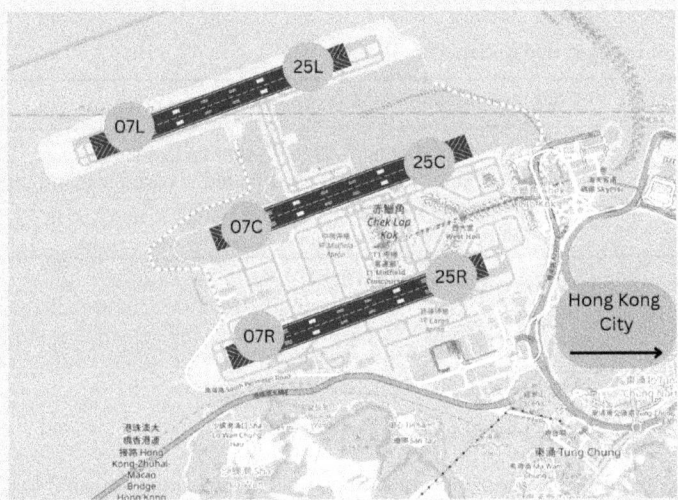

Figure 1.7b Chek Lap Kok Airport three-runway system
Source: adapted from OpenStreetMap (n.d.)

1.4.2 Climate change

Climate change refers to long-term shifts in temperature and weather patterns. The Intergovernmental Panel on Climate Change (IPCC) – the United Nations (UN) body for assessing the science related to climate change – states that human activities, principally through emissions of greenhouse gases, have unequivocally caused global warming, with global surface temperatures in 2011–2020, reaching 1.1°C above 1850–1900 temperatures (IPCC, 2023).

Chapter 6 will assess the impacts of global warming and climate change on the air transport industry, however the industry itself is by no means an innocent bystander in the causes and effects of climate change. According to the IEA (2023), in 2022, aviation accounted for 2% of global energy-related CO_2 emissions. Between 1987 and 2018, air transport industry CO_2 emissions doubled, whilst air transport passenger traffic quadrupled. On this metric, air transport emissions on a per passenger basis have significantly improved but the problem is growth rather than individual technologies – which have generally become much more efficient on a per unit basis.

In addition to CO_2, aircraft engines also emit oxides of nitrogen (NOx) which have a net warming impact, in the short-term, due to the formation of atmospheric ozone (Lee, 2018). In addition, water vapour and soot emissions from engines may form condensation trails (contrails) depending on atmospheric conditions (Padhra and Kurnaz, 2023).

ATG Did you know?

Scientists demonstrated the heat-trapping nature of carbon dioxide and other gases in the mid-19th century.

In 1824, Joseph Fourier calculated that an earth-sized planet, at our distance from the Sun, ought to be much colder. He suggested something in the atmosphere must be acting like an insulating blanket. In 1856, Eunice Foote discovered that blanket, showing that carbon dioxide and water vapor in earth's atmosphere trap escaping infrared (heat) radiation.

In the 1860s, physicist John Tyndall recognised earth's natural greenhouse effect and suggested that slight changes in the atmospheric composition could cause climatic variations.

In 1896, a seminal paper by Swedish scientist Svante Arrhenius first predicted that changes in atmospheric carbon dioxide levels could substantially alter surface temperature through the greenhouse effect (NASA, 2024).

The air transport industry has been successful in reducing its per unit emissions due to the following four areas (IATA, 2023):

- improvements in aircraft and engine technology
- service weight reductions
- operational improvements
- greener on-the-ground infrastructure.

If air transport wishes to continue to grow, and there is a forecast doubling of demand over the next 20 or so years (Boeing, 2023), further solutions have to be found to combat both CO_2 and non CO_2. emissions. At their 77th Annual General Meeting, member airlines of the International Air Transport Association (IATA) committed to achieving *net-zero carbon emissions* from their operations by 2050, which would bring the air transport sector in line with the objectives of the Paris Agreement of 2015, to limit global warming to well below 2°C. To succeed, it will require coordinated efforts from the entire air transport industry, the introduction of game-changing new technologies such as hydrogen propulsion, further developments in sustainable aviation fuels (SAFs), continuing operational and infrastructure improvements and carbon offsetting, as well as significant government support. Chapter 14 analyses the impacts of air transport on climate change and possible mitigation strategies.

1.4.3 Air transport noise

Noise is very much a local environmental problem. According to ICAO (n.d.), aircraft noise is the most significant cause of adverse community reaction related to the operation and expansion of airports. Factors such as annoyance, cognitive impairment, sleep disturbance and cardiovascular problems have all been stated as being possible negative impacts of noise. The noise footprint from each commercial aircraft movement has reduced massively since the introduction of the jet age, however aircraft still create noise.

Politically, community opposition to air transport because of noise can have a large impact on policies and regulations. Despite the reduction in individual aircraft noise over the years, one of the key issues is industry growth and that although individual noise events may be much quieter, cumulatively there are now many more noise events. Due to this increase in movements and with the planned growth in air transport movements, if the industry is to successfully undertake its planned growth, solutions will have to be found in many parts of the world to reduce the issue of aircraft and airport noise. Chapter 15 will assess the effects of air transport noise, how it is measured, regulatory factors and mitigation strategies.

1.5 Geography of transport

Before we move on to examine the air transport industry specifically in Chapter 2, it is important to understand the broader geographical concept of transport. According to Rodrigue (2020, p.1), transport geography is a "sub-discipline of geography concerned about the mobility of people, freight and information and its spatial organization considering attributes and constraints related to the origin, destination, extent, nature and purpose of movements".

There are four main types (or *modes*) of transport which can facilitate these movements:

- roadways
- railways
- waterways
- airways

In addition, there are also pipelines, cable transport and space transport.

1.5.1 Concept of space

The concept of **space** is fundamental to the study of transport geography. Spatial interaction between two places contains a "supplying area" – which has a surplus of a commodity – and a "generating area" – which has a demand for the commodity. This is known as *spatial differentiation* (Boniface, Cooper and Cooper, 2021) with transport linking the two areas. One of the seminal authors in the sphere of transport geography was Ullman (1980), who proposed that three main factors were necessary for spatial differentiation and transport development:

1. **Complementarity.** As places differ, there is the desire to travel from one place to another. Transport can fulfil this demand – from a shopping trip to the city by train ten miles away, to a holiday from Germany to Australia by air.
2. **Intervening opportunities.** There may be competing attractions. For example, rather than flying to Australia from Germany on holiday, Thailand is on the way and may compete.

3. **Transferability or the friction of distance.** This refers to the cost (in time and money) of overcoming the distance between two places. If this cost was too high, then even complementarity and a lack of intervening opportunities would not persuade movement to take place. High airline fares, hotel costs or a lack of available flight times would fit this category.

1.5.2   The elements of transport

Faulks (1990) stated that there were four basic physical elements in any transport system:

1. **The way.** The medium of travel, for example artificial means, such as roads and railways or natural means, such as air or water. For air transport, this would be controlled by organisations – in the UK, National Air Traffic Services (NATS) manages commercial airspace.
2. **The terminal.** This provides access to the way at both ends of the transport route – airports for example – or transferring between modes, such as a railway terminal allowing access to an airport for travellers.
3. **The carrying unit.** Such as aircraft, cars, trains and ships.
4. **Motive power.** This combines with the way and the carrying unit to determine characteristics such as speed and range. From the 1950s onwards, jet air transport made air travel much quicker than with propeller driven aircraft.

These characteristics vary in importance depending on the mode of transport and Chapter 2 will explain these in the context of air transport.

1.5.3   Core principles of transport geography

Geography trades space for time and money and therefore can be a constraint. Rodrigue (2020) stated that transport geography could be understood from eight core principles:

1. The spatial linking of a *derived demand*. Air transport is a derived demand for almost all journeys (except for those who solely fly for pleasure). So, to visit another country, air travel may be chosen as the mode, but the purpose may be to travel for business or go on holiday. Air travel is a means to an end.
2. *Distance* is a relative concept involving space, time and effort.
3. *Space* is the generator, supporter and a possible constraint for air travel. The medium of transport is the air, however bad weather can sometimes hinder this, either making it unsafe or necessitating longer routings, to avoid weather phenomena such as thunderstorms.
4. The relation between space and time can *converge* or *diverge*. As air transport has become faster, it has mostly converged. However, the withdrawal of supersonic travel, via the retirement of Concorde in 2003, has meant a divergence on routes served by this mode.
5. *Locations* can be central or intermediate. They are central when attractors or generators of movements. Locations are intermediate when used as transit. In air travel, complex route networks are often categorised into two main types: *point-to-point* and *hub-and-spoke* (Figure 1.8).
6. To overcome *geography*, transport must consume space. Airports function as both regional and global activities – from the local airports allowing airlines to connect communities, to the global airport hubs such as in Dubai, London Heathrow and Istanbul – servicing much of the world.

24 *Fundamentals of Global Air Transport Geography*

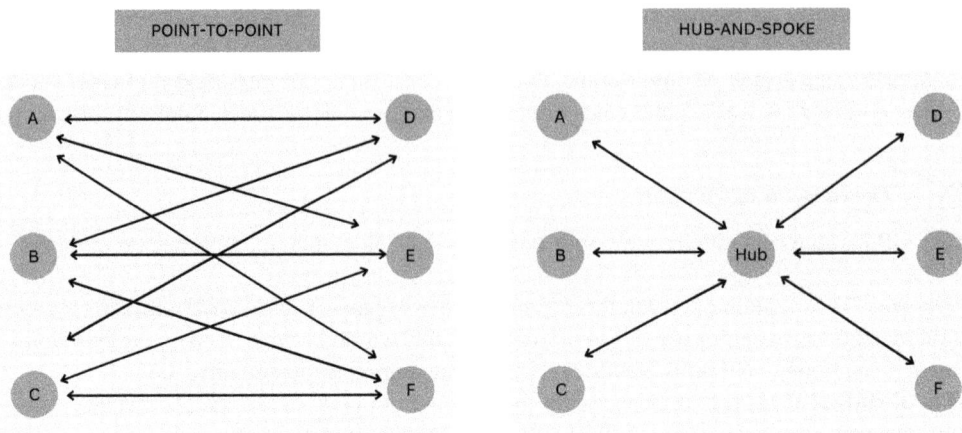

Figure 1.8 Point-to-point and hub-and-spoke networks
Source: author

7. Transportation seeks *massification* but is constrained by *atomisation*. Massification involves higher capacity – such as global airports and large aircraft such as the Airbus A380 double-decker. Atomisation may involve the first or last segment of transport. So before flying, an individual car journey may be required.
8. *Velocity* is concerned with the time taken for a passenger or unit of freight to move along the transport system and aspects such as scheduling are important. As Case Studies 1.4 and 1.5 illustrate, whilst air transport is usually the quickest form of transport, for other reasons it may not be the transport mode which is used.

ATG Case study 1.4

Transport mode choice – London to Paris

Eurostar – the high-speed rail service between London and Paris – commenced operations in 1994. Between 1996 and 2019, airline annual seat capacity declined from 4.8 million to 2.7 million seats. The number of scheduled flights reduced from 36,600 to 16,755 (although average aircraft size has increased) and the number of daily flights reduced from around 100 to 46 (OAG, 2019).

The main reason for this is that whilst flying times are generally quicker, door-to-door times were now much more competitive. For those travelling between the centres of Paris and London, the Eurostar avoids commutes to airports often on the city periphery and longer advance check-in times. The fact that there are still a considerable number of flights between these two cities demonstrates that modal choice is split between high-speed rail and air. In addition, there are also the options of car journeys between the two cities via ferries and the Channel Tunnel.

ATG Case study 1.5

Transport mode choice – France

In May 2023, new legislation was introduced in France effectively banning domestic flights, where the same journey could be made by train in less than 150 minutes and where a suitable rail service exists. This ban essentially ruled out air travel between Paris and Nantes, Lyon and Bordeaux, although transit flights were unaffected. The law specifies that the new train route must be "frequent, timely and well-connected" (Peters, 2023). The new legislation was introduced as part of France's 2021 Climate Law and is intended to offer more sustainable domestic travel and reduce carbon emissions. That said, routes between Paris and the South coast such as Marseilles and Nice are outside the scope and were not included.

Modal choice, therefore, can be directly influenced by government policies.

Transport can thus be described as an extremely important global activity – it allows us to overcome, or at least reduce, geographical constraints. Transport creates economic value and links people on the plus side, but environmental impacts can be very much a negative. Transport can have a huge influence *on* geography as well as being influenced *by* geography.

The importance of geography and the concept of space will be examined throughout the book, in terms of both physical and human geography. From a physical perspective, these constraints are one of the reasons why air transport has grown so much in the last 50 years, as technological advances have enabled this mode to significantly reduce the constraints of space. Physical constraints such as weather, climate, geomorphology, landforms, hydrology and biogeography, all have an influence on air transport systems and these will be considered in Part II, along with an overview of the air transport industry.

Chapter review questions

1.1 What are the key relationships between geography and air transport? Why is understanding geographical factors key to understanding different aspects of air transport?

1.2 What are the key sub-disciplines of physical geography which have an impact on the air transport industry? For any one of these, research your local airport. How do you think physical geography has influenced its location and development?

1.3 Explain the main aspects of human geography in the context of air transport. For your own nearest commercial airport, which airlines operate there and what are the routes offered? Is it more of an urban or rural airport? How important do you think the airport is in terms of the economic development of your region?

1.4 What are the main ways in which air transport creates environmental impacts? Which of these are local and which are more widespread? Which issues are created by your local airport?

1.5 In addition to air transport, what are the other available modes of transport? Provide examples as to how these may either complement or compete with one another.

1.6 Explain the core principles of transport geography. Choose any ONE transport mode and analyse how these principles compare to air transport.

1.7 What is the IATA three-letter and ICAO four-letter code for your local airport? Can you find logic in the choice of letters?

ATG trivia

Airport codes

Have you ever wondered what the three- or four-letter acronym on your plane ticket means, or how they are assigned? Airport codes are used to differentiate airports around the world, to ensure there is no confusion between countries and cities. The International Civil Aviation Organisation (ICAO) and the International Air Transport Association (IATA) are the two official organisations that issue airport codes, but their codes are different.

ICAO

Each airport – even small aerodromes – is assigned a unique code by ICAO. These airport codes are four-character alphanumeric codes assigned to airports worldwide. They are used for air traffic control and airline operations, making it easier to identify airports in communication and documentation.

For many ICAO codes, the first letter refers to a larger region. The second shows the country within that larger region, and the remaining two letters are a two-letter abbreviation. In Europe, codes often begin with either an E (northern Europe,) B (Greenland, Iceland and Kosovo) or L (southern Europe) (Figure 1.9). Barcelona's Josep Tarradellas Barcelona-El Prat Airport's ICAO code is LEBL – "L" denoting southern Europe, "E" for Spain (España) and finally "BL" for Barcelona.

Larger countries may have an initial letter dedicated to them, for example, Canadian airport codes start with "C". Montreal/Pierre Elliot Trudeau International Airport would be CYUL – the "C" is for Canada, and "YUL" is the specific airport code. Airport codes starting

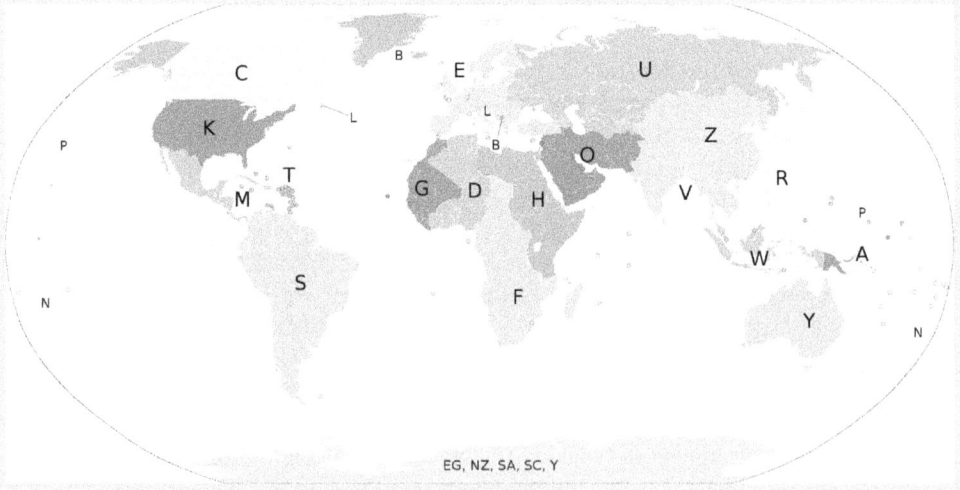

Figure 1.9 ICAO four-letter codes – first letter principle
Source: Hytar (2013)

with "K" designate US airports – so New York's John F Kennedy International airport is "KJFK" (Uniting Aviation, 2022).

IATA

IATA codes are three-letter location identifiers and are used for commercial airline purposes. They are generally those which are more passenger (and cargo) facing, such as on reservations and baggage tags. There are, therefore, far fewer IATA than ICAO codes in circulation. IATA also issues codes for non-airport locations, most commonly for train and ferry stations where there is an *intermodal connection*, for example "BPP" at Berlin Hauptbahnhof (Central Station).

A common approach for code allocation involves using the first letters of the city – e.g. MIA for Miami and MAD for Madrid – but in certain instances the code is already in use. San Diego International Airport was "SAN", therefore Santiago de Chile became "SCL". Another method would be to combine initial letters, such as DFW for Dallas Fort Worth.

So far, so good. Unfortunately, there are some airports where the codes appear to bear no resemblance to the location. AGP – Malaga (Spain) and NLU – Felipe Ángeles in Mexico City – are examples. Either way, using these codes is much better than having to type out the whole airport names, especially when some are as long as São Paulo/Guarulhos International Airport – Governador André Franco Montoro. Using the IATA code (GRU) or ICAO code (SBGR) is a little less time consuming.

On the basis that IATA codes are three letters, there are inevitably the occasional actual words thrown up and flying from DIE to HEL may not sound like the best idea, however flying between Arrachart Airport in Madagascar (DIE) and Helsinki Airport in Finland (HEL) sounds a lot more pleasant!

References

Aerotropolis (2024) *Aerotropolis*. Available at: https://aerotropolis.com/airportcity/index.php/about/.

Boeing (2023) *Commercial market outlook*. Available at: www.boeing.com/commercial/market/commercial-market-outlook/index.page.

Boniface, B., Cooper, R. and Cooper, C. (2021) *Worldwide destinations: the geography of travel and tourism*. 8th edn. Abingdon: Routledge.

Boyle, M. (2015) *Human geography: a concise introduction*. 2nd edn. Chichester, West Sussex: John Wiley and Sons.

Boyle, M. (2021) *Human geography*. 2nd edn. Chichester, West Sussex: John Wiley and Sons Ltd.

Cathay Pacific (n.d.) *The Hong Kong airport story: then, now and the future | Cathay, Cathay Pacific*. Available at: www.cathaypacific.com/cx/en_HK/inspiration/hong-kong/hong-kong-airport-story-now-future.html.

Climate.gov (2023) *El Niño & La Niña (El Niño-Southern Oscillation) | NOAA Climate.gov*. Available at: www.climate.gov/enso.

Dastrup, R.A. (2020) *Introduction to human geography: a sustainable development perspective*. CC BY-NC-SA 4.0 Deed: PressBooks. Available at: https://slcc.pressbooks.pub/humangeography/part/political-geography/.

Debbage, K.G. (2014) 'The geopolitics of air transport', in *The geographies of air transport*. Farnham: Ashgate (Transport and Mobilities Series), pp. 25–39.

Doganis, R. (2019) *Flying off course: airline economics and marketing*. 5th edn. Abingdon: Routledge.

Douglas, I. and Lawson, N. (2003) 'Airport construction: materials use and geomorphic change', *Journal of Air Transport Management*, 9(3), pp. 177–185. Available at: https://doi.org/10.1016/S0969-6997(02)00082-0.

Duram, L.A. (2018) *Environmental geography: people and the environment*. Santa Barbara: ABC-CLIO.

European Parliament (2019) *EU external aviation policy*. Available at: www.europarl.europa.eu/RegData/etudes/BRIE/2019/642221/EPRS_BRI(2019)642221_EN.pdf.

European Union (n.d.) *History of the European Union – 2000–09 | European Union*. Available at: https://european-union.europa.eu/principles-countries-history/history-eu/2000-09_en.

Faulks, R.W. (1990) *The principles of transport*. 4th edn. New York: McGraw-Hill.

flydenver (n.d.) 'About DEN', *Denver Airport*. Available at: www.flydenver.com/about-den/.

Fouberg, E.H. and Murphy, A.B. (2020) *Human geography: people, place and culture*. 12th edn. Wiley. Available at: https://ereader.perlego.com/1/book/3866011/24.

Google Maps (n.d.) *Google maps*. Available at: www.google.com/maps.

Hannam, K., Sheller, M. and Urry, J. (2006) 'Editorial: mobilities, immobilities and moorings', *Mobilities*, 1(1), pp. 1–22. Available at: https://doi.org/10.1080/17450100500489189.

Holden, J. (2011) *Physical geography: the basics*. Abingdon: Routledge.

Hytar (2013) *ICAO first letter*. Available at: https://commons.wikimedia.org/wiki/File:ICAO_FirstLetter.svg.

IATA (2023) *Net zero 2050: new aircraft technology*. Available at: www.iata.org/en/iata-repository/pressroom/fact-sheets/fact-sheet-new-aircraft-technology/.

ICAO (2018) *Annex 14 to the Convention on International Civil Aviation, Volume 1: Aerodromes*. 8th edn. Montreal: ICAO.

ICAO (n.d.) *Aircraft noise*. Available at: www.icao.int/environmental-protection/pages/noise.aspx.

IEA (2023) *Aviation, IEA*. Available at: www.iea.org/energy-system/transport/aviation.

Industry High Level Group (2019) *Aviation benefits report – 2019*. Montreal: ICAO. Available at: www.icao.int/sustainability/Documents/AVIATION-BENEFITS-2019-web.pdf.

IPCC (2023) *Climate change 2023 – synthesis report: summary for policymakers*. Intergovernmental Panel on Climate Change (IPCC). Available at: www.ipcc.ch/report/ar6/syr/downloads/report/IPCC_AR6_SYR_SPM.pdf.

Kaplan, D.H. and Holloway, S. (2014) *Urban geography*. 3rd edn. Hoboken, NJ: John Wiley & Sons.

Kasarda, J. (2008) *Airport cities: the evolution*. London: Insight Media.

Lee, D.S. (2018) *The current state of scientific understanding of the non-CO2 effects of aviation on climate*. Available at: https://assets.publishing.service.gov.uk/media/5d19c4fc40f0b609cfd97461/non-CO2-effects-report.pdf.

Marshall, T. (2015) *Prisoners of geography: ten maps that explain everything about the world*. London: Elliott and Thompson.

Marshall, T. (2021) *The power of geography: ten maps that reveal the future of our world*. London: Elliott and Thompson.

Met Office (n.d.) 'What is meteorology?', *Met Office*. Available at: www.metoffice.gov.uk/about-us/what/what-is-meteorology.

NASA (2024) *How do we know climate change is real?* Available at: https://climate.nasa.gov/evidence/#:~:text=In%201896%2C%20a%20seminal%20paper,temperature%20through%20the%20greenhouse%20effect.

National Geographic (n.d.) *Geography*. Available at: https://education.nationalgeographic.org/resource/geography-article.

Newbold, K.B. (2017) 'Population geography', in *International encyclopedia of geography*. John Wiley & Sons, Ltd, pp. 1–10. Available at: https://doi.org/10.1002/9781118786352.wbieg1040.

NOAA (2023) *What are El Nino and La Nina?* Available at: https://oceanservice.noaa.gov/facts/ninonina.html.

OAG (2019) *High speed rail vs air: Eurostar at 25, the story so far*. Available at: www.oag.com/blog/high-speed-rail-vs-air-eurostar-at-25-the-story-so-far.

OpenStreetMap (n.d.) *OpenStreetMap*. Available at: www.openstreetmap.org/copyright.

Pacione, M. (2011) *Population geography: progress & prospect (Routledge Revivals)*. Abingdon: Routledge.

Padhra, A. and Kurnaz, S. (2023) 'Aviation and climate change: becoming a climate-neutral industry', in *Challenges and opportunities for aviation stakeholders in a post-pandemic world*. Hershey, PA: IGI Global, pp. 84–108.

Peters, L. (2023) 'France begins short haul flight ban where trains are suitable', *Simple Flying*. Available at: https://simpleflying.com/france-begins-short-haul-flight-ban/.

Pijet-Migon, E. and Migon, P. (2018) 'Landform change due to airport building', in M.J. Thornbush and C.D. Allen (eds) *Urban geomorphology: landforms and processes in cities*. Amsterdam: Elsevier, pp. 101–111.

Rodrigue, J.-P. (2020) *The geography of transport systems*. 5th edn. Abingdon: Routledge.

Squire, R. and Jackman, A. (2024) *Political geography: approaches, concepts, futures*. London: Sage.

Ullman, E. (1980) *Geography as spatial interaction*. Washington: University of Washington Press.

UNDRR (2007) *Hazard*. Available at: www.undrr.org/terminology/hazard.

United Nations (2024) 'Population'. *United Nations*. Available at: www.un.org/en/global-issues/population.

Uniting Aviation (2022) 'You see airport codes every time you travel – do you know the story behind them?', *Uniting Aviation*, 5 August. Available at: https://unitingaviation.com/news/general-interest/you-see-airport-codes-every-time-you-travel-but-do-you-know-the-story-behind-them/.

US EPA (2016) *Drinking water & ground water kids' stuff: Water facts of life – Ride the water cycle with these fun facts*. Available at: www3.epa.gov/safewater/kids/waterfactsoflife.html.

US GAO (2019) *U.S. airlines: Information on DOT's oversight of and stakeholders' perspectives on foreign ownership | U.S. GAO*. Available at: www.gao.gov/products/gao-19-540r.

USGS (2002) *USGS digital orthophoto of Denver International Airport in Denver, Colorado, United States*. Available at: https://commons.wikimedia.org/wiki/File:Denver_airport_USGA_2002.jpg (Accessed: 20 February 2024).

World Bank (2024a) 'Urban development', *World Bank*. Available at: www.worldbank.org/en/topic/urbandevelopment/overview.

World Bank (2024b) *World Bank Open Data*. Available at: https://data.worldbank.org.

Zimmerer, K.S. and McSweeney, K. (2020) 'Environmental geography', in A. Kobayashi (ed.) *International encyclopedia of human geography*. 2nd edn. Oxford: Elsevier, pp. 183–192. Available at: https://doi.org/10.1016/B978-0-08-102295-5.10784-X.

2 The air transport industry

Chapter outcomes

At the end of this chapter, you will be able to:

- Understand the key regulatory developments relating to the growth of the commercial air transport sector.
- Explain the Five Freedoms of the Air and their importance to air transport.
- Describe the purpose and roles of ICAO.
- Understand the various interlinked segments within the aviation value chain.
- Understand and explain the key metrics for assessing commercial airline performance.
- Explain the key business models utilised by commercial airlines and the adaptation which has occurred within individual airlines over time.
- List the key historic commercial aircraft developments which have resulted in the types of aircraft flying today.
- Describe the main types of airliner aircraft and explain the geographical markets they serve.
- List the main aircraft manufacturers and explain the types of aircraft produced and the importance of aircraft variants and fleet commonality.
- Explain the concept of aircraft performance and payload-range.
- Describe the types of airports in operation today. Explain the different functions these serve and their key characteristics.
- Understand the different types and classifications of airspace.

2.1 Introduction

On 1 January, 1914, the world's first regularly scheduled heavier-than-air airline – the St. Petersburg-Tampa Airboat Line – took off from the Municipal Pier in St. Petersburg, Florida, USA, carrying a single passenger – Abe Pheil – and touched down 23 minutes later on the Hillsborough River in Tampa (IATA, 2023d). It would have been difficult to imagine that 109 years later, an estimated global air passenger volume of 8.6 billion people followed in Abe Pheil's footsteps in 2023 alone (ACI World, 2024).

Air transport is the movement of passengers and freight by any conveyance that can sustain controlled flight (Bowen and Rodrigue, 2020). It is the only rapid worldwide transportation network, making it essential for global business. It is a vital enabler in achieving economic growth and development around the world and facilitates integration into the global economy, providing vital local, national and international connectivity. It creates employment opportunities, helps to generate trade and develop tourism opportunities and has doubled in size approximately every 15 years. According to the Industry High Level Group (2019), prior to the impacts of the COVID-19 pandemic being felt from 2020, the air transport industry:

- carried an annual volume of 58 million tonnes of freight
- operated 38 million scheduled commercial flights
- flew 48,500 routes worldwide
- supported 65.5 million jobs
- created an economic impact of $2.7 trillion in direct, indirect, induced and tourism-connected benefits – around 3.6% of the world's gross domestic product (GDP).

It is forecast that the air transport industry will have recovered from the pandemic (on most metrics) in 2024 (ICAO, 2024a). However, this recovery will not be globally even and it still means a "missing" period of four to five years, to recover to 2019 levels.

ATG Did you know?

Air transport, aviation and aerospace

These are all related terms, but they have specific meanings:

- *Air transport* involves the movement of passengers, cargo or mail by air.
- *Aviation* is a broader term including all activities related to the design, development, production, operation and maintenance of aircraft.
- *Aerospace* is the most comprehensive term, encompassing both aeronautics and astronautics. It includes the design, development and production of aircraft and spacecraft, as well as the associated technologies, systems and infrastructure.

Aviation itself can be split into three main categories (Kearns, 2021):

1. *Airlines* are organisations which provide the commercial air transport of passengers and/or cargo. **Commercial** means the organisation charges fares for passengers and cargo movement and operates for profit and **air transport** means that people and goods are moved by aircraft.
2. *Military aviation* is the use of aircraft to support military activities.
3. *General aviation* refers to all operations that fall outside commercial airlines and military aviation, such as flight instruction, medical transport and recreational flying.

The term "air transport" is generally used in this book for consistency, as the overall focus is on commercial operations, however the words "aviation" and "aerospace" are also used where appropriate.

Whilst the movement of passengers, cargo or mail by air is a relatively straightforward definition of air transport, this can involve aircraft such as the A380, with 500+ seats, flying flights of 15 hours+, right down to small aircraft such as the DHC-6 Twin Otter, carrying less than 20 passengers to a range of remote global destinations.

Making all this possible is a broad range of industry sectors, in what can be defined as the **aviation value chain**. The commercial aviation value chain consists of a few interlinked segments, with airlines as the central node and "upstream" sectors such as aircraft and engine manufacturers, airports, air traffic services (airspace), ground services and financial institutions and "downstream" sectors such as global distribution systems (GDSs), travel agents, freight forwarders and cargo integrators. Following a discussion on the origins of air transport and the key regulatory frameworks, this chapter is then structured to analyse four of the key components of the aviation (or air transport) value chain – the "*airs*" – airlines, aircraft, airports and airspace.

2.2 The origins of air transport

After the end of World War I in 1918, the market for civil aviation began to develop. Airlines such as KLM from the Netherlands and Avianca from Columbia began operations in 1919. In this year, the first regular international passenger service began between London, England and Paris, France. Back in 1910, the International Air Navigation Conference, held in Paris, attempted to create multilateral legal principles related to air navigation. However, talks were not successful due to disagreements about sovereignty.

> **ATG Did you know?**
>
> **Sovereignty**
>
> Every state has complete and exclusive sovereignty over the airspace above its territory (ICAO, n.d.b). This is the cornerstone on which almost all air law is founded, through to the present day.

2.2.1 Paris Convention (1919)

Not only did the market for air transport begin to develop after World War I, but there was also a realisation that aircraft not only had tremendous, but also devastating, potential. For obvious reasons, aviation was a key subject at the Paris Peace Conference in 1919 and the most notable result from an aviation perspective was the Paris Convention, titled "Convention Relating to the Regulation of Aerial Navigation". The Paris Convention constituted the first successful attempt at common regulation of international air navigation and laid the foundations for air law. One of the key principles in drafting the Convention, which governed many of the 43 Articles which outlined agreements on technical, organisational and operational aspects of civil aviation, stated that: "Each nation has absolute sovereignty over the airspace overlying its territories and waters. A nation, therefore, has the right to deny entry and regulate flights (both foreign and domestic) into and through its airspace".

The Convention was ultimately ratified by 37 states, of which four countries (Bolivia, Chile, Iran and Panama) denounced it and therefore by 1940, it was in force for 33 states (ICAO, 2024b).

The International Commission for Air Navigation (ICAN) was also created, directed by the forerunner of the United Nations (the League of Nations). This was a fantastic help in drafting the annexes of the Chicago Convention, which was to come in 1944 (Kearns, 2021).

2.2.2 Chicago Convention (1944)

Following the Paris Convention in 1919, there were two further Conventions – Madrid (1926) and Havana (1928) – however there was little in the way of the advancement of international law, and the rapid growth of international air transport between the two World Wars, demonstrated the possibilities of civilian air transport.

In 1944, it seemed that World War II was coming to an end and that commercial aviation would be growing internationally. In September 1944, the United States invited governments to an international civil aviation conference in Chicago in November 1944, which 52 states attended. On 7 December, 1944 (three years to the day after the attacks on Pearl Harbor), 52 states signed the new convention, which set out as its primary objective: the development of international civil aviation "in a safe and orderly manner" such that international air transport services would be established on the basis of equality of opportunity and operated soundly and economically (ICAO, 2024b)

There were several key instruments in the Convention and of vital importance was the restating and reinforcement of the **principle of air sovereignty**:

1. *The Interim Agreement on International Civil Aviation*, which was a bridging mechanism to permit a beginning of the global effort while awaiting ratification of the Convention. This created the Provisional International Civil Aviation Organisation (PICAO) until the permanent organisation (ICAO) came into being.
2. *The International Air Services Transit Agreement* (or *"Two Freedom" agreement*), under which the aircraft of member states may:

    - Fly over each other's territory without landing (**First Freedom of the Air**).
    - Land in another country for non-traffic purposes (without picking up or dropping off passengers, cargo or mail), e.g. for refuelling or maintenance (**Second Freedom of the Air**).

3. *The International Air Transport Agreement* (or *"Five Freedoms" Agreement*). In addition to these two freedoms, a further three concerning commercial transport were established (Figure 2.1), being the right to:

    - Deliver paying passengers from a home country to a foreign country (**Third Freedom of the Air**).
    - Deliver paying passengers from a foreign country to a home country (**Fourth Freedom of the Air**).
    - Carry passengers from a home country to a foreign country, then drop off passengers, pick up new ones and carry them to a third, new country (**Fifth Freedom of the Air**).

   These Freedoms are often known as "traffic rights". The **Five Freedoms of the Air** were part of the multilateral Chicago Convention and there were a further four (to make the **Nine Freedoms of the Air**) which were not included; however, these can be agreed on a state-by-state basis through *bilateral agreements*. Chapter 9 will assess these in more detail.

4. *The drafts of 12 Twelve Technical Annexes* covered technical and operational aspects of international civil aviation, such as aircraft airworthiness, air traffic control and telecommunications. These have grown to 19 Annexes in contemporary times, some with multiple volumes, now including aspects such as Environmental Protection, which was not considered as a key aspect in 1944 (Table 2.1). Their key function is the production of standards and recommended practices (SARPs).
5. A standard form of *Bilateral Agreement* for the exchange of air routes was prepared and recommended by the Conference.

34 *Fundamentals of Global Air Transport Geography*

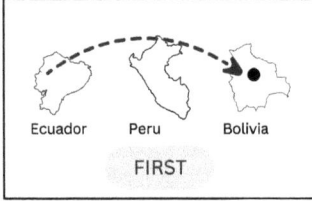
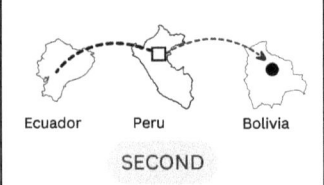
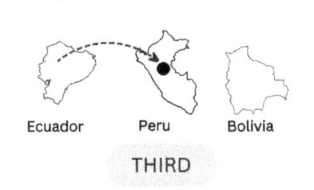
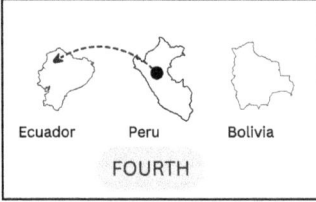
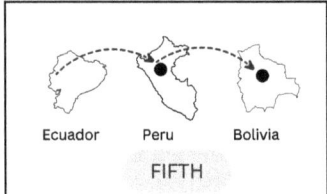

Figure 2.1 The Five Freedoms of the Air
Source: author

Table 2.1 Annexes 1–19 of the Chicago Convention

Annex 1	Personnel Licensing	**Annex 11**	Air Traffic Services
Annex 2	Rules of the Air	**Annex 12**	Search and Rescue
Annex 3	Meteorological Service for International Air Navigation	**Annex 13**	Aircraft Accident and Incident Investigation
Annex 4	Aeronautical Charts	**Annex 14**	Aerodromes
Annex 5	Units of Measurement to be used in Air and Ground Operations	**Annex 15**	Aeronautical Information Services
Annex 6	Operation of Aircraft	**Annex 16**	Environmental Protection
Annex 7	Aircraft Nationality and Registration Marks	**Annex 17**	Security: Safeguarding International Civil Aviation Against Unlawful Acts of Interference
Annex 8	Airworthiness of Aircraft	**Annex 18**	The Safe Transport of Dangerous Goods by Air
Annex 9	Facilitation	**Annex 19**	Safety Management
Annex 10	Aeronautical Telecommunications		

Source: author

2.2.3 ICAO

ICAO is a specialised agency of the United Nations, which came into being on 4 April, 1947, upon sufficient ratifications to the Chicago Convention being achieved. As a result, ICAO became the sole universal institution of international public aviation rights. For the first time in history, a single international organisation would standardise technical issues in aviation and harmonise practices between states. As outlined in Article 44, the purpose of ICAO was to develop the principles and techniques of international air navigation and foster the planning and development of international air transport so as to insure the safe and orderly growth of international civil aviation throughout the world (Convention on International Civil Aviation, 1944).

ATG Did you know?

ICAO Standards and recommended practices (SARPs)

SARPs are contained within the ICAO annexes and are applied universally, producing a degree of technical uniformity which has enabled civil aviation to develop in a safe, orderly and efficient manner.

Standards are:

> Any specification for physical characteristics, configuration, material, performance, personnel or procedure, the uniform application of which is recognised as **necessary** for the safety or regularity of international air navigation and to which Contracting States will conform in accordance with the Convention; in the event of impossibility of compliance, notification to the Council is compulsory under Article 38 of the Convention.
>
> (SKYbrary, n.d.)

Recommended Practices are:

> any specification for physical characteristics, configuration, material, performance, personnel or procedure, the uniform application of which is recognised as **desirable** in the interest of safety, regularity or efficiency of international air navigation, and to which Contracting States should endeavour to conform in accordance with the Convention.
>
> (SKYbrary, n.d.)

ICAO annexes do not have the same legally binding force as the articles of the Convention because the annexes are not international treaties subject to ratification. However, ICAO carries out audits to monitor member states' compliance with SARPs. Neither the operation of ICAO nor the adoption of SARPs contravenes the sovereignty of the signatory states. The latter can adopt more restrictive rules (European Parliament, 2016). Today, ICAO manages over 12,000 SARPs.

The Convention on International Civil Aviation was one of the most productive and successful conferences ever held (ICAO, 2024b) and the role of ICAO still underpins air transport in contemporary times, with a *Vision* to "achieve the sustainable growth of the global civil aviation system", and a *Mission* to:

> serve as the global forum of States for international civil aviation. ICAO develops policies and Standards, undertakes compliance audits, performs studies and analyses, provides assistance and builds aviation capacity through many other activities and the cooperation of its Member States and stakeholders.
>
> (ICAO, n.d.c)

ATG Did you know?

ICAO – Who is in?

According to Article 52 of the Chicago Convention, the Convention shall be open for adherence by members of the United Nations and states associated with them. States who are members of the ICAO are also named *contracting states* or *member states*.

Contracting states have usually established a Civil Aviation Authority (CAA) within their country as the regulatory body in the field of civil aviation primarily dealing with safety issues and responsible for the regulation of air transport services to/from/within the country and for the enforcement of civil air regulations, air safety and airworthiness standards. They coordinate all regulatory functions with ICAO by taking the policy decisions necessitated by the advances made in the field of *international standards and recommended practices* (SARPs) adopted in the ICAO Annexes.

As of April 2024, there were 193 contracting states, with only two non-contracting states – the Holy See and Liechtenstein (ICAO, 2024b).

2.3   The aviation value chain

The commercial air transport value chain consists of a number of interlinked segments which accounted for in excess of $1.4 trillion in revenue in 2022 and these worked together to enable approximately 4.5 billion passengers to travel in 2022 (IATA, 2024a). These are the key segments which make up the air transport industry and the remaining sections in this chapter will be structured using the key segments. Tretheway and Markhvida (2014) state that the aviation value chain can be broadly divided into upstream and downstream segments with airlines being the central node in the value chain (Figure 2.2).

The airline sector is the key node at the centre of the chain. In terms of revenue, it is by far the largest subsector (Figure 2.3) with a global revenue in 2022 of $732 billion.

The aviation value chain is not a collection of firms operating in isolation. As discussed in section 2.2.3, the development by ICAO of SARPs has led to standards and procedures being adopted globally by the various sectors, improving safety and efficiency – in areas such as airport and airfield design, navigation procedures and the facilitating of financial transaction clearing between sectors.

Profitability levels in the value chain do vary massively and despite the airline sector being the central node, it is also the sector with one of the worst historic levels of profitability. In 2022, this amounted to a net loss of $3.8 billion, although this was massively improved from a loss of $137.7 billion in 2020 during the COVID-19 pandemic and an estimated return to net profit of $23.3 billion in 2023 (IATA, 2023b). It is airlines we now turn our attention to.

2.4   The airline industry

From the earliest days of the first airlines over 100 years ago, the airline industry has transformed into a behemoth carrying an estimated 4.5 billion passengers per year, flying 38 million scheduled flights and 48,500 routes. Apart from being the means of travel from a to b, the industry is barely recognisable from its earliest days. That said, there are several airlines still operating today, which were operating under essentially the same name over 100 years ago (Table 2.2).

The air transport industry 37

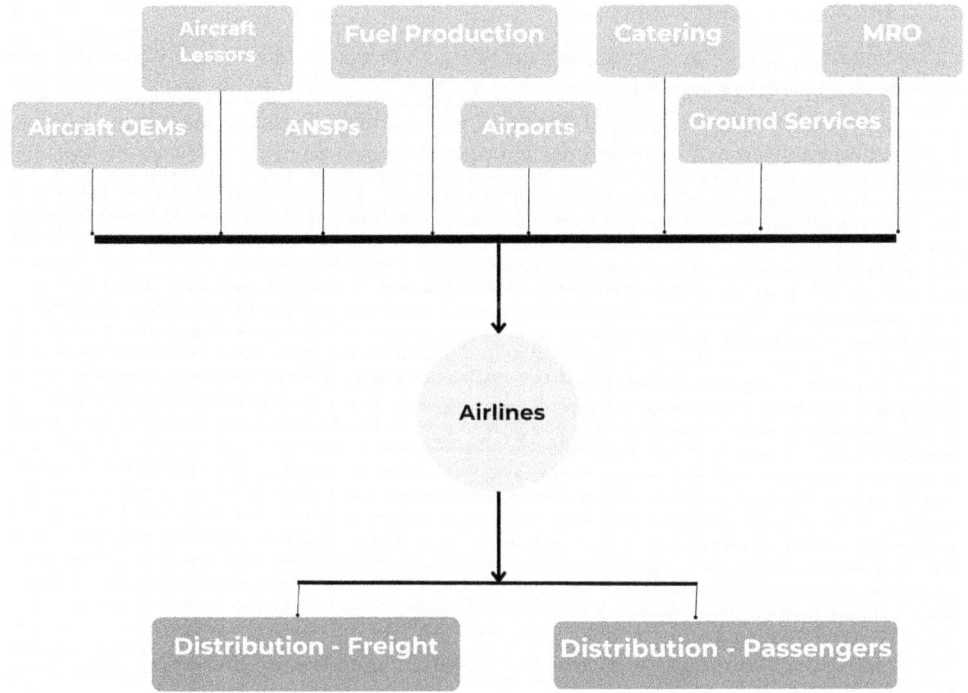

Figure 2.2 The commercial aviation value chain
Source: adapted from Tretheway and Markhvida (2014); IATA (2023a)

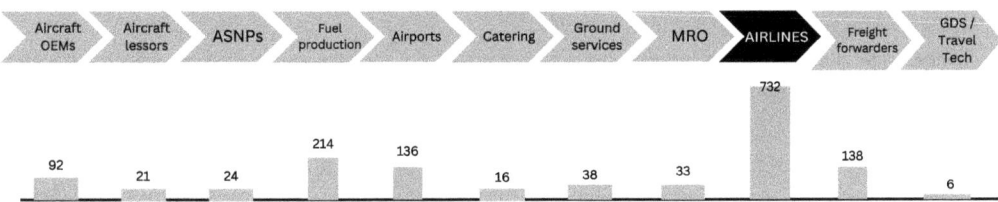

Figure 2.3 Revenues in the aviation value chain
Source: adapted from IATA (2023a)

Table 2.2 Example oldest airlines still operating

Airline	Year of foundation	Estimated 2023 passenger numbers (million)
KLM Royal Dutch Airlines (Netherlands)	1919	30.3
Avianca (Colombia)	1919	32.3
Qantas (Australia)	1920	46
Aeroflot (Russia)	1923	47.3
Finnair (Finland)	1923	11

Source: author from airline annual reports

2.4.1 Airline industry performance

The airline industry often has a reputation for being glamorous and exciting, which in some respects it is, however its financial performance equates to very slim profit margins (if at all) and a very cyclical and often very seasonal business. Willie Walsh, Director General of IATA, states that the airline industry was experiencing net profit margins more than 4% in the last decade 2010–2019. However, in 2020–2022, margins were negative following COVID-19, recovering to estimated net profit margins of 2.6% and 2.7% for 2023 and 2024 respectively (Airline Business, 2023).

The airline industry is the product of a *derived demand* – being that for almost all passengers, their ultimate purpose is to travel from a to b – and the aircraft is the means of doing that, rather than the purpose. As a result, the industry is especially open to global and national macro level factors, such as the performance of the wider economy, terrorist threats, wars, pandemics and political factors such as regulations. The economic performance of the airline industry – globally and regionally – will be analysed in Chapter 10.

> **ATG Did you know?**
>
> **Airline metric terminology**
>
> To analyse who are the "biggest" airlines in the world, the **revenue passenger kilometre (RPK)** and **cargo tonne kilometre (CTK)** metrics are often used. These are sometimes referred to as the airline's "*traffic*".
>
> **RPKs** illustrate the total number of kilometres travelled by paying passengers and is calculated as revenue passengers x distance travelled.
>
> **CTKs** illustrates the total number of kilometres travelled by cargo and is calculated as tonnes of cargo carried x distance travelled.
>
> The total number of passengers or cargo tonnage on their own can also be used to gauge the traffic carried by airlines, but as the geographic concept of distance is so important to the air transport industry, RPKs and CTKs are often used. Transporting 200 passengers 3,000 miles involves greater production than carrying 200 passengers 300 miles.
>
> To achieve these metrics, there must be a *carrying capacity*. For passengers, **available seat kilometres (ASKs)** are used – which is the number of seats x distance travelled.
>
> **Available tonne kilometres (ATKs)** are often used either for cargo airlines or where passenger airlines also carry cargo and is the available payload (cargo and passengers) x distance travelled.

The largest airlines in the world by RPKs (Table 2.3) carry more than 100 million passengers per annum – some are global (American Airlines), some are regional (Ryanair – mainly Europe) and some are mainly domestic (Southwest Airlines – mainly USA). The difference in using RPK rather than passenger numbers to compare traffic can be evidenced by comparing Emirates (number 5 in RPKs) with Ryanair (number 8 in RPKs). Despite having approximately four times the passenger numbers of Emirates, Ryanair has fewer RPKs. This is due to the significantly higher average flight length of Emirates over Ryanair. However, this does give Ryanair the potential to drive extra (or *ancillary*) revenues (such as in-flight drinks) from four times as many people!

Table 2.3 Top ten airline groups by passenger traffic (RPKs), 2022

Rank	Airline group	State/territory	Traffic (RPKs million)	Passengers (million)
1	American Airlines Group	USA	346,939	199.3
2	United Airlines Holdings	USA	332,736	144.3
3	Delta Airlines Group	USA	314,527	175.0 est
4	Air France-KLM Group	France	237,567	83.3
5	Emirates Airline	UAE	225,867	43.6
6	IAG	UK	215,749	94.7
7	Lufthansa Group	Germany	207,035	101.7
8	Ryanair	Ireland	207,000 est	168.6
9	Southwest Airlines	USA	199,263	157.0
10	Qatar Airways	Qatar	168,000 est	31.7

Source: adapted from Flight Global (2023)

Table 2.4 Top ten cargo airlines by CTKs, 2022

Rank	Airline	Scheduled CTK (million)	Year-on-year change (%)
1	Federal Express	19,547	−5.4
2	United Parcel Service	15,889	2.3
3	Qatar Airways	14,267	−11.4
4	Emirates	10,153	−14.3
5	Korean Air	9,518	−8.7
6	Atlas Air	8,675	2.8
7	Turkish Airlines	8,318	−9.8
8	Cargolux	7,971	−7.2
9	China Southern Airlines	6,915	−14.4
10	China Airlines	6,359	−15.4

Source: adapted from IATA WATS (2022) in Jeffrey (2023)

From a cargo perspective, the pandemic witnessed an upsurge in demand and, unlike for passengers, 2021 saw greater CTKs vs 2019, whereas RPKs were massively lower vs 2019. Although lockdowns meant fewer passengers could travel, the demand for goods was still high, and with many bricks and mortar shops being closed, the demand for online shopping was huge. For air cargo carriers, 2022 was a testing year after the pandemic boom period, with the commencement of the Ukraine–Russia war and a toughening of economic conditions in many parts of the world with trade flow disruptions and rising inflation. CTKs therefore declined by 8% year on year and by 1.6% compared to 2019.

2.4.2   Airline business models

A business model is essentially a representation of how an organisation intends to make money. It describes an organisation's purpose and corporate objectives, its value proposition, its key activities and resources, customer segments, cost structure and revenue streams. Airlines have adopted a range of business models (Figure 2.4) to try to attain commercial success and there have been many successes and failures from each of these, illustrating that there is no one best way – only the right way! There are also a few crossovers between business models – there are many airlines who incorporate some of the characteristics of one business model, within another.

40 *Fundamentals of Global Air Transport Geography*

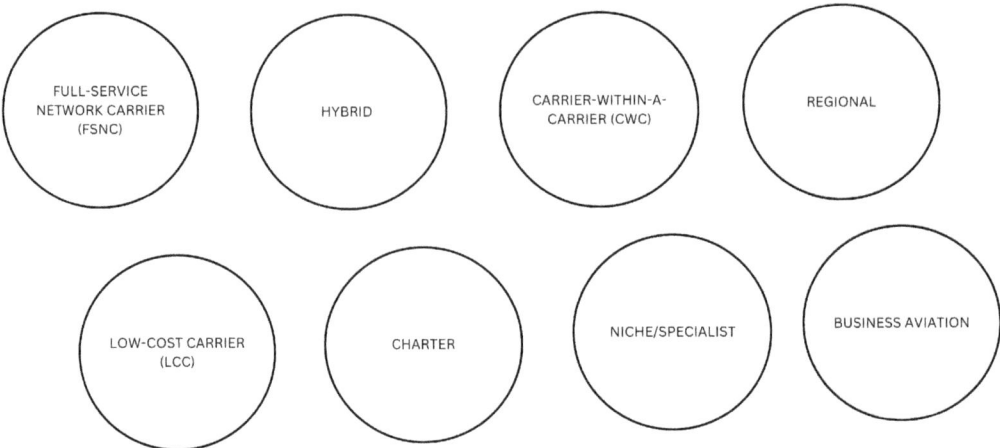

Figure 2.4 Airline business models
Source: author

Full-service network carriers (FSNCs)

These are sometimes referred to as "traditional" carriers and many of these airlines started their lives (and some still are) as so-called "flag" carriers, flying the national flag of a country – such as British Airways, Kenya Airways and Singapore Airlines – and often owned by the respective governments. From the late 1970s in the USA and 1990s in Europe, the industry in those parts underwent a period of deregulation, opening markets to increasing liberalisation, meaning greater opportunities for competition and starting new airlines. In 1987, British Airways was privatised by the UK government, removing government funding and meaning it now operated as a commercial entity.

These airlines traditionally offered a range of amenities within the ticket price, such as a free hold bag, in-flight entertainment (IFE) and on-board catering. Most operate what is called a "hub-and-spoke" network (Figure 1.8) – the "hub" being a central airport that flights are routed through and the "spokes" being routes out of the hub airport. This enables them to have greater efficiency and serve a large number of destinations (Edmunds, 2015). Table 2.5 illustrates the key characteristics of the "traditional" FSNC and LCC business models.

ATG Case study 2.1

FSNC – KLM Royal Dutch Airlines (Netherlands)

KLM was established in October 1919 by a group of investors and its first director Albert Plesman. The first flights began in May 1920, between London's Croydon Airport and Amsterdam. KLM was one of the first European airlines to use Douglas aircraft, operating the DC-2, DC-3, DC-4 and DC-6, and its first jet aircraft was the DC-8 in 1960, with the Boeing 747 introduced in 1971. It was the first European airline to launch a frequent flyer programme (FFP), known as the Flying Dutchman. This became the Flying Blue program in 2005 (Hayward, 2021).

In 2004, KLM joined the SkyTeam alliance and merged with Air France and together they carry more than 77 million passengers per year, operating over 500 aircraft in 118 countries. It is the oldest airline still operating under its original name (KLM, 2024).

Table 2.5 FSNC and LCC characteristics

Characteristic	FSNC	LCC
Length of flight	Short, medium and long-haul	Mostly short-haul
Aircraft types	Operate a varied fleet, enabling them to connect smaller towns and cities and fly long trans-continental flights	Generally one aircraft type or one family – e.g. Airbus A320/A321. Cost reductions through fleet commonality
Brand	Brand segmented to appeal to differing markets	Basic and easily recognisable
Route network	Hub-and-spoke with high number of connections	Point-to-point and no connections
Pricing	Flexible, higher fares, catering for business and corporate travellers, different classes of travel (economy, business and first-class) and traditionally round-trip fares	Simple fare structures, one-way tickets and only one price available at any one time
Distribution	Direct and via travel agents (corporate and high street) who charge commission	Direct sales – originally via telephone, then mainly via internet
In-flight service	Free in-flight food and drink, checked baggage, headsets, different classes	Charge for all ancillary products such as food, drink and baggage. One class
Frequent flyer programmes	Loyalty programmes to incentivise retained business, offering mileage, executive lounge access and other perks	None offered
Airports	Primary airports	Higher use of secondary airports to reduce airport charges and turnaround times
Market	Corporate travel, travel for business, leisure passengers prepared to pay higher fares for perceived level of service	Price sensitive and less time-sensitive leisure passengers and cost-conscious business passengers
Partnerships	Often partner with other airlines via interlining, codeshare and alliances	No partnerships to reduce complexity
Cargo	Carry cargo in the aircraft belly and some have dedicated cargo aircraft (such as Korean Air)	None carried to reduce complexity, cost and minimise turnaround times
Staff	Generally higher degree of unionisation and more "legacy" contracts	More outsourcing of jobs such as ground handling, to lower costs
Aircraft utilisation	Lower utilisation due to connecting flights at hubs, busier airports, time zone differences and crew rest periods	Higher utilisation, reduced turnaround times (commonly up to 30 minutes at many airports), aircraft generally at home base at the end of the day, reducing crew accommodation costs

Source: author

Low-cost carriers (LCCs)

Low-cost carriers (often referred to, especially by themselves, as low *fare* airlines) came into being because of deregulation in the USA in 1978 (this concept will be fully examined in Chapter 9) which essentially eased the conditions for market entry. This business model is now fully established in North America, Europe and much of the Asia Pacific region, as well as parts of Central and South America. Africa has been much slower to embrace the LCC concept, although in October 2023 the LCC share of the seat capacity within Africa had grown to around one in five seats (CAPA, 2023). Airlines such as Southwest Airlines in the USA, Ryanair and easyJet in Europe and Air Asia in the Asia Pacific exploited first mover advantage to grow rapidly, especially in the decade from 2000.

These airlines specialise in keeping their costs low to offer lower fares and either stimulate demand or take demand from other carriers with higher fares. They generally have single aircraft fleets to benefit from fleet commonality, reductions in crew costs, training, maintenance and operational flexibility and a simple business model where the fare includes the seat and much else, such as bags and food, must be paid for (Table 2.5).

According to OAG (2023), there were 811 scheduled airlines in 2023, of which 102 were LCCs. Between September 2022 and August 2023, 31% of global flights were on LCCs, 35% of seats and 27% of ASKs. The top ten LCCs account for over 50% of global LCC ASKs (Table 2.6).

ATG Case study 2.2

LCC – GOL Linhas Aéreas Inteligentes S.A (Brazil)

GOL was one of the first LCCs in South America. It operated its first flight in 2001 and is now the largest LCC in South America, operating from Brazil to markets in South America, the Caribbean and USA. In 2023 it had a nearly 40% market share of the domestic Brazilian market and carried almost 31 million passengers. It is an exclusively Boeing 737 operator and is the second largest airline in South America behind LATAM (CAPA, 2024b).

There are low-cost carriers and then there are **ultra-low-cost carriers (ULCCs)** such as Ireland's Ryanair, Hungary's Wizz Air and the US Spirit and Frontier Airlines. In many ways, the LCC and ULCC modus operandi is very similar, but the ULCC has bare minimum inclusions in the fare, often involving non-reclining seats, one small carry-on bag which must fit under the seat and a more significant number of add-on fees. That said, they often offer the lowest base fares in their market and for those passengers who are price sensitive, they are often a very good option.

Table 2.6 Global share of LCC ASKs, 12 months to August 2023

Airline	ASKs (million)	Global share (%)	State/territory
Southwest Airlines	258,787	11	USA
Ryanair	222,578	10	Ireland
Jetblue	125,968	6	USA
IndiGo	119,765	5	India
easyJet	111,439	5	UK
Spirit	92,025	4	USA
Wizz Air	73,574	3	Hungary
Volaris	60,537	3	Mexico
Frontier	57,215	3	USA
Lion Air	51,742	2	Indonesia

Source: adapted from OAG (2023)

ATG Case study 2.3

ULCC – Ryanair (Ireland)

Ryanair was founded in 1985 as anything but an ULCC. It began flying between Waterford in Ireland and London Gatwick with a 15-seat Embraer Bandeirante aircraft. The loss-making airline transformed its business model, from offering business class and a FFP, to an LCC, in the early 1990s. Its CEO – Michael O'Leary – oversaw the transformation from 1994, where it is now the largest airline in Europe by passengers, operating the most aircraft – over 500 – with mostly a single fleet type of Boeing 737 (excluding its subsidiary Malta Air, which operates A320s). Throughout the last 15 years, it has regularly been one of the most profitable airlines in Europe, with its relentless focus on cutting costs, offering low base fares and charging for ancillary products.

Hybrid carriers

A key point to note is that the airline industry is extremely dynamic and requires evolution – many airlines have historically adapted their business models depending on circumstances. As illustrated, the industry tends to categorise airlines by generic business model terms such as FSNC and LCC, but in reality, one size does not fit all, and many airlines adapt – to a lesser or greater extent – characteristics from other business models. The term "hybrid" is often used to describe such airlines (Table 2.7). This could mean changing business model entirely (such as Ryanair) from an FSNC to LCC or adapting characteristics from another (LCC Southwest offering two free hold bags in the ticket price and FSNCs American Airlines, United and Delta charging for hold bags in economy on their domestic flights).

ATG Case study 2.4

Hybrid – JetBlue (USA)

JetBlue began operations in 2000 as an LCC and whilst retaining many of these characteristics, it has increasingly adopted characteristics more akin to an FSNC, such as a Premium cabin (Mint) offering lie-flat beds on transatlantic flights between New York, Boston and London, free WiFi and a loyalty programme called TrueBlue. It competes on both short-haul leisure routes as well as long-haul business.

Charter carriers

Charter flights are unscheduled flights not part of an airline's published schedule and are utilised by organisations such as tour operators, as part of inclusive tour holidays and sports teams travelling to play matches. There are two types:

- *Ad hoc.* One-off charters, such as football teams and supporters travelling to a UEFA Champions League fixture.
- *Series.* These involve multiple flights, often for an entire season, such as a tour operator operating a weekly flight from Oslo to Tenerife from May to October.

Table 2.7 Airline differences from the "traditional" FSNC or LCC business models

Characteristic	Airline example	Change
Length of flight	JetStar (Australia)	A subsidiary of FSNC Qantas, the LCC operates long-haul routes, including Sydney to Honolulu (Hawaii) which takes just under 10 hours
Aircraft types	Norwegian (Norway)	B737 operator – LCC Norwegian – acquired B787 aircraft in 2013 to fly long-haul routes between Europe and North America/Asia. These left the fleet in 2021
Route network	Frontier (USA)	ULCC Frontier operated part of its routes as hub and spoke from Denver International Airport
Pricing	easyJet (UK)	In addition to its Standard fares, LCC easyJet also offers Flexi fares, which provide additional products such as a hold bag and speedy boarding
Distribution	GOL Linhas Aéreas Inteligentes S.A. (Brazil)	LCC GOL offers its tickets for sale via online travel companies such as Expedia
Product and service	Southwest, Delta, United, American (USA)	LCC Southwest allows two free hold bags in its ticket price, whereas on domestic flights, FSNC American, Delta and United charge for hold bags
Frequent flyer programmes	Air Asia (Malaysia)	LCC Air Asia has a points-based loyalty programme – Air Asia Rewards – with benefits available such as complimentary flights and discounts on hotels and travel attractions
Airports	easyJet (UK)	LCC easyJet has always had a major focus on primary rather than secondary airports, which are generally preferred by business passengers
Partnerships	Value Alliance (Asia)	Asian LCCs Cebu Pacific, Cebgo, Jeju Air, Nok Air and Scoot were part of the Value Alliance of LCC airlines
Cargo	Cebu Pacific (Philippines)	LCC Cebu Pacific has often generated over 5% of its total revenues from cargo
Aircraft utilisation	Allegiant (USA)	ULCC Allegiant has often operated low utilisation schedules

Source: author

It is possible for a charterer to not only book the entire aircraft, but perhaps an allocation shared amongst others. Monarch Airlines of the UK was an example, where many seats were sold to multiple tour operators on the same flight, such as Airtours, First Choice, Thomson and Cosmos – called consolidated charters.

Charter airlines were, in many ways, the original low-cost airlines. They grew initially in response to the restrictive regulatory environment in force at the time (see Chapter 9) in the 1960s and 1970s (Doganis, 2019), had high aircraft utilisation, often flew overnight, had short turnarounds and chargeable amenities such as food and drink. The advent of the LCC business model saw the importance of the charter airlines decline – largely because of the timing in Europe coinciding with the rise of the internet and the growth of self-packaging the various elements of a holiday and the opportunity to fly for flexible periods, rather than for the standard 7 or 14 days. That said, there are still many charter airlines flying today and it still retains important in many markets, especially in Northern Europe to destinations in Spain, Portugal and the Canary Islands.

ATG Case study 2.5

Charter – TUI Group (Germany)

TUI AG is one of the world's leading tour operator companies, owning over 400 hotels, 16 cruise ships, 1,200 travel agencies and five airlines – essentially covering the entire tourism value chain (TUI Group, 2024).

The Group's subsidiary airlines include TUIfly, TUIfly Nordic, TUI Airways, TUI fly Belgium and TUI fly Netherlands, operating around 130 aircraft including short-haul Boeing B737 and long-haul B787 aircraft and operates as essentially a "virtual airline". The total Group revenue for Fiscal year 2023 was €20.7 billion.

Regional carriers

Regional airlines, also known as feeder airlines or commuter airlines, are airlines that primarily operate short-haul flights connecting smaller airports to larger hub airports. They typically use smaller aircraft, such as turboprops or regional jets, and focus on providing service to smaller communities that may not have direct access to major airline hubs. These airlines often have agreements with major airlines, known as codeshare agreements or feeder agreements, allowing them to operate flights on behalf of the larger carrier under their brand. This allows major airlines to extend their reach to smaller markets without having to operate flights themselves with larger aircraft, which may not be economically viable.

Regional airlines play a crucial role in the aviation industry, by providing essential air service to communities that would otherwise have limited or no access to air travel. They often operate under the umbrella of larger airline alliances or as subsidiaries of major carriers, such as Republic Airways and Mesa Airlines.

ATG Case study 2.6

Regional – SkyWest Airlines (USA)

SkyWest is one of the largest regional airlines in the world. Founded in 1972, its original fleet included a two-seat Piper Cherokee, a four-seat Cherokee Arrow, a six-seat Piper Seneca and a six-seat Cherokee Six. In 2023, it carried 38.6 million passengers with a fleet of almost 500 Bombardier CRJ and Embraer ERJ aircraft. It operates flights on behalf of Delta, United, American and Alaska Airlines to a total of 237 destinations (SkyWest, 2024).

Carrier-within-a-carrier (CWC)

A CWC model refers to a subsidiary airline operating under the umbrella of a larger parent airline, often with its own branding and business model. These allow the parent airlines to diversify their operations, enter new markets or cater to different customer segments, without diluting their main brand.

Often these were established by FSNCs to counter the threat of the LCCs by having subsidiaries with lower cost bases, who could more easily compete on price. In most instances, in the USA and Europe, these did not work – for example Air Canada Tango, Continental Lite, Delta Song, Air Canada Zip, KLM Buzz and SAS Snowflake. These essentially operated as airlines with what amounted to a "low fare-high cost" model, which is extremely challenging.

There are some examples which have been successful, such as Qantas subsidiary Jetstar – established in 2004 and making an earnings before interest and tax (EBIT) profit in 2022/2023 of US$264.4 million (CAPA, 2024a).

ATG Case study 2.7

CWC – Go (UK)

To meet the changing demand for air travel in Europe and the growth in the LCC market from easyJet and Ryanair, in 1998 British Airways established their own low-cost airline – Go – to lure the budget traveller. Go operated from London Stansted (away from the main BA hub at London Heathrow) on flights generally to destinations in Europe, using Boeing 737 aircraft. After BA CEO Bob Ayling left in 2000 and amid concerns that its low fares were taking passengers from its own BA flights, Go was sold to private equity firm 3i in 2001 and subsequently bought (ironically) by easyJet in 2002, who absorbed the airline into its own operations.

Specialist/niche carriers

Whyte and Lohmann (2020) state that specialist operators undertake low density but vital services, such as flights flown as part of public service obligation (PSO) routes (see Chapter 11). These may often use specific aircraft (including helicopters) operating from short and unprepared runways. Loganair in the UK is an example of a regional airline which operates some PSO routes, including London Heathrow to Dundee and Glasgow to Barra, Tiree and Campbeltown in remote parts of Scotland, where these may not otherwise be commercially viable.

ATG Case study 2.8

Specialist – "The Mail Plane": remote air services subsidy (RASS) scheme – Australia

RASS is part of the Australian Government's Regional Aviation Access Program (RAAP) and subsidises a regular weekly air transport service for the carriage of passengers and cargo – such as educational materials, medicines and fresh foods – to communities in remote and isolated parts of Australia, where a regular air service offers the only reliable means of transport. Populations range from six to 200 and there are ten regions under the scheme. In 2023, there were subsidised flights to 269 remote communities, by air operators including Aviair, Chartair and Northern Territory Air Services (Australian Government, 2023).

Business aviation

The International Business Aviation Council (IBAC, 2024) defines business aviation as "that sector of aviation which concerns the operation of aircraft by companies for the carriage of passengers or goods as an aid to the conduct of their business, flown for purposes generally considered not for public hire..." There are four sub-definitions:

COMMERCIAL

The commercial operation or use of aircraft by companies for the carriage of passengers or goods, as an aid to the conduct of their business and the availability of the aircraft for whole aircraft charter.

CORPORATE

The non-commercial operation or use of aircraft by a company for the carriage of passengers or goods as an aid to the conduct of company business.

OWNER OPERATED

The non-commercial operation or use of aircraft by an individual for the carriage of passengers or goods as an aid to the conduct of their business.

FRACTIONAL OWNERSHIP

The operation or use of aircraft operated by an entity for a group of owners who jointly hold minimum shares of aircraft operated by the entity. Fractional ownership operations are normally non-commercial; however, the operation of the aircraft may be undertaken as a commercial operation in accordance with the air operators certificate (AOC) held by the entity.

ICAO states that business aviation sits somewhere between commercial air transport (such as air ambulance operations, charter operations and air taxis) and general aviation (corporate operations). Its importance is illustrated in Europe, where there are almost three times as many airports serviced by business aviation, as scheduled airlines, and represents 8% of European aviation traffic (EBAA, 2024).

ATG Case study 2.9

Business aviation – NetJets

NetJets owns and controls one of the largest fleets of private jets, with around 750 aircraft worldwide. They specialise in fractional ownership, jet leases and private jet card programmes. They divide their aircraft into five classes:

- Light (Embraer Phenom 300/E)
- Midsize (Cessna Citation XLS)
- Super-Midsize (Bombardier Challenger 350)
- Large (Bombardier Challenger 650)
- Long-range (Bombardier Global 7500)

Cargo

The air cargo industry is a trade facilitator that contributes to global economic development and creates millions of jobs. According to Doc 9626 of ICAO (n.d.a), air cargo (or freight) refers to any property, other than mail, stores and passenger baggage, carried on an aircraft. Categories of air cargo include dangerous goods, live animals and humanitarian aid, as well as perishable, temperature-sensitive pharmaceuticals, chemicals, food and ornamental plants.

Air cargo is generally inherently high-value, process critical, business-to-business and pre-consumer in the supply chain. According to IATA (2024b), air cargo transports over US$6 trillion worth of goods, accounting for approximately 35% of world trade by value. In 2017 in Colombia, air freight accounted for around 12% of Colombian international trade by value but only 0.48% by tonnage. Asia, North America and the USA account for many of the major trade routes.

There are many different types of specialist companies in the air cargo industry:

1. **Integrators.** These are companies such as FedEx and DHL, which provide customers with a door-to-door service for mainly time-sensitive and small shipments. They have fully integrated operations and a direct interface with customers. In less than 20 years, integrators have grown into the largest air cargo operators worldwide (Dresner and Zou, 2020).
2. **Passenger-cargo combi.** Many airlines, such as British Airways, Etihad and Lufthansa, carry belly freight in their holds, in addition to passengers in the cabin. Those routes with regular passenger flights can then be utilised, subject to payload limitations, for freight carriage.
3. **All-cargo operators.** These are dedicated cargo aircraft by operators such as Atlas Air, Cargolux and Polar Air Cargo. They work with *freight forwarders* to secure bookings and provide door-to-door pick-up and delivery.

Cargo often has a unidirectional demand flow, for example from China (where many goods are produced) to Europe.

ATG Case study 2.10

Cargo – Cathay Cargo (Hong Kong)

Established in 1946, Cathay Pacific Airways has grown to become one of the world's leading combi cargo carriers at one of the largest cargo hubs in the world – Hong Kong. Cathay operates a fleet of 20 B747, six B747-400 Extended Range and 14 B747–8F freighters – in addition to belly capacity on many passenger aircraft. In 2022, Cathay Cargo carried more than 1.1 million tonnes of cargo (Cathay Cargo, 2024).

2.5 Aircraft

When pilot Tony Jannus landed his Benoist Model XIV airboat on the Hillsborough River in Tampa, Florida in 1914, after a 23-minute flight from St. Petersburg, carrying one passenger, would he have foreseen that 110 years later there would be 28,400 commercial aircraft in the skies (Oliver Wyman, 2024) and approximately 8.6 billion passenger journeys (ACI World, 2024) on a commercial airline flight every year? When the Wright Brothers made their flight in

1903, did they envision aircraft flying faster than the speed of sound or carrying upwards of 500 people, each watching the latest movies on their IFE screens?

The purpose of transport is to overcome space, which is shaped by constraints such as distance and time. Since those early flights, aircraft have enabled us to fly much further and quicker, such that it is possible to now fly non-stop from Singapore to New York – 15,330km (9,520 miles), in less than 19 hours, carrying 161 passengers. Aircraft development has enabled a lowering of the cost (friction) of distance and helped in the development of globalisation – in terms of economic development, tourism and mobilities.

2.5.1 Aircraft and engine historical developments

Technological developments happened apace in the early 20th century and the requirements of aircraft production during World War I and World War II significantly facilitated aircraft technological developments and milestones in the commercial air transport sector (Table 2.8). The development of the Douglas DC-3 in the mid-1930s revolutionised commercial air travel. It was faster, more reliable and had greater range than previous aircraft. First flown in 1935, the Douglas DC-3 became the most successful airliner in the formative years of air transport and was the first to fly profitably without US government subsidy – 803 DC-3 for commercial use were produced (Smithsonian, n.d.) and many are still flying.

The introduction of the de Havilland Comet into commercial service in 1952 ushered in the era of the jet age, which marked a step change in commercial air transport, although it was jet aircraft built later in the 1950s and 1960s such as the Douglas DC-8 and the Boeing 707 – which was the first truly successful passenger jetliner – that changed the commercial air transport landscape, selling 1,010 of the type. It also marked the start of the successful 700 series of Boeing aircraft.

Since the 1950s, there have been fewer fundamental changes to aircraft design, but significant improvements in terms of power, size, range and efficiency. In terms of size, the Boeing 747 – a twin-aisle, widebody "jumbo jet" – was a true gamechanger following its introduction in 1970, and sold 1,574 in its variants. Pan American World Airways was a key driver of the aircraft development, requesting an aircraft 2.5 times the size of the B707. The 747 was not just a gamechanger in terms of size and range, but also economics and its lower costs per seat mile facilitated (when regulations allowed) lower fares, opening flying to many more people. The Airbus A380, introduced in 2007, became the largest commercial passenger airliner, carrying around 500 passengers, although only 251 were built and production ceased in 2021.

The introduction of the single-aisle, narrowbody Boeing 737 in the late 1960s, and the reduction from four (B707) to three (B727) to two engines, enabled lower costs and a wider cabin. The B737 has gone through four iterations over a 50+ year period and over 18,000 had been ordered by 2024. Its narrowbody competitor – the Airbus A320 – was not in service until 1988, however alongside its variants, orders also exceed 18,000 by 2024 and the two aircraft families have been the best-selling aircraft of all time. Table 2.9 illustrates the key aircraft developments from 1949 until 2023.

The space-time-distance relationship resulted in the iconic supersonic Concorde being built, entering service in 1976, cutting flight times between London/Paris and New York to around three hours. For commercial and safety reasons, Concorde was withdrawn from service in 2003. New manufacturers such as Boom – who developed a supersonic airliner – the prototype XB-1 making its maiden flight in March 2024 – hope to place an emphasis back on speed, within the economic and environmental constraints present in the 2020s.

50 *Fundamentals of Global Air Transport Geography*

Table 2.8 Key developments in commercial aircraft development, 1783–1947

Year	Event	Key trivia
1783	First flight carrying "cargo" and "passengers"?	The first hot air balloon – designed by the Montgolfier brothers – flew in 1783. The first passengers were not actually humans, but a sheep, duck and a rooster! Later that year, the first human passenger flew.
1852	First powered, controlled airship flight	Henri Giffard's steam-powered airship travelled around 6mph and flew almost 17 miles (27km) from Paris to Elancourt, France.
1903	First powered, sustained and controlled flight of a heavier-than-air aircraft	The first flight was by the Wright Brothers in Kitty Hawk, North Carolina. The Wright Flyer flew 120ft in 12 seconds. By 1905, the Wright Flyer III had flown a 24-mile flight in 39 minutes.
1906	First powered flight in Europe	Alberto Santos Dumont, in his *14-bis* aeroplane, flew 722 feet (220 metres) in 21 seconds.
1909	First flight across the English Channel	Louis Blériot made the first flight across the English Channel in a heavier-than-air aircraft – aType XI monoplane – from Calais to Dover, in 36 minutes.
1910	First cargo flight	The first cargo flight took place in the USA between Dayton and Columbus, Ohio. Philip Parmalee piloted a Wright Model B aeroplane 65 miles (105km) carrying a 200lbs (90kg) package of silk. A race was set up between the aircraft and an express train. The aircraft won. This was also the first intermodal transport, as the cargo was brought to the start by car.
1911	First airmail flight	The first "airmail" was transported between Allahabad and Naini in India by Henri Pequet, in a Humber-Sommer biplane, bringing 6,500 letters to the destination, 13km away. The flight lasted 13 minutes.
1914	First airline	The world's first regularly scheduled heavier-than-air airline took off from the Municipal Pier in St. Petersburg on New Year's Day 1914. The airline was known as the St. Petersburg–Tampa Airboat Line. The airboat was known as Benoist Airboat Model XIV, no. 43. After a 23-minute flight, including a brief landing on the bay to make adjustments to the propeller drive chain, pilot Tony Jannus and his single passenger, Abe Pheil, touched down on the Hillsborough River in Tampa.
1919	First non-stop transatlantic flight	John Alcock and Arthur Whitten Brown completed the first non-stop transatlantic flight in a Vickers Vimy biplane, in around 16 hours, from Newfoundland to County Galway. They were knighted by King George V the following week.
1927	First solo non-stop transatlantic flight	Charles Lindbergh flew his single-engine Spirit of St. Louis from the USA to France in 33hr 30min. Interest in aviation soared after this crossing and in many ways laid the foundation for future aviation developments.
1930	Jet engine invented	Frank Whittle filed a patent for a gas turbine to propel an aircraft directly by its exhaust – the preliminary design for the turbojet engine.
1935	Introdcution of the Douglas DC-3	The DC-3 was arguably the most successful airliner in the formative period of air transport.
1939	First jet aircraft	A German physicist – Hans von Ohain – worked for Ernst Heinkel and helped to develop the world's first jet plane – the turbojet powered Heinkel He 178.
1947	First person to fly faster than the speed of sound	A Bell X-1, piloted by Captain Charles E. "Chuck" Yeager, became the first aeroplane to fly faster than the speed of sound over the Mojave Desert in the USA.

Source: author

The air transport industry 51

Table 2.9 Key developments in commercial aircraft development, 1949–2023

Year	Event	Key trivia
1949	World's first commercial jet airliner	The de Havilland DH106 Comet prototype flew in 1949. BOAC inaugurated the world's first commercial jet service in 1952, carrying passengers and mail. Passengers could travel comfortably at 460mph (100+ mph faster than the fastest propeller-driven airliner) making it a revolutionary leap in air travel. The world was suddenly a lot smaller.
1952	First commercial turboprop	The Vickers Viscount went into service with BEA and became the first commercial turboprop aircraft to conduct passenger operations.
1957	Boeing 707	This was the first in the Boeing 700 series and with the Comet struggling with design issues and customer confidence, the B707 was in a different league to the Comet. The B707-320B (1962) could carry 189 passengers, fly 5,000NM at a cruise speed of 600mph and could arguably be the first truly successful passenger jetliner – with 1,010 sales.
1958	Douglas DC-8	The DC-8 sold 556 units and had comparable speeds and passsenger capacity to the B707 and both aircraft were instrumental in pushing passenger air travel forward in terms of speed and comfort.
1968	Boeing B737	With previous aircraft generally having three or four engines, there was a desire for more economical twin-engine aircraft. The B737 has gone through four generations of variants and is one of the most successful aircraft of all time.
1970	First widebody airliner	The Boeing B747 was the first widebody airliner, with four engines and a partial upper deck. The "jumbo jet" became an icon of the skies, selling over 1,500 units in its variants, with the last unit being delivered to Atlas Air in 2023. The success of the B707 led to Pan Am requesting a much larger (2.5 times) aircraft to operate longer flights, with more passengers and lower per seat mile costs. The main deck could also be fully used for freight.
1974	First widebody twin-engined commercial aircraft	The A300 was also Airbus's first commercial aircraft and its first revenue flights were between Paris and London Heathrow with Air France.
1976	First supersonic passenger-carrying commercial aeroplane	The BAe/Aerospatiale Concorde entered regular commercial service with British Airways and Air France in 1976. It was developed when a desire for speed was great and could fly between London/Paris and New York in around three hours. It cruised at around 1,350mph (Mach 2) at an altitude of 60,000ft – 20–25,000ft higher than most other commercial aircraft. Only 14 production aircraft were ever made and due to a combination of commercial and safety factors, it operated its last flight in 2003.
1977	First extended-range twin-engine operations performance standards (ETOPS)-compliant aircraft	In 1977, the A300B4 became the first ETOPS-compliant aircraft. The draw for this with customers was that it qualified for extended twin-engine operations over water, offering more versatility in routing.
1988	First civil aircraft to introduce fly-by-wire technology	The Airbus A320 went into commercial service with Air France in 1988 and along with its A319 and A321 variants has become one of the most successful families of aircraft.
1995	Most powerful in-service engine	In 1995, the GE90 engine entered service on the Boeing 777, powering a British Airways flight between London and Dubai. The engine generates a maximum thrust of 115,540lbf (pound force).

(Cotinued)

Table 2.9 (Continued)

Year	Event	Key trivia
2008	World's largest passenger airliner	The Airbus A380 entered commercial service with SIA in 2007. The full-length double-deck aircraft typically carries around 500 passengers, but is certified to carry up to 853.
2011	First airliner to benefit from primarily composite materials	The Boeing 787 Dreamliner was the first major airliner with majority composite material. The primary advantage of composite materials is the weight that they save compared to traditional structures. When used widely in the context of an entire aircraft, this can make a big difference to its weight and, therefore, its efficiency levels and operating costs.
2018	Longest-range jetliner	The A350-900ULR entered commercial service in 2018 and is capable of flying up to 18,000km (9,700NM)
2022	World's first commercial electric passenger aeroplane	Eviation Aircraft launched its "Alice" aircraft from Washington's Grant County International Airport, travelling to an altitude of 3,500 feet on an eight-minute flight. The aircraft is desgined to fly up to 250 miles with up to nine passengers.
2023	Worlds first transatlantic flight on 100% sustainable aviation fuel (SAF)	A Virgin Atlantic Boeing 787, powered by Rolls Royce Trent 1000 engines, flew between London Heathrow and New York JFK using 100% SAF.

Source: author

Since the 1960s (Concorde excepted), rather than speed, the focus has been more on economics, passenger comfort, reducing noise and aircraft emissions – making aircraft more efficient rather than faster and generally using more fuel-efficient twin-engine aircraft. Higher composite usage (such as on the B787 and A350) making aircraft lighter, alongside aerodynamic and wing design changes and improved high-bypass ratio engines, have facilitated this improved efficiency. According to IATA (2023c), the fuel efficiency of the current aircraft fleet is around 80% better than 50 years ago

ATG Did you know?

Extended-range twin-engine operations performance standards (ETOPS)

Have you ever looked out the window of a two-engined aircraft flying over the Atlantic or Pacific and wondered what would happen if one of the engines failed?

An ETOPS approval permits twin-engined aeroplanes to operate over a route that contains a point further than one hour flying time at the approved one-engine-inoperative cruise speed. This had a massive impact on the commercial air transport industry as prior to ETOPS, transoceanic flights remained the domain of three- and four-engine aircraft, as twin-engined aircraft had to remain no more than 60 minutes from a diversion airport.

In 1977, the Airbus A300B4 became the first "ETOPS compliant" aircraft – enabling it to fly for longer from a diversion airport (90 minutes – ETOPS-90). As a guide for how contentious the ETOPS issue was, the Federal Aviation Administration (FAA) was vehemently opposed to giving twinjets permission to cross the Atlantic and according to the *Journal of Air Law and Commerce* (in Loh and Pande, 2019), when Boeing Vice-President Dick Taylor approached the FAA for a higher ETOPS certification for the Boeing 767, their response was – "It'll be a cold day in hell before I let twins fly long-haul over-water routes". However, in 1985, the FAA increased the ETOPS to 120 minutes (ETOPS-120) which was a gamechanger for the twin-engine Boeing 767. Their first ETOPS flight was operated by TWA between Boston and Paris, leading to orders for B767 from a number of operators, especially as this opened up many more transatlantic routes.

In 1990, the FAA extended the ETOPS to 180 minutes for the Boeing 777 and fast forward to today, the B777-300ER has ETOPS-330 and the Airbus A350XWB has ETOPS-370 approval, essentially covering the entire globe, save a small part of Antarctica.

2.5.2 Types of airliner aircraft

There are many different types of airliners, each fulfilling a specific niche element of flying, however these can be broadly categorised into three types: widebody, narrowbody and regional aircraft.

Widebody aircraft

A widebody aircraft has a twin-aisle and can seat up to ten across. These include aircraft such as the A350, A380, B777 and B787, and many airlines configure these to carry 300+ passengers, with the A380 carrying around 500. These generally operate on medium- and long-haul routes, although there are some exceptions, where the volume of traffic warrants widebody services on shorter routes, such as between Tokyo and Osaka in Japan – a flight of 1hr 15mins – where the A350, B777 and B787 are all operated.

Narrowbody aircraft

A narrowbody aircraft has one aisle and up to six passenger seats across, and these constitute most commercial aircraft flying today. They include the two best-selling models of all time – the B737 and A320 – and are generally used on short- and some medium-haul routes. Passenger configurations generally range from 130 up to around 250, with many being around 180 seats (most of Ryanair's aircraft are B737–800, configured with 189 seats).

Regional aircraft

Regional aircraft are typically those with a seating capacity from 19 to 130 and generally operate on shorter routes. In 2020, regional aircraft accounted for more than 30% (9,300 units) of the world's commercial aircraft fleet (Clean Aviation, 2024). These can be split into two categories:

- *Regional jets*: These can fly higher and faster and therefore can be more fuel efficient on longer journeys. They are generally preferred for longer short-haul journeys where higher capacities are desired. Aircraft in this category include the Embraer 175, the Bombardier CRJ series and the Airbus A220-100.

54 *Fundamentals of Global Air Transport Geography*

- *Turboprop*: These are propeller-driven aircraft and are generally used on shorter, low-density routes and in airports where field performance is more limited in terms of length or runway surface and are often cheaper to operate. Aircraft include the ATR 42 and the DHC Dash 8.

ATG Case study 2.11

The "Kangaroo Route" – Qantas, Australia–London

The "Kangaroo Route" refers to the commercial passenger air routes between Australia and London. This is due to the way the aircraft "hopped" on the long distance between the two countries. Qantas cites the birthdate of this route as 1947, when they began operating the entire Kangaroo route independently.

The aircraft used on this route is a microcosm of the aircraft developments between the commencement of the route and the present day. In 1947, a Lockheed Constellation (the "Connie"), carrying 29 passengers (paying an adjusted for inflation A$35,000), flew between Sydney and London, utilising seven stops (Figure 2.5) with a travel time around 58 hours. This was still quicker than the ten days on a flying boat in 1938 and the 28 days on the first flight between the United Kingdom and Sydney in 1919 (Qantas, 2019). By 1965 this had dropped to 30 hours and five stops on the B707. In 1990 the B747-400 took 23–24 hours with only one stop (often via Singapore) and by 2024, the A380–800 (485 seats) was operating the route via Singapore in around 23–24 hours.

In 2017, Qantas announced plans to take on one of the final frontiers of commercial air transport – commercially viable, direct fights between London/New York and Sydney – dubbed "Project Sunrise". In 2022, A350–1000 aircraft were ordered, with a planned four-class configuration on the routes, with 238 seats. Entry into service is currently scheduled for mid-2026 (Qantas, 2024). Flights are expected to take around 19–20 hours.

2.5.3 Aircraft manufacturers

In the narrowbody and widebody market, there is essentially a duopoly between European manufacturer **Airbus** and US manufacturer **Boeing**.

Boeing

Boeing can trace its routes back to 1916 and has been at the forefront of commercial aircraft manufacture since the B247 in 1933 and the B707 in 1958. Other notable achievements include the first commercial aircraft to reach 1,000 orders (B727), the B737, B747, B757, B767, B777 and B787 – each of which has filled a niche and been successful in the markets it set out to serve. The B737 MAX, B777X and B787 are the key product lines still being sold today (Table 2.10).

Airbus

Airbus operates not just in the commercial aircraft manufacturing sphere, but also in the helicopter, defence and space sectors. It was founded in 1970 as a consortium of French and German

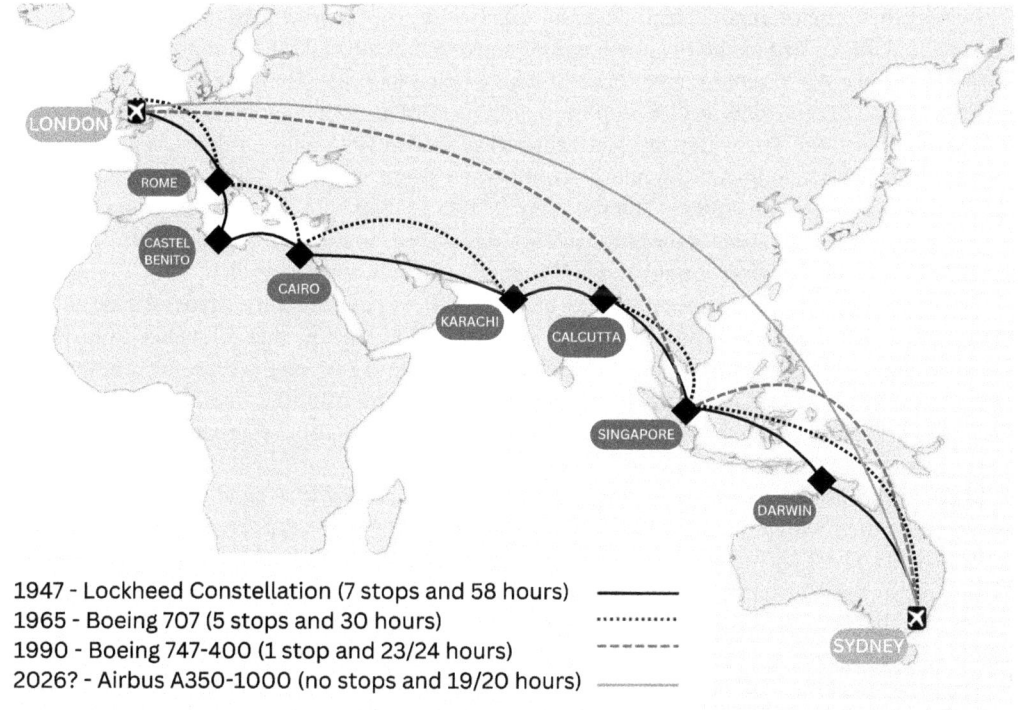

1947 - Lockheed Constellation (7 stops and 58 hours)
1965 - Boeing 707 (5 stops and 30 hours)
1990 - Boeing 747-400 (1 stop and 23/24 hours)
2026? - Airbus A350-1000 (no stops and 19/20 hours)

Figure 2.5 Qantas route developments between London and Sydney from 1947

Source: author from Qantas data

Table 2.10 Boeing aircraft currently in production

Type	Number of passengers (two-class)	Range NM (km)	Entry into service	Deliveries (as of March 2024)	Unfilled orders (as of March 2024)
B737 MAX 7	138–153	3,800 (7,040)	N/A		
B737 MAX 8	162–178	3,500 (6,480)	2017		
B737 MAX 9	178–193	3,300 (6,110)	2018		
B737 MAX 10	188–204	3,100 (5,740)	N/A		
Total 737 MAX				1,462	4,752
B767	N/A	3,255 (5,238)	1982	1,306	101
B777	317–392	7,370 (13,649)	1995	1,727	62
B777-8	395	8,745 (16,190)	N/A		
B777-9	426	7,285 (13,500)	N/A		
Total B777-8/9				0	453
B787-8	248	7,305 (13,530)	2011	396	43
B787-9	296	7,565 (14,010)	2014	626	581
B787-10	336	6,330 (11,730)	2018	96	170

Source: author compiled from Boeing (2024)

aerospace companies. Later, the conglomerate would be joined by British and Spanish aerospace firms. More than 50 years later, the French government (via SOGEPA) and German government (via GZBV) both own 10.9% each and the Spanish government (via SEPI) owns 4.1%, with the remaining shares owned by third party investors.

Airbus's first commercial aircraft development was the A300, a twin-engine widebody, entering service with Air France in 1974. A range of narrowbody aircraft was also produced, the most notable being the A320, entering service with Air France in 1988. The next in the "family" was the stretched A321, entering service with Lufthansa in 1994 and the shorter A319, entering service with Swissair in 1996. The widebody A340 entered service in 1993 with Lufthansa and Air France, and the smaller widebody A330 made its commercial debut with Air Inter in 1994. The largest commercial aircraft – the A380 – entered commercial service with Singapore Airlines in 2007. Since then, a new generation of aircraft has been developed, some from the same family – A319/A320/A321neo (new engine option) and A330neo plus the new ultra-long-haul A350 (Table 2.11).

One of the benefits for airlines of the Airbus fleet is fleet commonality, across the single-aisle A320 family and the widebody A330, A340 and A350. This means that pilots need only spend a limited amount of time upgrading from the A320 to the A380, and within the A320 family, flight crew members can perform single-fleet flying – flying multiple versions with the same type rating. Commonality benefits also extend to aircraft components, reducing maintenance costs (Edmunds, 2015).

In 2017, Airbus acquired a controlling share in a joint venture with Canadian aircraft manufacturer Bombardier, which included the C Series aircraft CS100 and CS300, which were rebranded as A220-100 and A220-300. These aircraft provided Airbus with a new niche at the smaller end of the market, as the A220-100 could be classified as a "regional" aircraft, seating between 100–130 and the A220-300 (130–160 passengers) was now in direct competition with its own A319neo, which has struggled for sales. The A220 has become a popular aircraft amongst airlines.

Since the inception of the A300 in 1974, and as of February 2024, Airbus has delivered 15,276 commercial aircraft, with orders for a total of 23,828 and a backlog of 8,552 aircraft. Of these total orders, 19,374 were for narrowbody aircraft.

Regional jet manufacturers

This market segment is dominated by Embraer of Brazil and Bombardier of Canada.

Embraer has two main aircraft types – the ERJ series and the E-Jets.

- The ERJ aircraft serves smaller demand markets, with aircraft between 30/37 seats and a range of 1,750NM (3,243km) with the **ERJ135** and 50 seats and 1,550NM (2,783km) with the **ERJ145**.

Table 2.11 Airbus aircraft currently in production

Type	Number of passengers (two-class)	Range NM (km)	Entry into service	Deliveries (as of March 2024)	Unfilled orders (as of March 2024)
A220-100	100–120	3,450 (6,390)	2016	59	43
A220-300	120–150	3,400 (6,297)	2016	263	549
A319neo	120–150	3,750 (6,950)	2022	17	45
A320neo	150–180	3,500 (6,500)	2016	1,898	2,224
A321neo	180–220	4,000 (7,400)	2017	1,282	4,887
A330-800neo	220–260	8,150 (15,094)	2020	7	5
A330-900neo	260–300	7,200 (13,334)	2018	118	166
A350-900	300–350	8,300 (15,372)	2015	505	412
A350-1000	350–410	8,700 (16,112)	2018	82	190

Source: author compiled from Airbus (2024)

- The Embraer E-Jet series focuses on the larger part of the regional market. The E-170 was introduced in 2004 seating around 72 and a range of 2,150NM (3,982km) through to the largest E195, seating around 116 and a range of 2,300NM (4,260km). An improved E2 programme was launched in 2013, with the E175-E2 seating around 88 and a range of 2,000NM (3,704km) through to the largest E195-E2, seating around 132 and a range of 2,600NM (4,815km). The E2 programme included new wing design, more fuel-efficient engines and updated avionics (Embraer, 2024).

Bombardier withdrew from commercial aircraft production in 2020, however there were 1,945 aircraft produced within the successful CRJ family.

Turboprop manufacturers

There are two main turboprop regional aircraft manufacturers – ATR and De Havilland Canada (DHC).

ATR has two main aircraft – the ATR42 and ATR72 (ATR, 2024). The two aircraft produced currently are the updated ATR42-600 (30–50 seats and a range of 726NM/1,345km) and ATR72-600 (44–78 seats and a range of 740NM/1,370km).

DHC produces the DHC Dash 8 (formerly the Q400 under previous manufacturer, Bombardier). The Dash 8–400 is the highest capacity turboprop in the market (around 90 seats and a range of 1,100NM/2,040km) and has sold around 620 units (De Havilland, 2024).

2.5.4 Aircraft performance and payload range

Performance is a term used to describe the ability of an aircraft to accomplish certain things that make it useful for certain purposes. For example, the ability of an aircraft to land and take off in a very short distance is an important factor to the pilot who operates in and out of short, unimproved airfields. The ability to carry heavy loads, fly at high altitudes at fast speeds and/or travel long distances is also essential for the performance of many commercial aircraft.

The primary factors most affected by performance are the take-off and landing distance, rate of climb, ceiling, payload, range, speed, maneuverability, stability and fuel economy. Some of these factors are often directly opposed: for example, high speed versus short landing distance, long range versus high payload and high rate of climb versus fuel economy. It is the preeminence of one or more of these factors that dictates differences between aircraft and explains the high degree of specialisation found in modern aircraft (FAA, 2023).

An aircraft's *payload* is the available weight of passengers, cargo and baggage, excluding fuel, and the *range* is how far it can fly. There is a natural trade-off between payload and range. If the maximum take-off weight (MTOW) of an aircraft is reached by the combined weight of:

- the aircraft itself,
- payload (passengers, baggage, cargo) and
- fuel

then if the range is increased beyond this, payload must be sacrificed for fuel. Aircraft can only fly further if they carry less passengers, cargo or baggage. Figure 2.6 illustrates the total typical number of two-class passengers versus the range of selected Boeing and Airbus aircraft.

58 *Fundamentals of Global Air Transport Geography*

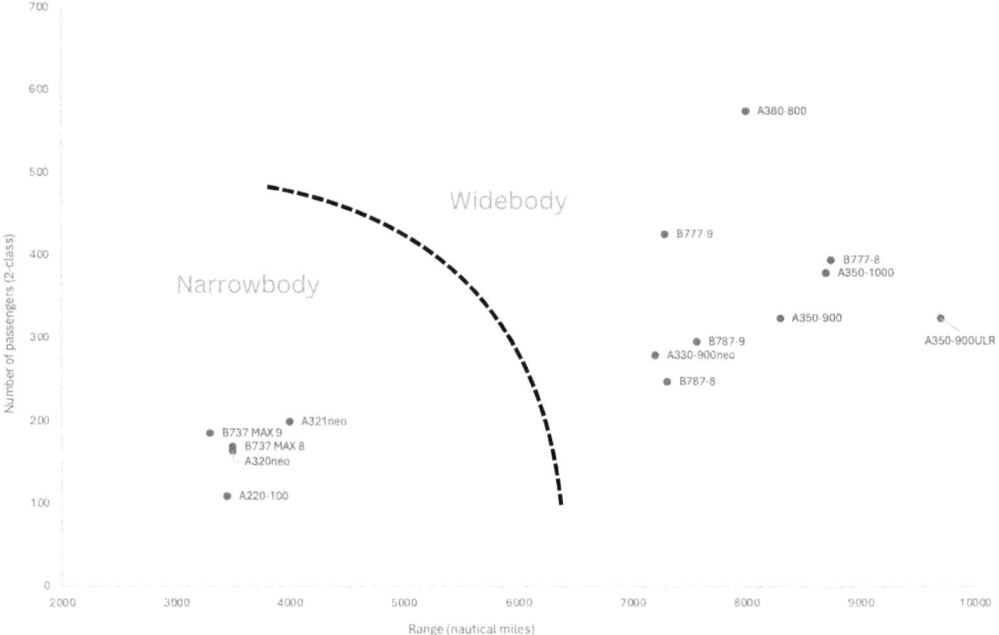

Figure 2.6 Airbus and Boeing aircraft on passengers versus range

Source: author from manufacturer specifications

2.5.5   Aircraft and the future

IATA (2023b) forecasts the demand for air transport will double by 2040, growing at an average annual rate of 3.4%. However, as with the air transport industry throughout history, it is unlikely this growth will be linear, or equally shared across the world's regions (Chapter 10). A variety of political, economic, social, technological, environmental and regulatory factors will likely conspire to impact future demand. However, as it always has, the industry will likely remain robust and able to adapt over the long term.

Aircraft demand

Every year, Boeing and Airbus produce a 20-year market forecast, in terms of the number of aircraft forecast to be required in 20 years' time – replacement plus growth. Figure 2.7 shows that the Boeing forecast is for the global aircraft fleet to double by 2042. Of this, 76% is expected to be narrowbody and 18% widebody. A key factor in achieving this will be to overcome the wide variety of environmental constraints the industry faces.

Aircraft design

To meet the IATA Net Zero 2050 emissions targets (Chapters 6 and 14), airlines are having to significantly adapt their operations. However, both IATA and ICAO believe it is unlikely the industry will be able to do this without radical new steps and key in this is expected to be related to aircraft and powerplant technologies. Aircraft with electric propulsion are in development, however due to the much higher energy density of aviation fuel versus electric power, it is unlikely that electric engines will power anything other than small, shorter-range

a) Aircraft demand: 2023-2042   b) New aircraft to 2042 by type

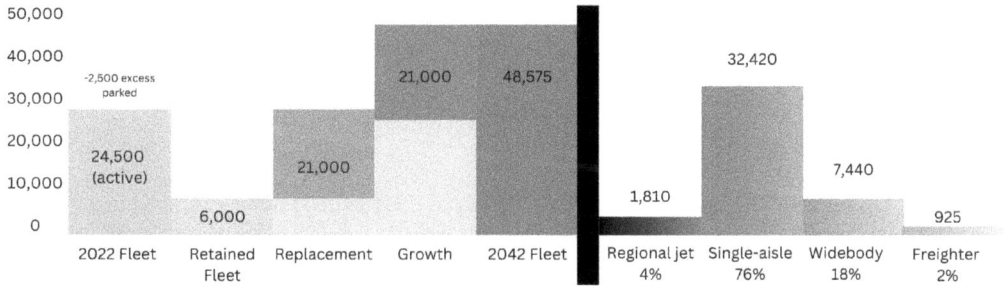

Figure 2.7 Boeing commercial aircraft forecast, 2023–2042
Source: adapted from Boeing (2023)

aircraft – at least in the short-to-medium term until the energy density of batteries improves markedly. Another possible solution in the long term is hydrogen as a fuel – either in combustion engines or fuel cells – as these reduce in-flight CO_2 emissions. Both Boeing and Airbus, as well as many other manufacturers, are working on designs to utilise these technologies.

If the air transport industry can manage its emissions footprint successfully, there is every likelihood that the aircraft we see in the skies in the future will look very different to those we see today.

2.6 Airports

At its core, an airport is a place where aircraft take off and land. They facilitate the transfer of passengers, baggage and cargo from the landside component to the airside. Airports are transport *nodes*, which act as a point of connection or intersection in a transport network. They act as an interchange between arriving at the airport by surface access (car, train, etc.) and departing by aeroplane. While the core purpose of an airport has not changed since the dawn of airports over 100 years ago, their variability in design, developments and architecture has led to the saying "if you have seen one airport … you have seen one airport".

2.6.1 Types of airports

Airports range in size from the smallest local airport to airports serving regions and/or national networks, through to large international and global megahub airports. According to Bowen and Rodrigue (2020), airports are embedded at several different *scales*:

- **Global**: Key nodes in the global economy and their importance as a function of:
  - *centrality* – the role as an origin and destination gateway to a surrounding region
  - *intermediacy* – the degree to which a node serves as an interchange between different regions. The key global airports enjoy centrality, intermediacy, or both, for example Dubai.
- **Regional/national**: Most flights do not cross international boundaries and around 80% are in the same region. Airports help to bind together nations and regions, for example the dense interregional network of flights in Southeast Asia.

- **Local**: Airports fulfil key community features, by allowing connection to other areas and producing beneficial impacts, such as job creation, but also negative ones in terms of environmental impacts.

Kearns (2021) states that airports can be categorised based on: route network, purpose, traffic characteristics or business model:

Route network

- **Hub airports**: Like the hub of a wheel, these airports are central nodes, with routes (spokes) feeding out to a number of small and large airports. They operate in waves, feeding in and out of one central airport, and the biggest airports in the world, such as Atlanta Jackson and Istanbul, are global hub airports.

    Table 2.12 illustrates the airports with the largest passenger numbers in 2023, each of which can be categorised as a hub airport, with five of the top ten based in the USA – the largest being Atlanta, with over 104 million passengers in 2023. Five of the top ten have still not recovered from their pre-COVID peaks in 2019, however airports such as Tokyo Haneda and London Heathrow have seen significant increases from 2022, as COVID restrictions were lifted and demand started to recover. Overall, 2023 represented a 27.2% increase from 2022 and a recovery of 93.8% from 2019, with an overall 2023 passenger volume of close to 8.5 billion (ACI World, 2024). Chapter 10 will analyse the factors behind these trends.

    In terms of aircraft movements, in 2023, on top spot again was Atlanta, with over 2,000 movements per day! Some of the largest airlines in the world have major hubs at these top airports – Delta at Atlanta and American and United Airlines at Chicago O'Hare. Eight of the top ten airports were in the USA (Table 2.13), which is as a result not only of the hub-and-spoke business model, but also the fact that the average size of aircraft at these airports is significantly smaller than those at airports such as Dubai (DXB) and London Heathrow (LHR). Many major US airports have significant regional (or commuter) feed, with smaller regional jets feeding from many smaller towns and cities. A significantly higher number of runways is required to accommodate this type of operation and a lack of runway capacity is a key factor constraining growth at airports such as LHR.
- **Reliever airports**: These airports are smaller than hub and generally accept overflow traffic from a hub, for example Fullerton Municipal Airport (FUL) in California is a reliever airport for Los Angeles (LAX).

Purpose

- **International**: Requiring facilities such as customs and immigration.
- **Regional**: These generally have shorter runways, general aviation and either feed larger airports or serve local communities, such as Newquay in Cornwall, UK.
- **Local**: These support general aviation activities and are often called aerodromes or airstrips.

Traffic characteristics

- **Origin-destination**: Those which serve over 70% of passengers who begin or end their journey at that airport (such as London Luton Airport), requiring more landside facilities such as car parking.

Table 2.12 Top ten airports for passenger numbers, 2023

2023 rank	2022 rank	2019 rank	AIRPORT	2023 passengers	% change vs 2022	% change vs 2019
1	1	1	Atlanta, USA (ATL)	10,46,53,451	11.7	−5.3
2	5	4	Dubai, UAE (DXB)	8,69,94,365	31.7	0.7
3	2	10	Dallas/Fort Worth, USA (DFW)	8,17,55,538	11.4	8.9
4	8	7	London Heathrow, UK (LHR)	7,91,83,364	28.5	−2.1
5	16	5	Tokyo Haneda, Japan (HND)	7,87,19,302	55.1	−7.9
6	3	16	Denver, USA (DEN)	7,78,37,917	12.3	12.8
7	7	28	Istanbul, Turkey (IST)	7,60,27,321	18.3	45.7
8	6	3	Los Angeles, USA (LAX)	7,50,50,875	13.8	−14.8
9	4	6	Chicago O'Hare, USA (ORD)	7,38,94,226	8.1	−12.7
10	9	17	New Delhi, India (DEL)	7,22,14,841	21.4	5.4

Source: adapted from ACI World (2024)

Table 2.13 Top ten airports for aircraft movements, 2023

2023 rank	2022 rank	2019 rank	AIRPORT	2023 movements	% change vs 2022	% change vs 2019
1	1	2	Atlanta, USA (ATL)	7,75,818	7.1	−14.2
2	2	1	Chicago, USA (ORD)	7,20,582	1.3	−21.7
3	3	3	Dallas/Fort Worth, USA (DFW)	6,89,569	5	−4.2
4	4	5	Denver, USA (DEN)	6,57,218	8.1	4
5	5	8	Las Vegas, USA (LAS)	6,11,806	5.3	10.6
6	6	4	Los Angeles, USA (LAX)	5,75,097	3.3	−16.8
7	7	7	Charlotte, USA (CLT)	5,39,066	6.6	−6.8
8	10	54	Istanbul, Turkey (IST)	5,05,968	18.8	53.4
9	9	21	New York, USA (JFK)	4,81,075	7.2	5.5
10	18	19	Tokyo, Japan (HND)	4,64,910	19.8	1.4

Source: adapted from ACI World (2024)

- **Transit airports**: More than 70% of passengers who transit (or gateway) through the airport on their way to another destination, requiring more transit lounges and less landside facilities (such as Atlanta Jackson).
- **Alliance hubs**: Strategically used as hubs for an airline alliance group, such as Star Alliance (Singapore), Oneworld (Dallas Fort-Worth) and Skyteam (Amsterdam Schiphol).

Business model

- **Cargo hub airports**: Focus primarily on cargo traffic (such as Anchorage, Alaska) but many also support passenger traffic (such as Shanghai, China) and the busiest of these in Hong Kong facilitated the carriage of over 4.3 million tonnes in 2023 (Table 2.14).
- **Business airport**: Catering specifically to business aviation, such as LeBourget in Paris, France.
- **Low-cost airports**: To facilitate the desire for LCC airlines to reduce their airport charges, a number of airports were developed to cater for these needs, such as Frankfurt Hahn, Germany and Brussels Charleroi, Belgium. Other airports developed specific terminals, such as Singapore Changi and Kuala Lumpur, Malaysia and some developed dedicated airside areas, such as piers in Amsterdam Schiphol, Netherlands.

Table 2.14 Top ten airports for cargo, 2023

2023 rank	2022 rank	2019 rank	AIRPORT	2023 metric tonnes	% change vs 2022	% change vs 2019
1	1	1	Hong Kong, Hong Kong (HKG)	43,31,976	3.2	−9.9
2	2	2	Memphis, USA (MEM)	38,81,211	−4	−10.2
3	4	3	Shanghai, China (PVG)	34,40,084	10.4	−5.3
4	3	6	Anchorage, USA (ANC)	33,80,374	−2.4	23.1
5	6	5	Seoul Incheon, Korea (ICN)	27,44,136	−6.9	−0.7
6	5	4	Louisville, USA (SDF)	27,27,820	−11.1	−2.2
7	8	12	Miami, USA (MIA)	25,25,591	1	20.7
8	11	8	Doha, Qatar (DOH)	23,55,503	1.5	6.3
9	9	13	Los Angeles, USA (LAX)	21,30,835	−14.9	1.9
10	7	9	Taipei, Taiwan (TPE)	21,12,988	−16.8	−3.2

Source: adapted from ACI World (2024)

2.6.2 An evolving industry

Developments in the air transport industry generally affect airlines, which in turn affect airports and nowhere is this more evident than in the regulatory sector, especially following deregulation and the liberalisation of certain air transport markets. *Flexibility* is key for the airport planner.

Hub-and-spoke networks

Post-deregulation in the USA, domestic major carriers transformed their crisscross domestic networks into radial hub-and-spoke networks – except the Delta hub at Atlanta which already existed before deregulation (Burghouwt, Mendes De Leon and De Wit, 2015). An additional impact has been that by creating *hub dominance*, an incumbent's most effective defensive tactic in a liberalised market, it offers the possibility of pre-empting or controlling competition at a particular airport (Goetz and Graham, 2004).

ATG Case study 2.12

Ethiopian Airlines and Addis Ababa Bole International Airport

Ethiopian Airlines is the largest airline in Africa (Finlay and Memon, 2024) with a fleet of 154 aircraft as of April 2024, including a mix of narrowbody B737 and DHC-8 and widebody A350, B767, B777 and B787 (Ethiopian Airlines, n.d.). Its primary hub is in Addis Ababa, which was the sixth largest airport in Africa for passengers (6,656,516), and third largest for aircraft movements (108.244) in 2022 (ACI Africa, 2023), as well as ranking fourth for cargo tonnage (226,417 tonnes).

Ethiopian Airlines is a hub-and-spoke carrier, geographically well-located in East Africa, targeting markets in the Middle East, Asia, Europe and North and South America, as well as within Africa and domestically within Ethiopia. Through its mixed fleet, it is more able to right size between demand and supply and operate shorter domestic flights, as well as long-haul inter-continental flights, connecting in "waves" via Addis Ababa.

Longer-range aircraft

With the advent of aircraft such as the Boeing 787 and Airbus A350, these aircraft can now fly extremely long routes (Singapore to New York is over 15,300km) and at lower unit costs than previous generation aircraft, meaning longer distance direct flights are now possible (such as London to Perth with Qantas) and more regional–hub flying, rather than just hub–hub, for example Glasgow to New York, rather than just Glasgow–London–New York. This changes route dynamics as well as airport geographies.

LCCs and the use of secondary airports

The growth of LCCs such as Ryanair has also created a growth in secondary airports, generally further away from major centres, where fees are lower, congestion is less and turnarounds can be quicker. Secondary airports such as London Stansted, Rome Ciampino and Bangkok Don Mueang have grown around this business model.

Airline consolidation

Airline mergers can have significant effects on airports, especially for hub airports which are no longer required by the new parent airline. Following the Delta and Northwest Airlines merger in 2008, Cincinnati Airport saw Delta make large cuts in flight capacity.

ATG Did you know?

Airport slots

A slot is permission granted for an airline to land or take off and use an airport's infrastructure (terminal, apron, gates, runway, etc.) on a particular day and within a specific time frame. Slots are used when an airport is constrained in terms of the demand from airlines exceeding available supply, sometimes at peak times, or at other airports (such as London Heathrow) for most of the day.

2.6.3 Airport constraints

The potential growth in air transport may be constrained by supply restrictions, many of which are related to airports:

- **Site location**: Airports have specific site requirements (Chapter 7), especially in relation to runways, but the overall space required and distance from urban areas are also important factors. Available land in many countries, such as the United Kingdom, is at a premium and even extending existing airports, let alone building new ones, can be difficult, as seen in the long-running London Heathrow Third Runway proposals.
- **Airport slots**: Many of the world's airports are slot-constrained – meaning supply cannot meet demand, especially at peak times – and therefore airlines are limited in terms of the

routes they are able to introduce or frequency increases. Paris CDG, Frankfurt Main and Mexico City Airports are examples.
- **Noise**: Noise restrictions have constrained many of the world's airports, in terms of types of aircraft allowed to operate and the times of airport operation, often resulting in night closures to minimise noise impacts to local residents (Chapter 15).
- **Emissions**: Airports have a key role to play in the reduction of emissions (Chapters 6 and 14). Although aircraft technology developments are the responsibility of manufacturers, the provision of storage and distribution at airports, for key factors such as future electric and hydrogen supply, are key factors in the viability of future aircraft technologies.

ATG Case study 2.13

Airports in China

Since China began to open up and reform its economy in 1978, its GDP growth has averaged 9% per year and has had the fastest sustained expansion by a major economy in history (World Bank, 2024a). In terms of air transport passengers, between 2000 and pre-COVID 2019, China experienced a growth of 965% (Figure 2.8a). Compare this to a mature market like the USA, which witnessed a growth of 39% over the same period and the scale of growth is apparent. In terms of airports, the number of civil airports has grown from 139 in 2000 to 254 in 2022 (Figure 2.8b). According to the Civil Aviation Administration of China (CAAC), China is aiming to have 400 civilian airports by the end of 2035, which is an average of ten new airports per year.

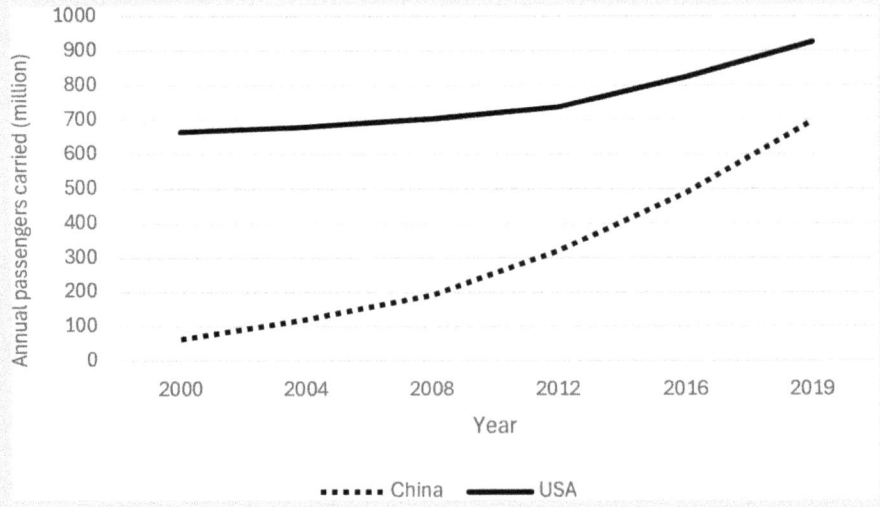

Figure 2.8a Total annual airline passengers carried, 2000–2019: China and USA

Source: adapted from World Bank (2024b)

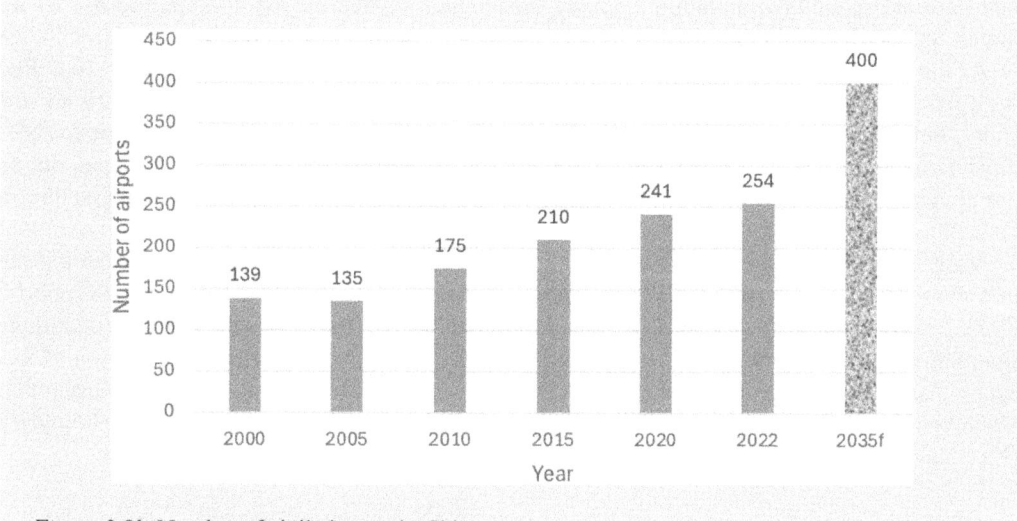

Figure 2.8b Number of civil airports in China
Source: adapted from Statista (2024)

2.7   Airspace

Airspace is that portion of the atmosphere controlled by a country above its territory. It is the medium through which aircraft fly. Think of it as a three-dimensional road or rail network in the sky, covering the entire earth in longitude, latitude and altitude. It may be invisible, however it is very strictly regulated and controlled for safety and efficiency reasons and can have an impact on an airline's cost structure, depending on how it is organised and which country's airspace the airline is flying through.

2.7.1   *Airspace horizontal boundary*

The Chicago Convention in 1944 defined sovereign national airspace boundaries, then strengthened by the United Nations Convention on the Law of the Sea (UNCLOS) in 1982. A state has complete and exclusive sovereignty over the airspace above its territory and this is defined horizontally as extending to 12 nautical miles (22.2km) from a nation's coastline. Airspace outside any country's territorial limit is classed as *international airspace* or *Oceanic*, as with the "*high seas*" in maritime law.

Subject to international agreement, specific countries may assume responsibility for managing this international airspace, for example – Portugal, United Kingdom, Canada, USA and Iceland managing airspace over the North Atlantic (Budd, 2020).

2.7.2   *Airspace vertical boundary*

There is no international agreement on the vertical limit of state sovereignty or the boundary between territorial airspace and outer space and a debate has raged for decades. The Outer Space Treaty was agreed by the UN in 1966 and provided the basic framework on

66 *Fundamentals of Global Air Transport Geography*

international space law, including: "Outer space shall be free for exploration and use by all States" (UNOOSA, 2024).

As the laws governing airspace and outer space are different, it is important to try to define the vertical extent of a territory's airspace. If you fly a satellite 55 miles above a nation's territory, but space does not begin until 60 miles, then there could be problems, particularly if those two nations are not on the friendliest of terms. It is commercially important also, due to the developing business model of **commercial space travel**, with companies such as Virgin Galactic.

Many experts say that space starts at the point where orbital dynamic forces become more important than aerodynamic forces, or where the atmosphere alone is not enough to support a flying vessel at suborbital speeds (National Geographic, 2018). The Fédération Aéronautique Internationale (FAI) states that the Kármán line at 62 miles (100km) is the upper limit of airspace. According to NASA (2019), 99.99997% of earth's atmosphere lies below this point. However, the US FAA, US Air Force and NOAA generally use 50 miles (80km) as the boundary (National Geographic, 2018).

ATG Did you know?

Kármán line

The Kármán line is named for Hungarian American aerospace engineer, Theodore von Kármán, who asked a simple question: At what altitude does the speed need to keep an aircraft aloft through aerodynamic lift become so high that it exceeds orbital velocity? Kármán calculated this figure, which is now widely used as the start of outer space – 62 miles (100km).

Commercial jet aircraft generally fly between 30,000 and 42,000ft, whilst supersonic aircraft such as Concorde flew around 55,000 to 60,000ft. In the UK, NATS (2024) defines the top height of upper airspace as flight level (FL) 600 (60,000ft).

2.7.3   *Flight information regions*

To facilitate the safe and orderly flow of air traffic, via ICAO, the earth is sub-divided into nine air navigation regions. These are defined in Appendix 1 to the *Directives to Regional Air Navigation Meetings and Rules of Procedure for their Conduct (DOC 8144-AN/874)*:

1. Africa-Indian Ocean (AFI) Region.
2. Asia (ASIA) Region.
3. Caribbean (CAR) Region.
4. European (EUR) Region.
5. Middle East (MID) Region.
6. North American (NAM) Region.
7. North Atlantic (NAT) Region.
8. Pacific (PAC) Region.
9. South American (SAM) Region.

All airspace in the world is divided further into flight information regions (FIRs), by international agreement through ICAO, and each FIR is managed by a controlling authority which has responsibility for ensuring air traffic services are provided for the aircraft flying within. Each of the nine regions contains a varied number of FIRs, which can vary in size. Some smaller countries have one FIR in their airspace, others have many. Figure 2.9 illustrates many of the global FIRs. For example, in the far west of the map is Tahiti FIR (NTTT) and in the far east is Auckland Oceanic (NZZO).

International airspace is managed, subject to international agreement, by individual countries. For example, the North Atlantic (NAT) region and the FIRs are managed by:

- Norway (Bodo Oceanic)
- Iceland (Reykjavik Oceanic)
- UK (Shanwick Oceanic)
- Portugal (Santa Maria Oceanic)
- USA (New York Oceanic East)
- Canada (Gander Oceanic)

2.7.4 Types of airspace

Airspace in each FIR is sub-divided into either controlled or uncontrolled, as well as special-use airspace (military airspace, airspace adjacent to nuclear power stations, etc.).

Controlled airspace

Controlled airspace is a defined sector of airspace where air traffic control (ATC) services are provided. This type of airspace is typically found around busy airports and in areas where there

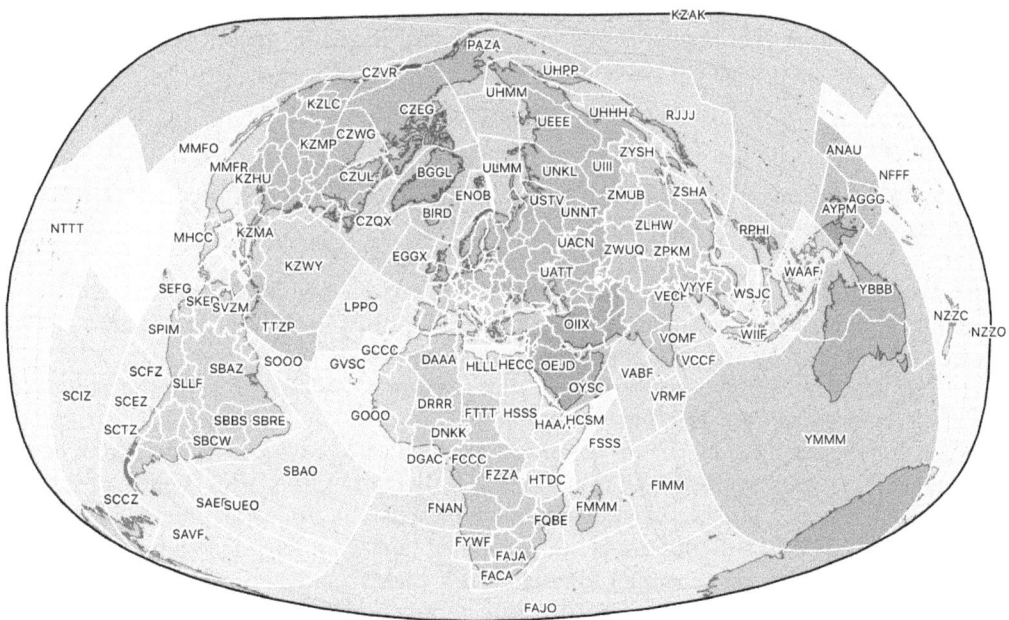

Figure 2.9 Flight information regions
Source: Olive and Spinielli (2022)

is significant air traffic. In controlled airspace, pilots must communicate with ATC and follow their instructions. ATC provides separation services between aircraft, issues clearances for flight operations, and manages the flow of air traffic to ensure safety and efficiency.

> **ATG Did you know?**
>
> The rules of the air: visual flight rules (VFR) and instrument flight rules (IFR)
> These are two sets of regulations used by pilots for navigating and operating aircraft in different weather and airspace.
>
> - **VFR** is a set of regulations where the weather conditions allow a pilot to visually separate themselves from other aircraft and terrain. There are regulated minimum weather requirements, such as a lack of clouds, and precise requirements may vary depending on country, airspace and time of day. These will then dictate minimum visibility. VFR flights typically operate in airspace where pilots can see and avoid other traffic and where ATC may not be required.
> - **IFR** is a set of regulations and procedures where pilots operate an aircraft by reference to instruments in the cockpit – flight instruments such as attitude indicators, altimeters and navigation instruments. These flights can operate in clouds and safe separation is generally the responsibility of ATC.

Uncontrolled airspace

This is where flights are not supervised by ATC and therefore no clearance is required to enter uncontrolled airspace. ATC may provide advisory information, but the pilot is responsible for maintaining separation from aircraft and terrain.

Airspace classification

ICAO defines airspace classifications in *Annex 11: Air Traffic Services, Chapter 2,* Section 2.6. This classifies airspace into seven types:

Class A. IFR flights only are permitted, all flights are provided with air traffic control service and are separated from each other. This is where air traffic is at its busiest and most complicated and users are generally major airlines and business jets.
Class B. IFR and VFR flights are permitted, all flights are provided with air traffic control service and are separated from each other.
Class C. IFR and VFR flights are permitted, all flights are provided with air traffic control service and IFR flights are separated from other IFR flights and from VFR flights. VFR flights are separated from IFR flights and receive traffic information in respect of other VFR flights.
Class D. IFR and VFR flights are permitted and all flights are provided with air traffic control service, IFR flights are separated from other IFR flights and receive traffic information in respect of VFR flights, VFR flights receive traffic information in respect of all other flights.
Class E. IFR and VFR flights are permitted, IFR flights are provided with air traffic control service and are separated from other IFR flights. All flights receive traffic information as far as is practical. Class E shall not be used for control zones.

The air transport industry 69

Class F. IFR and VFR flights are permitted, all participating IFR flights receive an air traffic advisory service and all flights receive flight information service if requested.

Class G. IFR and VFR flights are permitted and may receive flight information services if requested.

The classifications may sound complicated, however they are designed for safety and efficiency and to ensure that, for example, a large airliner departing a major airport does not get too close to a student pilot in a small training aircraft.

2.7.5   Conclusion

The large growth in air traffic has meant a corresponding increase in operational complexity within airspace regions. The reality is that not all aircraft fly the most direct, or most efficient, flight path and these inefficiencies cost substantial amounts in extra fuel and hence costs. Just as it is extremely frustrating to be driving down the motorway and becoming stuck in a traffic jam, air transport is no different. Whether it involves aircraft being held on the ground until more capacity is available, slowed down in-flight to avoid bottlenecks over large airports or diverted to less efficient routings to avoid congestion, the effective management of airspace is critical to safe and efficient air transport operations. This will also apply in the longer term, to ensure that forecast air transport growth can be met and in the most efficient way possible.

Chapter review questions

2.1 What were the key regulatory developments in the growth of the air transport sector? What are the Five Freedoms of the Air? Can you find routes which represent the rights granted on each of these Five Freedoms?

2.2 What are the key roles and responsibilities of ICAO? Research the SARPs of ICAO in 1947 – which of these are still relevant today? Which SARPs have been introduced since then? When and why?

2.3 List the key segments which make up the aviation value chain. Choose any THREE of these segments. How are they interlinked and how important are these linkages in the successful operation of the air transport industry?

2.4 Explain the key metrics used to analyse airline commercial performance. For any TWO airlines, research their annual reports for these key metrics. What are the differences and why?

2.5 List and explain the wide variety of airline business models in operation today. For any ONE FSNC or LCC airline and using Table 2.5, analyse to what extent the airline conforms to these key characteristics. Which are different and have these changed over time? Why?

2.6 What were the key aircraft developments in the commercial air transport industry? For any ONE of these developments, research its timeline and success, from beginning to end. Would you class it as ultimately successful? What influence has it had on contemporary aircraft?

2.7 What are the different types of airliner aircraft and aircraft categories? Which markets do they serve? Research any ONE airline. From their annual report or another web source, which aircraft do they have? Can you find which routes they use these aircraft on, using a flight search engine?

2.8 Who are the main aircraft manufacturers and what are their key aircraft product developments? For any ONE aircraft type, how many variants has it had? Which are still operating? If so, what is their future potential?

2.9 Explain the concept of aircraft performance and payload range. From Tables 2.10, 2.11 or manufacturer data, choose ONE aircraft type. What is its range? Using an online tool such as Great Circle Mapper (www.gcmap.com) what are the routes it can operate?

2.10 What are the different types and categories of airports? Research your local airport. Which type and category does it fit into? Are there any developments planned and what benefits and possible environmental issues may they bring?

2.11 What are the main types of airspace and how is this classified? Research the airspace around your own area – which type is it, what type of aircraft can fly, and under what conditions?

ATG trivia

Great circle route

Have you wondered why, if you've watched airline moving maps on aircraft IFE systems, the curved route seems to be taking you the long way home? Surely the shortest distance between two points is a straight line (also known as the *rhumb line*)? When you look at a standard map, as in an atlas, north is north, south is south and countries are relatively in the correct place, however as the earth is a spheroid, cartographers need to distort the map to provide a two-dimensional representation of a three-dimensional globe.

ATG Did you know?

Mercator projection

Have you ever seen a map of the world in an atlas or classroom? The chances are you have seen the work of Gerardus Mercator, a 16th-century Flemish cartographer. The Mercator projection flattens a spherical globe into a two-dimensional map, with latitude and longitude lines drawn in a straight grid. Mercator's view of the world is one that has endured through the centuries and is still widely used today (National Geographic, 2023).

The shortest distance between two points on a globe is an arc called a *great circle*. Imagine splitting the earth into two equal hemispheres and cutting in any direction, the line would be the earth's circumference, which is the Great Circle Route. Great circle routes are vital to understand how aeroplanes fly between two points. Figure 2.10 shows the difference between the Great Circle and Rhumb line between New York and Madrid. While they are both on similar latitudes, the northerly curved route is shorter.

One of the key geographical concepts in the air transport industry is the *hub-and-spoke network* (Chapter 1). Some of the key global hubs of today (such as Dubai) owe as much to their central geographical location as any other factor. A good example of this is the development of Anchorage, Alaska Airport (ANC) as a key global cargo hub. A glance at a Mercator map will show its location as being a little out of the way, however according to Flightradar24 (2023), it was the fourth busiest airport in the world for cargo throughput and in May 2023 – more than 8,000 cargo flights arrived and departed from ANC.

Figure 2.11 illustrates that the most direct (Great Circle) route between Louisville, USA (UPS cargo hub) and Tokyo, Japan, is almost directly over ANC. This central location between the key cargo regions of North America and Asia facilitated the development of ANC as a global hub, especially when many aircraft were unable to fly direct due to weight restrictions and therefore had to stop for fuel. Carrying less fuel also enabled airlines to carry much more cargo.

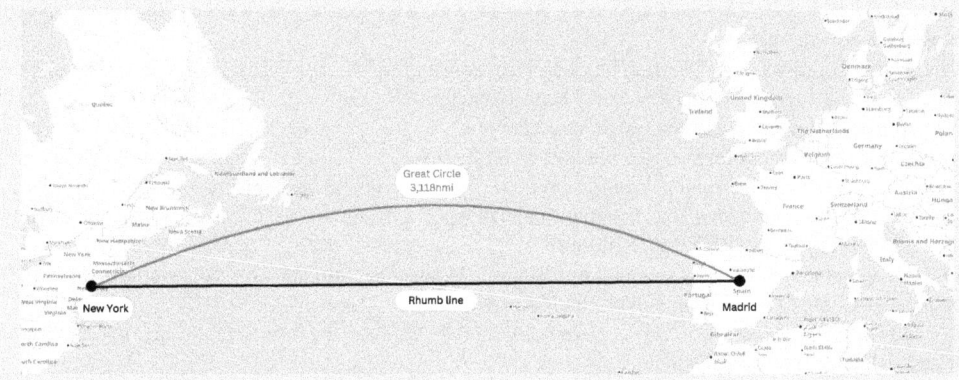

Figure 2.10 Great Circle Route between New York and Madrid
Source: author adapted using OpenStreetMap (n.d.)

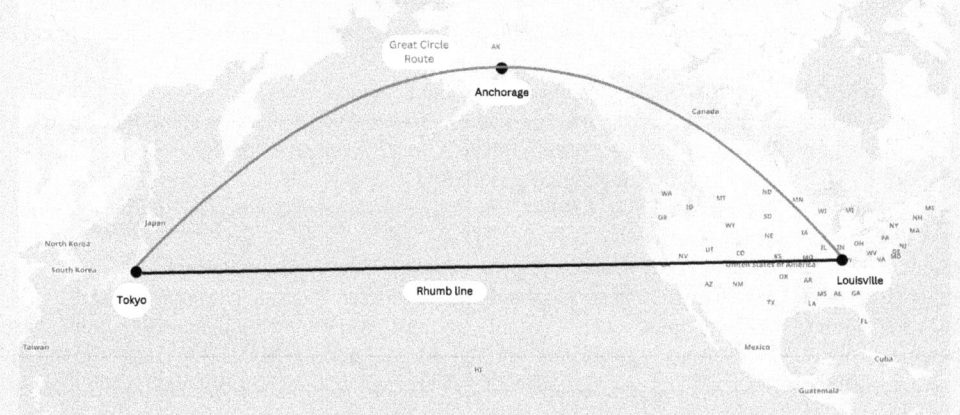

Figure 2.11 Great Circle Route between Tokyo and Louisville, via Anchorage
Source: author adapted using OpenStreetMap (n.d.)

References

ACI Africa (2023) *Air traffic performance for Africa 2022*. Available at: www.aci-africa.aero/files/Africa-Air-Traffic-Performance-2022-EN-Final-270423.pdf.

ACI World (2024) *Latest air travel outlook reveals 2024 to be a milestone for global passenger traffic*. Available at: https://aci.aero/2024/04/14/top-10-busiest-airports-in-the-world-shift-with-the-rise-of-international-air-travel-demand/.

Airbus (2024) *Orders and deliveries | Airbus*. Available at: www.airbus.com/en/products-services/commercial-aircraft/market/orders-and-deliveries.

Airline Business (2023) *Why airline industry focus is on the margins of profitability*, Flight Global. Available at: www.flightglobal.com/airline-business/why-airline-industry-focus-is-on-the-margins-of-profitability/156134.article.

ATR (2024) *Home, ATR*. Available at: www.atr-aircraft.com/.

Australian Government (2023) *Remote Air Services Subsidy Scheme*. Department of Infrastructure, Transport, Regional Development, Communications and the Arts. Available at: www.infrastructure.gov.au/infrastructure-transport-vehicles/aviation/regional-remote-aviation/remote-air-services-subsidy.

Boeing (2023) *Commercial market outlook*. Available at: www.boeing.com/commercial/market/commercial-market-outlook/index.page.

Boeing (2024) *Boeing commercial airplanes*. Available at: www.boeing.com/content/theboeingcompany/us/en/company/about-bca.

Bowen, J. and Rodrigue, J.P. (2020) 'The rise of air transportation', in J.P. Rodrigue (ed.) *The geography of transport systems*. 5th edn. New York: Routledge, pp. 181–195.

Budd, L. (2020) 'Airspace and air traffic management', in *Air transport management: an international perspective*. 2nd edn. Abingdon: Routledge, pp. 253–269.

Burghouwt, G., Mendes De Leon, P. and De Wit, J. (2015) *EU Air Transport Liberalisation Process, Impacts and Future Considerations*. International Transport Forum Discussion Papers 2015/04. Available at: https://doi.org/10.1787/5jrw13t57flq-en.

CAPA (2023) *Out of Africa – will things ever change?*. Available at: https://centreforaviation.com/analysis/reports/out-of-africa--will-things-ever-change-664099.

CAPA (2024a) *Financials for Jetstar Group*. Available at: https://centreforaviation.com/data/profiles/airlines/jetstar-airways-jq/financial.

CAPA (2024b) *Fleet for GOL*. Available at: https://centreforaviation.com/data/profiles/airlines/gol-g3/fleet.

Cathay Cargo (2024) *About Cathay Cargo – Cathay Cargo*. Available at: www.cathaycargo.com/en-us/aboutcathaypacificcargo.aspx.

Clean Aviation (2024) *Regional aircraft*. Available at: www.clean-aviation.eu/regional-aircraft.

Convention on International Civil Aviation (1944) *Convention on International Civil Aviation done at Chicago on the 7th Day of December 1944*. Available at: www.icao.int/publications/Documents/7300_orig.pdf.

De Havilland (2024) *Home*. Available at: https://dehavilland.com/en.

Doganis, R. (2019) *Flying off course: airline economics and marketing*. 5th edn. Abingdon: Routledge.

Dresner, M. and Zou, L. (2020) 'Air cargo and logistics', in *Air transport management: an international perspective*. 2nd edn. Abingdon: Routledge, pp. 287–307.

EBAA (2024) *About Business Aviation, EBAA – European Business Aviation Association*. Available at: www.ebaa.org/about-business-aviation/.

Edmunds, J. (2015) *Introduction to airline and airport management*. Harlow: Pearson.

Embraer (2024) *Our aircraft*. Available at: www.embraercommercialaviation.com/our-aircraft/.

Ethiopian Airlines (n.d.) *Ethiopian Airlines*. Available at: https://corporate.ethiopianairlines.com/aboutethiopian/overview?_gl=1*sf4mpx*_gcl_au*MTUxODIxMzM2MC4xNzMyNTQ3MDEx*_ga*NDIyMTAxMDE4LjE3MzI1NDcwMTI.*_ga_N20335F3KZ*MTczMjU0NzAxMi4xLjEuMTczMjU0NzA4My4yMS4wLjA.

European Parliament (2016) *The International Civil Aviation Organization*. Available at: www.europarl.europa.eu/RegData/etudes/ATAG/2016/593483/EPRS_ATA(2016)593483_EN.pdf.

FAA (2023) *Pilot's handbook of aeronautical knowledge*. Oklahoma City: Dept of Transportation. Available at: www.faa.gov/regulations_policies/handbooks_manuals/aviation/faa-h-8083-25c.pdf.
Finlay, M. and Memon, D.O. (2024) 'The 5 largest airlines in Africa', *Simple Flying*. Available at: https://simpleflying.com/largest-airlines-africa/.
Flight Global (2023) *Digital issues of Airline Business*. Available at: www.flightglobal.com/news/airline-business/digital-issues-of-airline-business.
Flightradar24 (2023) 'Anchorage: the world's cargo hub', *Flightradar24 Blog*, 18 July. Available at: www.flightradar24.com/blog/anchorage-worlds-cargo-hub/.
Goetz, A.R. and Graham, B. (2004) 'Air transport globalization, liberalization and sustainability: post-2001 policy dynamics in the United States and Europe', *Journal of Transport Geography*, 12(4), pp. 265–276. Available at: https://doi.org/10.1016/j.jtrangeo.2004.08.007.
Hayward, J. (2021) 'Top 10: the oldest airlines in the world', *Simple Flying*. Available at: https://simpleflying.com/10-oldest-airlines/.
IATA (2023a) *Global outlook for air transport – highly resilient, less robust*. Available at: www.iata.org/en/iata-repository/publications/economic-reports/global-outlook-for-air-transport----june-2023/.
IATA (2023b) *Industry statistics fact sheet December 2023*. Available at: www.iata.org/en/iata-repository/pressroom/fact-sheets/industry-statistics/.
IATA (2023c) *Net zero 2050: new aircraft technology*. Available at: www.iata.org/en/iata-repository/pressroom/fact-sheets/fact-sheet-new-aircraft-technology/.
IATA (2023d) *The story of the world's first airline*. Available at: www.iata.org/en/about/history/flying-100-years/firstairline-story/.
IATA (2024a) *Aviation value chain*. Available at: www.iata.org/en/iata-repository/publications/economic-reports/aviation-value-chain/.
IATA (2024b) *Value of air cargo*. Available at: www.iata.org/en/programs/cargo/sustainability/benefits/.
IBAC (2024) *IBAC definitions of business aviation, International Business Aviation Council*. Available at: https://ibac.org/about-ibac/resources-and-links/ibac-definitions-of-business-aviation.
ICAO (2024a) *Passenger air traffic surpasses pre-pandemic levels*. Available at: www.icao.int/Newsroom/Pages/Passenger-air-traffic-surpasses-pre-pandemic-levels.aspx.
ICAO (2024b) *The postal history of ICAO*. Available at: https://applications.icao.int/postalhistory/annex_16_environmental_protection.htm.
ICAO (n.d.a) *Air cargo*. Available at: www.icao.int/sustainability/economic-policy/Pages/default.aspx.
ICAO (n.d.b) *USAP principles*. Available at: www.icao.int/Security/USAP/Pages/USAP-Principles.aspx.
ICAO (n.d.c) *Vision and mission*. Available at: www.icao.int/about-icao/Council/Pages/vision-and-mission.aspx.
Industry High Level Group (2019) *Aviation benefits report – 2019*. Montreal: ICAO. Available at: www.icao.int/sustainability/Documents/AVIATION-BENEFITS-2019-web.pdf.
Jeffrey, R. (2023) 'Top 25 air cargo carriers: cargo airlines tackle tough times', *Air Cargo News*. Available at: www.aircargonews.net/data/top-25-air-cargo-carriers-cargo-airlines-tackle-tough-times/.
Kearns, S.K. (2021) *Fundamentals of international aviation*. 2nd edn. Abingdon: Routledge (Aviation Fundamentals).
KLM (2024) *Air France KLM – KLM United Kingdom*. Available at: www.klm.co.uk/information/corporate/about-air-france-klm.
Loh, C. and Pande, P. (2019) 'What are ETOPS rules and why do they matter?', *Simple Flying*. Available at: https://simpleflying.com/what-are-etops-rules/.
NASA (2019) *Earth's atmosphere: a multi-layered cake, climate change: vital signs of the planet*. Available at: https://climate.nasa.gov/news/2919/earths-atmosphere-a-multi-layered-cake.
National Geographic (2018) *Where, exactly, is the edge of space? It depends on who you ask*, Science. Available at: www.nationalgeographic.com/science/article/where-is-the-edge-of-space-and-what-is-the-karman-line.
National Geographic (2023) *Gerardus Mercator*. Available at: https://education.nationalgeographic.org/resource/gerardus-mercator.
NATS (2024) *Introduction to airspace*. Available at: www.nats.aero/ae-home/introduction-to-airspace/.

OAG (2023) *Low-cost carriers in the aviation industry: what are they?* Available at: www.oag.com/blog/what-are-low-cost-carriers-aviation.

Olive, X. and Spinielli, E. (2022) *Flight information regions / Open Aviation. CC.BY.4.0.*, Observable. Available at: https://observablehq.com/@openaviation/flight-information-regions.

Oliver Wyman (2024) *Oliver Wyman Global Fleet and MRO Market Forecast 2024–2034: key trends*. Available at: www.oliverwyman.com/our-expertise/insights/2024/feb/global-fleet-and-mro-market-forecast-2024-2034.html.

OpenStreetMap (n.d.) *OpenStreetMap*. Available at: www.openstreetmap.org/copyright.

Qantas (2019) *Qantas and the UK – a timeline*. Available at: www.qantasnewsroom.com.au/wp-content/uploads/2019/11/8151-CORP-A-History-of-the-Kangaroo-Route-Timeline-D3-3.pdf.

Qantas (2024) *Project sunrise*. Available at: www.qantas.com/gb/en/about-us/our-company/fleet/new-fleet/project-sunrise.html.

SKYbrary (n.d.) *Standards and recommended practices (SARPS) | SKYbrary Aviation Safety*. Available at: https://skybrary.aero/articles/standards-and-recommended-practices-sarps.

SkyWest (2024) *SkyWest Airlines » Fact Sheet*. Available at: www.skywest.com/about-skywest-airlines/facts.

Smithsonian (n.d.) *Douglas DC-3 | National Air and Space Museum*. Available at: https://airandspace.si.edu/collection-objects/douglas-dc-3/nasm_A19530075000 (Accessed: 27 March 2024).

Statista (2024) *China: number of civil airports 2022*. Available at: www.statista.com/statistics/258207/number-of-civil-airports-in-china/.

Tretheway, M.W. and Markhvida, K. (2014) 'The aviation value chain: economic returns and policy issues', *Journal of Air Transport Management*, 41, pp. 3–16. Available at: https://doi.org/10.1016/j.jairtraman.2014.06.011.

TUI Group (2024) *About TUI Group*. Available at: www.tuigroup.com/en-en/about-us/about-tui-group.

UNOOSA (2024) *The Outer Space Treaty*. Available at: www.unoosa.org/oosa/en/ourwork/spacelaw/treaties/introouterspacetreaty.html.

Whyte, R. and Lohmann, G. (2020) 'Airline business models', in *Air transport management: an international perspective*. 2nd edn. Abingdon: Routledge, pp. 129–144.

World Bank (2024a) *China, World Bank*. Available at: www.worldbank.org/en/country/china.

World Bank (2024b) *World Bank open data*. Available at: https://data.worldbank.org.

Part II
Physical geography and air transport

3 Meteorology and air transport

The atmosphere

> **Chapter outcomes**
>
> At the end of this chapter, you will be able to:
>
> - Explain the underlying processes that enable aircraft to fly and to be controlled whilst in flight.
> - Identify the components of the atmosphere and the key gases involved in the functioning of the atmosphere.
> - Distinguish between the five layers of the atmosphere and how each of these impacts the weather we experience.
> - Explain the concept of the International Standard Atmosphere and its importance for pilots and aircraft.
> - Understand the principles of temperature, its geographical variations and roles within air transport.
> - Analyse the effects of altitude on atmospheric pressure and its effect on aircraft performance.
> - Explain the importance of altimetry in safe flying.

3.1 Introduction

Atmospheric science has its sub-fields in meteorology and climatology. These are similar in terms of the scientific principles and phenomena being analysed, but differ in terms of approach, time scale and application (Coleman and Law, 2015).

Meteorology is the study of the earth's atmosphere and its constituent parts and processes, such as temperature, precipitation, air pressure, humidity and movement of air, over short time scales (minutes to weeks) and enables weather forecasts to be compiled. In the context of aviation, a knowledge of meteorology is vital for pilots, who encounter the vagaries of meteorological factors before they have even left the runway and throughout the duration of the flight, as well as airline flight and route planners. Safety critical weather conditions, such as thunderstorms, right through to passenger comfort during flight for passenger airlines, necessitates a thorough analysis of meteorology. Airport operations are also impacted by factors such as rain, wind and snow, which can all have impacts from both a safety perspective and the ability to adequately, and in a timely manner, service aeroplanes during turnarounds and whilst aeroplanes are taxiing on the ground.

Climatology involves analysing long-term weather patterns and trends. It studies long-term mean atmospheric conditions in a particular area as well as extremes, within both regional climates and global climate systems. The study of climatology aims to answer questions about climate variability and the impact of these variations on the earth's different systems. Increasingly, the study has focused on climate change – both the natural climate variations and understanding how human activities, such as greenhouse gas emissions, contribute to climate change. The influence of climate change on air transport will be discussed in Chapter 6 and the influence of air transport on climate change will be discussed in Chapter 14. To quote from A.J. Herbertson, the first Professor of Geography at the University of Oxford: "climate is what on average we may expect, weather is what we actually get" (Herbertson, 1913, p.118).

This chapter will focus on the atmosphere, its components, layers and atmospheric principles, including temperature, pressure and density. Chapters 4 (Water vapour) and 5 (Wind) will focus on these key atmospheric phenomena and their influence on air transport. It is important to note that atmospheric phenomena cannot be studied in isolation – each is significantly inter-related with many others and the structure within these three chapters is for categorisation, rather than separate processes. Chapter 6 will focus on Climatology as a separate concept, explaining global atmospheric circulation, world climate zones and the impact of climate on airline operations, and will conclude with a discussion on climate change and its potential influence on air transport.

3.2 Principles of flight

One of the key factors in understanding the geography of air transport is in the use of the word "air". According to Bowen and Rodrigue (2020, p.181), "air transportation is the movement of passengers and freight by any conveyance that can sustain controlled flight". Thus, it takes place (predominantly) in the air and therefore the influence of atmospheric phenomena can be considerable. To enable a more thorough analysis of atmospheric variables and how these influence air transport, an understanding of the principles of flight is extremely useful. For a more detailed explanation and of aircraft aerodynamics in general, publications such as those by Kundu, Price and Riordan (2016) and Sadraey (2016) should be consulted.

3.2.1 Newton's laws of motion

In 1665, Sir Isaac Newton proposed three laws of motion, which revolutionised science (NASA, n.d.a) and these help to explain how an aeroplane is able to fly.

As far as aircraft are concerned, *inertia* would occur when the pilot changes the throttle setting of an engine, *force* relates to an aircraft's motion resulting from aerodynamic forces, aircraft weight and thrust and finally, the motion of lift from an airfoil, when the air is deflected downward by the airfoil's *action and in reaction*, the wing is pushed upward.

Table 3.1 Newton's laws of motion (NASA, n.d.a)

Law 1	An object at rest remains at rest and an object in motion remains in motion at constant speed and in a straight line, unless acted on by an unbalanced force	INERTIA
Law 2	The acceleration of an object depends on the mass of the object and the amount of force applied	FORCE
Law 3	Whenever one object exerts force on another object, the second object exerts an equal and opposite force on the first	ACTION and REACTION

Meteorology and air transport: the atmosphere 79

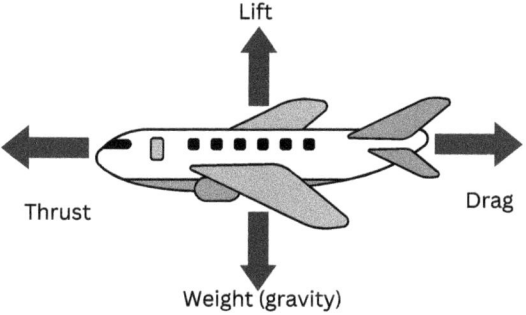

Figure 3.1 The four forces of flight
Source: author

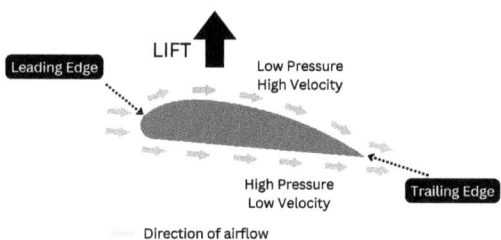

Figure 3.2 Aircraft wing and lift
Source: author

3.2.2 The four forces of flight

An aircraft can fly because of four forces: lift, weight, thrust and drag (Figure 3.1).

The wings of the aircraft are designed with the front edge slightly tipped up (a slight curvature of the upper half of the wing), which means that air flows more quickly over the top surface of the wing and more slowly over the bottom when an aircraft moves through the air (Kearns, 2021). This creates low pressure in the air above the wing and high pressure in the air below the wing (Figure 3.2).

- **Lift** is a mechanical force generated by a solid object moving through a fluid (NASA, n.d.a). The flow is turned in one direction and the lift is generated in the opposite direction, as per Newton's third law in Table 3.1. It is the force which allows an aircraft to overcome gravity and obtain (and maintain) flight. The difference between the high- and low-pressure areas creates lift. Most of the lift on a normal aeroplane is generated by the *wings*.

ATG Did you know?

Without air, there is no lift generated by an aircraft's wings. Space is nearly a vacuum; therefore, the Space Shuttle cannot stay in flight by lift from its wings – rather it is due to orbital mechanics and its speed.

- **Drag** is the force which opposes an aircraft's movement through the air. Imagine putting your hand out a moving car's window and feeling it pull backwards. It is essentially caused by:

  a) *Aerodynamic friction.* Drag occurs when the surfaces of an aircraft meet the air. Smooth aircraft surfaces produce less friction than roughened surfaces.
  b) *Form drag.* This depends on the shape of the object. Formula One car designers spend huge amounts of time and money trying to reduce form drag, as indeed do aircraft designers!
  c) *Induced drag.* This is a byproduct of generating lift. It is highest at the wingtips and lowest at the wing roots. This creates *wingtip vortices*, creating a rearward pulling force, which causes drag. It can also have safety implications, which will be analysed in Chapter 5.

- **Thrust** is the force which moves an aircraft through the air (NASA, n.d.a). Aeroplanes generate thrust through Newton's third law – the principle of action and reaction. Thrust is generated by the *engines* of the aircraft and when thrust is stronger than drag, the aircraft accelerates in the direction of the net force. There are four main propulsion systems: propeller, turbine (jet) engine, ramjet and rocket.

- **Weight** is the force generated by the gravitational attraction of the earth on the aeroplane (NASA, n.d.a). For an aircraft to fly, the force of the lift needs to be equal to or greater than the aircraft's weight. We are all familiar with the concept of weight, it is something we can measure when we step on the bathroom scales in the morning! Unlike lift and drag, which *are aerodynamic (mechanical) forces*, weight is a *gravitational force.* Control over weight distribution is critical for stability and control and this is generally planned for using the Load Control/Flight Dispatch function within airlines, using the principles of weight and balance. How heavy the aeroplane is versus certified weights – such as the maximum take-off weight (MTOW) and where is the centre of gravity/centre of lift – are critical if the aeroplane is even able to take off!

ATG Did you know?

Winglets

Many modern aeroplanes use *winglets* to reduce induced drag. Winglets are vertical extensions of wingtips and were designed as small airfoils. An engineer at the NASA Langley Research Centre – Dr. Richard Whitcomb – took the concept of winglets from the drawing board in the 1970s to flight testing, when in 1979–1980, a military version of the B707 recorded an increased fuel mileage rate of 6.5% (NASA, n.d.b). Since then, many commercial aircraft now incorporate wingtip technology. These can improve an aircraft's payload-range and reduce both its fuel burn and its environmental footprint.

3.2.3   Aeroplane flight controls

The marvel that is aircraft flight is one thing, however once airborne, the next challenge is to control the aircraft. Figure 3.3 illustrates the key aeroplane parts and flight control surfaces.

Ultimately, this control is achieved in a three-dimensional plane, around **three axes of rotation** – *pitch, roll and yaw*. This is achieved using **three flight control surfaces** – *elevators, ailerons and the rudder* – respectively (Figure 3.4).

- **Elevator (pitch).** The elevator is usually located on the tail of the aircraft, on the horizontal part of the tailfin. It serves two purposes. The first relates to producing a downward force

Meteorology and air transport: the atmosphere 81

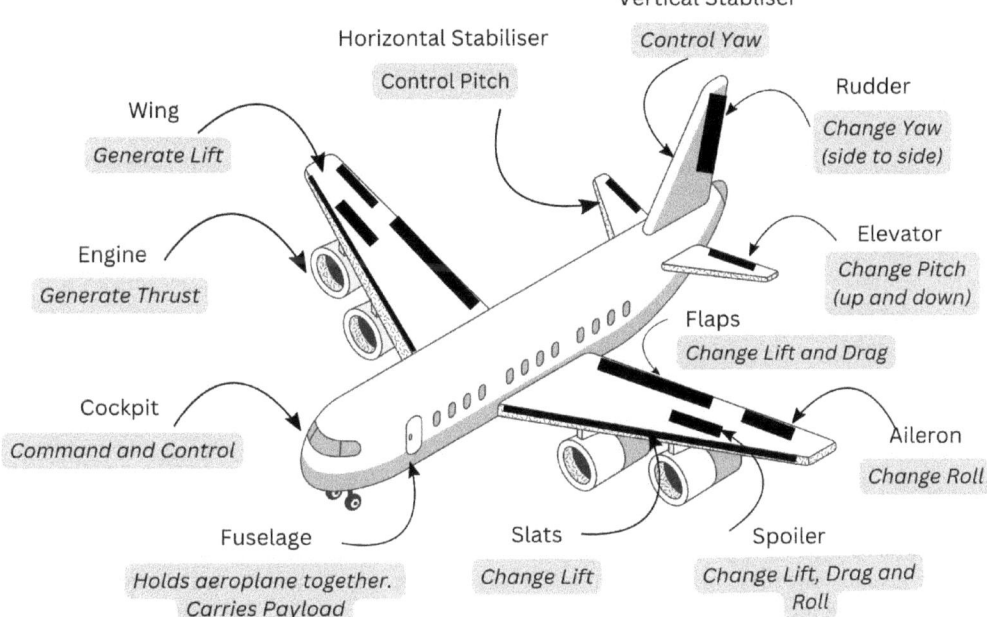

Figure 3.3 Aeroplane parts, flight control surfaces and functions
Source: author

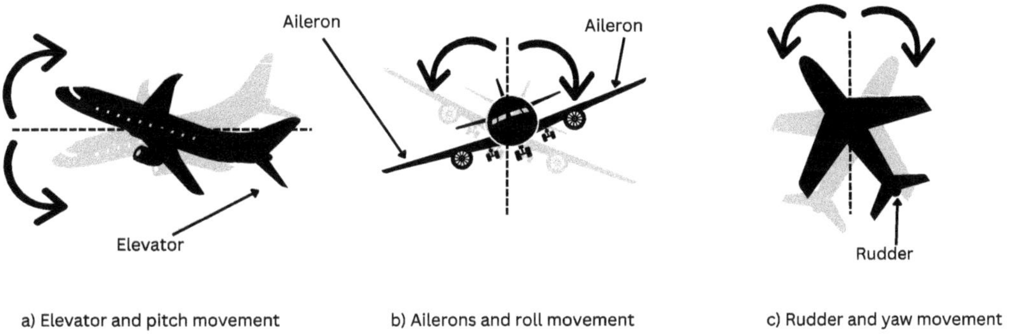

Figure 3.4 Flight control surfaces and the three axes of rotation
Source: author

on the tail to provide stability. Second, movement of the elevator up or down, to make the aircraft climb or descend.
- **Ailerons (roll).** These are located on the outer part of each wing and move in opposite directions. When a pilot moves the yoke left or right, one aileron moves up and the other downwards. This increases the lift on one wing and reduces lift on the other. This rolls the aircraft sideways and means it can turn (much like a car steering wheel).
- **Rudder (yaw).** This is located on the vertical tailfin (stabiliser) and works in the same way as a rudder on a boat – except in air rather than water – steering the aircraft nose left and right. Its primary mechanism is to counteract the drag caused by the ailerons in a turn – termed *coordinated flight*. It is also used on the ground to align the aircraft during takeoff and landing during crosswinds.

82 *Fundamentals of Global Air Transport Geography*

3.2.4   Centre of gravity (CG)

The centre of gravity, also known as CG, is the effective point whereby all weight is concentrated. It is the point where the axes of flight meet. It is not fixed but can be determined by the aircraft weight distribution – for example, passengers, fuel and cargo. It is an extremely safety critical concept, as it is vital to maintain the CG to ensure the aircraft does not go out of trim – either too nose-heavy or tail-heavy and can retain stable flight characteristics. Different aircraft types have different CG limits and handling characteristics and pilots will be given training specific on the aeroplane type they are flying.

3.3   Components of the atmosphere

Air is the main constituent of the atmosphere. As a result, understanding the properties of air is vital to understand how the atmosphere influences aircraft performance. Aircraft certification will require aircraft to be tested in a variety of atmospheric conditions, to measure performance in a range of weather scenarios. Part of the certification for the Airbus A350 involved hot weather testing in the United Arab Emirates, cold weather testing in Iqaluit, Canada and high-altitude evaluations in La Paz, Bolivia (Airbus, 2021).

The atmosphere is a layer of gas and suspended solids extending from the earth's surface up many thousands of miles, becoming increasingly thinner with distance, but always held by the earth's gravitational pull (NOAA, n.d.). It provides a number of vital functions, such as providing air to breathe, shielding us from harmful ultraviolet (UV) radiation from the Sun, trapping heat to warm the planet and preventing extreme temperature differences between day and night (UCAR, 2021). It is the protective bubble we live in. It consists of a few gases, each having its own physical properties, in which varying quantities of tiny solid particles and water droplets are suspended. These vary over time and from place to place (Sadraey, 2016).

Table 3.2 illustrates the four main gases within the atmosphere.

- **Nitrogen** is the most common. Living things need nitrogen to make proteins and it also dilutes oxygen, preventing rapid burning at the earth's surface.
- **Oxygen** is used by anything living and is essential for respiration and in aviation terms, for combustion (burning) which is a key component in the process of aircraft propulsion.
- **Argon** is used in incandescent light bulbs to stop oxygen corroding the filament, in fluorescent tubes and low-energy bulbs, as well as double-glazed windows (Royal Society of Chemistry, n.d.).
- **Carbon dioxide** is used by plants to make oxygen. It also acts as a blanket that prevents the escape of heat into outer space. It is earth's most important greenhouse gas (Lindsey, 2024). By adding more carbon dioxide to the atmosphere – via the burning of fossil fuels such as

Table 3.2 Chemical makeup of the atmosphere EXCLUDING water vapour

Gas	Content (%)
Nitrogen	78.084
Oxygen	20.947
Argon	0.934
Carbon dioxide	0.035
Others – trace	0.002

Source: adapted from NOAA (n.d.)
Note: the proportion varies day by day.

coal and oil – humans are in effect supercharging the natural greenhouse effect, causing temperatures to rise. We will return to this vital concept for contemporary geography in Chapters 6 and 14 as it has a vital role in changing global atmospheric conditions, hence has a big effect on the air transport industry, not to mention the planet as a whole.

These gases make up the DRY atmosphere. However, by far and away the most important component that makes up our daily weather is water vapour (Walmsley, 2021). This influences a wide range of atmospheric phenomena – for example, if it is a lovely, sunny day, or a muggy day with a threat of thunderstorms. It is the source of all clouds and precipitation (Sadraey, 2016). Hence, water vapour is a key component for aircraft performance, as will be explained in Chapter 4.

In the earth's desert regions, when dry winds are blowing, the water vapour is near zero. This climbs to near 3% on hot and humid days. The upper limit, approaching 4%, is found in tropical climates (Table 3.3).

Not only is water vapour an important raw material for clouds and precipitation, but it is also a vehicle for the transfer of heat energy and as a regulator of the earth's temperatures through absorption and emission of radiation (FAA, 2023).

3.4 Layers of the atmosphere

According to ICAO (in FAA, 2023), the earth's atmosphere is sub-divided into five concentric layers, based on the vertical profile of average air temperature changes, chemical composition, movement and density (Figure 3.5). Each of these is topped by a "pause", where the maximum changes in thermal characteristics, chemical composition, movement and density occur (Figure 3.6).

The boundary heights for these regions can vary considerably. Although aviation generally occurs in the troposphere and lower stratosphere, an understanding of what takes place above is important, as this can directly impact the weather that is experienced.

3.4.1 Troposphere

This is the lowest and most important layer for air transport, as most of the earth's weather occurs here. It makes up around 80% of the atmosphere's mass. It begins at the earth's surface and extends to about 11 kilometres (36,000ft) high (FAA, 2023). The air becomes thinner as the gases decrease with height, therefore the temperature also drops with height, at a rate on average of 1.98°C per 1,000 feet. This part of the atmosphere is densest and temperature drops from about +15°C to −56°C (Sadraey, 2016).

According to Walmsley (2021), the vertical limit for most small (unpressurised) aircraft is around 10,000 feet and the typical cruising level for a commercial airliner is 30,000 feet,

Table 3.3 Chemical makeup of the atmosphere including water vapour

Water vapour (%)	Nitrogen (%)	Oxygen (%)	Argon (%)
0	78.084	20.947	0.934
1	77.30	20.70	0.92
2	76.52	20.53	0.91
3	75.74	20.32	0.90
4	74.96	20.11	0.89

Source: adapted from NOAA (n.d.)

84 *Fundamentals of Global Air Transport Geography*

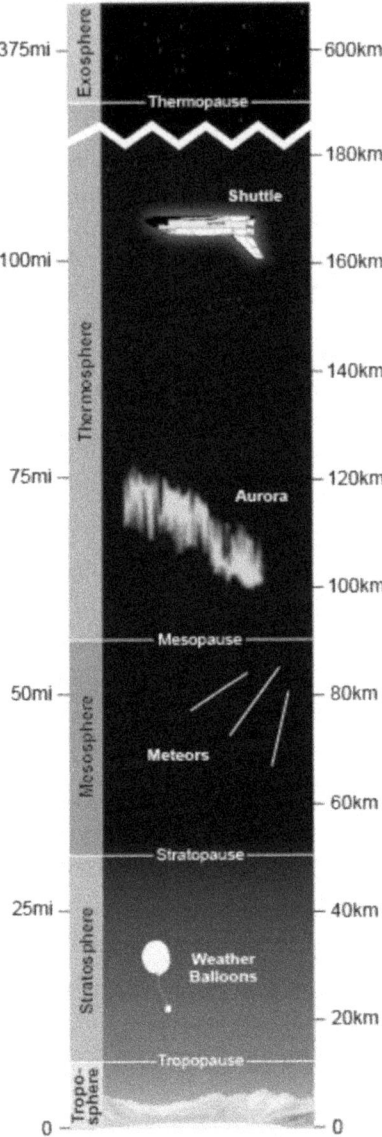

Figure 3.5 The basic layers of the atmosphere
Source: NOAA (2023)

therefore much of air transport flight will occur in the troposphere. An altitude of 18,000 feet is the altitude after which oxygen is not enough for breathing and any aircraft flying beyond this altitude must be equipped with an air pressure system (Sadraey, 2016).

At the top of this layer is the *tropopause*, where temperatures are very cold and various weather-related phenomena, such as jet streams and clear air turbulence (Chapter 5), take place.

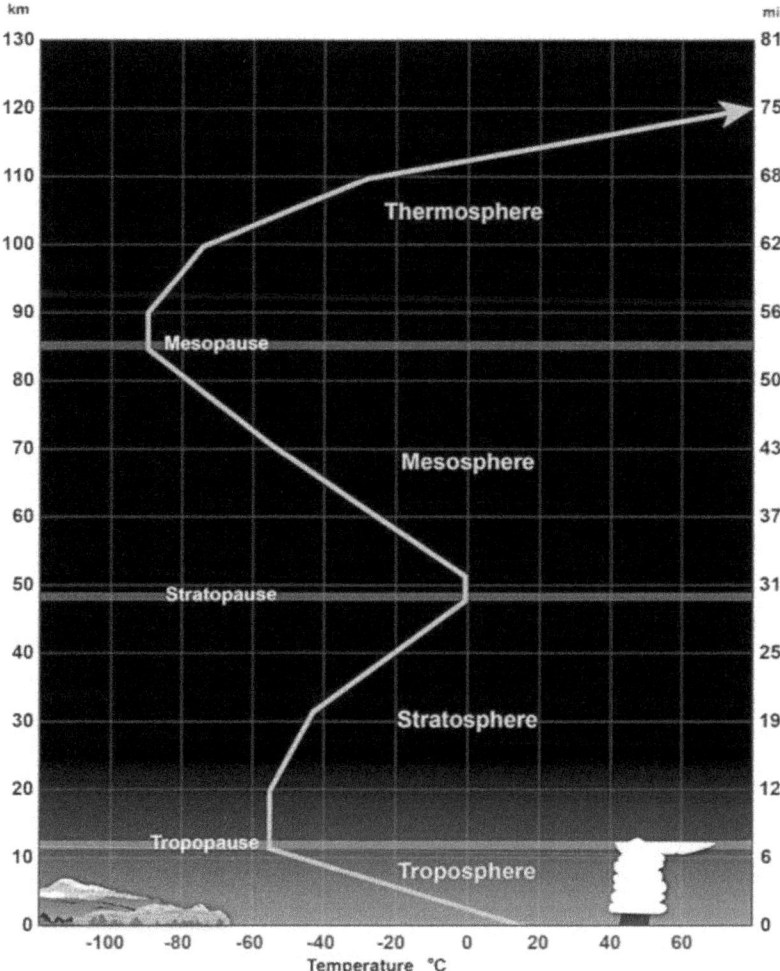

Figure 3.6 Average temperature profile for the lower layers of the atmosphere
Source: NOAA (2023)

3.4.2 Stratosphere

Above the tropopause, the temperature increases throughout the stratosphere due to the ozone layer (Kearns, 2021). This plays an important role in filtering UV light and only allowing a limited amount to reach earth's surface. Human-produced chemicals such as chlorofluorocarbons (CFCs) have a depleting impact on the ozone layer.

The stratosphere extends to about 50km high, and it is dryer and less dense than the troposphere. The temperature starts around −56.5°C and then increases gradually to −3°C in the upper section. The stratosphere contains about 19.9% of the total mass of the atmosphere, therefore 99% of air is located in the troposphere and stratosphere (Sadraey, 2016).

Modern jet transport aircraft often cruise in the lower stratosphere, where there is less "weather", avoiding atmospheric turbulence and convection in the troposphere (FAA, 2023).

The supersonic aircraft, Concorde (retired in 2003) flew much higher than subsonic aircraft – up to 60,000 feet – which is well into the stratosphere.

3.4.3 Mesosphere

Above the stratosphere, the temperature drops again through the mesosphere, with the lowest atmospheric temperatures at the mesopause (280,000ft/85km) (Kearns, 2021).

3.4.4 Thermosphere

Located between about 80–700km above earth's surface, it is known as the upper atmosphere (FAA, 2023). The gases become increasingly thin compared to the mesosphere. The temperature rises again because of solar radiation (Kearns, 2021) and can reach as high as 2,000°C (3,600°F) near the top of this layer.

3.4.5 Exosphere

Located between about 700–10,000km (NASA, 2019), this is the uppermost atmospheric layer where gas density is very low and some air particles can escape from earth's gravitation. This is generally considered space. There is no weather in this region – the *aurora borealis* and *aurora australis* are sometimes seen in its lowest part.

> **ATG Did you know?**
>
> **Meteor showers**
>
> Have you ever seen a *meteor shower* – sometimes called *shooting stars*? These burn up in the *mesosphere*. They make it through the exosphere and thermosphere as these don't have much air, however when they reach the mesosphere, there are enough gases to cause friction and create heat – leaving fiery trails in the sky.

All aeroplanes are certified by their manufacturers to fly up to a maximum altitude (Table 3.4) and this is discovered during production and flight-testing and approved by the relevant regulatory authorities, such as the Federal Aviation Administration (FAA) in the USA and the European Union Aviation Space Agency (EASA). For operational and safety reasons, airlines normally choose cruising altitudes below these ceilings – for jet aircraft between 31,000–38,000 feet (although this can vary daily due to various factors). However, for reasons explained in this Chapter and Chapter 4, flying "high" is preferred by airlines (and pilots!) for a variety of reasons.

3.5   International Standard Atmosphere (ISA)

The atmosphere is in a constant state of flux and aircraft are expected to operate safely and efficiently in a variety of atmospheric conditions, such as different temperatures (see section 3.6) and atmospheric pressure (see section 3.7). It is unlikely that the atmosphere will be "standard" on any given day and that aircraft performance may not be "normal".

Table 3.4 Aeroplane service ceilings

Aircraft type	Approximate service ceiling (ft)
Concorde	60,000
Airbus A380–800	43,100
Airbus A350–900	43,100
Boeing 787–9	43,100
Airbus A350–1000	41,450
Boeing B787-10	41,100
Boeing 737 MAX	41,000
Airbus A320 family	39,100 to 41,000
Boeing 737-300	37,000
ATR 42 (turboprop)	25,000
Twin Otter DHC-6 (turboprop)	25,000
Cessna 172 (single engine piston)	14,000

Source: author, from manufacturer specifications

Table 3.5 ISA values at sea level

	Sea-level values	Reduces by
Temperature	+15°C (+59°F)	1.98°C (3.6°F) per 1000ft until tropopause is reached at 36,090ft and −56.5°C
Pressure	1013.25mb/hPa	1mb/hPa per 30ft (up to about 5000ft)
Density	1,225 grm/m3	In accordance with ISA temperature and pressure values

Source: author compiled from Wickson (2015); Walmsley (2021); SKYbrary (n.d.b)

Table 3.6 ISA values at a range of altitudes

Height ft (km)	Temperature (°C)	Pressure (mb)	Relative density (%)
45,000 (13.7)	−56.5	148.2	19.5
40,000 (12.2)	−56.5	188.2	24.7
36,090 (11.0) tropopause	−56.5	228.2	29.7
30,065 (9.2)	−44.4	300.0	36.8
18,289 (5.6)	−21.2	500.0	56.4
10,000 (3.0)	−4.8	696.8	73.8
4,781 (1.5)	+5.5	850.0	87.3
Mean sea level (MSL)	+15	1013.25	100.0

Source: adapted from Wickson (2015)

However, to provide a common basis for comparing performance features of different aircraft, a standard is required. ICAO has developed a unique, internationally accepted atmospheric condition, called the *International Standard Atmosphere* (ISA). This provides a standard against which to compare the actual atmosphere at any point and time throughout the year, for the entire world. ISA values are calculated at sea level for temperature, pressure and density (Table 3.5) and assumes values decrease with an increase in height (Table 3.6) (SKYbrary, n.d.b). It assumes standard temperature lapse rates for the atmosphere and a constant tropopause height. These are roughly equivalent to the mean values in temperate latitudes. Weather-related processes are generally referenced to the standard atmosphere (FAA, 2023).

A pilot can then compare their actual weather conditions on the day, to determine if their aircraft will perform better or worse than the ISA. Aircraft performance relies heavily on the atmospheric parameters of *temperature*, *pressure* and *density*. These will now be analysed in turn, but as a reminder, with most weather-related concepts, these are inter-related.

3.6 Temperature

It can be said that temperature is the controlling factor in meteorology. It is certainly one of the fundamental components in aviation weather, as large variations occur all around the world – by latitude, altitude, season, day, night, even one hour to the next.

Temperatures are generally higher near the equator and colder at the poles, but the effect of altitude can also mean colder temperatures near the equator. Quito in Ecuador is located next to the equator, at an altitude of around 9,350ft (2,850m) and has a mean annual average temperature of 11.3°C (52.4°F), whereas Singapore in Southeast Asia – also located close to the equator – has most of its topography under 50ft (15m) and has an average temperature of 26.7°C (80°F) (Climate-data.org, 2023). Changes of even a few degrees can have a big impact on atmospheric phenomena and hence aircraft performance.

3.6.1 Defining temperature

According to the FAA (2023), temperature is a numerical value representing the average kinetic energy of the atoms and molecules within matter and depends on the energy of molecular motion. Higher (warmer) temperatures indicate a higher average kinetic energy of molecular motion due to molecular speeds. Temperature is an indicator of the internal energy of air. The amount of heat required to raise the temperature of air, or lost when cooled, is termed *heat capacity*. The importance here is the energy released or absorbed during the process, as we will see in this section.

3.6.2 Temperature measurement and units

There are three main units of *temperature measurement*: Celsius (°C), Fahrenheit (°F) and Kelvin (K) scale. The technical conversions for these are beyond the scope of this book.

- **Kelvin (K)** scale is a thermodynamic (absolute) temperature scale, where absolute zero, the theoretical absence of all thermal energy, is 0K. Nothing can be colder than absolute zero, so the Kelvin scale contains no negative numbers.
- **Fahrenheit (°F)** is rarely used now, although it is still used for many everyday non-aviation temperature measurements in the USA.
- **Celsius (°C)** is the most used temperature scale worldwide and in meteorology. The scale is approximately based on the freezing point (0°C), and boiling point (100°C), of water under a pressure of one standard atmosphere – approximately sea level (FAA, 2023). The key for aviation is that as per ICAO Annex 5, degrees Celsius (°C) is the temperature unit which should be used in international civil aviation air and ground operations (ICAO, 2010) and therefore is the unit used within this book.

Accurate temperature measurement is critical for aircraft performance calculations, such as the *take-off distance required*. These are usually calculated using thermometers inside a box such as a *Stevenson screen* (Figure 3.7) and are normally located at the aerodrome, standing about 1.25m (4ft) above the ground. It is a white box with slats, to allow air to flow through. These face north, which, combined with the colour and slats, provide the best measure of current temperature (Met Office, n.d.).

3.6.3 Solar and terrestrial radiation

The majority of the energy which warms the planet is provided by the Sun and is called *solar radiation*. Solar radiation has a short wavelength and high frequency. As the sun is extremely

Figure 3.7 Stevenson screen
Source: Bidgee (2011)

90 *Fundamentals of Global Air Transport Geography*

hot, this results in the solar radiation travelling through the troposphere – some being absorbed by the ozone layer, some reflected back to space by clouds and the atmosphere and some warming the earth's surface. This in turn warms the air above, becoming heat energy, raising the temperature of land and water surfaces.

The earth and the atmosphere absorb most of the energy reaching its surface (approximately 70%) and around 30% is reflected back to space (WMO, 2019). The earth radiates energy at wavelengths longer than the Sun because it is colder. Part of this is absorbed by greenhouse gases, trapping heat in the atmosphere. Without this natural greenhouse effect, the earth's average temperature would be −18°C (0°F), instead of today's 14°C (59°F). The burning of fossil fuels can exacerbate this.

Current season and latitude determine the amount of solar radiation a particular area receives – peaking on 21 June in the northern hemisphere and 21 December in the southern hemisphere on the Summer Solstice. Due to the curvature of the earth, solar radiation at higher latitudes, such as the poles (90 degrees latitude), is spread out, limiting the amount of energy absorbed. The equator (0 degrees latitude) has solar radiation concentrated into a smaller area, leading to warmer temperatures (Figure 3.8). The angular elevation of the sun is therefore reduced, with the result that the ground area being radiated is increased, so the degree of *insolation* reduces with the increase in latitude (Wickson, 2015).

ATG Did you know?

Radiation

Have you ever stood in front of a fireplace? The side of the body nearest the fire warms, while the other side remains unaffected by heat. This is the heat transfer called *radiation*. Radiation is the transfer of heat energy through space by electromagnetic radiation. Once the earth's surface absorbs the radiation, it will release its own radiation, called *terrestrial radiation*. We can thus say that *the atmosphere is warmed from below*.

Once the earth's surface warms up, it can now transfer this energy through three main processes: *conduction, convection* and *advection*.

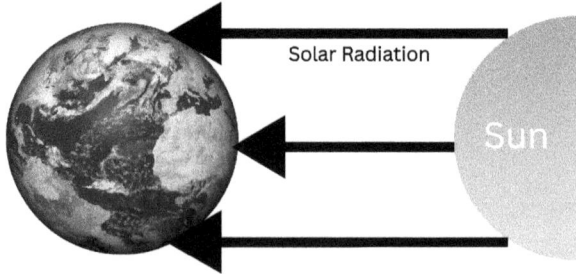

Figure 3.8 Solar radiation at the high and low latitudes
Source: author

3.6.4 *Conduction*

Conduction is the transfer of energy (including heat) by molecular activity from one substance to another in contact with or through a substance (FAA, 2023). Since air is a poor conductor, most energy transfer by conduction occurs right near the earth's surface. It therefore only affects air temperature a few centimetres into the atmosphere. During the day, sunlight heats the ground, which then heats the air directly above via conduction. At night, the ground cools and the heat flows from the warmer air directly above to the cooler ground via conduction (UCAR, 2018).

3.6.5 *Convection*

Convection is the transport of heat within a fluid, such as air or water, via motions of the fluid itself (FAA, 2023). Because air is a poor thermal conductor, convection plays a very important role in the heat transfer within the atmosphere, and it helps transfer energy higher in the atmosphere. When a warm surface heats the air above via conduction, the air expands and rises. It then carries its heat higher in the atmosphere – think of a hot air balloon rising: the air inside the balloon is heated, making it less dense than the air around it (Figure 3.9). This can cause thermal upward-moving currents or "thermals" – the knowledge of these being especially important for glider pilots!

3.6.6 *Advection*

Advection is the transport of a substance or quantity by bulk motion of a fluid. It is the horizontal movement of energy and is important in many weather conditions. The horizontal movement is typically caused by the wind. There are two main types of advection. *Cold advection* occurs when cold air travels horizontally towards warmer air – this can cause weather phenomena such as thunderstorms. *Warm advection* occurs in reverse and can create phenomena such as low clouds and fog.

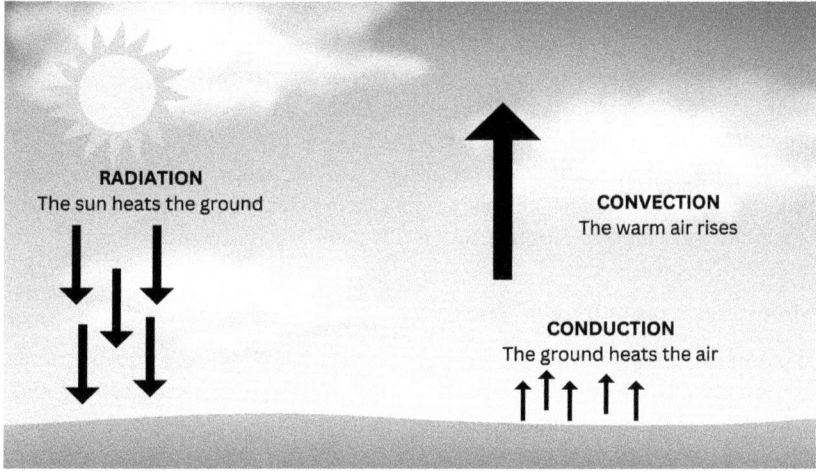

Figure 3.9 Radiation, conduction and convection

Source: author

92 *Fundamentals of Global Air Transport Geography*

3.6.7 *Surface temperature variations*

In addition to latitude and seasons, there are a few other factors which influence temperatures at or near the surface.

Time of day – diurnal temperature range

From sunrise, as the angle of the sun increases, surface air temperatures will also increase. The maximum insolation is received at Noon, but temperatures continue to rise until about 1500. The earth is also experiencing terrestrial radiation, which after 1500 is greater than heat gained, hence temperatures drop. The coldest time of the day is around sunrise (when solar and terrestrial radiation are roughly in balance).

Effect of cloud and wind on temperature

Local temperature variations are largely influenced by cloud, wind and surface types. Chapter 4 (specifically Section 4.5 on clouds) and Chapter 5 (Wind) will provide a detailed analysis of these key weather phenomena and their influence on air transport:

- **Cloud** results in warmer night-time but cooler daytime temperatures. During the day, clouds will reflect incoming solar radiation, but at nights clouds act like a blanket, trapping in energy. This is important for the phenomena of frost and fog, key in aviation operations, for example a cloudless night allows cooler temperatures at night due to the release of terrestrial radiation.
- **Wind** can often even out large temperature variations. Wind causes *turbulence*. During the day, the warmer air will then be mixed with cooler air from above and it reduces the time air is in contact with the warm land surface, both resulting in lower daytime temperatures. On a windless day, limited mixing occurs, meaning higher daytime temperatures.

Nature of the surface

Different surfaces require a different amount of energy to heat up. The amount of energy reflected by a surface is called *albedo*. Dark colours (such as aircraft runways) have an albedo close to zero, which means little energy is reflected, whereas pale colours have an albedo closer to 100%, meaning nearly all energy is reflected. For example, forests have an albedo of about 15%, meaning 15% of sunlight is reflected out to space, whereas fresh snow has an albedo of 90%, meaning 90% of sunlight is reflected (UCAR, n.d.).

This is extremely important in understanding climate change – if warming causes ice and snow to melt, darker coloured surfaces become more exposed, less solar energy is reflected into space and the planet warms even more.

3.7 Atmospheric pressure and density

Section 3.5 explained ISA and the relationship between temperature, atmospheric pressure and air density. A more thorough explanation of the concepts of atmospheric pressure and air density is required, along with their implications for air transport.

In simple terms, **atmospheric pressure** is the weight of the air above a certain point. It can be defined as force per unit area (Lankford, 2001). It is the sum of all the air molecules above a specific point. **Density**, which is directly related to pressure, is a property of the atmosphere, which can be used by pilots to help determine how their aircraft will perform at various altitudes

(FAA, 2023). It is the number of air molecules in a particular area and the more of these air molecules per unit of air, the higher the air density.

3.7.1 Measurement and units

Atmospheric pressure is measured using a *barometer* and accurate measurement is vital for the safety of aircraft. Mercury barometers were designed by Evangelista Torricelli in the 17th century. In 1844, Lucien Vidi invented the aneroid barometer, which is the technology commonly used by meteorologists and the aviation community today (FAA, 2023). These are commonly found in Stevenson screens on the ground and on aircraft, using a sensor mounted outside the aircraft, called a *static port*.

Atmospheric pressure is usually measured in hectopascals (hPa) (ICAO, 2010) and can also be expressed in inches of mercury (inHg) or millibars (mb). Like temperature, pressure has a *standard*, which forms part of the ISA (Table 3.5). At sea level, *atmospheric pressure* is 1013.25hPa (1013.25mb/29.92inHg). Near sea level, atmospheric pressure reduces by about 1hPa for every 30ft and reduces to about 700hPa by 10,000ft. This is called the *pressure lapse rate*. Higher in the atmosphere, pressure decrease slows. Around 18,000ft, around half sea level pressure (500hPa) can be found – now the lapse rate is around 1hPa per 50ft.

Oxygen levels are at roughly 50% of sea level values at this point (Mount Everest Base Camp is around this height – at 17,598ft/5,364m). The summit of Everest (29,035ft/8,849m), based on the ICAO Standard Atmosphere, is around 313hPa, approximately 30% oxygen values at sea level. Oxygen content is so low at this height because its abundance is directly proportional to atmospheric air pressure within the troposphere, which falls exponentially with increasing elevation (Matthews *et al.*, 2020).

Average *air density* at sea level is 1225g/m3. This reduces in accordance with ISA temperature and pressure values.

> ### ATG Did you know?
>
> **Aeroplane ear**
>
> Have you ever felt like your ears feel clogged up when taking off or landing in an aeroplane or going up quickly in an elevator? Ear barotrauma (aeroplane ear) happens when your middle ear is affected by sudden changes in air pressure. When aircraft change altitude, air pressure changes (as aircraft increase in height, pressure decreases) and this can happen faster than your eustachian tubes (which run from the back of your middle ear to the back of your throat and help to maintain equal pressure on both sides of your eardrum) can react. Congestion can make this feel worse. This is usually a temporary issue which goes away when air pressure changes back (Cleveland Clinic, 2023).

3.7.2 High and low pressure

As explained in Section 3.5, the atmosphere is rarely standard and atmospheric pressure is constantly varying, mainly due to passing pressure systems.

If there is a pattern of high atmospheric pressure – a *"high"* or *"anticyclone"*, for example, 1024hPa – this generally results in good weather conditions, with little wind. A High signifies that there is a greater mass of air over a given point.

Conversely, when a pattern of low atmospheric pressure is around, for example 995hPa, this is termed a *"low"* or *"depression"*, which often brings poor weather, rain and stronger winds. A Low signifies there is less overlying air over a given point.

Winds specifically will be looked at in Chapter 5. These Lows and Highs can create very different conditions for aeroplane flight and therefore accurate measurement is very important.

3.7.3 Density's effects on pressure

In general, the density of an air parcel can be changed by changing its mass, pressure or temperature (FAA, 2023).

Density is directly related to pressure. Assuming constant mass and temperature, an air parcel with a higher pressure is denser than an air parcel with a lower pressure. Air pressure decreases with height in the atmosphere, therefore so does density. The greatest effect is in the vertical direction (Figure 3.10).

3.7.4 The effects of temperature and water vapour on density

Density is inversely related to *temperature*. Assuming constant mass and pressure, an air parcel with a higher temperature is less dense than an air parcel with a lower temperature. The warmer air occupies a larger volume. In the atmosphere, these changes are reflected in latitudinal differences – e.g. Helsinki, Finland and Cairo, Egypt. Temperature has the greatest effect on density. Density is also inversely related to its quantity of *water vapour*. This is since dry air molecules have a larger weight than water vapour molecules and density is directly related to mass.

For an aircraft waiting to take off, the ideal conditions would be a cold day with high pressure. The less-than-ideal conditions would be a warm day with lower pressure. Low density conditions can occur around the world leading to aircraft being unable to take off or perhaps having to reduce their weight (payload) to take off. High temperatures affect how the air pushes

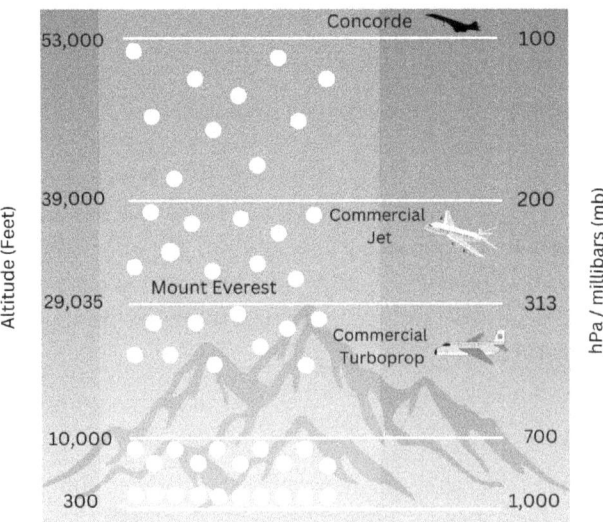

Figure 3.10 Pressure's effects on density in the atmosphere
Source: author

back. A hot day means the density of the air above the runway drops. High temperatures thin the atmosphere and oxygen molecules are spread further apart. Fewer air molecules beneath the wings mean the aircraft may not generate enough force to take off or need a longer take-off roll (Figure 3.11).

Aerodromes in the Southwestern United States – in California and Arizona – can experience temperatures above 49°C (120°F) in summer, leading to flight cancellations.

3.8 Altimetry

Altimeters measure the difference in pressure between a particular pressure surface and the pressure at the aircraft level. The altimeter is basically an aneroid barometer, however rather than units of pressure, the altimeter measures increments of *altitude*. The pressure altimeter is one of the most important instruments a pilot uses during flight, as it enables them to know how high they are flying, especially in relation to terrain and relevant aerodromes. These devices are essentially inaccurate (as we will see) and therefore their use must be undertaken with care as the ramifications can be extremely safety critical.

In simple terms, altitude means the vertical elevation of an object above the earth, however there are many different meanings of altitude in aviation – *indicated altitude*, *true altitude*, *absolute altitude*, *pressure altitude* and *density altitude* (FAA, 2008).

3.8.1 Types of altitude

- **Indicated altitude** is the reading on the altimeter when set to the local barometric pressure. Mainly used for aircraft separation.
- **True Altitude** is the vertical distance above mean sea-level – it is the indicated altitude corrected for non-standard temperature and pressure.
- **Absolute Altitude** is the height above ground level.

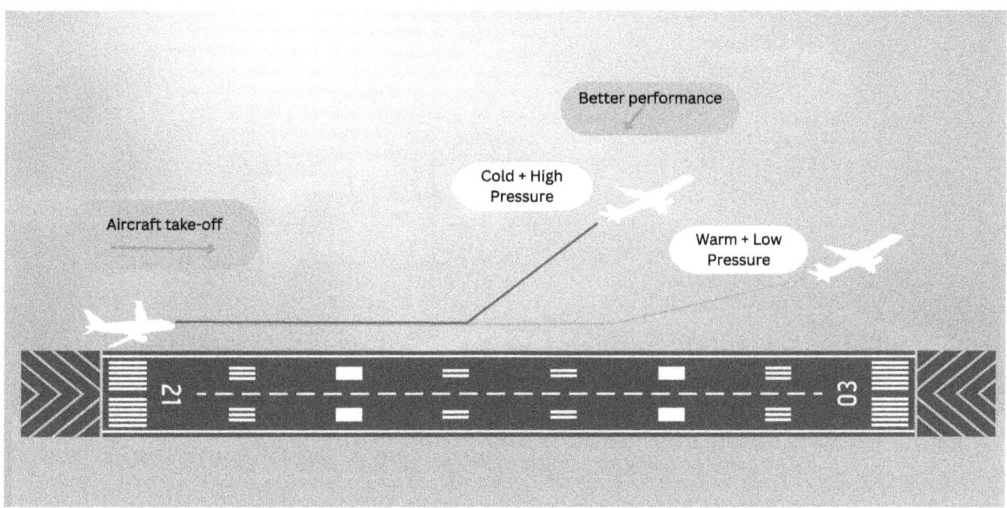

Figure 3.11 Pressure, temperature and aircraft performance
Source: author

96 *Fundamentals of Global Air Transport Geography*

- **Pressure Altitude** is the indicated altitude when an altimeter is set to 1013.2hPa, mainly used in aircraft performance calculations and in high-altitude flight.
- **Density altitude** is the pressure altitude corrected for non-standard temperature variations.

The altimeter measures atmospheric pressure and then displays an altitude based on calibration with ISA. Therefore, at sea level, on a "standard" day, the pressure will be 1013hPa; when the aircraft climbs to 10,000ft it will be 700hPa and climbing to 18,000ft it will be 500hPa. As we have seen, the atmosphere is rarely "standard" and therefore the altimeter has an adjustable subscale to set the correct subscale pressures – known as altimeter "*Q*" codes. Three references for barometric pressure are commonly used – *QNH*, *QFE* and *Standard Pressure* (Figure 3.12).

ATG Did you know?

Datum

The word *datum* can mean two different things in aviation. First, regarding weight and balance, it is an imaginary vertical plane from which horizontal measurements are taken – essentially it could be viewed as a vertical line perpendicular to the aircraft's longitudinal axis. Second, it is also used through a collection of specific points on the earth with known heights above or below a defined reference surface – such as mean sea level (MSL) and "Q" codes.

3.8.2   QNH

QNH is the pressure set on the subscale of the altimeter so that the instrument indicates *height above sea level* (SKYbrary, n.d.a). This is useful as many maps provide terrain heights in

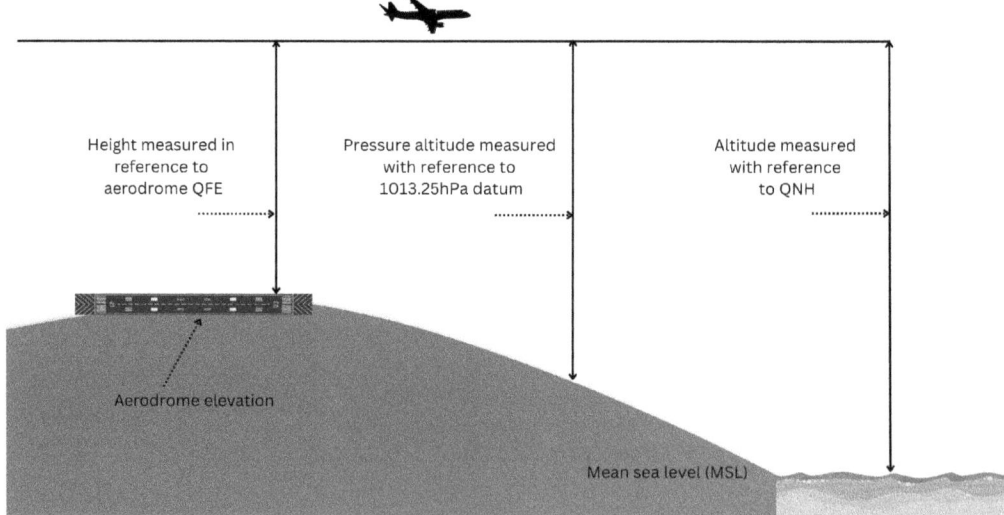

Figure 3.12 Altimeter pressure settings
Source: author

reference to sea level, for example a mountain at 10,000ft in height. It is easily calculated at an aerodrome. For example, when the pressure at an aerodrome elevation of 2,000ft is 940hPa, then the sea-level pressure would be 1,006hPa (pressure increases by 1hPa for every 30ft, therefore 2,000ft divided by 30ft per hPa would be 66hPa difference).

3.8.3 QFE

QFE indicates *height above the reference elevation being used*. It is the pressure at a datum point, which would normally be the aerodrome. The altimeter will therefore give an indication of height above the aerodrome (at the highest point on the runway) during flight. In the previous example, QFE would therefore by 940hPa (the aerodrome elevation). This is a useful datum when the flight remains close to the aerodrome – circuit flying for example. It is of less use far from the aerodrome, especially with mountainous terrain en route.

3.8.4 Standard pressure

QNE is different in that it is an altitude, not a pressure. With standard pressure of 1013.2 set, an aircraft altimeter indicates Pressure Altitude (Flight Level) and is used by all aircraft operating above the transition altitude, which can vary from airport to airport. It is a fixed value published on airport documentation. In the USA and Canada, the transition altitude is fixed at 18,000ft and the airspace above is known as the Standard Pressure Region (SKYbrary, n.d.a).

3.8.5 Altimeter errors and corrections

During a flight, atmospheric pressure regularly changes, therefore it is important the altimeter is kept up to date, according to the nearest surface weather reporting station (FAA, 2023).

- **Pressure.** Flying from areas of low to high pressure without adjusting the altimeter would result in the altimeter under-reading. In other words, the pilot would think they are lower than they are. Conversely, flying from high to low pressure without adjusting the altimeter will result in over-reading – in other words the pilot thinks they are higher than they are – which can be a serious problem! Pilots use phrases such as "high to low, look out below" here.
- **Temperature.** As a pilot flies from warm to cold air, the altimeter reads too high – the pilot is lower than the altimeter indicates. This is especially an issue in cold weather over mountainous areas, where the pilot needs to know the difference between indicated and true (or approximately true) altitude. Phrases such as "from hot to cold, don't be bold" can be used here!
- **Density.** *Density altitude* is a significant indicator of aircraft performance. A "high" density altitude means air density is reduced, which has a negative impact on performance. Three important factors contribute to high density altitude. The higher the (1) **altitude** and (2) **temperature**, the less dense the air (FAA, 2008). As shown in Figure 3.11, this can have a negative impact on aircraft performance – airports such as Phoenix Sky Harbor, Arizona, where the Northern Runway is at a MSL of 1,135ft above MSL (Sky Harbor, 2023) and temperatures regularly reach into the 40s°C during summer. (3) **Humidity** can result in issues regarding engine efficiency rather than aerodynamic efficiency.

Chapter review questions

3.1 How do the four forces of flight enable an aircraft to become and to stay airborne? What are the flight control surfaces used and how do these mechanisms work in achieving safe flight? In the context of the principles of flight, explain the reasons for the accident involving Air Midwest Flight 5481 on 8 January, 2003. Use the NTSB.gov website to locate the Aircraft Accident Report.

3.2 Nitrogen accounts for approximately 78% of the chemical composition of the atmosphere. Carbon dioxide (CO_2) accounts for a fraction of 1%. Why is CO_2 therefore so important to our atmosphere? In what ways do some of the other gases influence aircraft flight?

3.3 What are the key differences between the troposphere and the stratosphere? What are the reasons commercial jet aircraft generally fly in the upper troposphere and lower stratosphere? Are there are any reasons you can think why this flying altitude may not always be best?

3.4 What are the key atmospheric conditions which form part of the International Standard Atmosphere? Based on the values in Table 3.5, research the "standard" temperature and pressure at your local airport.

3.5 Can you explain the differences between radiation, convection and conduction? What are the factors which influence temperatures at (or near) the earth's surface?

3.6 Can you explain the different ways temperature and atmospheric pressure affects aircraft performance? Analyse the role played by density altitude.

3.7 What are the different types of altimeter "Q" codes? Why is each important and in what scenarios would each be used?

ATG trivia

Hot and cold

According to the WMO (2023a), the years 2015–2022 were the warmest eight years on record globally, fueled by ever-rising greenhouse gas concentrations and accumulated heat. However, the warmest officially recorded temperature was well over 100 years ago, in Furnace Creek, Death Valley, California, of 56.7°C (134°F), on 10 July, 1913 (WMO, 2023b). Conversely, the coldest ever recorded temperature was taken at the Vostok high-altitude weather station in Antarctica, of −89.2°C (−128.6°F), on 21 July, 1983 (WMO, 2020).

That is on our own planet earth, however. As for the Solar System, planetary surface temperatures tend to get colder the farther a planet is from the Sun, with the coldest being the dwarf planet Pluto, with a mean temperature of −225°C/−375°F. It may be expected that Mercury (mean temperature of 167°C/333°F) would be the warmest, being the closest to the Sun, however it is actually Venus (464°C/867°F), because of its dense atmosphere acting as a greenhouse (NASA, 2022).

References

Airbus (2021) *Test and certification: approvals for entry-in-service*. Available at: www.airbus.com/en/products-services/commercial-aircraft/the-life-cycle-of-an-aircraft/test-and-certification.

Bidgee (2011) *Automatic Weather Station (AWS) stevensonscreen at the Airport Meteorological Office (AMO) at Wagga Wagga Airport*. Available at: https://commons.wikimedia.org/wiki/File:AWS_stevenson_screen_at_Wagga_Airport.jpg.

Bowen, J. and Rodrigue, J.P. (2020) 'The rise of air transportation', in J.P. Rodrigue (ed.) *The geography of transport systems*. 5th edn. New York: Routledge, pp. 181–195.

Cleveland Clinic (2023) *Cleveland Clinic*. Available at: https://my.clevelandclinic.org/health/diseases/17929-airplane-ear.

Climate-data.org (2023) *Climate data for cities worldwide*. Available at: https://en.climate-data.org/.

Coleman, J. and Law, K. (2015) 'Meteorology', *Reference module in earth systems and environmental sciences* [Preprint]. Available at: https://doi.org/10.1016/B978-0-12-409548-9.09492-6.

FAA (2008) *Density altitude*. FAA-P-8740–2. AFS-8 (2008). Federal Aviation Administration.

FAA (2023) *Aviation weather handbook*. Newcastle, Washington: Aviation Supplies and Academics Inc.

Herbertson, A.J. (1913) *Outlines of physiography, an introduction to the study of the earth*. 3rd rev edn. London: Arnold. Available at: http://archive.org/details/outlinesofphysio00herb.

ICAO (2010) *Annex 5 — units of measurement to be used in air and ground operations*. 5th edn. Available at: https://elibrary.icao.int/reader/268467/&returnUrl%3DaHR0cHM6Ly9lbGlicmFyeS5pY2FvLmludC9lHBsb3JlO3NlYXJjRleHQ9YW5uZXglMjA1O21haW5TZWFyY2g9MTt0aGVtZU5hbWU9Qmx1ZS1UaGVtZQ%3D%3D?productType=eBook&themeName=Blue-Theme.

Kearns, S.K. (2021) *Fundamentals of international aviation*. 2nd edn. Abingdon: Routledge (Aviation Fundamentals).

Kundu, A., Price, M. and Riordan, D. (2016) *Theory and practice of aircraft performance*. Chichester: John Wiley & Sons.

Lankford, T.T. (2001) *Aviation weather handbook*. New York: McGraw Hill.

Lindsey, R. (2024) 'Climate change: atmospheric carbon dioxide', *Climate.gov*. Available at: www.climate.gov/news-features/understanding-climate/climate-change-atmospheric-carbon-dioxide.

Matthews, T. *et al.* (2020) 'Into thick(er) air? Oxygen availability at humans' physiological frontier on Mount Everest', *iScience*, 23(12), p. 101718. Available at: https://doi.org/10.1016/j.isci.2020.101718.

Met Office (n.d.) *How we measure temperature*. Available at: www.metoffice.gov.uk/weather/guides/observations/how-we-measure-temperature.

NASA (2019) *Earth's atmosphere: a multi-layered cake, climate change: vital signs of the planet*. Available at: https://climate.nasa.gov/news/2919/earths-atmosphere-a-multi-layered-cake.

NASA (2022) *Solar system temperatures, NASA solar system exploration*. Available at: https://solarsystem.nasa.gov/resources/681/solar-system-temperatures.

NASA (n.d.a) *Dynamics of flight*. Available at: www.grc.nasa.gov/www/k-12/UEET/StudentSite/dynamicsofflight.html.

NASA (n.d b) *NASA – NASA Dryden technology facts – winglets*. Brian Dunbar. Available at: www.nasa.gov/centers/dryden/about/Organizations/Technology/Facts/TF-2004-15-DFRC.html.

NOAA (2023) *Layers of the atmosphere*. Available at: www.noaa.gov/jetstream/atmosphere/layers-of-atmosphere.

NOAA (n.d.) *Atmosphere | National Oceanic and Atmospheric Administration*. Available at: www.noaa.gov/jetstream/atmosphere.

Royal Society of Chemistry (n.d.) *Argon – element information, properties and uses | Periodic Table*. Available at: www.rsc.org/periodic-table/element/18/argon.

Sadraey, M.H. (2016) *Aircraft performance: an engineering approach*. Boca Raton: CRC Press. Available at: https://doi.org/10.1201/9781315366913.

Sky Harbor (2023) *Airport facts, Phoenix Sky Harbor International Airport*. Available at: www.skyharbor.com/about-phx/history-economic-development/airport-facts/.

SKYbrary (n.d.a) *Altimeter pressure settings | SKYbrary Aviation Safety*. Available at: www.skybrary.aero/articles/altimeter-pressure-settings.

SKYbrary (n.d.b) *International Standard Atmosphere (ISA)*. Available at: www.skybrary.aero/articles/international-standard-atmosphere-isa.
UCAR (2018) *Conduction | Center for Science Education*. Available at: https://scied.ucar.edu/learning-zone/earth-system/conduction.
UCAR (2021) *What is the atmosphere? | Center for Science Education*. Available at: https://scied.ucar.edu/learning-zone/atmosphere/what-is-atmosphere.
UCAR (n.d.) *The energy budget | Center for Science Education*. Available at: https://scied.ucar.edu/learning-zone/how-climate-works/energy-budget.
Walmsley, S. (2021) *Aviation weather for the private pilot*. Aviation Books Private Pilot Series.
Wickson, M. (2015) *Meteorology for pilots*. Marlborough, Wiltshire: Airlife. Available at: www.perlego.com/book/3157975/meteorology-for-pilots-pdf?queryID=7c7fff055d06d98c38f49c9e9bf14cbd&index=prod_BOOKS&gridPosition=6.
WMO (2019) *The Sun's impact on the earth*. Available at: https://public.wmo.int/en/sun%E2%80%99s-impact-earth.
WMO (2020) *WMO verifies −69.6°C Greenland temperature as northern hemisphere record*. Available at: https://public.wmo.int/en/media/press-release/wmo-verifies-696%C2%B0c-greenland-temperature-northern-hemisphere-record.
WMO (2023a) *Past eight years confirmed to be the eight warmest on record*. Available at: https://public.wmo.int/en/media/press-release/past-eight-years-confirmed-be-eight-warmest-record.
WMO (2023b) *Simultaneous heatwaves hit northern hemisphere in summer of extremes*. Available at: https://public.wmo.int/en/media/news/simultaneous-heatwaves-hit-northern-hemisphere-summer-of-extremes.

4 Meteorology and air transport
Water vapour

Chapter outcomes

At the end of this chapter, you will be able to:

- Explain the importance and stages of the hydrologic cycle.
- Identify the different types of humidity and explain their importance in flight operations.
- Distinguish between the different types of atmospheric stability and explain how they influence conditions for aircraft flight.
- Identify the different types of clouds, understand how these are formed and their influence on aircraft operations.
- Understand the different types of precipitation and their formation.
- Explain the types of visibility hazards and their risk to air transport.
- Analyse the different types of icing hazards and how these can affect air transport both in-flight and on the ground.

4.1 Introduction

Water vapour is the foundation for much of the weather we experience. It is the gaseous form of water and one of the most important of all constituents of the atmosphere (FAA, 2023). We may not always see it, but it is almost always in the atmosphere. Even on a clear sunny day, there is likely to be water vapour present. Water vapour is a key component in terms of the impact of weather on air transport – from its interaction with other atmospheric parameters such as temperature, through to the formation of different cloud types and the phenomena we experience such as rain and snow, as well as the severe events like thunderstorms, hurricanes and tropical cyclones.

As we saw in Section 3.3, water vapour is a key factor in the chemical make-up of the atmosphere. It can vary from close on 0% in some of the earth's desert regions (30°N/S) when dry winds are blowing, to near 3% on very hot/humid days. The upper limit around 4% is found in tropical climates (NOAA, n.d.a).

The importance of temperature in the presence and quantity of water vapour cannot be overstated, both in terms of how much water vapour can be held in the air but also changing states – into liquid or solid (ice). Understanding why and how these occur is vital to understanding much of the weather the air transport industry will encounter.

DOI: 10.4324/9781003405351-6

102 *Fundamentals of Global Air Transport Geography*

Water vapour is also the most important greenhouse gas in the atmosphere. Heat radiated from the earth's surface is absorbed by water vapour molecules in the lower atmosphere, which then radiate heat in different directions, including back to the surface, therefore providing earth with an additional source of warmth as well as the sun (NASA, 2023). This is key to understanding the warming of our planet and the impacts of this on air transport, such as decreasing aircraft performance.

Water is constantly cycling through the atmosphere and understanding these processes is important in understanding weather phenomena and their impacts on air transport, particularly on flight operations. The best place to start, therefore, is with an analysis of the hydrologic (water) cycle, before discussing concepts such as humidity and dewpoint.

4.2   Hydrologic (water) cycle

The hydrologic cycle involves the continuous circulation of water in the earth–atmosphere system (Figure 4.1).

Fundamentally, the water cycle is the motion of the water from the ground to the atmosphere and back again. There are many processes involved in this cycle and each has its own importance.

4.2.1   Evaporation

Evaporation is the change in state of a liquid to a gas. For meteorology purposes, water is the most important. There are a number of different sources, such as oceans, rivers and lakes, of which the ocean is the primary one, annually about 120cm (47in) of water being evaporated into the atmosphere (FAA, 2023). The process can also occur as water droplets fall from clouds but evaporate before reaching the ground. Evaporation generally occurs quicker in hotter and drier conditions, as this air will have greater capacity.

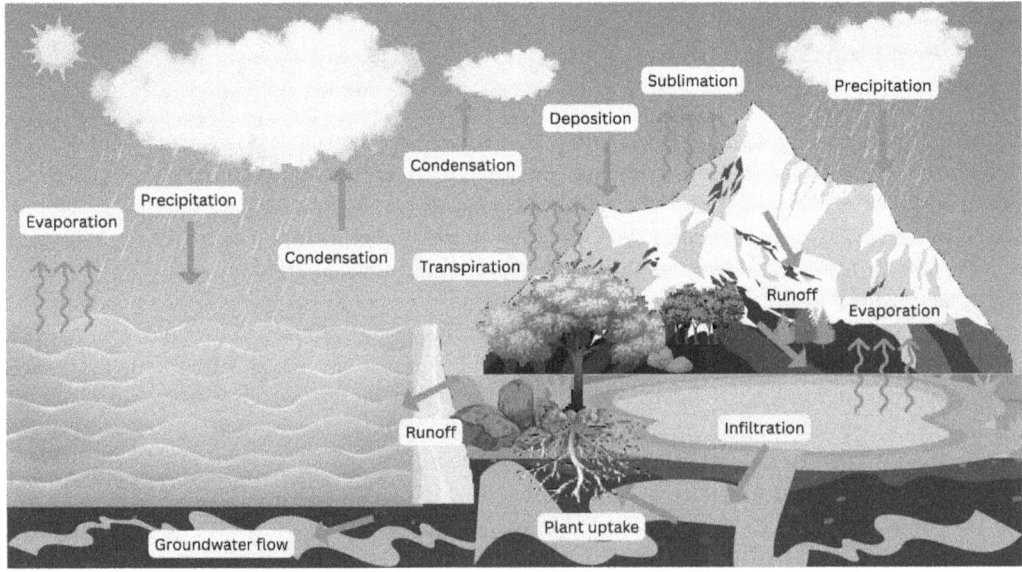

Figure 4.1 The hydrologic cycle
Source: author adapted from NOAA (n.d.b)

Everyone has experienced evaporation in their own lives. For example, the body can heat up due to external temperature increases or via sweat from exercise, with water leaking onto the skin. The body uses its own heat to evaporate the liquid and cool the body – think about stepping out of a swimming pool and the feeling of cold as the bodily heat is removed by evaporating water on the skin.

4.2.2 Transpiration

Transpiration is the evaporation of water from plants through stomata. These are small openings found on the underside of leaves connected to vascular plant tissues (NOAA, n.d.b). In most plants it is a relatively passive process, largely controlled by the humidity of the atmosphere and the moisture content of the soil. Only 1% is used in the growth process of a plant, the remaining 99% is passed into the atmosphere (FAA, 2023).

4.2.3 Condensation

Condensation is the phase transition by which a gas (water vapour) is changed into a liquid. It is the opposite process of evaporation. In the atmosphere, condensation may appear as clouds, fog, mist, dew or frost (FAA, 2023). Think of the water appearing on the side of an uninsulated cold drinks can or bottle. It requires condensation nuclei, tiny particles – typically in the form of dust or pollutants – that water vapour can bind to.

4.2.4 Precipitation

Precipitation occurs when small condensation particles, via coalescence and collision, grow too large for the rising air to support and therefore fall to earth – in the form of rain, hail, snow or sleet. On average, the world receives about 980mm (38.5in) each year over both land mass and the oceans (NOAA, n.d.b).

4.2.5 Runoff

Runoff occurs when there is excessive precipitation which causes the ground to become saturated. The excess liquid flows across the surface of the land into nearby streams, rivers and lakes and some returns to the atmosphere. Runoff can be a major cause of water pollution as it picks up pollutants during its flow. For lakes with no outlet, evaporation is the only means for the water to return to the atmosphere. Evaporation thus begins the hydrologic cycle over again.

4.2.6 Infiltration

Infiltration is the movement of water into the ground from the surface.

4.2.7 Groundwater flow

This is the underground flow of water in aquifers. This may return to the surface in springs or eventually seep into the oceans.

4.2.8 Plant uptake

This is water taken from the groundwater flow and soil moisture and allows transpiration to take place in plants.

4.2.9 Sublimation

This is the phase by which a solid is changed into a gas (water vapour) without passing through the liquid stage. In the atmosphere this occurs when ice and snow (solids) change into water vapour.

4.2.10 Deposition

This is the phase transition by which water vapour is transferred into a solid without passing through the liquid stage – for example, a below freezing point cloud changing into ice crystals.

> **ATG Did you know?**
>
> Water fast facts:
>
> - There are 336,000,000,000,000,000,000 gallons of water on the earth.
> - 71% of the earth's surface is covered in water.
> - 97% of this water is in the oceans (therefore salt water).
> - Less than 3% of all water is freshwater (and suitable for drinking).
> - Of this 3%, more than two-thirds is locked in glaciers and icecaps.
> - 0.001% of water is in the atmosphere.
>
> Source: USGS (2019); NOAA (2023); NASA (n.d.)

4.3 Humidity

As we have seen, one of the key ingredients of air is water vapour. With the sun shining over water sources, such as oceans, seas, rivers and lakes, some of this water evaporates into the atmosphere and this produces *humidity*. Humidity can also affect air pressure and air density and including the significant role temperature plays, these atmospheric parameters can have a major impact on aircraft performance. One of the main effects of high humidity is a reduction in engine thrust; water vapour weighs less than an equal volume of dry air and this can decrease aircraft take-off and climb performance by approximately 7% on a very humid day (Sadraey, 2016).

Air can only hold a certain amount of water vapour before it becomes "full" – *saturation* – given a certain temperature and pressure. At this point condensation may occur.

There are several different methods to define the amount of water vapour in the air – and therefore measuring humidity. Some are more directly useful for the air transport industry, and pilots in particular, than others.

4.3.1 Absolute humidity

The amount of water vapour in a unit volume of air is called the absolute humidity. This is generally measured in grams (g) per cubic metre (m3). The maximum amount will be dependent on the temperature – more water vapour can be held in air in higher temperatures.

- In isolation, absolute humidity is of little direct use to a pilot, as the figure – say 12g/m3 – may result in the air being saturated if the weather is colder and the air capacity is only 12g/m3.

However, on a warmer day, the air capacity may be 24g/m3, therefore the air is only half full, yet the amount of absolute humidity is the same. In the first example, there may be low clouds and fog but in the second there may be clear skies.

ATG Did you know?

Measuring humidity

Instruments for measuring humidity are called *hygrometers*. Most commonly, these use two thermometers mounted side-by-side: a normal mercury thermometer and another where the bulb of the instrument is kept continuously wet by means of a wet cloth. A comparison between the two readings can be used as a measure of humidity.

However, according to the Royal Meteorological Society (2013) a more simple hygrometer can be used containing human hair! The principle being that when relative humidity is high, hair gets longer.

On a slightly different humidity theme, research by James, MacDonald and Hart (2012) seems to have disproved many a cricket lover's belief – their research indicating no direct or indirect manner in which humidity can significantly affect the ability of the bowler to make the cricket ball swing!

4.3.2 Relative humidity

This is a much more practical measure for flight operations which gives a percentage value for how full the air is of water vapour, based on the particular pressure and temperature. It is a measure of how close the air is to becoming saturated (FAA, 2023). A parcel of air with 100% relative humidity (RH) is *saturated*, but anything less than 100% is called *unsaturated*.

- In the previous example, the cold day (12g/m3 by 12g/m3) had a RH of 100%, but on the warmer day the RH (12g/m3 by 24g/m3) was only 50%.

The effects of temperature on RH are illustrated in Figure 4.2. In this example, an air parcel at sea level at a temperature of 30°C has the capacity to hold 27g of water vapour. If it held 8g its RH would be 30%, If the temperature decreased to 20°C, its water storage capacity decreases to 15g (53%) and at 10°C, the capacity decreases to the amount of water vapour (8g), therefore the air parcel becomes saturated. The water vapour content did not change.

Diurnal variation of relative humidity

Temperature changes throughout a 24-hour period and therefore so does RH. As the air cools, RH increases and as the air warms, RH decreases, therefore low cloud and fog is more likely at night and in the early morning (Walmsley, 2021) around sunrise when the temperatures are generally at their lowest and RH is generally at its lowest around 1500 when the temperature is at its highest (FAA, 2023).

The amount of water vapour present in a parcel of air will generally be higher over water – lakes, seas, etc. – than over desert regions, due to evaporation.

106 *Fundamentals of Global Air Transport Geography*

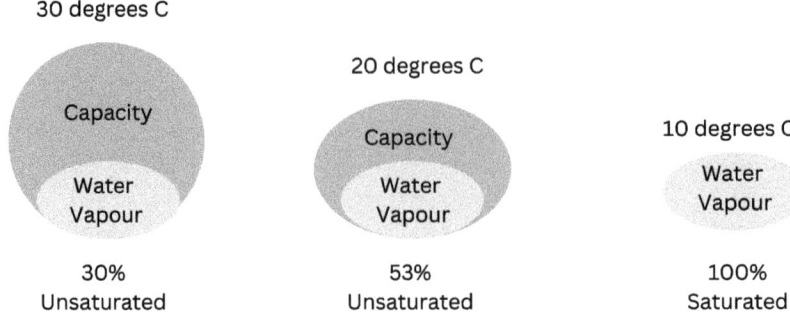

Figure 4.2 Temperature effects on relative humidity
Source: author

4.3.3   Dewpoint

A key measurement given to pilots as part of their weather reports is *dewpoint*. This is the air temperature at which a sample of air would reach 100% humidity based upon its current degree of saturation (SKYbrary, n.d.a). Once the RH of a mass of air becomes 100%, then if the temperature falls it cannot hold all the water vapour and therefore condenses into clouds (or at low levels, *fog*; or in contact with objects at or near the ground, *dew*). When the temperature is below 0°C (32°F) it is sometimes called the *frost point*.

ATG Did you know?

Weather reports – METAR

METAR stands for Meteorological Terminal Aviation Routine Weather Report (or Meteorological Aerodrome Report). These reports are issued for flight planning purposes and summarise the current weather at aerodromes in near real time (Met Office, n.d.b). They provide a snapshot of current weather conditions, including wind speed and direction, visibility, weather, cloud, temperature, dewpoint and pressure.

As per ICAO (2018) Chapter 4, a state must establish aeronautical meteorological stations at aerodromes within its territory, as it determines to be necessary. Section 4.3.1 states that "At aerodromes, routine observations shall be made throughout the 24 hours of each day … such intervals shall be made at intervals of one hour or … at intervals of one half-hour". Through ICAO, the coded format is highly standardised.

Figure 4.3 illustrates the types of information present in METAR weather reports. Regarding temperature and dewpoint (highlighted), at London Heathrow Airport, the example temperature is 12°C as is the dewpoint, RH is 100% and therefore condensation is likely to occur. However, at Phoenix Sky Harbor Airport, the temperature is 30°C but the dewpoint is only 5°C, RH is low and it is extremely unlikely condensation will occur near the surface.

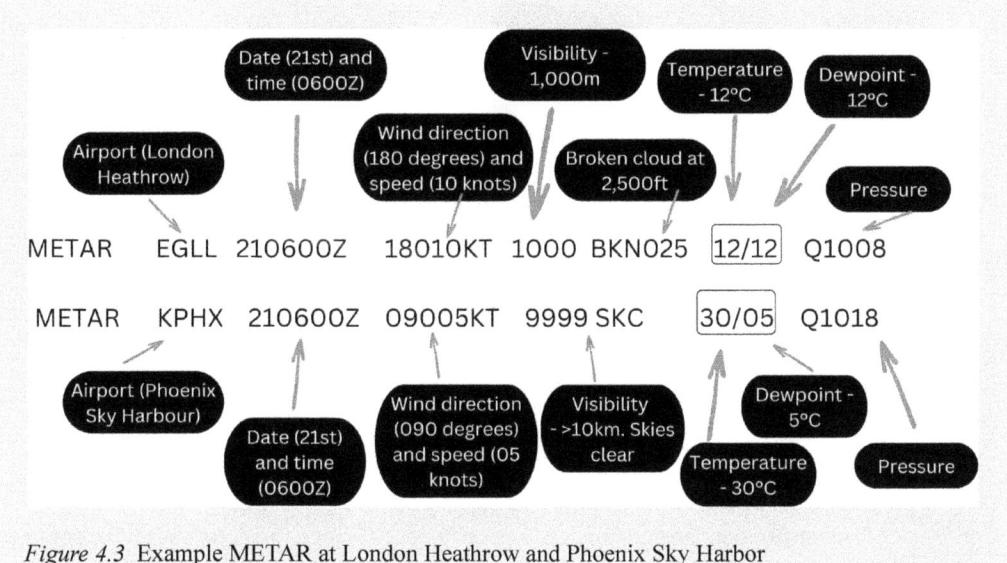

Figure 4.3 Example METAR at London Heathrow and Phoenix Sky Harbor
Source: author

It is also important that the pilot is aware as to which direction the temperature is heading. For example, if it is mid-morning, it is likely that the temperature is increasing, therefore all other things being equal, the RH should drop, and the risk of condensation will also drop. However, if it were in the early evening, the temperatures are likely to drop further, with the likelihood of RH increasing and condensation occurring, depending on how close the dewpoint is to the temperature.

4.3.4 Latent heat

Every time a change of phase occurs, latent heat/energy is involved. When evaporation occurs, heat is required. The temperature of the water vapour is not increased, it is heat which is hidden. The latent heat being released increases the temperature of the surrounding air (sensible heat). Latent heat is hidden heat which is absorbed or released without a change of temperature when water changes state.

Figure 4.4 illustrates the process of latent heat transactions. During condensation and freezing, latent energy is released, and this is important for cloud formation and the icing of aircraft. Latent energy is required during melting and evaporation.

ATG Did you know?

Latent heat exchange

The annual global energy consumption is estimated to be approximately 580 million trillion joules. According to the FAA (2023), an average hurricane releases 52 million trillion joules per day as water vapour condenses into clouds and precipitation. The amount of energy associated with latent heat exchange should not be understated!

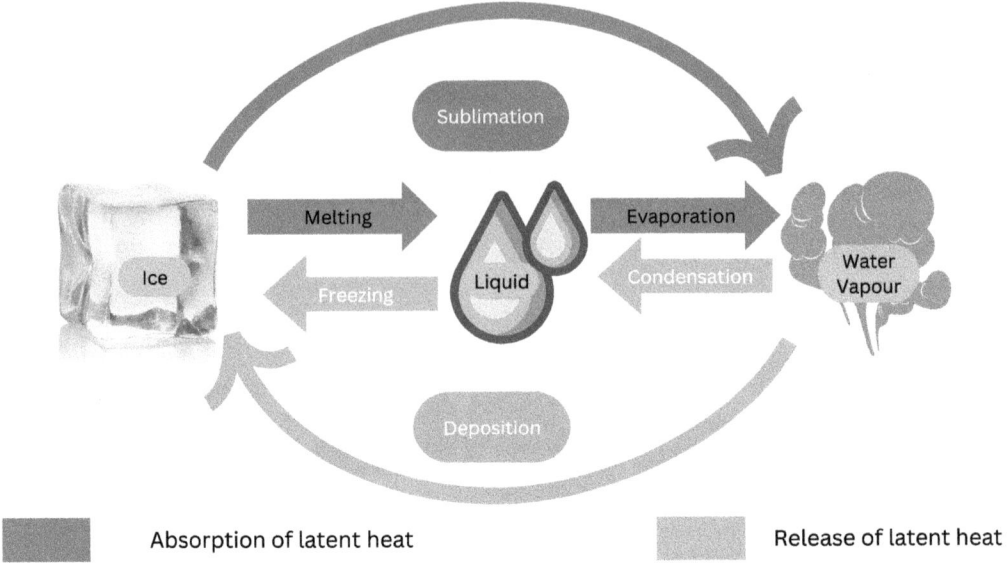

Figure 4.4 Latent heat transactions
Source: author

4.4 Atmospheric stability

Looking up at a bright, sunny, blue sky with very few clouds and a question to be asked could be why are there so few clouds on a day like today? This is down to the concept of *atmospheric stability*. Stability is defined as the ability of a mass (parcel) of air to remain in equilibrium – its ability to resist displacement from its initial position (Lankford, 2001). A parcel of air can be displaced for a number of reasons, called **triggers**, such as: warming from below (*convection*), air forced over higher surfaces (*orographic*), convergence or *frontal lifting* (Walmsley, 2021).

A parcel of air is what carries water vapour higher in the atmosphere and hence is important for the formation of clouds and any subsequent precipitation. Stability is the property of the ambient air that either enhances or suppresses *vertical motion* of air parcels. This concept is crucial in determining weather conditions and pilots require a sound understanding of the concept.

4.4.1 Adiabatic lapse rate

How a parcel of air behaves (whether it rises, sinks or stays the same) is due to a combination of temperature change in the environment (environmental lapse rate – ELR) and the internal temperature change within the parcel of air (adiabatic lapse rate). In Chapter 3, we saw that the atmosphere cools at around 1.98°C per 1,000ft. This is the overall temperature change in the atmosphere – the ELR, but with no movement of air up or down. As with the fact that sea level is not always 15°C and air pressure at sea level is not always 1013.25mb/hPa, temperature change in the atmosphere will also be variable.

Dry adiabatic lapse rate (DALR)

If a parcel of air is forced to rise it will cool, often at a different rate to the ELR – this is due to the *adiabatic process*. As the air rises, the pressure decreases, allowing the air parcel to expand.

Meteorology and air transport: water vapour 109

This uses energy which means it may cool at a higher rate than the ELR, at 3°C per 1,000ft (Wickson, 2015), as long as the air is not saturated. This is called the *dry adiabatic lapse rate*. This affects the temperature inside the air parcel, not surrounding it. A good analogy is the air inside a hot air balloon.

Saturated adiabatic lapse rate (SALR)

As the air rises, it may eventually reach an RH of 100% (saturation point), which results in condensation and cloud formation. During this process, *latent energy* is released, warming the air surrounding the parcel, resulting in two competing processes of both the DALR inside the parcel and warming outside. This type of air parcel will cool slower and is called the *saturated adiabatic lapse rate*. A typical figure here in temperate latitudes may be 1.5°C per 1,000ft (Wickson, 2015) although this can vary both geographically by latitude and at differing altitudes – the higher the altitude, the less moisture and less latent energy released, therefore the SALR is often higher than at lower altitudes (Walmsley, 2021).

4.4.2 Types of atmospheric stability

When the parcel of air is clear of its "trigger", the distribution of parcel stabilities within each column of air is key. Atmospheric stability generally falls into four unique categories: absolute stability, neutral stability, absolute instability and conditional instability (Lankford, 2001; FAA, 2023). Table 4.1 illustrates the typical characteristics of atmospheric stability.

Absolute stability

Absolute stability is when an air parcel resists vertical displacement whether it is saturated or unsaturated (Lankford, 2001). An air parcel rising would be colder and denser than the surrounding environment and would therefore sink back down again. If air is blown against a building and lifted, when it has moved away from the buildings and is on the "lee" side, it sinks back to the ground as the atmosphere is stable. If the ELR is less than 3°C per 1,000ft for dry air and less than 1.5°C for saturated air, the atmosphere is stable (Wickson, 2015).

A stable atmosphere results in limited vertical motion, hindering cloud development (Section 4.5) and generally meaning better conditions for aircraft flight.

Absolute instability

An atmosphere is unstable if, when the lifting force is removed, the rising air continues upwards. This occurs when vertical displacement of an air parcel is spontaneous. Its temperature is always

Table 4.1 Characteristics of atmospheric stability

Stable	Conditionally unstable	Unstable
Ceiling poor (where one exists)	Ceiling poor	Ceiling good
Visibility poor	Visibility poor	Visibility good
Smooth	Smooth below/turbulent above	Turbulent
Precipitation steady and light to moderate	Steady/embedded heavy showers	Precipitation showery and heavy

Source: adapted from Lankford (2001)

warmer than the surrounding air and the cooler, more dense (hence lighter), air forces it upwards. If the ELR is greater than 3°C for dry air or 1.5°C per 1,000ft for saturated air, the air is unstable. This can create significant cloud formations, having a more negative impact on aircraft flight.

Conditional instability

When the ELR is between 1.5°C and 3°C per 1,000ft, the atmosphere is termed conditionally unstable. In other words, it is stable if dry and unstable if saturated. As the average ELR is around 2°C per 1,000ft, this is the most likely condition in general. According to Lankford (2001), high moisture content at low levels and dry air at higher levels favours instability whilst dry air in low levels and high moisture content in higher levels favours stability.

Neutral stability

This is the state of a column of air in the atmosphere in which an ascending or descending parcel of air always has the same temperature and density as the surrounding environmental air (FAA, 2023). If the column of air is unsaturated, neutral stability will exist when the temperature lapse rate equals the DLR. If the air is saturated, then neutral stability exists when the temperature rate equals the SALR. Hence, the parcel of air is dry and the ELR is exactly 3°C per 1,000ft or the air is saturated and the ELR is exactly 1.5°C. The parcel of air remains at rest and would neither rise nor fall after its trigger – for example, when clear of the building, it will remain at the height of the building.

In practical terms, pilots will not regularly be calculating ELR values. However, in terms of weather reports such as METARs, the terms stable and unstable are important and can therefore be linked with different weather conditions which may be experienced during the flight – most noticeably via types of cloud formation.

4.5 Clouds

In meteorological terms, clouds are our signposts in the sky and are collections of water droplets or ice crystals, or a combination of both. Clouds are incredibly important for aviation operations, potentially resulting in rain, sleet, snow or hail, through to thunderstorms and potentially life-threatening events such as hurricanes. The signposts can tell the pilot a lot about current weather and that coming up in the near future – whether the visibility is going to be poor, if it is likely to be turbulent and the type of precipitation which can happen.

4.5.1 Cloud classification

There are ten major cloud types – "genera" (World Meteorological Organization, 2017; Met Office, n.d.a) and these are generally classified based on both their:

a) **Level in the atmosphere** (high, middle or low) and
b) **Stability** (stable or unstable) (Figure 4.5).

Cloud levels in the atmosphere

The approximate heights of cloud bases vary dependent on latitude, and Table 4.2 provides approximate altitudes for High, Middle, and Low cloud bases. Note that clouds such as cumulonimbus generally have their bases at low levels, however they are also classed as clouds with vertical motion and according to the FAA (2023), they can have their tops at over 60,000ft of altitude – as high as Concorde flew!

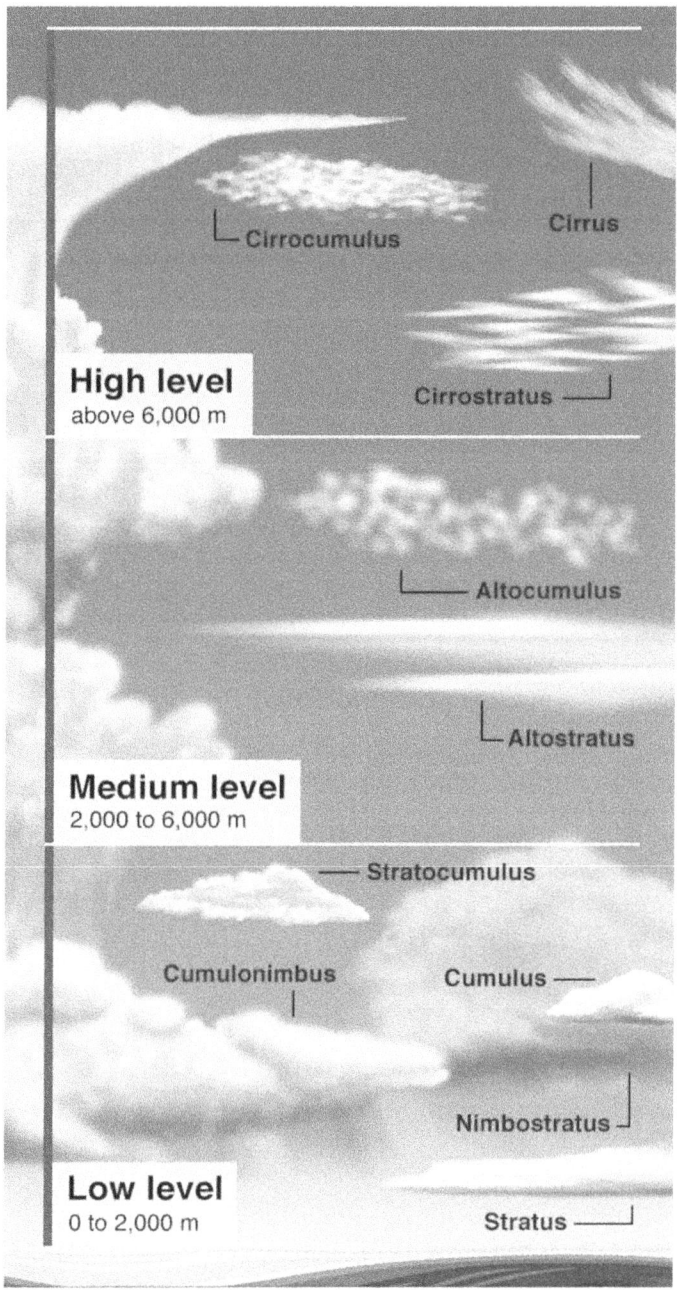

Figure 4.5 Cloud types
Source: © Crown copyright, Met Office

Table 4.2 Approximate height of cloud bases above the surface

Level	Polar regions	Temperate/mid-latitude regions	Tropical regions
High Clouds	10,000–25,000ft (3–8km)	16,500–40,000ft (5–13km)	20,000–60,000ft (6–18km)
Middle Clouds	6,500–13,000ft (2–4km)	6,500–23,000ft (2–7km)	6,500–25,000ft (2–8km)
Low Clouds	Surface–6,500ft (0–2km)	Surface–6,500ft (0–2km)	Surface–6,500ft (0–2km)

Source: adapted from FAA (2023)

Cloud stability

A second cloud categorisation is based on cloud stability, which falls into two types: *Stratiform* and *Cumuliform*.

1. **Stratiform** clouds are clouds with extensive horizontal development associated with stable air and generally consist of small water droplets. Generally, stable clouds have little turbulence due to the limited vertical movement of air found within. Clouds include:

    - stratus
    - stratocumulus
    - nimbostratus
    - altostratus
    - cirrostratus

2. **Cumuliform** clouds are characterised by extensive vertical development, in the form of rising shapes such as domes or towers, and are associated with an unstable air mass. Due to upward-moving currents, they generally contain large water droplets. Turbulence is often moderate to severe due to the vertical development of air. These include:

    - altocumulus
    - cirrocumulus
    - cumulus
    - cumulonimbus

ATG Did you know?

Cloud names

The classification of clouds dates to the early 19th century, courtesy of Luc Howard, an amateur meteorologist. These were based on Latin names and most clouds contain Latin prefixes and suffixes, which gives an indication of the cloud's character:

- Stratus/strato: flat/layered and smooth
- Cumulus/cumulo; heaped up/puffy
- Cirrus/cirro: feathers/wispy

- Nimbus/nimbo: rain-bearing
- Alto: mid-level (though Latin for high!)

Source: World Meteorological Organization (2017)

Although there are ten main classifications, there are about 100 combinations of "species" and "varieties" of clouds, including human-induced forms such as **contrails**, which will be discussed in Chapter 14.

An important point to note is that the classification of clouds is generally related to altitude – from a particular vantage point. For example, at sea level, we might look up and see nimbostratus clouds. However, were we on top of a mountain, we may see these clouds as fog.

4.5.2   Cloud types – high

These clouds are almost always made up of ice crystals due to their height in the atmosphere (above 16,500ft at mid-latitudes). The icing risk here for aircraft is generally low as the air is so cold – rather than sticking to the aircraft they bounce off it. Figure 4.6 illustrates two examples of high clouds – cirrus (a) and cirrocumulus (b).

Cirrus (Ci)

- Latin *cirrus* = ringlet or curling lock of hair.

Cirrus is a common high cloud, composed of detached cirriform elements of filaments. They have a fibrous appearance and often trail downward in wisps called "mares tails".

Cirrocumulus (Cc)

Indicate vertical motion at high levels, thus unstable and may indicate turbulence. They may be composed of highly supercooled water droplets as well as small ice crystals, or a mixture. Cirrocumulus clouds are not very common, forming ripples which may resemble honeycomb.

Figure 4.6 High-level clouds
Source: photos courtesy of John Irish

114 *Fundamentals of Global Air Transport Geography*

Cirrostratus (Cs)

These are generally very thin, high, stable clouds, appearing as sheets or layers of cirrus. They often barely look like clouds, more like a whitening of the blue sky. When these appear within a few hours after cirrus in midlatitudes, it is often the probability of an approaching front and hence poor weather. These sometimes contain a halo caused by the light interacting with the ice crystals.

4.5.3   Cloud types – middle

These are formed between 6,500 and 23,000ft at midlatitudes. Figure 4.7 illustrates altostratus mid-level clouds – in (a) the lower cloud in the sky and (b) forming ahead of a thunderstorm. They are normally formed by liquid water, though part may be supercooled (Frey, 2017) liquid water droplets and/or ice crystals (FAA, 2023). Due to the presence of supercooled water droplets, there is an increased risk of icing if an aircraft is flown through these clouds.

Altocumulus (Ac)

Altocumulus is a mid-level unstable cloud due to its vertical motion. Lumpy in appearance but precipitation is unlikely, although there may be light–moderate turbulence. A thickening band of altocumulus below a band of cirrus may indicate the approach of a cold front.

Altostratus

Altostratus is a cloud type in the form of a grey or bluish sheet or layer, often covering a considerable amount of the sky and according to the FAA (2023), it may cover an area of several thousand square miles. As with altocumulus, these may indicate a front approaching and therefore light precipitation. These can sometimes produce mild turbulence and light–moderate icing in the supercooled water regions.

4.5.4   Cloud types – low

There are a wide variety of clouds found at low levels – some small with insignificant effects on the weather, whereas others are huge, stretching into the stratosphere and bringing very poor

Figure 4.7 Mid-level clouds
Source: photos courtesy of John Irish

weather. As they are closer to the surface, much of the clouds contain water droplets, but in winter in some latitudes, there may be ice crystals. Figure 4.8 illustrates some low-level cloud types – cumulus (a and b), nimbostratus (c) and a late-stage cumulonimbus (d) – although this last type is sometimes referred to as a multi-level cloud, as it can extend high into the sky.

Stratus (St)

Stratus indicates a stable air mass and is a common cloud in the early morning, formed commonly with the cold surface cooling the air above. It is a grey layer with a uniform base. It does not normally produce precipitation but when it does, this will be drizzle or snow grains. Stratus produces little or no turbulence, but when the temperatures are near or below freezing, icing risk can be hazardous.

Nimbostratus (Ns)

This is sometimes classed as a mid-level cloud and is one of the two main rain-producing clouds. It is a grey cloud layer, often dark, rendered diffuse by more or less continuously falling rain, snow or ice pellets (FAA, 2023). It is generally formed in a front with the lifting of different air masses or lowering of altostratus. With the large water content, in colder temperatures, icing is a significant risk.

Cumulus (and towering cumulus)

Often called fair-weather cumulus, it is a common cloud in the mid-afternoon in summer. These develop vertically, often appearing as rising mounds, domes or towers. The top parts sometimes resemble a cauliflower. When a surface warms during the day, a shallow unstable layer may be created and may extend a few thousand feet. There should be little or no icing, but some turbulence may be experienced. However, for cumulus of moderate to strong development (towering cumulus) very strong turbulence may be experienced and some clear icing above the freezing level.

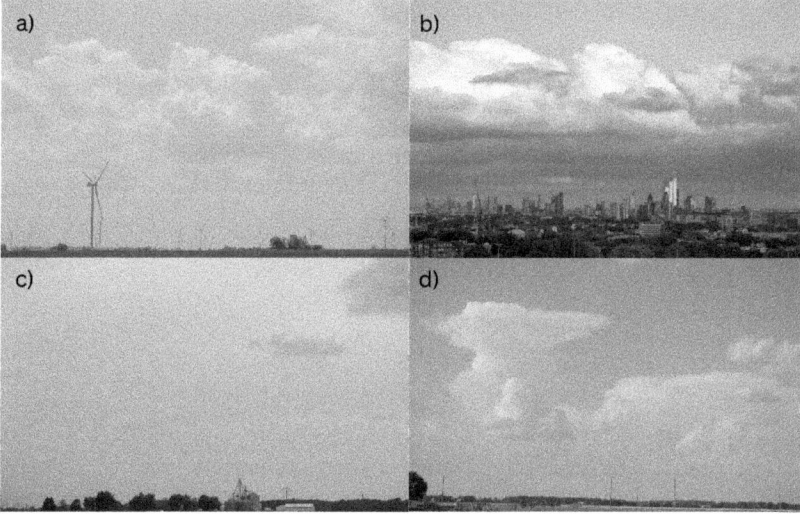

Figure 4.8 Low-level clouds

Source: photos courtesy of John Irish

Stratocumulus

As the two components of the name indicate, there is a mixture of stable and unstable air. The air is weak vertically, making the clouds rounded, but in more of a semi-continuous layer than cumulus. Generally low precipitation and maybe some turbulence, with possible icing at sub-freezing temperatures. From a plane's eye view above the cloud, they can look like a landscape of rolling cloud valleys and hills.

Cumulonimbus

These are sometimes also referred to as clouds with vertical development, rather than as low-level, even though that is where their bases are. These are very large, vertically developed clouds produced by very unstable conditions. They can appear as mountains or huge towers. Sometimes the tops spread out in an anvil shape. Precipitation is often heavy and showery in nature and sometimes accompanied by thunder and lightning. Ultimately, these clouds contain almost the whole spectrum of flying hazards, including severe turbulence and a serious potential hazard they cause for air transport – thunderstorms.

ATG Did you know?

Why is the sky blue?

When the sun's light reaches the earth's atmosphere it is scattered by the tiny molecules of gas (mainly nitrogen and oxygen) in the air. These molecules are much smaller than the wavelength of visible light. Shorter wavelengths such as blue and violet are scattered most strongly, so more of the blue light is scattered towards our eyes than the other colours (Met Office, n.d.c).

4.5.5 Cloud formation

It is important to know not just the types of clouds pilots may encounter, but also the processes by which clouds are made/formed. When air rises and the condensation point of 100% RH is reached, the water vapour will condense onto *hygroscopic nuclei*.

As we have seen, air being lifted to form clouds requires a trigger. There are several potential lifting mechanisms/triggers (Figure 4.9) for example:

- orographic lifting
- frontal lifting
- convection

Orographic lifting

This occurs when a parcel of air is lifted by a geographical feature – such as hills or mountains. Some of the air will have no choice but to rise, expand and cool and if the air is moist, clouds can form. Windward slopes of mountain ranges will therefore tend to have higher rainfall then the leeward slopes and can produce both stable and unstable air. Dry conditions form on the leeward side as the air descends, warms up and reduces its RH and can often extend hundreds of miles in a region known as the rain shadow. Figure 4.10 illustrates a stable orographic cloud.

Meteorology and air transport: water vapour 117

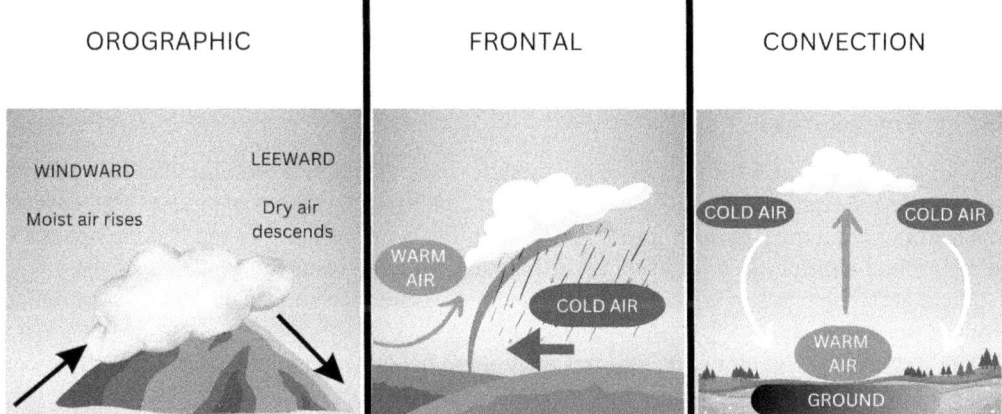

Figure 4.9 Cloud trigger mechanisms
Source: author

Figure 4.10 Orographic cloud on Athos, Greece
Source: author

Frontal lift

Frontal lift occurs when cold, denser air wedges under the warm, less dense air, pushing it upwards or alternatively the warmer less dense air will attempt to rise over the colder heavier air. The surface where the two air masses meet is termed a *front*. Due to this lifting mechanism, many cloud types can form over a wide area.

Convection

Air heated over a land surface rises and if cooled below dewpoint produces cloud. If conditions allow, this air can continue to rise, cooling as it goes, forming cumulus clouds. Stronger

118 *Fundamentals of Global Air Transport Geography*

convection can result in even larger clouds developing as the air rises higher before it is cooled, sometimes even producing cumulonimbus clouds and possibly thunderstorms.

4.6 Precipitation

Precipitation is any of the forms of water particles, whether liquid or solid, that fall from the atmosphere and reach the ground (FAA, 2023). Precipitation can take many guises and these can be hazardous to aircraft, for example reducing visibility (see Section 4.7) and making operations on the ground hazardous, in terms of aircraft landing and day-to-day ground operations.

There are three types of water formation in clouds: water, ice and supercooled water. All can produce precipitation.

The formation of precipitation requires three ingredients: water vapour, sufficient lift to condense the water vapour into clouds and a growth process to allow cloud droplets to grow large and heavy enough to fall as precipitation. The heavier the precipitation, the thicker the clouds are likely to be.

4.6.1 Precipitation growth

Only some clouds will produce precipitation, yet all contain water. These water droplets or ice crystals are too small and light to fall to the ground as precipitation – sometimes 0.01–0.02mm – whereas a typical raindrop may be 1–2mm. Cloud droplets falling from the sky from a cloud base of 2,300ft would take 48 hours just to reach the ground(!) (FAA, 2023). Evaporation will likely occur long before this, called *virga*. There are two main processes by which droplets can grow large enough to reach the ground as precipitation:

- coalescence
- deposition

Coalescence

These generally occur in warmer clouds (those containing liquid) whereby cloud droplets collide and merge in a process known as *collision-coalescence*. The greater the mixing, the larger the droplets, the more likely they are to be too heavy to remain suspended in the air and therefore fall to the ground as precipitation.

Deposition

This process occurs in colder clouds (colder than 0°C/32°F), mainly made up of ice crystals. The ice crystal process occurs when both ice crystals and water vapour are present and it is easier for water vapour to deposit directly onto the ice crystals, which then grow and may eventually become heavy enough to fall. If it is cold near the surface, it may snow; otherwise, the snow may melt to rain.

4.6.2 Types of precipitation

Pilots can come across a range of different types of precipitation on any one flight. The amount is generally due to the amount of mixing in a cloud as well as the temperature.

Drizzle

Drizzle is very small raindrops. According to the FAA (2023), drizzle is defined as having a diameter of between 0.2–2mm – leaving very little impact on a water surface. This may be from clouds such as stratus.

Rain

Rain is a drop of water between 2–5.5mm in diameter, where the surface impact can be seen. This would often be from cumulus or cumulonimbus clouds.

Snow

Snow is ice crystals that reach the ground and if it is cold enough can occur from most cloud types. It can be in the form of *grains*, *needles* or *snowflakes*. Snow can reach the ground still in its frozen state with a surface temperature up to 4°C.

Sleet

Sleet is a mixture of rain and snow.

Hail

This is ice in the form of balls or pellets and can grow to large sizes. They generally fall from cumulonimbus clouds. Hail can be more common at higher elevations because the stones begin to melt when they fall below the freezing level.

ATG Did you know?

Largest and heaviest hailstones

According to the US Department of Commerce (2010), the largest hailstone ever recorded by diameter landed in Vivian, South Dakota, USA, on 23 July, 2010, measuring 8.0 inches in diameter, 18.625 inches in circumference – almost as large as a ten-pin bowling ball!

The heaviest hailstone recorded fell in April 1986 in the Gopalganj District, Bangladesh, weighing 1.02kg (2.25lb) (World Meteorological Organization, n.d.)!

4.7   Hazards – visibility

Visibility is a measure of the distance at which an object or light can be clearly discerned (SKYbrary, n.d.b). It is the furthest horizontal distance that a dark object can be seen by an observer with normal eyesight and is a measure of atmospheric clarity. How far a pilot can see can have a significant impact on their ability to fly their aircraft safely.

There are three aspects of visibility which may apply to aviation:

- Flight visibility – the horizontal visibility from the aircraft.
- Slant visibility – from the air to the ground. Care needs to be taken between thin and thick layers of cloud, especially descending coming into land.
- Ground visibility – reported from ground stations.

There are two main reasons for reduced visibility:

- water and ice crystals in the air
- solid particles such as smoke, sand and dust – this can be called *haze* – and can be especially common around desert regions, especially in windy conditions

4.7.1   Obstructions to visibility

There are many obstructions to visibility and each can be hazardous to air transport. This is especially the case for pilots flying in what is called visual flight rules (VFR) flight, with aircraft flying in visual meteorological conditions (VMC).

4.7.2   Precipitation

The type and intensity of precipitation can influence visibility – this can be in the form of rain, drizzle or snow. Heavy rain can reduce visibility, whereas drizzle and snow can reduce visibility even more. Blowing snow can reduce visibility much less and may even be essentially zero with heavy or blowing snow in high winds – often called a whiteout.

4.7.3   Dust storm and sandstorm

Dust storms originate over regions when fine-grained soils, rich in clay and silt, are exposed to strong winds and taken airborne. Aircraft operation in a dust storm can be extremely hazardous and visibility can reduce to low very quickly and this is especially an issue coming into land and being able to pick out the airfield. These can also cause technical issues in terms of dust clogging engine air intakes and problems with electrical systems.

Sandstorms are particles of sand carried along by a strong wind. These are generally lower (rarely higher than 50ft above the ground) – sand particles are generally heavier and larger than dust. Hazards are especially high in the last phase of landing or at take-off.

Haboobs are dust or sandstorms that may extend horizontally for more than 60 miles (100km) and rise vertically to the base of the thunderstorm.

4.7.4   Fog

Fog is basically cloud at ground level. The cooling process for cloud formation occurs at the surface. This can be extremely hazardous, especially for aircraft and ground vehicles moving around the airport.

There are numerous types of fog, which are formed according to their formation mechanism. Regardless of fog type, pilots should be aware of the temperature and dewpoint, especially at night. If they are close together in the evening, then the possibility of fog increases. Figures 4.11 and 4.12 illustrate some of the common types of fog.

Radiation fog

This is one of the most common types of fog. It forms over land, especially during the colder months of the year. It generally occurs at night and peaks, then clears after sunrise. Valleys can be particularly prone to radiation fog. The ideal conditions are a clear sky, light wind and a high RH.

Meteorology and air transport: water vapour 121

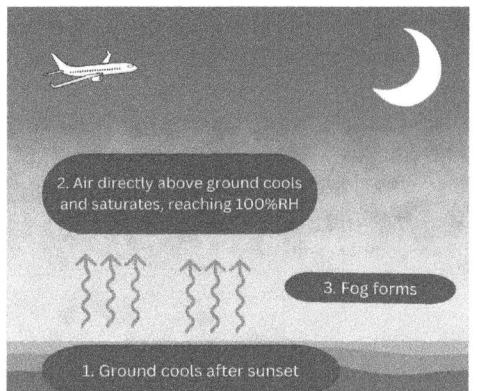
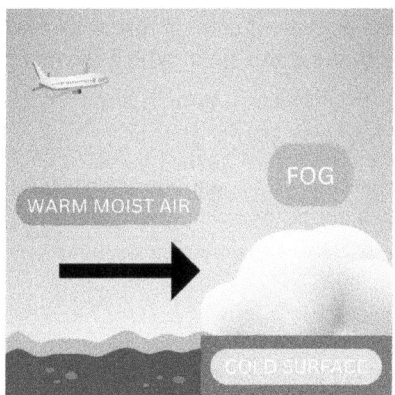

Figure 4.11 Radiation and advection fog
Source: author

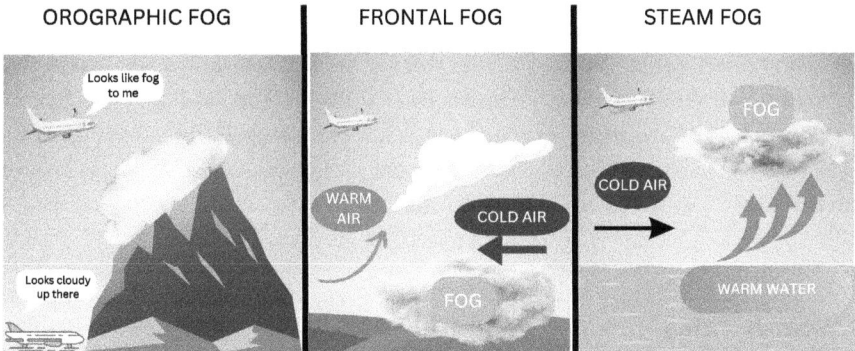

Figure 4.12 Orographic, frontal and steam fog
Source: author

Advection fog

This can form over both land and ocean. It is movement fog and occurs when moist, warm air flows over a cold surface. The cold surface must be at a temperature lower than the dewpoint of the moving air for saturation and condensation to occur. In the UK and Northern Europe, advection fog is most likely over land in winter and over sea in late spring or early summer when the water is colder.

Steam fog

This occurs when very cold, dry air moves across relatively warm water. There may be enough water evaporating from the water surface to produce saturation. It is commonly observed over lakes and streams on cold autumn mornings and over the ocean during winter and in higher latitudes, such as Greenland and Northern Canada. Steam fog is often very shallow and can look like smoke. It is associated with unstable air and therefore there may be convective turbulence encountered.

Frontal fog

This occurs at warm fronts and occlusions. It can cover 200–300NM in front of the warm front and occurs when moisture falls into cooler air below, which can become saturated. It travels with the warm front, which moves relatively slowly and can linger for a long period of time.

Orographic/hill fog

In simple terms, a mountain at 10,000ft viewed from sea level may have clouds at the top, however flying over the mountain, it could be classed as fog (see Section 4.5.5). Pilots must take extreme care when near mountainous terrain, especially when clouds/fog are present.

> **ATG Case study 4.1**
>
> **The hazard of fog: crash of KLM 4805 and Pan Am 1736 at Tenerife Los Rodeos Airport, 27 March, 1977**
>
> This was the worst aviation disaster of all time, between two Boeing 747s and involved miscommunication and fog. Due to fog, the captain of the departing KLM aircraft and the air traffic control tower could not see that the Pan Am aircraft, which had just landed, was still on the runway and the KLM aircraft commenced its take-off run, leading to the crash which killed 583 people. The investigation concluded that it was communication confusion that led the KLM Captain to initiate take-off without proper clearance. A contributing factor was the presence of dense fog (ICAO, 1978; Kearns, 2021).

4.8   Hazards – icing

Ice will form on an aircraft if liquid water hits a part of the airframe which has a temperature below freezing. According to the World Meteorological Organization (2020), there are three strategies to cope with aircraft icing:

1. Aircraft must be certified for icing (which commercial aircraft will be).
2. Aircraft must be cleaned of ice prior to take-off (clean wing principle).
3. Aircraft should be equipped with de-icing equipment.

Icing may occur in different locations:

- in-flight or at the surface
- ground icing

Icing can also be categorised into:

- airframe icing
- engine icing

4.8.1 Airframe icing

At very low temperatures, ice generally forms near the wing leading edges and is generally quite visible. At temperatures just below zero, the ice spreads back and can be less visible covering more of the airframe (EASA, 2015).

Airframe icing normally occurs when the ambient air temperature is below 0°C and supercooled water droplets are present. However, it may also occur when the temperatures are higher, and the aircraft have descended quickly from a high altitude – "cold soak". There are several different types of airframe icing:

Clear ice

If the supercooled water droplets are large, then they may immediately form ice when in contact with an airframe. This can then release latent heat and raise the temperature of the rest of the droplet, which can then delay the freezing process until it has flowed back over the wing or tailplane. This is very tough, dense and smooth and adheres strongly to aircraft skin. It usually occurs at warmer temperatures (>−10°C) and/or high liquid water contents.

Rime ice

This is generally smaller supercooled water droplets and occurs at low (<−10°C) temperatures and/or low liquid water contents; due to cold temperatures, it can freeze instantly. It has low latent release and therefore quick ice formation. This can cause compaction on the leading edge by airflow, but generally breaks away relatively easily and can cause a loss of airfoil shape and effect air intakes.

Mixed ice

Occurs around the crossover point/temperature between rime and clear ice – impingement of supercooled water and ice. Generally, the greatest icing severity occurs between −5 and −15°C (Walmsley, 2021) as there are still plenty of water droplets at a reasonably "cool" temperature. Higher in the atmosphere, there is less liquid that can freeze on contact with the aircraft surface, despite the colder temperatures.

Hoar frost

This is a thin coating, occurring in the absence of rain or cloud, usually when the aircraft is parked outside on cold nights. The water vapour goes straight to ice. It is vitally important to remove this before flying and to be vigilant that it does not return between removal and take-off, especially if the aircraft is being held at a busy airport before being cleared to leave stand and with a potentially long taxi time.

4.8.2 Engine icing

The risks can differ for both piston and turbine engines, although the risk of fuel freezing occurs in both.

- *Piston* engine icing occurs in the engine intake area and is comparable to airframe icing. This may restrict the flow of air to an engine. There is also the risk of fuel icing, caused by water in the fuel freezing in pipe bends, reducing fuel flow to an engine.

- *Carburetor* icing is caused by a lowering of the temperature inside the carburetor, so ice can form. This generally affects the general aviation sector more, due to having aircraft types with a carburetor.

4.8.3 Aviation hazards associated with icing

- Reduction in aerodynamic properties.
- Change in flight performance.
- Increase in weight and uneven loading.
- Engine intakes become blocked.
- Undercarriage retraction/extension problems.
- Control surfaces jam or become stiff.
- Pitot tubes become blocked.
- Communications affected.
- Vision impaired.

Source: World Meteorological Organization (2020)

4.8.4 Aircraft ground de-/anti-icing

This is a vital part of airport ground operations, where atmospheric conditions mean the risk of snow and ice is sufficient to warrant investment in equipment to prevent or remove ice from aircraft prior to take-off, to ensure safe operation (Figure 4.14). This is especially important on the aerodynamic surfaces, control surfaces and sensors/vents.

The aircraft should be positioned in a suitable area, with adequate drainage to try to eliminate ground pollution and to reduce the risk of human harm via any chemicals. De-icing occurs at three possible places:

- On stand prior to any loading beginning. This is a riskier time as depending on outside air temperatures, there is a risk of refreezing. However, it means the aircraft can depart quicker after loading is complete.
- On stand after loading is complete and prior to taxi. There is a lower risk of refreezing, but it takes time and will slow the aircraft departure time.
- At a designated de-icing area enroute from stand to the runway. The airport needs to provide a dedicated area for this.

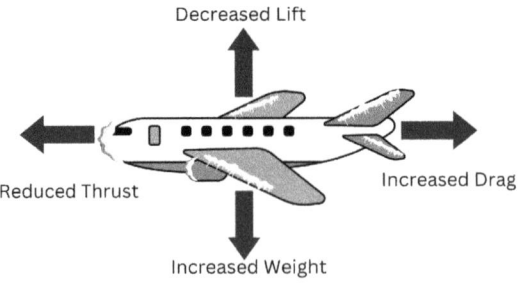

Figure 4.13 Hazards to the four forces of flight due to icing

Source: author

Meteorology and air transport: water vapour 125

Figure 4.14 De-icing of aircraft
Source: Canva.com

ATG Case study 4.2

The hazard of ice at the airport: serious incident involving ATR72–212A on departure from Manchester Airport, UK on 4 March, 2016

The aircraft arrived at Manchester Airport (MAN) from Guernsey (GCI) and was on the ground for more than an hour while it was snowing, and the temperature was 0°C. The flight crew decided no de-/anti-icing was required, and the aircraft departed. From rotation onwards, the flight crew found that manual forward control column input beyond trim capability was necessary to maintain controlled flight. The aircraft was then diverted to East Midlands Airport (EMA). The investigation concluded that ice contamination affected the tailplane and caused pitch control difficulty after departure. The evidence indicated that this would have been avoided if the aircraft had been de-/anti-iced and then inspected carefully before flight (AAIB, 2017).

ATG Case study 4.3

The hazard of ice in-flight: accident of British Airways Boeing 777-236ER at London Heathrow Airport (LHR) on 17 January, 2008

Whilst on approach to LHR from Beijing (PEK), the aircraft touched down with a loss of airspeed, 330m short of the paved surface at runway 27L. The investigation identified that a reduction in thrust occurred, which was due to restricted fuel flow to both engines. It determined that the probable causal factors included:

- Accreted ice from within the fuel system released, causing a restriction to the engine fuel flow.
- Ice had formed within the fuel system, from water that occurred naturally in the fuel.

Source: AAIB (2010)

Chapter review questions

4.1 Explain the different stages of the water cycle. Expand by analysing how each affects the weather.
4.2 What are the methods used to define humidity? Choose a location you are familiar with and research the annual RH values found there. How do these potentially influence flight operations at the location?
4.3 What are the different types of atmospheric stability? Can you find examples of any ONE of these and how they may influence aircraft flight?
4.4 Explain the different cloud types. What are the different weather conditions each produces? For any ONE cloud type, research an incident where this impacted aircraft operations.
4.5 Can you identify the different types of precipitation? Choose any ONE location and analyse how precipitation caused hazardous conditions for air transport.
4.6 Explain the main types of visibility hazards. Using fog as a case study, research any ONE incident or accident where fog played a contributory role. Evaluate any mistakes which were made and what could be done to reduce the likelihood of reoccurrence of the incident/accident.
4.7 Analyse the main ways icing can create hazards for air transport. For either in-flight or ground icing, research any ONE incident/accident and explain the role played by icing.

ATG trivia

Cabin humidity

We have seen the importance of RH to air transport. However, what about the RH inside aircraft cabins?

As a context, humid Malaysia – near the equator – has an average annual RH of around 83–84%. London, in the UK, has an average around 70%. Arizona, one of the driest states in the USA, averages around 35% and Ahaggar, in the Saharan Desert in Algeria, only averages around 26% RH (WorldData, n.d.).

According to the World Health Organization (2020), aircraft cabins have an RH of often less than 20%. Aircraft get fresh air from their jet engines, with air which does not enter the combustion chamber, and is routed to the aircraft cabin via high-efficiency particulate air (HEPA) filters. This is regularly recycled to make the air, according to Airbus and Boeing, cleaner than the air breathed in homes and offices.

Unfortunately, the air is generally very dry, as a low RH is good for metal aircraft structures. Recent aircraft, such as the B787 and A350, are increasingly using composite structures which are not as susceptible to corrosion. Hopefully this will mean less dehydration for passengers on flights in the future!

References

AAIB (2010) 'Aircraft Accident Report 1/2010. Report on the accident to Boeing 777-236ER, G-YMMM, at London Heathrow Airport on 17 January 2008'. Available at: www.gov.uk/aaib-reports/1-2010-boeing-777-236er-g-ymmm-17-january-2008.

AAIB (2017) 'AAIB Bulletin 2/2017'. Available at: https://skybrary.aero/sites/default/files/bookshelf/3889.pdf.

EASA (2015) *In flight icing*. European General Aviation Safety Team. Available at: www.easa.europa.eu/en/document-library/general-publications/egast-leaflet-ga-10-flight-icing.

FAA (2023) *Aviation weather handbook*. Newcastle, Washington: Aviation Supplies and Academics Inc.

Frey, R.P. (2017) *Whether the weather: aviation meteorology from A to Z*. Norderstedt, Germany: Books on Demand GmbH.

ICAO (1978) *Circular 153-AN/56*. Montreal.

ICAO (2018) *Annex 3 — Meteorological Service for International Air Navigation*. Available at: https://elibrary.icao.int/reader/264207/&returnUrl%3DaHR0cHM6Ly9lbGlicmFyeS5pY2FvLmludC9teS1saWJyYXJ5L3Byb2R1Y3QtZGV0YWlscy8yNjQyMDc%3D?productType=ebook&themeName=Blue-Theme.

James, D., MacDonald, D. and Hart, J. (2012) 'The effect of atmospheric conditions on the swing of a cricket ball', *Procedia Engineering*, 34(2012), pp. 188–193.

Kearns, S.K. (2021) *Fundamentals of international aviation*. 2nd edn. Abingdon: Routledge (Aviation Fundamentals).

Lankford, T.T. (2001) *Aviation weather handbook*. New York: McGraw Hill.

Met Office (n.d.a) *Cloud names and classifications*. Available at: www.metoffice.gov.uk/weather/learn-about/weather/types-of-weather/clouds/cloud-names-classifications.

Met Office (n.d.b) *Pilot resources*. Available at: www.metoffice.gov.uk/services/transport/aviation/regulated/pilot-resources.

Met Office (n.d.c) *Why is the sky blue?* Available at: www.metoffice.gov.uk/weather/learn-about/weather/optical-effects/why-is-the-sky-blue.

NASA (2023) *Water vapor*. NASA Earth Observatory. Available at: https://earthobservatory.nasa.gov/global-maps/MYDAL2_M_SKY_WV.

NASA (n.d.) *Ocean worlds*. Available at: www.nasa.gov/specials/ocean-worlds/index.html.

NOAA (2023) *Learning lesson: water, water everywhere*. Available at: www.noaa.gov/jetstream/ll-water.

NOAA (n.d.a) *Atmosphere*. Available at: www.noaa.gov/jetstream/atmosphere.

NOAA (n.d.b) *The hydrologic cycle*. Available at: www.noaa.gov/jetstream/atmosphere/hydro.

Royal Meteorological Society (2013) *What is humidity and how is it measured?* Available at: www.rmets.org/metmatters/what-humidity-and-how-it-measured.

Sadraey, M.H. (2016) *Aircraft performance: an engineering approach*. Boca Raton: CRC Press. Available at: https://doi.org/10.1201/9781315366913.

SKYbrary (n.d.a) *Dew point*. Available at: https://skybrary.aero/articles/dew-point.

SKYbrary (n.d.b) *Visibility*. Available at: https://skybrary.aero/articles/visibility.

US Department of Commerce (2010) *Record setting hail event in Vivian, South Dakota on July 23, 2010*. NOAA's National Weather Service. Available at: www.weather.gov/abr/vivianhailstone.

USGS (2019) *How much water is there on Earth?* Available at: www.usgs.gov/special-topics/water-science-school/science/how-much-water-there-earth#overview.

Walmsley, S. (2021) *Aviation weather for the private pilot*. Aviation Books Private Pilot Series.

Wickson, M. (2015) *Meteorology for pilots*. Marlborough, Wiltshire: Airlife. Available at: www.perlego.com/book/3157975/meteorology-for-pilots-pdf?queryID=7c7fff055d06d98c38f49c9e9bf14cbd&index=prod_BOOKS&gridPosition=6.

World Health Organization (2020) *Air travel advice*. Available at: www.who.int/news-room/questions-and-answers/item/air-travel-advice.

World Meteorological Organization (2017) *Classifying clouds*. Available at: https://wmo.int/world-meteorological-day-2017/classifying-clouds#:~:text=High%2Dlevel%20clouds%20typically%20have,m%20(6%20500%20feet).

World Meteorological Organization (2020) *Aviation | Hazards | Icing*. Available at: https://community.wmo.int/en/activity-areas/aviation/hazards/icing.

World Meteorological Organization (n.d.) *World Meteorological Organization's world weather & climate extremes archive*. Available at: https://wmo.asu.edu/content/world-heaviest-hailstone.

WorldData (n.d.) *Worlddata: the world in numbers*, *Worlddata.info*. Available at: www.worlddata.info/.

5 Meteorology and air transport
Wind

Chapter outcomes

At the end of this chapter, you will be able to:

- Understand how wind direction and speed are defined and measured.
- Explain and analyse the wind-related factors influencing runway location and orientation.
- Identify the four main forces influencing wind creation and explain the main types of wind these create and how they influence air transport operations.
- Discuss the main types of local winds and how these influence aircraft flight.
- Explain the different types of turbulence and how these can pose a hazard to aircraft.
- Identify the ingredients required for thunderstorm formation and explain the life cycle of thunderstorms.
- Analyse the reasons why thunderstorms can be extremely hazardous to air transport.

5.1 Introduction

Looking up at the sky, on a still, sunny day, it can be hard to imagine that at 30,000ft, aircraft may be in the process of being buffeted by strong winds. Conversely, when a gale is blowing at ground level, that same aircraft and passengers can be flying along in a calm, clear flight with no idea how blustery it is on the ground. That is the nature of wind – both low-level, surface winds and upper air winds – which may be causing the same or very different weather conditions on any given day.

From events such as kite flying in Ancient China and hot-air ballooning in 18th-century France, the importance of *wind* in flight has been recognised. On 17 December, 1903, the first powered, sustained, controlled flight of a heavier-than-air, manned flying machine took place, by the Wright Brothers (Smithsonian Institute, n.d.). This is a date which is etched as one of the key events in the history of air transport. The first flight covered 120 feet; however, this was shorter than the 300 feet the Wrights had suggested was the point at which an aviator had achieved sustained flight. How could it have been classed as the "first flight"? Orville Wright took off from the sand flats of Kitty Hawk, North Carolina, and flew into a headwind gusting from 24–27mph. So, while the distance on the ground was only 120 feet and took 12 seconds, the true distance was calculated at 540 feet, due to the headwind, and was greater than the 300-feet minimum (Crouch, 2014). The rest, as they say, is history!

DOI: 10.4324/9781003405351-7

130 *Fundamentals of Global Air Transport Geography*

For contemporary air transport, wind can be both a blessing and a curse. Headwinds enable aircraft to take off and land in shorter distances, whilst tailwinds can allow aircraft to fly much quicker in the cruise, with less fuel burn. However, wind can also be very dangerous, creating turbulent flight conditions, crosswinds and windshear, as well as serious large-scale weather events such as thunderstorms. An understanding of the wind is vital – for operational reasons, for safety critical reasons and for economic reasons.

5.2   Wind direction and speed

Wind is the term used to describe the large-scale flow of atmospheric air (SKYbrary, n.d.d). It is essentially the movement of air (McCabe, 2023), and is either responsible for the formation of weather or its redistribution. Before explaining the key processes behind wind, it is important to understand some terminology and key principles.

Wind velocity is a vector, defined by two parameters – direction and speed. Wind velocity is critical for air transport, both in terms of the take-off and landing phases, and the cruise. The direction the wind is blowing from can reveal what sort of weather to expect, due to air mass theory, which will be examined in detail in Chapter 6 which focuses on climatology.

5.2.1   *Defining and measuring wind direction*

Wind direction is always given as the direction *from* which the wind is blowing. It is given as degrees true in written form, and verbally from the air traffic control tower (in degrees magnetic) for assessment relative to the runway to be made. There are 36 specific azimuth degrees expressed in intervals of 10 degrees. The points of the compass are used in aviation to represent the direction the wind is blowing from – for example, a Northerly wind would be from 360° and a Westerly wind from 270° (Figure 5.1).

There are also 16 cardinal compass directions relative to the wind, with four primary ones: north (N), east (E), south (S) and west (W). There are four intermediate points: northeast (NE), southeast (SE), southwest (SW) and northwest (NW); as well as eight sub-divisions: north-northeast (NNE), east-northeast (ENE), east-southeast (ESE), south-southeast (SSE), south-southwest (SSW), west-southwest (WSW), west-northwest (WNW) and north-northwest (NNW) (Figure 5.1).

- **Wind direction** is measured using a *wind vane* (Figure 5.2) consisting of a thin horizontal arm carrying a vertical flat plate at one end with its edge to the wind and at the other end, a balance weight, which also serves as a pointer (Met Office, n.d.b).

5.2.2   *Defining and measuring wind speed*

- **Wind speed** is measured using an *anemometer* (Figure 5.2), which should be placed in an open space (so as not to be influenced by obstacles). Readings should be taken at 10m (33ft) above ground level (ICAO, 2021). There are different types of anemometers, but wind speed is normally measured by a *cup anemometer*, consisting of three or four cups, conical or hemispherical in shape, mounted symmetrically about a vertical spindle. The wind blowing into the cups causes the spindle to rotate (Met Office, n.d.b). The anemometer and wind vane are usually located together.

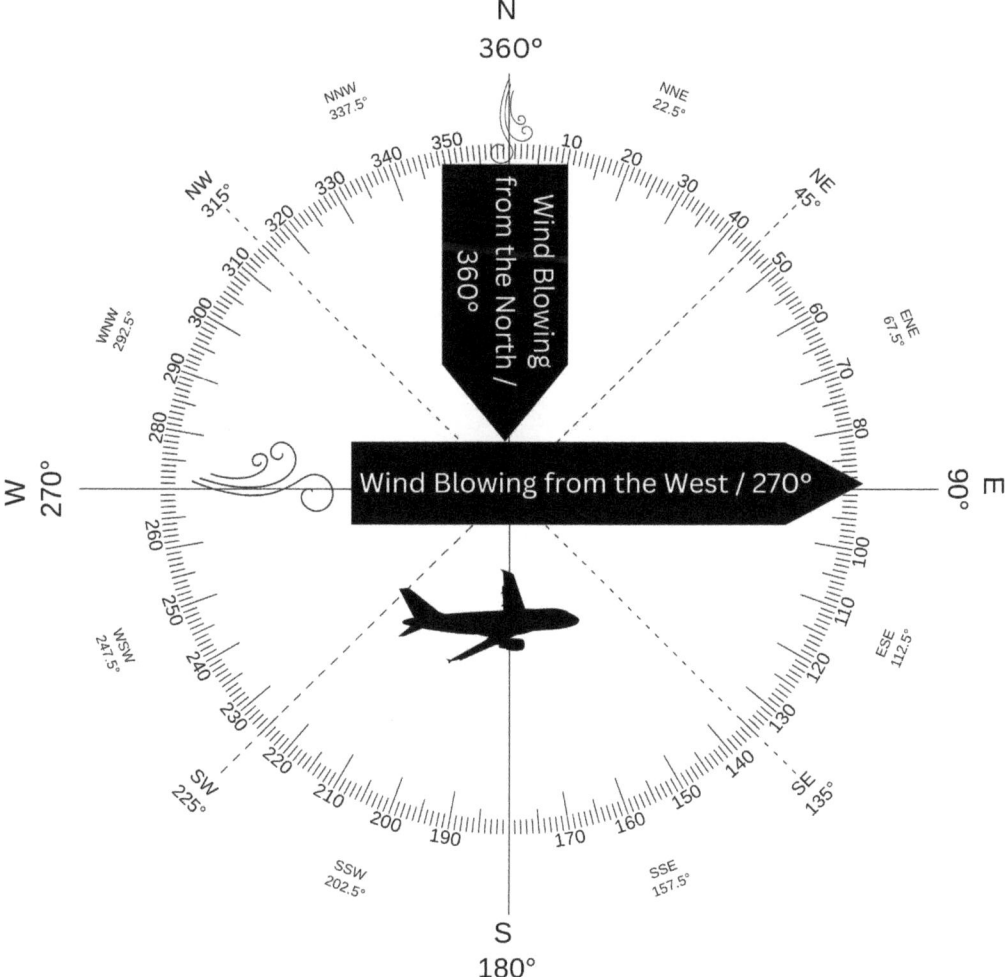

Figure 5.1 Wind directions in air transport
Source: author

- **Windsocks** are often used as a basic guide to both *wind direction and speed* (Figure 5.2). The wind direction is the opposite of the direction the windsock is pointing – so a windsock pointing east indicates a wind blowing from the west. The speed is indicated by the angle of the windsock in relation to the mounting pole. There are sometimes alternating stripes of orange and white, with each stripe adding up to 3 knots on to the wind speed, therefore a fully extended, horizontal windsock suggests a wind speed of at least 15 knots.

Beaufort wind force scale

The Beaufort scale is an empirical measurement for wind intensity based on observed sea or land conditions (Table 5.1). It can still be used today when no instrumentation is available.

132 *Fundamentals of Global Air Transport Geography*

Figure 5.2 Windsock and anemometer at Old Warden Aerodrome, Shuttleworth, Bedfordshire
Source: author

Table 5.1 The Beaufort wind force scale

Wind force	Wind speed (kts)	Wind speed (km/h)	Description	Probable wave height (metres)
0	<1	<1	Calm	–
1	1–3	1–5	Light air	0.1
2	4–6	6–11	Light breeze	0.2
3	7–10	12–19	Gentle breeze	0.6
4	11–16	20–28	Moderate breeze	1.0
5	17–21	29–38	Fresh breeze	2.0
6	22–27	38–49	Strong breeze	3.0
7	28–33	50–61	Near gale	4.0
8	34–40	62–74	Gale	5.5
9	41–47	75–88	Strong gale	7.0
10	48–55	89–102	Storm	9.0
11	56–63	103–117	Violent storm	11.5
12	64+	118+	Hurricane	14+

Source: adapted from Royal Meteorological Society (n.d.)

ATG Did you know?

The Beaufort scale

Devised by Irish hydrographer Francis Beaufort in 1805, whilst serving in the UK Royal Navy on HMS Woolwich. In the early 19th century, naval officers made regular weather observations, but there was no standard scale and thus could be subjective – one person's fresh breeze may be another's strong breeze. A scale was therefore standardised and helped sailing ships operate under a variety of wind and sea conditions.

Meteorology and air transport: wind 133

5.2.3 Headwinds

A headwind is a wind which blows against the direction the aircraft is travelling in (Figure 5.3). This is desirable for:

- **Take-off**: Headwinds increase the flow of air and therefore lift can be achieved earlier and at lower speeds. As a result, less runway will be required for the aircraft to take off and this can have benefits for airlines and airport operators, as will be illustrated in Chapters 6 and 7.
- **Landing**: As before, an airfoil moving into a headwind generates more lift than with no air or a tailwind, therefore less runway is required to land. This again can benefit airlines in terms of their choice of airports to land, based on aircraft type.

5.2.4 Tailwinds

A tailwind is a wind which blows in the same direction as the aircraft is moving (Figure 5.3). This is especially desirable:

- In the **cruise** phase, the aircraft can travel faster and complete its journey quicker than with still air or a headwind, which has many economic benefits, especially relating to lower fuel burn (due to reduced drag) and arriving at destinations quicker. Judicious flight planning can potentially mean more sectors in the day/week if flights are planned to take advantage of prevailing cruise tailwinds, such as flying from west to east over the Atlantic from the USA to Europe, especially overnight, due to the jet streams. An issue may occur if the airport cannot plan for early arrivals and the aircraft ends up in holding stacks prior to landing.
- According to IATA (2023), the cost of jet fuel accounts for between 25–30% of the total operating costs of airlines, therefore any reduction can improve historically thin airline profit margins.
- The less fuel burned, the less environmental impact caused by the air transport industry.

Take-off and landing in tailwinds, due to decreased lift, is generally discouraged, due to longer runway requirements to perform safe operations. Longer stopping requirements will also be required following rejected take-offs as speeds will be higher. Take-offs in tailwinds exceeding certain values (usually 10 knots) are often avoided (SKYbrary, n.d.c) but other factors such as runway conditions, presence of rain, etc., may influence this.

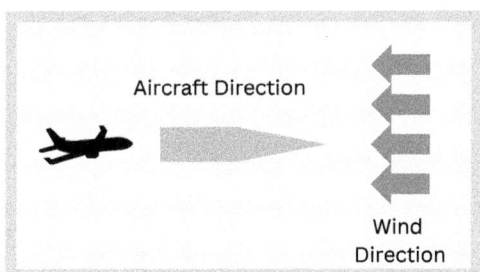
HEADWIND

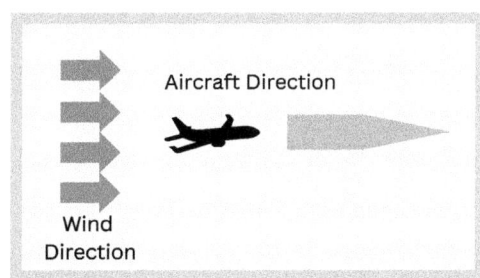
TAILWIND

Figure 5.3 Headwinds and tailwinds
Source: author

ATG Case study 5.1

The hazard of tailwinds: crash of Raytheon Hawker 800XP at Aspen, Colorado, 21 February, 2022

When ATC provided take-off clearance to the aircraft, they reported the wind as from 160° at 16kts, gusting to 25kts. They also reported the "instantaneous" wind as from 180° at 10kts. The maximum allowable tailwind for this aircraft is 10kts. The runway used was Runway 33 (330°), hence there was a tailwind.

During the take-off roll, after the aircraft would not become airborne, the captain aborted the take-off, and the aircraft departed the end of the runway and sustained substantial damage to the right wing and fuselage. The NTSB concluded that the term "instantaneous wind" was not used in any FAA publication and that pilots who operate infrequently at the airport are likely not familiar with the definition and potential operational impact. Multiple wind reports for 30 minutes before the attempted take-off were significantly above the tailwind limitation.

The NTSB (2023) attributed the probable cause of the accident as "the flight crew's improper decision to take-off in tailwind conditions that exceeded the airplane's performance capabilities, which resulted in a runway overrun following an aborted take-off".

5.2.5 Crosswinds

A crosswind is a wind which blows from the side versus the direction the aircraft is travelling. It blows in perpendicular components to the direction of travel and can be extremely hazardous, especially in the landing phase. Poorly executed crosswind landings can cause runway excursions.

5.2.6 Wind direction and runway orientation

The runway(s) are one of the most critical aspects involved in airport design. Runway location and orientation are paramount to airport safety, efficiency, economics and environmental impact. According to the FAA (2022), when designing an airport – and its runways – there are several factors which need to be considered:

- runway length
- runway threshold and location of obstructions
- capacity
- orientation and wind
- airspace analysis
- environmental factors
- topography
- wildlife hazards

These aspects will be dealt with in more detail in Chapter 7, when airport site selection is examined. As far as runway orientation and wind are concerned, the direction of the prevailing winds is critical to consider when deciding on the orientation of a runway and the primary orientation should be in the direction of the prevailing wind (Kazda and Caves, 2015; Prather, 2015; FAA, 2022).

Usability factor

According to ICAO (2018) in Annex 14, Volume 1 – Aerodromes, the *usability factor* is key in runway orientation, siting and number of runways. This is determined by the wind distribution.

- Point 3.1.1 states: "The number and orientation of runways at an aerodrome should be such that the usability factor of the aerodrome is not less than **95%** for the aeroplanes that the aerodrome is intended to serve".
- Point 3.1.3 also states that "it should be assumed that landing or take-off … is, in normal circumstances, precluded when the *crosswind* component exceeds:
    - 37km/h (20kt) in the case of aeroplanes whose reference field length is 1,500m or over.
    - 24km/h (13kt) in the case of aeroplanes whose reference field length is 1,200m or up to, but not including, 1,500m.
    - 19km/h (10kt) in the case of aeroplanes whose reference field length is less than 1,200m.

The crosswind component of wind direction and velocity is the resultant vector acting at a right angle to the runway (FAA, 2022).

Wind Rose

To try to calculate the usability factor and hence decide the orientation of runways, wind data over a long period is required. Annex 14, Section 3.1.4 (ICAO, 2018) states:

> The selection of data … should be based on reliable wind distribution statistics that extend over as long a period as possible, preferably not less than **five years**. The observations used should be made at least eight times daily and spaced at equal intervals of time.

The FAA (2022), in section B.4, states that: "The FAA recommends a data period covering at least the last **ten** consecutive years of wind observations … Use up to 30 years of consecutive data if needed to assess long-term trends in weather".

The use of *Wind Roses* is common to ascertain the long-term trends in wind speed and direction. These are graphical charts which characterise the speed and direction of winds at a location. They are presented in circular format and the length of each "spoke" indicates the amount of time the wind blows from a particular compass point direction. Colours indicate the wind speed categories (NOAA, n.d.).

ATG Case study 5.2

Single runway orientation: Mariscal Sucre International Airport (UIO) – Quito, Ecuador

Mariscal Sucre is the busiest airport in Ecuador, situated about 18km (11 miles) east of Quito and opened in 2013. The Wind Rose in Figure 5.4a illustrates that the most common wind direction is from the north and there are very few winds blowing from east to west, therefore the airport diagram in Figure 5.4b shows that the single runway was designed with a north–south orientation (18/36).

136 *Fundamentals of Global Air Transport Geography*

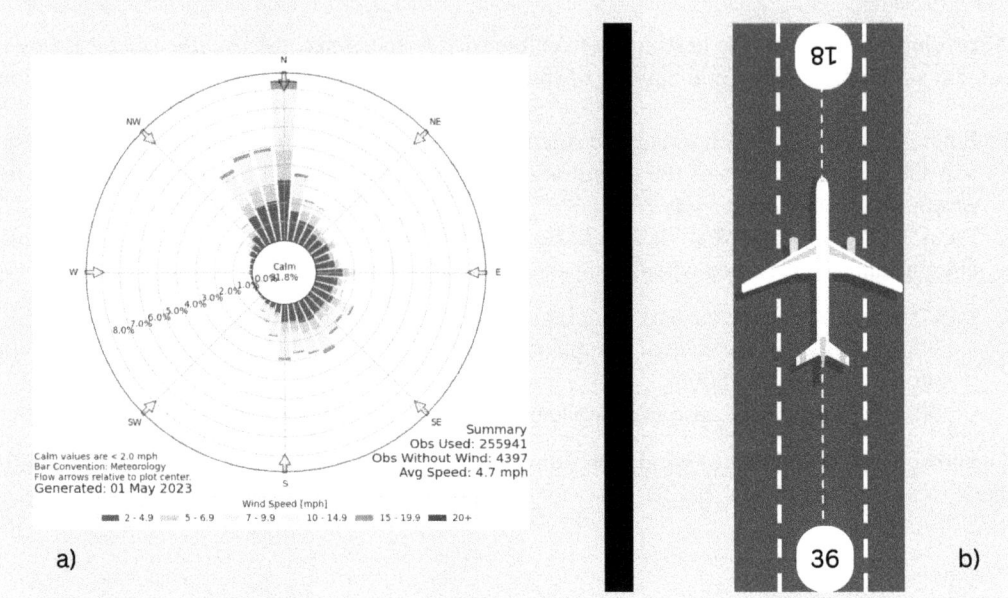

Figure 5.4a Wind Rose for Mariscal Sucre International Airport, Quito, between 30/6/1977 and 15/11/2015

Source: Iowa State University (2023b)

Figure 5.4b Mariscal Sucre runway orientation

Source: author

ATG Case study 5.3

Multiple/crosswind runway orientation: Newark Liberty International Airport (EWR), New Jersey, USA

Newark International is one of the airports which serves the New York metropolitan area along with John F. Kennedy International and La Guardia Airports and acts as a hub for both FedEx Express and United Airlines. It has three runways – the two main runways are parallel runways, running approximately SW–NE (04L/22R and 04R/22L). However, there is also a runway which is oriented 11/29 and this is often used when there are strong crosswinds on the main runways. Figure 5.5 illustrates that winds often come from both the SW and NNE directions (04/22 runways), but there are also a reasonable percentage of winds blowing roughly from the NW, hence the use of the 11/29 runway.

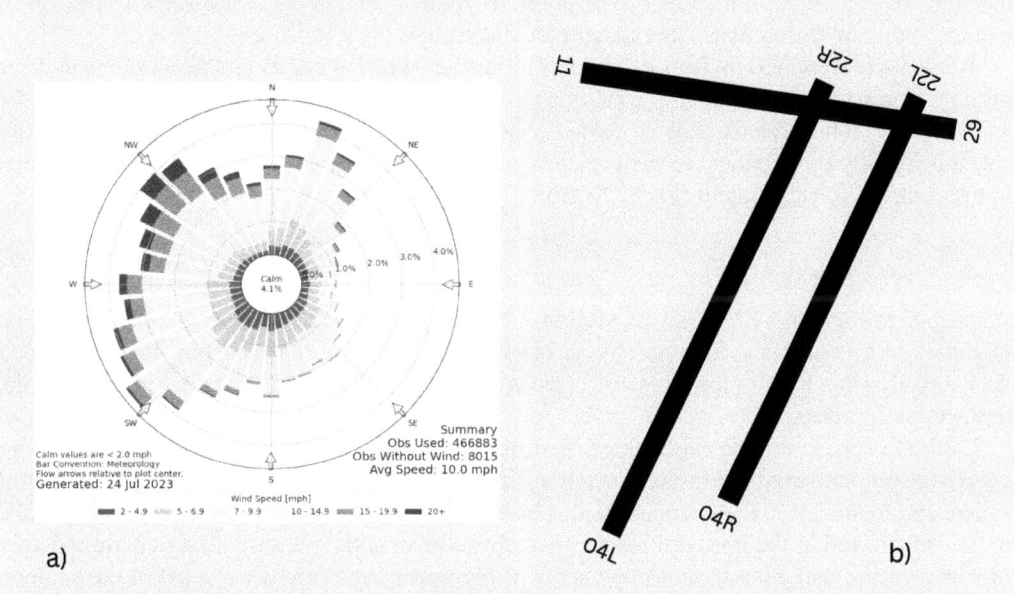

Figure 5.5a Wind Rose for Newark Liberty International Airport, between 01/01/1970 and 24/07/2023

Source: Iowa State University (2023a)

Figure 5.5b Newark Liberty runway orientation

Source: author

5.3 Forces affecting wind

The atmosphere is continually in motion. At the largest scale, there are relatively constant wind belts at the global level circling the earth. These will be covered in detail in Chapter 6. Understanding the immediate controls on atmospheric motion is a good starting point, and the earth has both massive storm systems covering thousands of kilometres and lasting days, through to small-scale, short-term local winds. According to Barry and Chorley (2010), there are four main controls on the horizontal movement of air near the earth's surface:

- pressure gradient force
- Coriolis force
- centripetal acceleration
- friction

5.3.1 *Pressure gradient force*

The most important factor influencing wind creation is the pressure gradient. The pressure gradient force is the rate of change of pressure over a given distance. Whenever a pressure difference

138 *Fundamentals of Global Air Transport Geography*

develops over an area, the force of the pressure gradient makes the wind blow to attempt to equalise pressure differences – the magnitude determines the wind strength.

These are represented by isobar gradients on surface weather charts and assist pilots in determining wind strength – for example isobars closer together can signify strong winds and further apart signify lighter winds (Figure 5.6). Air tends to move from high to low pressure. Wind speed is directly proportional to the pressure gradient force, which is also directly proportional to the contour/isobar gradient (FAA, 2023b).

5.3.2   Coriolis force

According to Newton's First Law of Motion, a body in motion at a constant velocity will remain in a straight line unless acted upon by an outside force (NASA, 1996). Thus, the wind would flow directly from high to low pressure (across the isobars). In reality, this does not happen due to the Coriolis effect.

Earth is a sphere and the earth rotates at different speeds – the Coriolis force acts at right angles to the pressure gradient force, causing wind to be deflected to the right in the northern hemisphere and to the left in the southern hemisphere (Figure 5.7). The wind therefore flows parallel to the isobars and in the northern hemisphere, flows in an anticlockwise direction around areas of low pressure and clockwise around areas of high pressure. This is reversed in the southern hemisphere.

The Coriolis force is zero at the equator and at its maximum at the poles. The earth rotates 360° in around 24 hours and therefore the earth travels further and thus quicker at the equator than the poles.

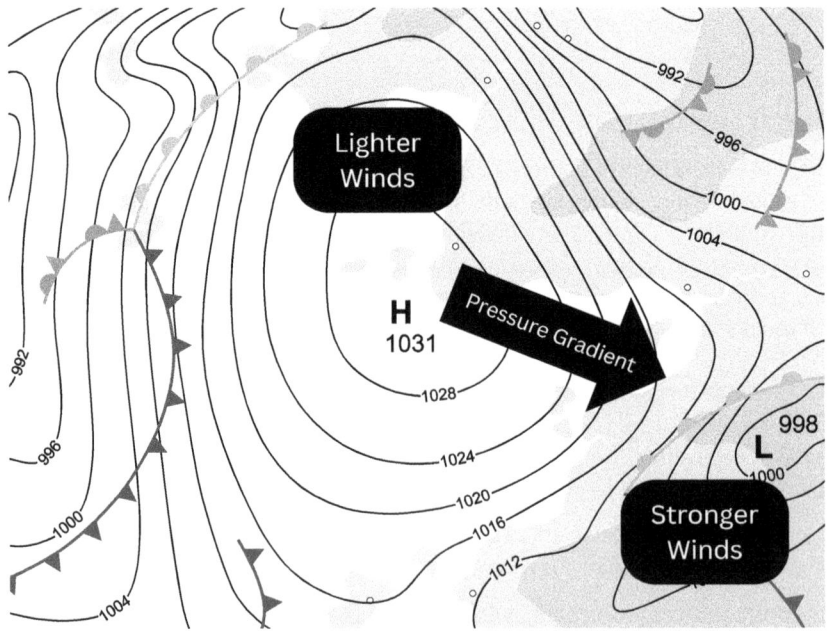

Figure 5.6 Isobars, pressure gradient and wind strength

Source: author, adapted from Canva.com

Meteorology and air transport: wind 139

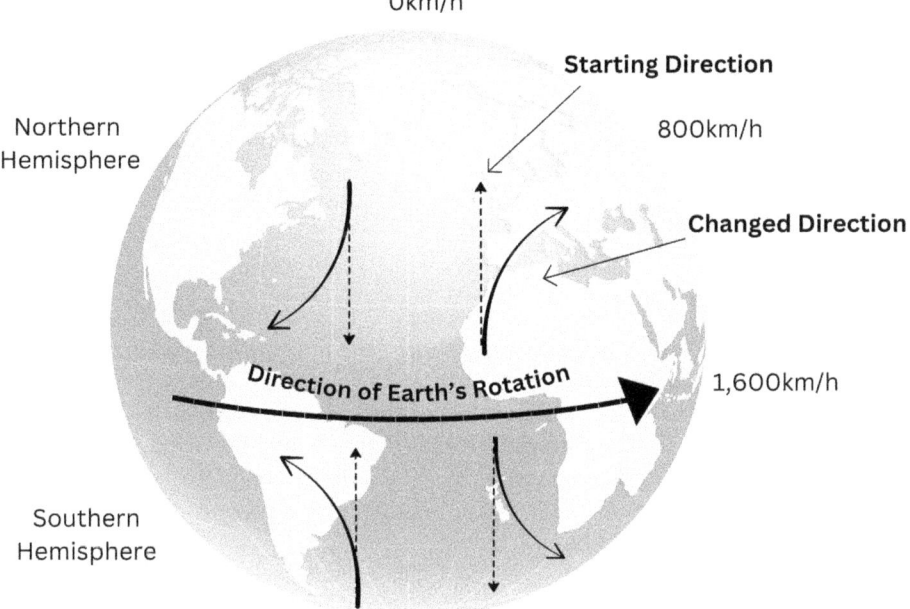

Figure 5.7 Coriolis effect
Source: author

ATG Did you know?

Who or what was Coriolis?

The Coriolis effect was described by a French physicist and mathematician Gustave-Gaspard de Coriolis – in 1835. He studied fluid dynamics through waterwheels and realised the same theories could be applied to the motion of fluids on the earth's surface (Ross, 1995).

Imagine a playground roundabout. Standing in the centre whilst it is spinning anticlockwise, if you throw a ball, it will appear to curve to the right – but to anyone standing watching, it is travelling in a straight line!

5.3.3 *Centripetal acceleration*

This is defined as the property of the motion of an object traversing a circular path. For a body to follow a curved path, there must be an acceleration inwards towards the rotational centre. The centrifugal effect due to rotation has resulted in a slight bulge of the earths mass in low latitudes and flattening near the poles (Barry and Chorley, 2010).

5.3.4 *Friction force*

The influence of terrain is extremely important in wind. The rougher the terrain the higher the effect of friction, which always acts as on opposing force to the direction and speed of the wind.

140 *Fundamentals of Global Air Transport Geography*

The influence of friction generally decreases with height above the lowest few thousand feet (FAA, 2023b). The main effect of the friction force is to reduce the wind closer to the surface, which will also reduce the Coriolis force. The friction layer has a particular impact on flight during the landing and take-off phases.

5.4 Wind types

Wind types include geostrophic, cyclostrophic and gradient.

5.4.1 Geostrophic wind

This refers to a wind produced when the Coriolis force and the pressure gradient force are in balance and acting in opposite directions (Frey, 2017). As an air mass moves, it is deflected to the right (northern hemisphere) or left (southern hemisphere) by the Coriolis force. This increases until it is balanced by the pressure gradient force (Figure 5.8a). The wind will then be blowing parallel to the isobars – called a *geostrophic wind* (Hong Kong Observatory, 2010). These generally only occur when centrifugal force and friction are not present and hence is only in the upper air, for example in the form of *jet streams*.

The geostrophic wind is an "ideal" situation. The effects of friction at lower levels (up to around 2,000 feet, except in mountainous areas) can have an impact. Figure 5.8b illustrates that in the northern hemisphere, when the pressure gradient force, frictional force and Coriolis force are in equilibrium, the wind direction will be pointing at a small angle towards the low-pressure side.

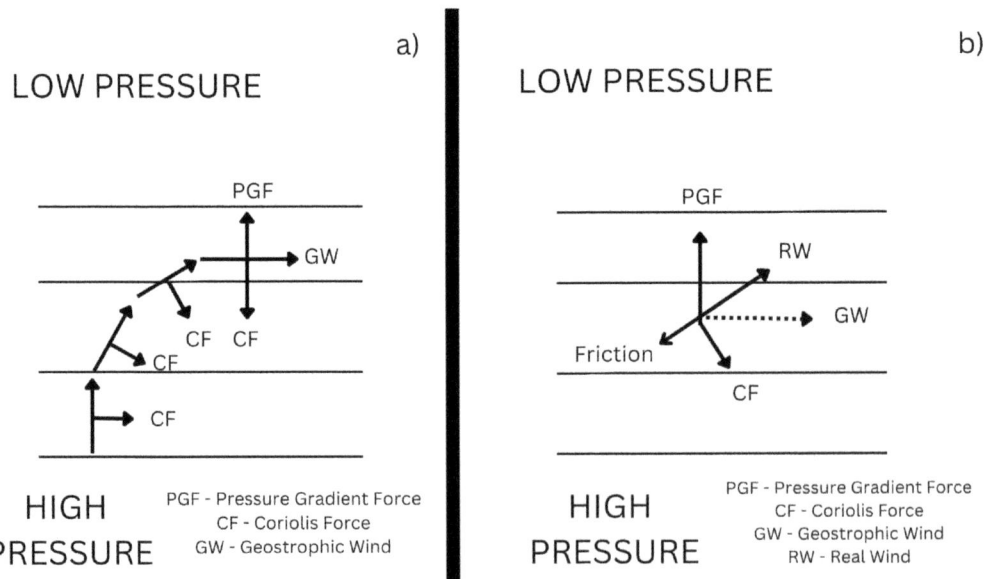

Figure 5.8a Geostrophic wind formation
Source: author

Figure 5.8b Real winds due to surface friction
Source: author

5.4.2   Cyclostrophic wind

When isobars are densely plotted or extremely curved, then the centrifugal force is much stronger than the Coriolis force. Therefore, an equilibrium exists between the pressure gradient force and the centrifugal force and this is called a cyclostrophic wind (Frey, 2017). One example would be in a tornado.

5.4.3   Gradient wind

Geostrophic winds exist in locations where there are no frictional forces and the isobars are straight. This is very rare, however. The isobars are almost always curved and rarely have completely even spacing. This means the winds are in *gradient wind balance* (University of Illinois, 2010). They still blow parallel to the isobars and in the northern hemisphere, circulation flows clockwise around highs and anticlockwise around lows (vice-versa in the southern hemisphere) but are no longer balanced by only the pressure gradient and Coriolis forces, and do not have the same velocity as geostrophic winds. As it considers the Coriolis, gradient and centrifugal forces, it defines relatively accurately the real wind. The missing parameters would be *terrain* and *friction* (Frey, 2017).

> **ATG Did you know?**
>
> **Buys Ballot's Law**
>
> Buys Ballot's Law was named for Dutch meteorologist, C.H.D. Buys Ballot, in 1857. It states that if a person stands with their back to the wind (in the northern hemisphere), the atmospheric pressure is low to the left and high to the right. The law implies that winds blow anticlockwise round a depression in the northern hemisphere and clockwise round an anticyclone. The process is reversed in the southern hemisphere.
>
> This can be useful in possibly determining the type of weather to come – a high to the west in the northern hemisphere indicates that settled weather may be approaching.

5.5   Local winds

You are flying along on a beautiful, clear day, with a weak pressure gradient and high pressure, yet you look down and see a windsock at 90-degree angles, indicating at least 15–20-knot winds. Why? The answer likely lies in the presence of *local winds*. These are small-scale wind systems, generally driven by diurnal heating or cooling of the ground (FAA, 2023b) which can occur near water or where mountainous terrain is present.

Low-level winds develop in the direction of the pressure gradient force and as they are small in scale and have a short lifespan, the Coriolis force plays no real role. As such, the wind generally blows from a high-pressure cooler surface to a low-pressure warmer surface.

5.5.1   Sea and land breezes

It can be common for airports to be near water, so pilots should be aware of the possibility of sea and land breezes. Land and sea breezes develop because of differential heating and cooling

142 *Fundamentals of Global Air Transport Geography*

of adjacent land and water surfaces (Yan 2005) and can be at both the sea and large lakes, such as the Great Lakes in the USA.

Sea breezes

The sun's radiation heats the air over the ground more than the air over the sea (Frey, 2017). When the process begins in the morning, the land warms up quicker than the water and the air rises through convection, creating a small local low pressure over the land (Walmsley, 2021). The air then travels at higher levels over the water before descending and creating a High, which forms a pressure gradient at the surface and the sea breeze flows from water to the land. This usually blows on calm, sunny days (Figure 5.9a) and there is the possibility of light to moderate turbulence, mainly over the land.

Land breezes

This is a coastal breeze blowing from the land to sea caused by temperature differences when the land surface is cooler than the water surface. These usually occur at night and early morning, as the land cools down quicker after sunset, whereas the water can remain warmer and this means the wind flows from land to water (Figure 5.9b).

5.5.2 *Valley and mountain winds*

Valley winds are more specific terms for *anabatic* winds and mountain winds are more specific terms for *katabatic* winds.

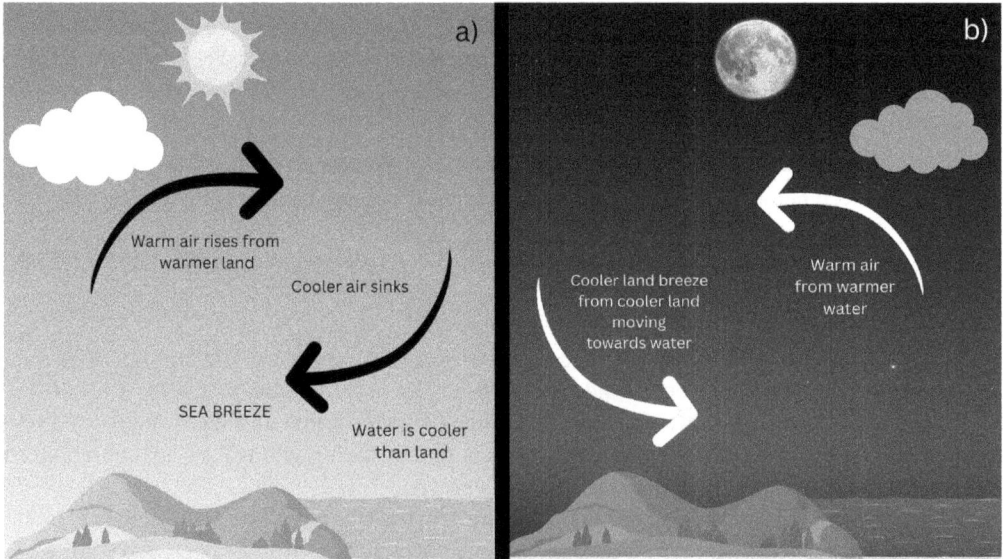

Figure 5.9a Sea breeze
Source: author

Figure 5.9b Land breeze
Source: author

Valley/anabatic winds

These are winds which ascend a mountain valley during the day. Air in a valley heats up more during the day than the air at the same height in the flat lands, therefore the pressure level rises slightly during the day (Frey, 2017). Winds develop in the direction of the pressure gradient force and the air rises up the heated hillsides and colder air from other valleys flows into the valley floor, creating a breeze. Clouds and precipitation may develop (Figure 5.10a).

Mountain/katabatic winds

These are nightly downslope winds commonly experienced in mountain valleys. During and after sunset, particularly with clear skies, air at and near the top of elevated surfaces cools faster than air at lower altitudes. As air cools it becomes denser and therefore heavier. The colder air then flows down the side of the mountain, resulting in a katabatic flow (or wind) (Figure 5.10b). These could be as strong as 100 knots (SKYbrary, n.d.b) and can pick up momentum as they proceed down the mountainside.

Pilots should take care when flying on the shady valley sides and this can impact aircraft performance, even during the day. Cold air can "pool" near the valley floor, and this can create very low temperatures. From an aircraft performance perspective, descending into the cold pool may increase the performance due to the cold air, however climbing out through the cold pool may experience inversions and this could degrade performance, which is a safety-critical risk.

The **venturi effect** could also occur if there are obstructions which funnel the air through smaller gaps, and this can cause the wind speed to increase even further.

Figure 5.10a Anabatic winds
Source: author

Figure 5.10b Katabatic winds
Source: author

ATG Case study 5.4

Tenzing-Hillary Airport, Lukla, Nepal

Tenzing-Hillary Airport is situated at 2,846m (9,337ft) above sea level in the Himalayas and is used by many Mount Everest climbers who arrive there at the beginning of their trek. It is noted for:

- Its height, as airports at high altitude present safety dangers due to the low pressure affecting aircraft performance.
- Its location, with mountains all around, between 6,000–7,000m high.
- Being sited in a valley between the mountains.
- The runway is only 527m (1,729ft) in length and sits on a narrow mountain shelf. There are no go-around procedures and at the end of runway 06 there is a mountain!
- Add all this to the strong winds often present and the airport presents a challenge to air transport operators.

5.5.3 Foehn winds

Foehn winds result in contrasting weather conditions on either side of a mountain range. The *windward* side of the mountain is the side which the prevailing wind is coming from, whereas the *lee* (or *leeward*) side faces away from the wind (National Oceanic and Atmospheric Administration, 2023). Foehn winds are warm, dry, downslope winds that occur on the lee side of topographic barriers (Wolfe, 2022). The windward side may experience cloud and precipitation prior to reaching the summit (Figure 5.11). The temperature difference can be huge – with a 15–20°C (27–36°F) difference sometimes experienced (Walmsley, 2021).

Foehn winds occur when moist and stable air is forced up the mountain, quickly resulting in saturated air, low cloud and precipitation. In the example in Figure 5.11, the sea level temperature is 18°C and cooling at the DALR (−3°C per 1,000ft) the dewpoint is 12°C at 2,000ft. The air then cools at the SALR (−1.5°C per 1,000ft) to the mountain top. On the way down it will adiabatically warm at the SALR, but the air being drier, it will become unsaturated quickly, in our example, with a dewpoint of 3°C. It will then warm at the DALR, meaning at sea level the temperature will be 27°C, which is 9°C warmer than on the windward side.

There are four mechanisms which combine to create the *foehn effect* (Met Office, n.d.a):

1. Condensation and precipitation
2. Draw-down of air
3. Turbulent mixing
4. Radiative warming

Pilots flying near mountains need to be extremely vigilant for the weather conditions caused by foehn winds and other meteorological phenomena. On the windward side, poor visibility due to low clouds is hazardous and at the top, the air accelerates rapidly as it is squeezed over the peak which can cause significant turbulence. On the lee side, the conditions will be clearer but may also be turbulent.

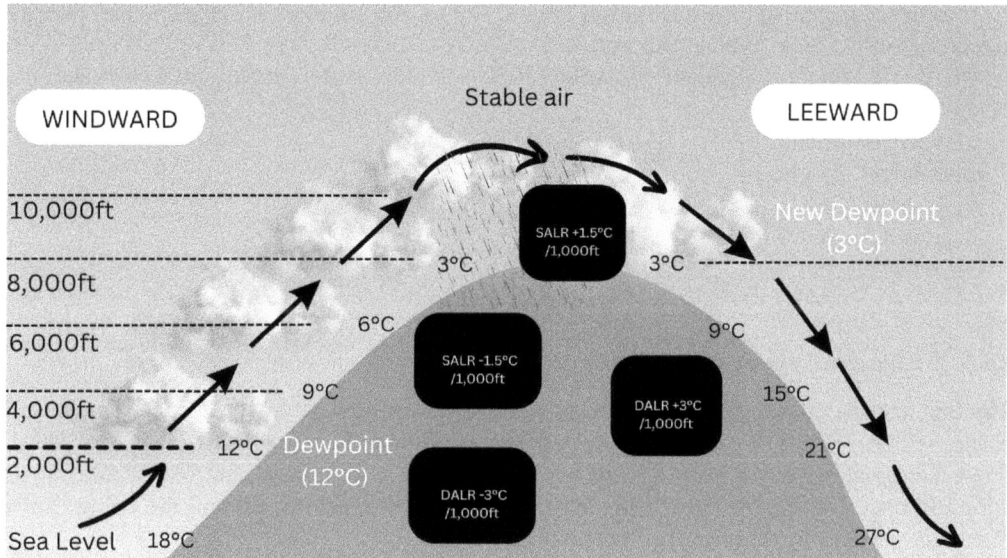

Figure 5.11 Foehn winds
Source: author

ATG Did you know?

When is a foehn wind not a foehn wind?

The term "foehn wind" originated from studies of orographic effects in the Alps in Central Europe. The term is from the German "föhn", which means hairdryer. However, this term is not always used in different parts of the world, for example:

- Chinook – Rocky Mountains, North America
- Zonda – Andes Mountains, Argentina, South America
- Helm – Cross Fell, Cumbria, England
- Canterbury Nor'wester, South Island, New Zealand

In January 1972, in Montana, USA, a foehn chinook event was responsible for the greatest temperature range over a 24-hour period ever recorded in the USA, rising 57°C: from −48 to 9°C(!) (Met Office, n.d.a).

So, foehn, chinook, zonda, helm and Canterbury nor'wester are essentially the same type of wind but are recognised with different names in different parts of the world.

5.6 Hazards: turbulence

According to the FAA (2023b), "aircraft turbulence is irregular motion of an aircraft in flight, especially when characterised by rapid up-and-down motion caused by a rapid variation of atmospheric wind velocities". Turbulence is encountered on many flights, much of which is small scale, but there are some turbulent events which can produce serious hazards.

Turbulence is graded on a relative scale by ICAO for the purposes of reporting and forecasting, and its effect on a "typical" aircraft.

The World Meteorological Organization (2020) identifies different types of turbulence:

- convective turbulence
- mechanical turbulence
- wind shear
- clear air turbulence (CAT)
- wake turbulence

5.6.1 Convective (thermal) turbulence

Daytime heating causes rising air currents which produce convective turbulence. This usually occurs within 7,000ft of the surface in stable or conditionally unstable air (Lankford, 2001). These are most active on warm summer afternoons when winds are light (FAA, 2023b). Air normally rises at irregular rates, especially when different types of surfaces are being heated. The *eddies* caused by this rising air create turbulence.

Pilots should be aware of several factors due to the uneven surface heating. Rocky terrain, plowed fields and paved areas produce upward currents, which may push the aircraft above the glide path causing overshooting of the touchdown point. Trees, rivers, lakes and green fields produce predominantly downward currents which can cause undershooting (Lankford, 2001) Often both types are present, meaning several corrections may be required on final approach (Figure 5.12).

5.6.2 Mechanical turbulence

An object in the way of the flow of air will provide an obstacle, causing changes in wind direction. The turbulence is not caused by anything meteorological, but rather the obstruction itself. Small-scale mechanical turbulence can occur with objects such as buildings and trees and can be a particular hazard during take-off and landing. When the wind exceeds about 15–20 knots, eddies can form on the lee side of buildings and turbulence can be significant (Figure 5.13a).

Hills and mountains cause particular hazards (Figure 5.13b). One example of mechanical turbulence around hills is at Gibraltar Airport, situated next to a large rock 426m (1398ft) high, and when the winds are blowing around the rock from certain directions, landing is extremely challenging.

Table 5.2 Turbulence classifications

Turbulence classification	Description
Light	Slight erratic changes in altitude and/or attitude.
Moderate	Change in altitude and/or attitude, but the aircraft remains in positive control at all times.
Severe	Large abrupt changes in altitude and/or attitude. Aircraft may be momentarily out of control.
Extreme	Aircraft is violently tossed about and practically impossible to control. May cause structural damage.

Meteorology and air transport: wind 147

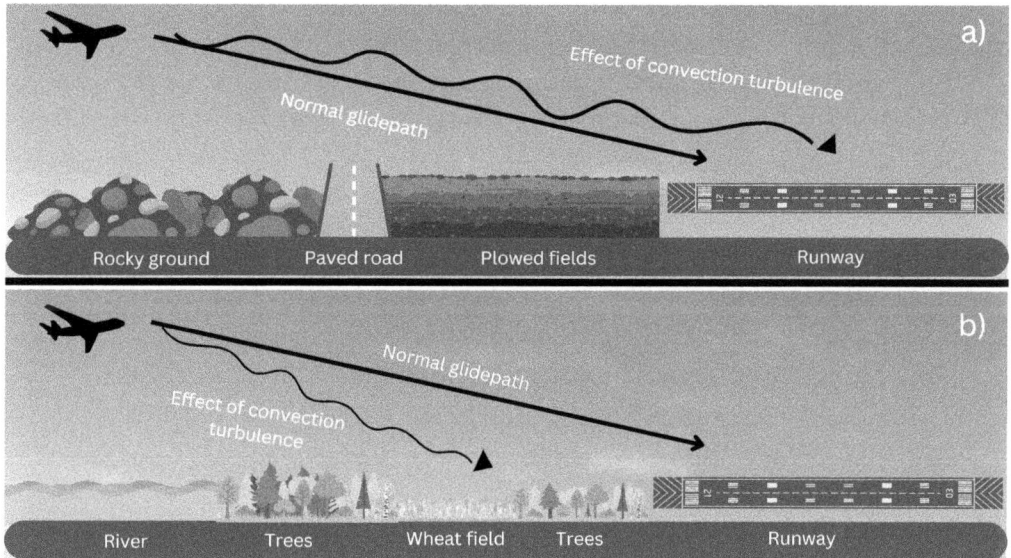

Figure 5.12 Thermal/convective turbulence: (a) surfaces creating rising currents; (b) surfaces creating descending currents
Source: author

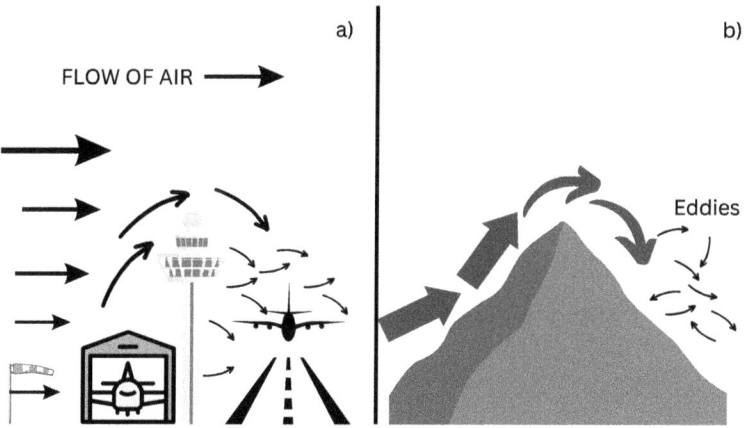

Figure 5.13 Mechanical turbulence: (a) over buildings at an airport; (b) over mountains
Source: author

ATG Case study 5.5

Kai Tak Airport, Hong Kong (closed 1998)

Kai Tak Airport was renowned as one of the most difficult airports at which to land. As Hong Kong developed, the airport was surrounded by high-rise developments and the approach over the densely populated city and hills was dramatic and challenging. The pilots flew towards a large orange and white chequerboard (known as Checkerboard Hill), made a 47-degree turn to the runway, with around two nautical miles left to fly and a height less than 1,000ft (Figure 5.14). The risk of strong winds and turbulent conditions was high, meaning pilots had to be specially trained to land here.

Figure 5.14 The old Kai Tak Airport and Checkerboard Hill location
Source: adapted from Google Earth (n.d.)

5.6.3 Wind shear

Wind shear is a sudden, drastic change in wind speed and/or direction over a small area and can be classified as either *vertical* or *horizontal*. It can occur at any altitude – at higher levels near jet streams or at lower levels near the ground due to convection. It can subject aircraft to sudden violent updrafts and downdrafts as well as sudden horizontal changes. It is often classed as a silent aviation hazard (FAA, 2023b). Low-level windshear can result in sudden reductions in airspeed and lift which can be hazardous on approach – on landing especially.

ATG Did you know?

Low-level wind shear alert system (LLWAS)

Some airports, mainly in the USA, have an LLWAS, which is a ground-based system used to detect wind shear and associated weather phenomena, close to an airport. A few anemometers are strategically placed around an aerodrome and the information can be passed to pilots via ATC, to warn of the presence of wind shear.

Meteorology and air transport: wind 149

5.6.4 Clear air turbulence

Clear air turbulence (CAT) is defined as sudden severe turbulence occurring in cloudless regions that causes violent buffeting of aircraft (SKYbrary, n.d.a). It is commonly applied to higher altitude turbulence, normally above 15,000ft, and generally occurs in patches and is transitory in nature (Lankford, 2001). CAT is a particular problem as it is often experienced unexpectedly without visual clues.

Structural damage could be caused, flight crew performance impaired and/or physical injury to crew/passengers may occur. For example, on 28 December, 1997, a Boeing 747-122 of United Airlines flew into an area of CAT over the Pacific Ocean, despite weather forecasts indicating no turbulence in the area. Fifteen passengers and three crew members received serious injuries and one passenger was killed (NTSB, 2001).

5.6.5 Wake vortex turbulence

This is turbulence which is generated by the passage of an aircraft in flight. A wing producing lift generates a disturbance (or wake) caused by a pair of counter-rotating vortices which trail the wingtips (Figure 5.15). The greatest strength of vortex occurs when the generating aircraft is:

- heavy
- slow
- clean (landing gear and flaps retracted)

(FAA, 2023a)

This can be a factor in the en route phase and is serious on take-off and landing when lighter aircraft follow heavier aircraft. Vortices tend to decay slowly, meaning they can be a hazard for several minutes.

One of the main safety precautions to prevent issues with wake vortex is separation minimum standards – keeping the aircraft spaced apart appropriately. Table 5.2 illustrates the radar wake turbulence separation minima set by ICAO for aircraft behind the A380–800, for the approach and departure flight phases.

Figure 5.15 Wake vortex turbulence
Source: author, via Canva.com

150 Fundamentals of Global Air Transport Geography

Table 5.3 A380–800 radar wake turbulence separation minima

Following aircraft	Wake turbulence radar separation minima
A380–800 (560,000kg)	Not required
Non A380–800 Heavy (>136,000kg)	11.1km (6.0NM)
Medium (>7.000–136,000kg)	13km (7.0NM)
Light (7,000kg or less)	14.8km (8.0NM)

Source: ICAO (2008)

5.7 Hazards: thunderstorms

A thunderstorm can be described as one or more sudden electrical discharges, manifested by a flash of light (lightning) and a sharp rumbling sound (thunder). Thunderstorms are associated with convective clouds and are generally (but not always) accompanied by precipitation at the ground (Met Office, 2007). Thunderstorms form from a well-developed cumulonimbus cloud.

Thunderstorms need three ingredients for their formation (Frey, 2017; FAA, 2023b):

1. **Sufficient water vapour**
   This can provide a key energy source for the cloud.
2. **Unstable air**
   This needs to be for a significant vertical portion of the atmosphere, where it can rise to the tropopause and this means clouds can grow to 50,000–60,000ft near the equator.
3. **A lifting mechanism (trigger)**
   This could be via air being forced up a mountain, rising through convection or frontal systems.

ATG Did you know?

Thunderstorms

It is estimated that there are approximately 2,000 thunderstorms in progress at any one time globally, with as many as 40,000 thunderstorm occurrences each day – this works out at 14.6 million occurrences annually(!) (NOAA, 2023b). The average thunderstorm is 15 miles wide and lasts around 30 minutes, and a thunderstorm cloud could contain 1.1 million tons of water!

5.7.1 Thunderstorm life cycle

A thunderstorm cell is the convective cell of a cumulonimbus cloud containing thunder and lightning. It undergoes three distinct stages (Figure 5.16):

1. **Developing/growth (towering cumulus) stage**
   The initial lifting mechanism forces the unstable air to rise and continue to rise due to the release of latent energy. All movement is upwards – called *updraughts* – and cloud droplets begin to form. There is usually little or no rain due to the small size of the droplets, but there may be occasional lightning. Updraughts can exceed 3,000ft per minute.

Meteorology and air transport: wind 151

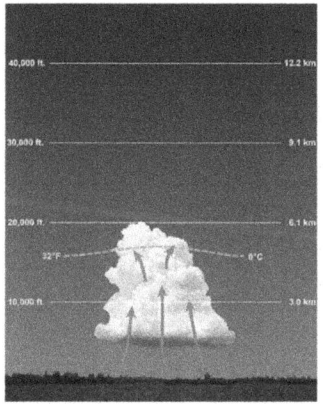
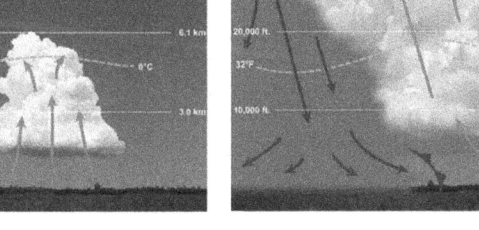
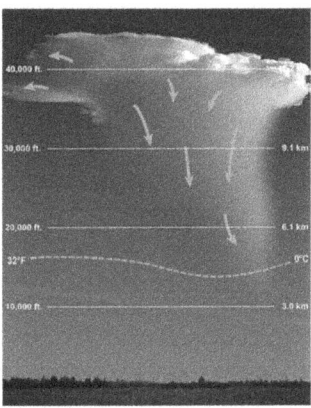

a) Towering cumulus stage   b) Mature stage   c) Dissipating stage

Figure 5.16 The three stages in a thunderstorm life cycle
Source: adapted from NOAA (2023a)

2. **Mature stage**
   When precipitation begins to fall. This stage contains both updraughts and *downdraughts*, of falling precipitation. Frequent thunder, lightning and heavy precipitation is likely. Strong winds are usually experienced, caused by the downdraughts forming gust fronts. The thunderstorm can reach as high as the tropopause, creating an *anvil*. This is the most dangerous stage.
3. **Decaying/dissipating stage**
   This occurs when the falling precipitation stops the updraughts, therefore all movement is downwards, meaning no more moisture is being introduced. There may still be heavy rain, strong winds and lightning, but these are reducing in intensity.

5.7.2   Thunderstorm hazards to aviation

Thunderstorms are especially hazardous to aviation for a few reasons:

- **Turbulence can be violent, both inside and around the cloud(s).**
  Pilot reports have noted up and down drafts exceeding 6,000ft per minute and turbulence exceeding the capability of many aircraft are present in thunderstorms.
- **Strong winds and wind shear.**
  Thunderstorms can cause very strong winds and extreme changes in wind speed and direction, which is especially problematic in the critical stages of flight.
- **Stressing the airframe.**
  In severe conditions, there is an increased possibility of stressing the airframe.
- **Loose articles in the cabin.**
  Any loose article can become a projectile during turbulence, even in-flight trolleys if the turbulence is severe enough.
- **Very heavy rain and/or hail.**
  Visibility significantly reduced and risk of engine flame out. This is also a potential hazard on landing the aircraft, where aquaplaning is a risk due to water accumulation and stopping

distances will be increased. Runways at most large airports have grooves cut into the surface and angled to the sides which helps drain water from the centre.
- **Increased risk of icing.**
Thunderstorms are in part driven by liquid to ice changes of state, therefore airframe icing is a risk. Clear icing, caused by larger supercooled drops, can be common.
- **Static interference on radio equipment.**
This could result in communication issues and increased pilot or air traffic workloads.
- **Lightning.**
Lightning is an electrical discharge caused by imbalances between storm clouds and the ground, or within the clouds themselves. It can cause temporary blindness to the pilots and aircraft damage, such as puncture holes in the radomes (nose) and tailfins of aircraft and control mechanism damage.

ATG Did you know?

Lightning and thunder

Contrary to the popular saying, lightning *can* strike the same place twice!

Cloud-to-ground lightning bolts are common – about 100 strike earth's surface every second. Each bolt can contain up to one billion volts of electricity! Central Africa is the area of the world where lightning strikes most frequently (National Geographic, 2009).

Intracloud lightning is the most common type of discharge (Royal Meteorological Society, 2017) and refers to lightning embedded within a single storm cloud, jumping between different charge regions within the cloud.

Technically lightning does not have a temperature; however, resistance to the movement of the electrical charges causes the materials that the lightning is passing through to heat up. Lightning can heat the air and it passes through to 50,000°F (five times hotter than the surface of the sun!) (National Weather Service, n.d.).

This heating causes the air to explode outward. The pressure in the shock wave decreases rapidly with increasing distance and within ten yards or so has become small enough to be perceived as *thunder*. Thunder can be heard up to 25 miles away from the lightning discharge.

Light travels through air much faster than sound. Count the number of seconds from the time you see a lightning flash until you hear thunder. Sound travels approximately 1/5 of a mile per second (1/3km p/s), dividing the number of seconds by five gives the number of miles to the flash (divide by three for km) (NSSL, n.d.).

The FAA (2008) states to never go closer than 5 miles to any visible storm cloud with overhanging areas and to strongly consider increasing that distance to 20 miles or more.

Ultimately, aircraft of all sizes should try to steer well clear of any thunderstorm activity. As a pilot saying goes, "It's better being on the ground wishing you were up than being up and wishing you were on the ground"!

ATG Case study 5.6

Aircraft accident involving Dornier DO 228-202 at Bodo Airport, Norway, 4 December, 2003

On approach to Bodo Airport, the aircraft experienced a powerful lightning strike, which struck the aircraft's nose area and passed to the tail, which damaged the aircraft such that the elevator was uncontrollable. The pilots were blinded for approximately 30 seconds and lost control of the aircraft for a period. They were able to regain limited control of the nose position using electric pitchtrim and were able to bring the aircraft in for landing. The two pilots were seriously injured and both passengers suffered minor injuries. The report concluded that there was reason to believe that the total amount of energy in the lightning exceeded the values of the aircraft construction specification and that up to 30% of the wiring in essential bondings in the tail may have been defective before lightning struck (NSIA, 2007).

Chapter review questions

5.1 How are wind speed and direction defined and measured in aviation? Which wind direction from headwind, tailwind and crosswind is generally preferred by airline pilots and why?

5.2 Choose an airport local to you. Can you find out the orientation of the runway(s)? Can you locate a wind rose for the airport or the local area? Based on the principles of runway orientation, analyse whether you believe the runway direction(s) are the best based on the prevailing winds.

5.3 What are the main forces influencing wind creation and how can each of these influence air transport operations?

5.4 Analyse how each of the various local winds can influence aircraft flight and research if you can find airports which may encounter each of these wind types?

5.5 What are the main types of turbulence? For each of these, how can they pose a hazard to air transport operations? For any ONE of these, research an aviation accident where turbulence played a major role and what recommendations the accident investigators suggested to make future scenarios safer.

5.6 Explain the different life cycle stages of a thunderstorm. For each of these, assess the risk to air transport operations.

5.7 Research the website for your local accident investigation bureau (NTSB, AAIB, etc.). Locate an accident/incident where a primary cause was thunderstorms. Evaluate the event and propose suggestions for how the air transport industry could reduce the future risks.

ATG trivia

How hot do aircraft get?

There are two factors which determine this – outside air temperature (OAT) and speed. We have seen that based on ISA values, sea level temperatures are calculated at 15°C and OAT at a cruising altitude of 35,000ft at around −54°C. Flying at that same 35,000ft, with a Mach speed of 0.85 and depending on prevailing winds, speed could be around 500 knots.

So how hot is the aircraft itself? As an aircraft flies, it compresses the air and this causes the air temperature to rise and the maximum temperature would be where the air has completely stopped, such as the leading edges. The temperature rises (called the *ram rise*) and the temperature could heat up to a total air temperature of approximately −22°C.

However, according to the National Research Council (1996) this all changes when the aircraft moves from subsonic to supersonic (>Mach 1), because of aerodynamic heating. By Mach 2.0 – the approximate cruising speed of Concorde (twice the speed of sound) – the skin temperature would increase dramatically, to nearly 100°C. At Mach 2.4 it would reach 150°C.

Extreme temperatures remain a challenge for aircraft designers and supersonic military aircraft often have shorter life spans due to thermal challenges and the aircraft structure. The increasing use of composite rather than aluminium alloys may enhance the life span of supersonic jets (Memon, 2022).

References

Barry, R. and Chorley, R. (2010) *Atmosphere, weather and climate*. 9th edn. Abingdon: Routledge.
Crouch, T. (2014) *First flight?*, *Smithsonian Institute, National Air and Space Museum*. Available at: https://airandspace.si.edu/stories/editorial/first-flight.
FAA (2008) *FAA-P-8740-12. AFS-8. Thunderstorms – don't flirt...skirt 'em*. Available at: www.faasafety.gov/files/gslac/library/documents/2011/Aug/56397/FAA%20P-8740-12%20Thunderstorms[hi-res]%20branded.pdf
FAA (2022) *Advisory circular: airport design; 150/5300–13B*. Federal Aviation Administration.
FAA (2023a) *Aeronautical information manual*. Available at: www.faa.gov/air_traffic/publications/atpubs/aim_html/chap_7.html (Accessed: 13 September 2023).
FAA (2023b) *Aviation weather handbook*. Newcastle, Washington: Aviation Supplies and Academics Inc.
Frey, R.P. (2017) *Whether the weather: aviation meteorology from A to Z*. Norderstedt, Germany: Books on Demand GmbH.
Google Earth (n.d.). Available at: www.google.co.uk/earth/.
Hong Kong Observatory (2010) *Geostrophic wind*. Available at: www.hko.gov.hk/en/education/weather/meteorology-basics/00010-geostrophic-wind.html.
IATA (2023) *Airfares recover amid soaring jet fuel costs and inflation*. Available at: www.iata.org/en/iata-repository/publications/economic-reports/airfares-recover-amid-soaring-jet-fuel-costs-and-inflation/.
ICAO (2008) *Guidance on A380–800 wake vortex aspects*. Available at: https://skybrary.aero/sites/default/files/bookshelf/160.pdf.
ICAO (2018) *Annex 14 to the Convention on International Civil Aviation, Volume 1: Aerodromes*. 8th edn. ICAO.
ICAO (2021) *Doc 8896 – manual of aeronautical meteorological practice*. 13th edn. ICAO.

Iowa State University (2023a) *IEM: site wind roses*. Available at: https://mesonet.agron.iastate.edu/sites/windrose.phtml?station=EWR&network=NJ_ASOS.

Iowa State University (2023b) *IEM: site wind roses*. Available at: https://mesonet.agron.iastate.edu/sites/windrose.phtml?station=SEQU&network=EC__ASOS.

Kazda, A. and Caves, R.E. (2015) *Airport design and operation*. 3rd edn. Bingley, UK: Emerald. Available at: https://ereader.perlego.com/1/book/387013/75.

Lankford, T.T. (2001) *Aviation weather handbook*. New York: McGraw Hill.

McCabe, K. (2023) *What is wind?*, *Royal Meteorological Society*. Available at: www.rmets.org/metmatters/what-wind.

Memon, D.O. (2022) 'How hot does the skin of a supersonic aircraft get during flight?', *Simple Flying*. Available at: https://simpleflying.com/how-hot-does-supersonic-aircraft-skin-get/.

Met Office (2007) *National Meteorological Library and Archive Fact Sheet 2 – Thunderstorms*. Available at: https://skybrary.aero/sites/default/files/bookshelf/447.pdf.

Met Office (n.d.a) *Foehn effect*. Available at: www.metoffice.gov.uk/weather/learn-about/weather/types-of-weather/wind/foehn-effect.

Met Office (n.d.b) *How we measure wind*. Available at: www.metoffice.gov.uk/weather/guides/observations/how-we-measure-wind.

NASA (1996) *1st&2nd laws of motion*. Available at: www.grc.nasa.gov/www/k-12/WindTunnel/Activities/first2nd_lawsf_motion.html (Accessed: 6 September 2023).

National Geographic (2009) *Lightning facts and information, environment*. Available at: www.nationalgeographic.com/environment/article/lightning.

National Oceanic and Atmospheric Administration (2023) *What do leeward and windward mean?* Available at: https://oceanservice.noaa.gov/facts/windward-leeward.html.

National Research Council (1996) *Accelerated aging of materials and structures: the effects of long-term elevated-temperature exposure*. Washington, D.C.: National Academy Press. Available at: https://doi.org/10.17226/9251.

National Weather Service (n.d.) *How hot is lightning?* NOAA's National Weather Service. Available at: www.weather.gov/safety/lightning-temperature.

NOAA (2023a) *Life cycle of a thunderstorm*. Available at: www.noaa.gov/jetstream/thunderstorms/life-cycle-of-thunderstorm.

NOAA (2023b) *Thunderstorms*. Available at: www.noaa.gov/jetstream/thunderstorms.

NOAA (n.d.) *Wind roses – charts and tabular data*. Available at: www.climate.gov/maps-data/dataset/wind-roses-charts-and-tabular-data.

NSIA (2007) *Report on the aircraft accident at Bodo Airport on 4th December 2003. 2007/23 eng*. Available at: www.nsia.no/Aviation/Published-reports/2007-23-eng.

NSSL (n.d.) *Lightning basics*. Available at: www.nssl.noaa.gov/education/svrwx101/lightning/.

NTSB (2001) *NTSB Aviation Accident Final Report – DCA98MA015*. Available at: https://skybrary.aero/sites/default/files/bookshelf/368.pdf.

NTSB (2023) *Aviation Investigation Final Report, Accident Number CEN22LA130*. Available at: Report_CEN22LA130_104676_8_27_2023%2010_39_19%20AM.pdf.

Prather, C.D. (2015) *Airport management*. Newcastle, Washington: Aviation Supplies and Academics Inc.

Ross, D. (1995) *Introduction to oceanography*. New York: HarperCollins.

Royal Meteorological Society (2017) *Types of lightning*. Available at: www.rmets.org/metmatters/types-lightning.

Royal Meteorological Society (n.d.) *The Beaufort wind scale*. Available at: www.rmets.org/metmatters/beaufort-wind-scale.

SKYbrary (n.d.a) *Clear Air Turbulence (CAT)*. Available at: https://skybrary.aero/articles/clear-air-turbulence-cat.

SKYbrary (n.d.b) *Katabatic wind*. Available at: https://skybrary.aero/articles/katabatic-wind.

SKYbrary (n.d.c) *Tailwind*. Available at: https://skybrary.aero/articles/tailwind.

SKYbrary (n.d.d) *Wind*. Available at: https://skybrary.aero/articles/wind.

Smithsonian Institute (n.d.) *1903 Wright Flyer | National Air and Space Museum*. Available at: https://airandspace.si.edu/collection-objects/1903-wright-flyer/nasm_A19610048000 (Accessed: 27 August 2023).

University of Illinois (2010) *Gradient wind: non-geostrophic winds which blow parallel to isobars, Department of Atmospheric Sciences at Urbana-Champaign*. Available at: http://ww2010.atmos.uiuc.edu/(Gh)/guides/mtr/fw/grad.rxml.

Walmsley, S. (2021) *Aviation weather for the private pilot*. Aviation Books Private Pilot Series.

Wolfe, S.A. (2022) '7.08 – Cold-Climate Aeolian Environments', in J. (Jack) F. Shroder (ed.) *Treatise on geomorphology*. 2nd edn. Oxford: Academic Press, pp. 195–219. Available at: https://doi.org/10.1016/B978-0-12-818234-5.00036-5.

World Meteorological Organization (2020) *Aviation | Hazards | Turbulence and wind shear*. Available at: https://community.wmo.int/en/activity-areas/aviation/hazards/turbulence.

Yan, Y.Y. (2005) 'Land and sea breezes', in J.E. Oliver (ed.) *Encyclopedia of world climatology*. Dordrecht: Springer Netherlands (Encyclopedia of Earth Sciences Series), pp. 446–446. Available at: https://doi.org/10.1007/1-4020-3266-8_121.

6 Climatology and air transport

Chapter outcomes

At the end of this chapter, you will be able to:

- Understand global atmospheric circulation processes and the three-cell circulation model and explain how these translate into global weather patterns.
- Describe the main global air masses and understand the typical weather conditions each of these brings.
- Explain the weather principles of different types of fronts and the impacts these can have on aircraft.
- Understand the locations and importance of jet streams and their impacts on air transport operations.
- Explain the processes involved in tropical cyclone formation and their related hazards.
- Understand the processes involved in climate change and the contemporary global situation.
- Explain the impacts of climate change on the air transport industry.

6.1 Introduction

The air that surrounds us has a massive effect on our day-to-day and long-term decision making. Where do we want to live? What do we wear today? Do we take an umbrella? Where do we go on holiday? What leisure pursuits do we engage in? These will involve predictions which influence our future behaviour, based on the *weather* and *climate.*

Climate is the long-term pattern of weather in a particular area. A region's weather patterns, usually tracked for a period of at least 30 years, are considered its climate (Goodwin, 2014).

The difference between climate and weather is temporal. Weather is the conditions in the atmosphere over a short period of time and climate is how it behaves over relatively long periods of time (NASA, 2005; Kearns, 2021). Put another way, climate is what you expect, weather is what happens!

Climatology, being the scientific study of earth's climate, is an interdisciplinary science, combining meteorology, geography, geology and physics (Frey, 2017). Climate is an extremely important factor for the air transport industry. It will affect which destinations we want to fly to, can fly to, how we land, take off and the routes we choose from A to B.

158 *Fundamentals of Global Air Transport Geography*

Chapters 3–5 studied many of the key factors which make up weather and climate – broadly in the context of the atmosphere, water and wind. This chapter will now focus on the larger scale and longer-term aspects which make up our climate and will provide us with a broader understanding of the weather patterns which are expected in different parts of the world, over the course of the year and their impact on air transport.

The first part of the chapter will focus on the broad range of factors which influence and create the earth's climate, including global atmospheric circulation processes, air masses and their associated weather conditions, weather fronts, jet streams and extreme weather conditions included within tropical cyclones.

Due to our changing climate, its study is vital in contemporary times, with rising global temperatures and sea levels, changing precipitation, changing wind patterns, reductions in biodiversity and desertification. Each of these is extremely important to the air transport industry and the second part of the chapter will focus on the impacts of changing climate on the air transport industry.

ATG Did you know?

The word "climate"

The word "climate" derives from the Greek *klinein* – meaning to lean. In ancient times, geographers believed that the world could be divided into distinct zones, based on the slope or inclination of the northern celestial pole, moving northward from the equator. The Greek word *klima* means slope or inclination – thus the zones were called *klimata*.

Each region of the world has a unique set of weather conditions, influenced by several factors, for example:

- local terrain: forest, mountains, coastal, flat plains
- latitude and distance from the equator
- global atmospheric circulation

6.2 Global atmospheric circulation

Global atmospheric circulation is the movement of air around the planet and consists of wind systems which have annual and seasonal variations and is one of the principle factors determining climatic zone distribution. According to Lockwood (2012), there are two major causes:

1. Inequalities in the distribution of solar radiation over the earth's surface (with more being received at the equator).
2. The rotation of the earth.

Without the earth's rotation, tilt relative to the sun and surface water, global circulation would be simple. With the sun directly overhead at the equator, this would heat up this area more than other parts of the earth. The hot air would rise into the upper atmosphere and create a low-pressure belt at the equator. This would then move directly north to the poles, becoming

Climatology and air transport 159

very cold and thus sinking, causing a high-pressure area, which would then circulate south back to the equator.

The effects of latitude on solar radiation were studied in Chapter 3 and the rotation of the earth creating the Coriolis effect was studied in Chapter 5. These concepts are extremely important in understanding the global climate. The solar radiation received mainly at the equator and radiated back out provides the energy to drive the global atmospheric circulation, while the earth's rotation determines its shape (Lockwood, 2012).

The earth's tilt means the Sun is not always directly over the equator. The earth rotates on its axis relative to the Sun every 24 hours, with an inclination of 23.45 degrees (NASA, n.d.a).

ATG Did you know?

What if the axial tilt was zero?

The days are longer in the summer and shorter in the winter due to the axial tilt, which also creates the seasons as one hemisphere gets more sunlight during its summer and less during winter. If the tilt was zero, the days and nights would be the same length and there would be no seasons.

As with most things with the earth's climate, the axial tilt does not stay constant. According to Buis (2020), the angle is slowly decreasing in a cycle which spans about 41,000 years. When the tilt is smaller, summers are milder and winters are warmer. The repeating cycles of increasing and decreasing tilt plays an important role in the glacial cycle climate shifts over the last million years.

6.2.1   The three-cell circulation model

Our global circulation provides a natural air conditioning system to stop the poles becoming colder and the equator becoming hotter. According to the Met Office (n.d.a), this global circulation can be described as the worldwide system of winds by which the necessary transport of heat from tropical to polar latitudes is accomplished.

In each equator there are three cells – Hadley, Ferrel and Polar cells – in which air circulates the entire depth of the troposphere (Figure 6.1). The vertical and horizontal movements of air within these cells create consistent areas of ascending and subsiding air, which is a key component in the formation of regular high and low pressure systems.

Hadley cell

Air rises near the equator where there is a heat surplus. It then flows poleward in the upper atmosphere and sinks about 30°N and S and returns to the equator at low levels. It is restricted to tropical regions by the earth's rotation (Lockwood, 2012).

Ferrel cell (mid-latitude)

From 30°N and S, at low-level, air converges and moves poleward rising at around 50–60°N and S. At this point, **frontal depressions** are created, where the tropical and polar air masses meet and this creates poor weather, with rain and wind. The upper-level winds are mainly from the west and therefore travel towards the east, bringing regular poor weather, often called a

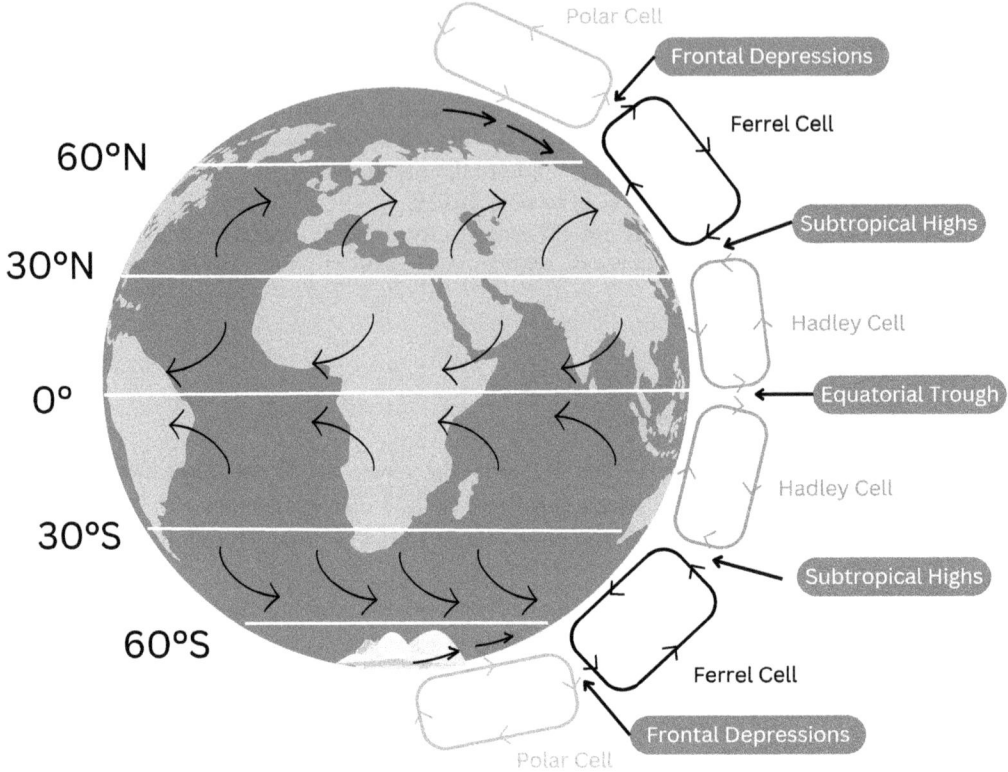

Figure 6.1 Global atmospheric circulation
Source: author

westerly situation. Some air will then return at high levels towards 30°N and S. It is essentially the net effect of air motions from all the storms or "depressions" that occur in the mid-latitudes (Royal Meteorological Society, n.d.).

Polar cell

This cell extends from between 60 and 70°N and S (Met Office, n.d.b) to the poles. At higher latitudes, air rises and travels towards the poles. Once over the poles, the air sinks, forming areas of high atmospheric pressure called the polar highs. At the surface, air moves outward from the polar highs, creating east-blowing surface winds called polar easterlies. It is the smallest and weakest of the cells.

6.2.2   The effects of the three-cell circulation

High and low pressure

Between each of the cells are surface-level bands of high and low pressure. The low-pressure bands are found at the equator and around 50–60°N and S, where the *polar front* occurs and *frontal depressions* develop, whereas the high-pressure band is located around 30°N and S – this is where the *subtropical highs* are found (Figure 6.1). Usually, fair and dry/hot weather is associated with high pressure, with rainy and stormy weather associated with low pressure.

The equatorial trough and inter-tropical convergence zone

The **equatorial trough** (Figure 6.1) is an almost continuous low-pressure belt found near the equator, where the **trade winds** of the northern and southern hemispheres come together. The humidity is so high that slight variations in *stability* cause major variations in weather (American Meteorological Society, 2012). It can produce large areas of cumulonimbus clouds extending into the upper troposphere and lower atmosphere, producing heavy precipitation.

The inter-tropical convergence zone (ITCZ) lies in the equatorial trough, only a few hundred kilometres in width and here, the conditions are favourable for ascending air motion, condensation of water vapour, cloudiness and high precipitation rates (Lockwood, 2012). The trade winds move from the subtropical anticyclones either side of the **heat equator** and there is a greater movement of the ITCZ north and south over land than the oceans (Wickson, 2015). It undergoes a regular seasonal shift in its position during the year (Quantick, 2001). In July and August, the ITCZ lies well north of the equator over Africa, Asia and Central America before moving south into South America, central Africa and Northern Australia by January and February (Met Office, n.d.c). The effect of the ITCZ determines the weather patterns over a significant portion of the globe.

From an air transport perspective, an active ITCZ can present several hazards to pilots:

- icing
- turbulence
- lightning
- wind shear

ATG Case study 6.1

Nigeria and the ITCZ

Nigeria's climate is characterised by the hot and wet conditions of the ITCZ, due to its location just north of the equator. When the ITCZ is south of the equator, the northeast winds prevail, producing the dry season. When it moves into the northern hemisphere, the SW wind prevails and brings rainfall during the wet season.

Lagos

Lagos averages around 1,740mm of rain annually, with four observed seasons:

1. Long rainy season from March to the end of July.
2. Short dry season, in August for three to four weeks, when the ITCZ moves to the north of the region.
3. Short rainy season, from early September to mid-October as the ITCZ moves south again.
4. Long dry season, from late October until early March.

Kano

Kano averages around 825mm of rain annually, with two main seasons:

1. Long dry season from October to mid-May, when the ITCZ is in the southern hemisphere.
2. Short rainy season from June to September.

Both the total annual rainfall and number of rain days decreases from south to north (NOAA, 2023c).

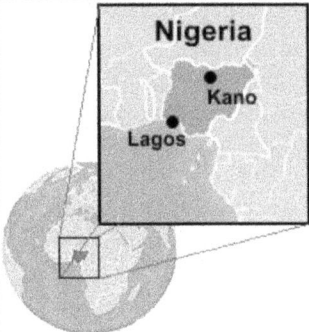

Figure 6.2a Location of Lagos and Kano, Nigeria
Source: author adapted from NOAA (2023c)

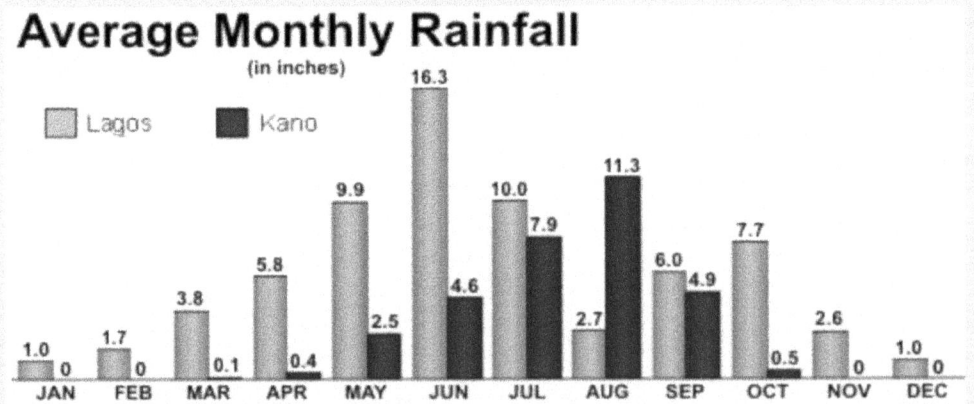

Figure 6.2b Monthly normal rainfall for Lagos and Kano
Source: author adapted from NOAA (2023c)

6.3 Air masses

The term "air mass" is given to a body of air that has a very large horizontal extent and in which its potential temperature and moisture content are similar through most of the troposphere (McClatchey, 2012). An air mass develops over a source region where it has remained for a period of days (Barry and Chorley, 2010) and the longer it stays over its source region, the more likely it will acquire the properties of the surface below (NOAA, 2023a).

Air masses are given a two-part name that describes the humidity and temperature characteristics of the region where they form. The first part describes the *humidity* – over the ocean this is **maritime** and over land it is **continental**. The second part describes the *temperature* of the air mass, which determines the latitude. Air masses from near the equator or in the tropics are called **tropical** (or equatorial) and those from more northerly latitudes are called **polar**.

According to McClatchey (2012), there are four basic types of air mass according to source region. Figure 6.3 illustrates these for Europe.

1. **Maritime Tropical** (mT) – A warm and extremely moist air mass originating from very warm tropical oceans, such as the mid-Atlantic and Indian Oceans. These contain warm, moist air bringing cloud, rain and mild weather.
2. **Continental Tropical** (cT) – A warm and dry air mass originating from warm land masses. In Europe, this originates from North Africa and brings warm, dry weather in summer.
3. **Maritime Polar** (mP) – A cold and slightly moist air mass originating from cold oceans, such as the Southern oceans around New Zealand and Australia. In Europe, these come from Greenland and the Arctic Sea, with wet, cold air bringing showery weather.
4. **Continental Polar** (cP) – These are cold dry air masses originating from cold land masses. In Europe these come from Central Europe, bringing dry summers and cold weather and snow in winter.

There are also extreme versions of polar air masses called continental or **maritime Arctic (cA/mA)** or Antarctic air masses (McClatchey, 2012; NOAA, 2023a). These originate over the Arctic or Antarctic and are thus very cold.

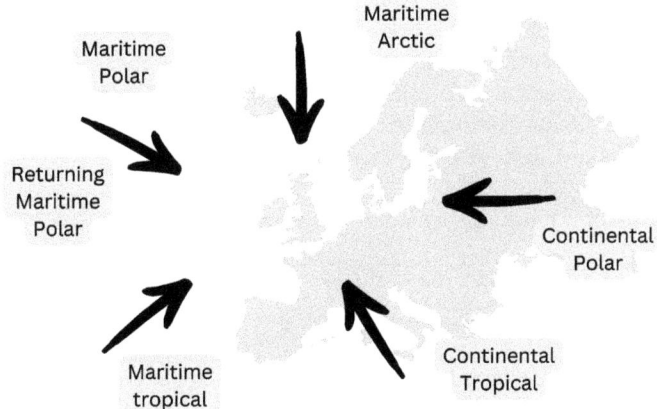

Figure 6.3 Air masses affecting the UK and mainland Europe
Source: author

164 *Fundamentals of Global Air Transport Geography*

There is also a **returning maritime polar** air mass in Europe, where maritime polar air has moved well to the south and then approaches the UK and France from the southwest or west, with moist, mild and unstable air bringing cloud and rain showers and possible thunderstorm activity.

These all play a key part in air transport operations, based on the atmospheric principles discussed in Chapters 3–5, in relation to temperature, humidity, clouds, precipitation, etc., and their impact on aircraft performance, as well as desired tourist destinations.

> ### ATG Did you know?
>
> ### Air masses
>
> An air mass can be compared to a football team. Players will all wear the same jersey and they will be competing against players from another team wearing a different jersey. In the case of air masses, we are talking about air not players and the "uniform" of air includes characteristics such as humidity and temperature. Like many football teams, when two air masses meet, there can be turbulence, some minor, some a little more major!

6.4   Fronts

A front is the name given to the boundary between two air masses – when two air masses meet, a front must be established and this will mean a change in the weather (UCAR, 2023). Many cause events such as turbulence, thunderstorms, gusty winds, icing and even tornadoes – each of which can have a significant impact on air transport operations. As air masses are global in nature, there will therefore be global fronts:

- **ITCZ** – a region of extensive unstable weather near the equator.
- **Mediterranean front** – the boundary in Europe, between polar continental/maritime air to the north and tropical continental air from North Africa to the south. It is a winter front when the sea is a region of low pressure where depressions form.
- **Polar front** – the frontal surface between polar and tropical air masses. This can extend across the Atlantic and Pacific Oceans. A series of waves forms on this front causing *depressions*.
- **Arctic front** – the boundary between arctic and polar air masses, at higher latitudes than the polar front. Depressions form affecting regions greater than 65°N.

There are four different types of weather fronts – *cold*, *warm*, *stationary* and *occluded* (Figure 6.4).

6.4.1   *Cold fronts*

A cold front occurs when cold air advances on warm air. As the colder air is heavier, it pushes underneath the warmer air, thus replacing the warm air at the surface. The lifted warm air ahead produces cumulus or cumulonimbus clouds and thunderstorms. They can move up to twice as fast as warm fronts (UCAR, 2023). The air pressure will drop, wind speeds will increase, temperatures drop and there is often heavy rain. The slope of cold fronts, being steeper, forces air up quicker and the band of cloud and precipitation is narrow.

Climatology and air transport 165

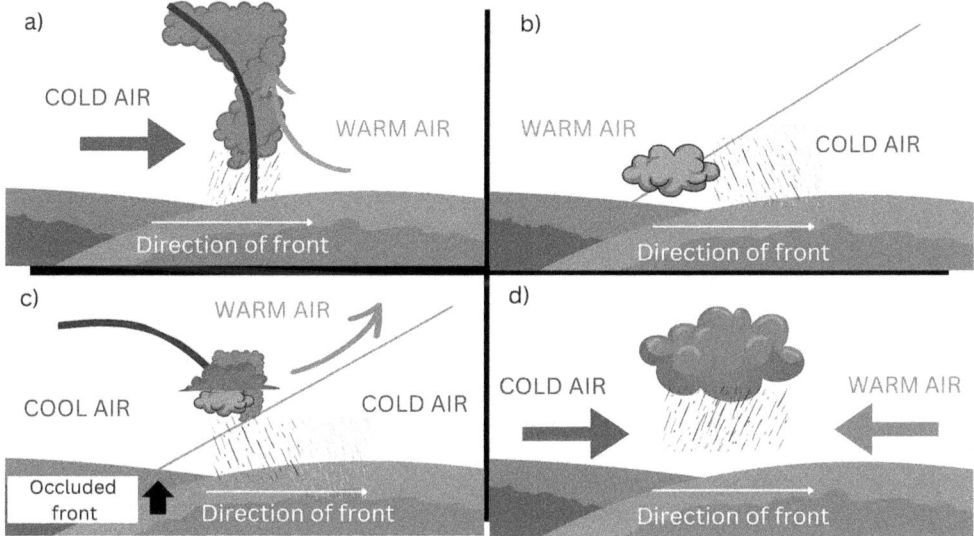

Figure 6.4 Weather fronts: (a) cold front, (b) warm front, (c) occluded front, (d) stationary front
Source: author

From a pilot perspective, this is extremely hazardous, due to the thunderstorms, turbulence, poor visibility and icing risk. Once the cold front passes, the air pressure starts to increase, the rain stops and cumulus clouds are replaced by stratus and stratocumulus.

6.4.2 *Warm fronts*

Warm fronts occur where warm air is advancing and rising over cold air, as warm air is lighter and warm fronts tend to travel relatively slowly. Clouds will typically be cirrus and cirrostratus initially, followed by mid-level clouds such as altostratus, perhaps bringing light rain and drizzle. When the main part of the front arrives, clouds such as heavy nimbostratus may be present with plenty of precipitation. The gentle slope of the warm front means a broader area of rising air. As the main part of the warm front passes, air pressure slowly reduces and wind speeds increase. As with cold fronts, very poor visibility and a risk of icing is present to pilots. When the warm front passes, temperatures rise as does the air pressure.

6.4.3 *Stationary fronts*

These form when a cold or warm front stops moving, so when both are pushing against each other, neither is strong enough to move the other – like equally balanced sumo wrestlers! Winds blowing parallel to the front instead of perpendicular help it to stay in place. They can still produce significant clouds and precipitation and in the same place due to its static nature. When the wind direction changes, the front will start moving again.

6.4.4 *Occluded fronts*

The word occluded means "hidden" and these occur when the cold front "catches up" with the warm front, which can happen as the cold front moves quicker. The warm air is then lifted from the surface and therefore hidden. An occluded front can be thought of as having the characteristics of both warm and cold fronts. The conditions for pilots are like the cold and warm fronts – wide areas of cloud, precipitation, turbulence and poor visibility.

6.5 Jet streams

According to the World Meteorological Organization (1958), "a jet stream is a strong narrow current, concentrated along a quasi-horizontal axis in the upper troposphere or in the stratosphere, characterised by strong vertical and lateral wind shears and featuring one or more velocity maxima".

Jet streams are relatively narrow bands of strong wind in the upper atmosphere, typically occurring around 30,000ft (9,100m) in elevation (NOAA, 2023b). The winds generally blow from west to east (due to the earth's rotation) but often the flow meanders southward and northward in waves (FAA, 2023). The winds in the jet stream are variable and may reach 500km per hour (310 miles per hour). In winter, the average speed is 160kmh (100mph) and in summer 80kph (50mph) and they may move equatorward and poleward from week to week (NASA, 2006). This movement helps to steer fronts at the surface and can therefore have a large impact on surface weather. A jet stream can be several thousand miles long, but only a few hundred miles wide and a few thousand feet in depth (SKYbrary, n.d.).

ATG Did you know?

Jet streams

The jet stream was virtually unknown until World War II when US B-29 pilots flying at high altitudes reported turbulence and very strong winds. It was recognised as a meteorological phenomenon in 1946 (Lankford, 2001; Maksel, 2018). A Japanese meteorologist named Wasaburo Ooishi is often credited with the first discovery back in the 1920s when he used weather balloons to track upper-level winds near Mount Fuji (McCabe, 2022).

Jet streams are caused by sharp temperature gradients associated with frontal depressions (Lockwood, 2012). The three-cell process explained in Section 6.2 helps to circulate the air within the troposphere and as such the jet streams largely exist because of a difference in heat, which in the northern hemisphere means cold air on the northern side of the jet stream and warm air to the south, reversed in the southern hemisphere (Met Office, n.d.d). In winter there is more of a temperature difference between the equator and poles, so the jet stream is generally stronger in winter and is why wetter weather is often experienced at this time in mid-latitude areas such as the UK.

The regions around 30°N/S and 50–60°N/S (as illustrated in Figure 6.1) are where the wind in the upper atmosphere is strongest, between the cells. There are generally two different jet streams in each hemisphere (Figure 6.5):

- The polar jet stream is located between the 50–60°N/S latitude lines.
- The subtropical jet stream is located around the 30°N/S latitude lines (NOAA, 2023d).
- In addition, there is also an equatorial easterly jet stream (Lockwood, 2012).

Climatology and air transport 167

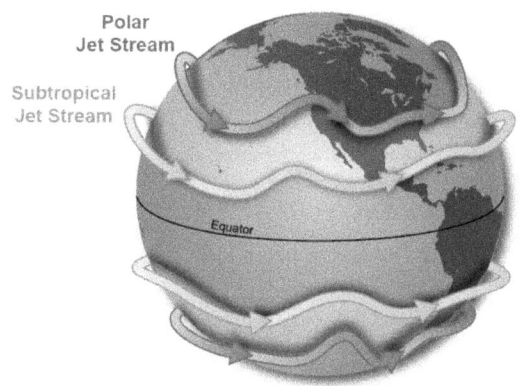

Figure 6.5 Location of jet streams
Source: NOAA (2023d)

6.5.1   Jet stream impacts on air transport

Ground speed

Aircraft flying west to east will try to utilise the jet stream to increase ground speed, arrive at their destination quicker and save on fuel burn. A tailwind of 100+mph can result in sizeable time savings on longer flights. An issue here may be if the flight arrives in the arrival airspace early around a busy airport, there may be significant holding, increasing fuel burn and workload of air traffic control. Flying against the jet stream may result in much more fuel burn and significantly longer journey times.

Indicated airspeed

When descending into stronger winds from calmer winds, an increased airspeed may occur and therefore care needs to be taken not to overspeed the aircraft when flying at maximum speeds.

Clear air turbulence (CAT)

Wind shear is possible flying in the vicinity of a jet stream. Passing over mountains may produce waves which could increase the turbulence.

ATG Did you know?

Speeding along in the jet streams

In February 2019, a Virgin Atlantic Boeing 787–9 reached an amazing speed of 801mph on a flight from Los Angeles to London. A year later, in February 2020, a British Airways Boeing 747-400 achieved an even greater speed of 825mph (1,328km/h) flying from New York to London. When you consider that the typical cruise speeds for these aircraft would be around

550mph, that is quite an increase! Flight times from New York to London are usually around seven hours, but this flight took less than five (Hardiman and Joshi, 2022).

To be clear, the aircraft did not break the sound barrier as this was a ground speed measurement rather than airspeed, with a tailwind more than 200mph. Step forward the polar jet stream!

6.6 Tropical cyclones

Tropical cyclone is a general term for any low which originates over tropical oceans (FAA, 2023), at least 5–30° latitude north and south of the equator (Met Office, n.d.a). They are rapid rotating storms, which draw their energy from the tropical oceans, with a diameter typically of around 200–500km, but can reach 1,000km (World Meteorological Organization, 2020). These are one of the biggest threats to life and property even in their formative stages of development. From an air transport perspective, flying in or around these presents potentially catastrophic risks. The risks are also huge on the ground for airports. According to the World Meteorological Organization (2020), in the previous 50 years, 1,942 varying types of disasters have been attributed to tropical cyclones, which killed 779,324 people, causing US$1,407.6 billion in economic losses.

6.6.1 Conditions required for tropical cyclone formation

According to Bryant (2005), at least **eight** conditions are required to develop a tropical cyclone:

- Efficient convective instability is required causing a near-vertical uplift of air throughout the troposphere. Ocean temperatures exceed 26–27°C.
- Convergence of surface air into warm waters, which the formation of numerous thunderstorms provides.
- A trigger mechanism is required to begin surface air rotation converging into the uplift zone.
- The Coriolis force must be sufficient to establish a vortex, hence why cyclones rarely form within 5° of the equator. Cyclones also cannot cross the equator as they rotate in opposite directions in each hemisphere (anticlockwise in the north and clockwise in the south).
- Cyclones cannot be sustained at temperatures above 24°C because they are dependent upon transfer of heat from the surface to the upper atmosphere.
- They must form an "eye" structure.
- They cannot develop if there are substantial winds in the upper part of the troposphere.
- The central pressure of a tropical cyclone must be below 990hPa.

6.6.2 Location of tropical cyclones

Figure 6.6 illustrates the seven regions around the world where tropical cyclones are likely to form. None are within 5° latitude of the equator and all have sea surface temperatures of at least 27°C.

Climatology and air transport 169

Location	Time of activity
a) Northeast Pacific basin (Mexico to about the dateline)	A broad peak with activity beginning in late May or early June, occurring until late October, early November
b) North Atlantic Ocean, Gulf of Mexico and the Caribbean Sea.	Hurricane season officially from 1 June to 30 November, peaking in early–mid September
c) North Indian basin (including the Bay of Bengal and the Arabian Sea)	From April to December, with a double peak in May and November. The severe cyclonic storms (>74mph/119kmh) occur from April to June and in late September–early December.
d) Northwest Pacific basin (from the dateline to Asia, including the South China Sea)	The main season runs from July to November, although occur regularly all year
e) Southwest Indian basin (From Africa to about 100°E)	Late October/early November to May, with a double peak in activity in mid-January and mid-February/early March
f) Southeast Indian/Australian basin (100° to 142°E)	Late October/early November to May, with a double peak in activity in mid-January and mid-February/early-March. A lull in activity in February is a bit more pronounced than the Southwest Indian basin's lull.
g) Australian/Southwest Pacific basin (142°E to about 120°W)	Begins in late October/early November, reaches a single peak in late February/early March and then fades out in early May

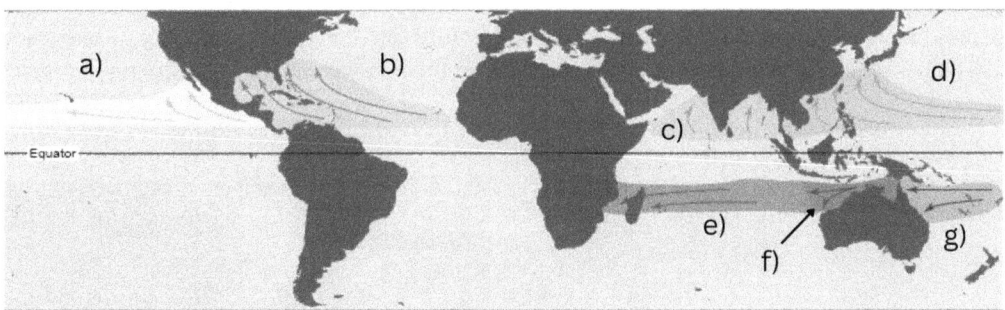

Figure 6.6 Global tropical cyclone formation basins
Source: adapted from NOAA (2023e)

ATG Did you know?

Q: When do tropical cyclones become hurricanes? When do hurricanes become typhoons?
A: They always are! According to the World Meteorological Organization (2020), they are different terms for the same phenomenon.

- In the Caribbean Sea, Gulf of Mexico and the eastern and central North Pacific Ocean, they are called **hurricanes**.

- In the western North Pacific, they are called **typhoons**.
- In the Bay of Bengal and Arabian Sea, they are called **cyclones**.
- In the western South Pacific and southeast Indian Ocean, they are called **severe tropical cyclones**.
- In the Southwest Indian Ocean, they are called **tropical cyclones**.

6.6.3 Classification of tropical cyclones

Tropical cyclones (Figure 6.7) are classified differently based on their region and intensity levels, although they begin when the maximum sustained wind exceeds 116km/h (63 knots) (Table 6.1).

6.6.4 Tropical cyclones and their related hazards

There are many hazards associated with tropical cyclones, such as destructive winds, torrential rainfall, flash-flooding, storm surges and landslides. Their potential for destruction is exacerbated by the length and width of affected areas, their intensity, frequency of occurrence and vulnerability of the impacted areas. Table 6.2 places these hazards into three categories.

For the air transport industry, hurricanes are a huge issue, from the physical impact of damage to airports, to the possibly thousands of flights being cancelled and large detours to flights to avoid the weather, adding to journey times, fuel bills and emissions. In theory, if hurricanes don't have a high vertical extent, aircraft can fly over them, however this is not common practice

Figure 6.7 Hurricane Maria (24 September, 2017), a CAT 5 hurricane, with winds as high as 174mph, devastating parts of the Northeastern Caribbean.

Source: NASA (n.d.c)

Table 6.1 Classification of tropical cyclones (based on data from World Meteorological Organization, 2023)

km/h	Southwest Indian Ocean	Arabian Sea and Bay of Bengal	Northwest Pacific	North Atlantic and Northeast Pacific	Southwest Pacific and Southeast Indian Ocean
118				Hurricane CAT 1 (up to 153kmh)	
	Tropical cyclone (up to 165kmh)	Very severe cyclonic storm (up to 166kmh)	Typhoon	Hurricane CAT 2 (154–177kmh)	Severe tropical cyclone/ Category 3 (up to 159kmh)
				Hurricane CAT 3 (178–208kmh)	
	Intense tropical cyclone (166–212kmh)	Extremely severe cyclonic storm (167–221kmh)		Hurricane CAT 4 (209–251kmh)	Tropical cyclone/ Category 4 (160–199kmh)
	Very intense tropical cyclone (213+kmh)	Super cyclonic storm (222+kmh)		Hurricane CAT 5 (252+kmh)	Tropical cyclone/ Category 5 (>200kmh)

Table 6.2 Hazards related to occurrence of tropical cyclones

Storm surge	Wind	Rain
Flooding of coastal lowland	Disruption to transport	Severe flooding
Saline intrusion	Death, injuries	Contaminated water supplies
Earthquakes	Urban fire	Spread of disease
Coastal erosion	Loss of crops and livestock	Landslides
Damage to structures		Death from drowning and disease
Death from drowning		

Source: adapted from Bryant (2005)

and is risky. Forecasting groups, such as the NOAA, help to predict cyclone tracks several days in advance, although there will always be unpredictability.

On the ground, hurricanes can cause significant damage and disruption to airports:

- **Typhoon Jebi, Kansai International Airport, Japan, September 2018.** Parts of the airport flooded and access to the island containing the airport cut off, stranding thousands of passengers and staff, who were evacuated by ferries and speedboats.
- **Cyclone Idai, Beira Airport, Mozambique, March 2019.** Airport closed and around 90% of Beira damaged or destroyed.
- **Typhoon Odette, Mactan-Cebu International Airport, Philippines, December 2021.** Substantial structural and systems damage, impassable roads and fuel shortages.

ATG Case study 6.2

U.S. National Weather Service (NWS) StormReady® Programme: Tampa, USA

The NWS has established a Storm ready programme using a grassroots approach to help communities, including airports, to develop plans to handle all types of extreme weather, including hurricanes. Tampa International Airport was the first StormReady Airport in Florida, the criteria for which was to:

- establish a local 24-hour warning point and an Emergency Operations Centre
- develop a formal hazardous weather plan
- conduct periodic drills/exercises
- have multiple ways of receiving NWS warnings and alerting the public
- promote the importance of public readiness through community seminars and presentations.

Source: Tampa Airport (2017)

Florida is under hurricane watch from approximately June until November each year and these are the types of programmes which can assist airports to prepare for these events.

6.7 Climate change

The importance of the climate to aviation has been discussed and cannot be overstated. Any changes to climate are widespread, affecting almost every part of the globe. In almost every human activity there is a climate footprint and therefore the work of all industries and their organisations is becoming more scrutinised (Padhra and Kurnaz, 2023). Any changes to climate will have serious impacts for the planet as a whole and focusing on the air transport industry specifically, many individual challenges will have to be faced.

6.7.1 Climate change and greenhouse gases

Climate change is linked to greenhouse gases (GHGs) in the earth's atmosphere (Kearns, 2021). These both absorb and emit heat, causing a warming process called the *greenhouse effect*. Four GHGs occur naturally in the atmosphere – water vapour, carbon dioxide, methane and nitrous oxide. The total greenhouse effect is the combination of natural and human induced GHGs.

The presence of natural GHGs is beneficial, capturing heat to maintain life on earth. Through photosynthesis, plants remove carbon dioxide (CO_2) from the atmosphere and much is absorbed by seawater in the oceans. Since the late 1800s, however, fossil fuel burning and other industrial and agricultural activities have significantly increased the amount of GHGs such as CO_2 into the atmosphere, with more heat being captured, creating global warming.

The rate of release and long lifetime in the atmosphere (tens of thousands of years) has meant that CO_2 has become the main focus in the effort to reduce greenhouse gas emissions (Padhra and Kurnaz, 2023).

6.7.2 Climate change history

To reiterate, climate is the average weather conditions over a long period of time, often 30 years or more. To measure the human impact on the climate, a very long time period is used, with global average temperatures often being compared to a baseline from 1850–1900 – before the dramatic increases in emissions from industrial activity and fossil fuel use (Kearns, 2021).

It is important to point out that changes to earth's climate due to natural causes such as volcanic eruptions have been occurring for millions of years, but it is the influence of human activity, since the industrial revolution, on climate change which is the focus, essentially supercharging the natural greenhouse effect.

In the late 19th century, Nobel Laureate Svante Arrhenius proposed that higher concentrations of CO_2 in the atmosphere would result in warmer atmospheric temperatures (Arrhenius, 1896). Many scientists and policymakers have since agreed that global warming beyond 2°C above the pre-industrial average would pose large and escalating risks to human life as we know it and many governments used this as an organising principle (MIT, 2021). Even back to the 1970s, William Nordhaus, an economist at Yale, suggested if this figure of 2°C were to be exceeded, global conditions would be past any point that human civilisation had experienced (Nordhaus, 1977).

In 1988, NASA climate scientist Dr James Hansen told the U.S. Senate that he was 99% certain the earth was warmer than at any time since measurements began, with a clear cause and effect of greenhouse gas emissions from humans (Hansen *et al.*, 2006).

ATG Did you know?

The Intergovernmental Panel on Climate Change (IPCC)

The IPCC is the United Nations body for assessing the science related to climate change, its impacts and future risks and options for adaptation and mitigation. It was founded in 1988 by the World Meteorological Organization (WMO) and United Nations Environment Programme (UNEP), with the objective of providing governments at all levels with scientific information which they can use to develop climate policies. IPCC reports are also a key input into international climate change negotiations. As of 2023, it had 195 members.

In 2015, the **Paris Climate Agreement** was negotiated under the UN unit, the Framework Convention on Climate Change, and was a treaty that brought the world's peoples into a common effort to combat climate change (MIT, 2021). The parties agreed to hold the rise in global temperature "well below" 2°C above pre-industrial levels and pursue efforts to limit the temperature increase even further to 1.5°C (UNFCCC, 2023). The target of 1.5°C was deemed a more desirable goal as it reduces risk for the worst outcomes of climate change in most of the world – 1.5°C or less is now used as a target in much IPCC documentation (IPCC, 2018).

6.7.3   Contemporary climate change

Atmospheric carbon dioxide

The modern record of atmospheric carbon dioxide levels began with observations recorded at Mauna Loa Observatory in Hawaii (Figure 6.8). From 1958, when records began, the level of CO_2 in the atmosphere has risen from around 317 parts per million (ppm) (NASA, n.d.b) to 418.56ppm in 2022 (Climate.gov, 2023). According to NASA (n.d.b), crossing the threshold of 400ppm is a signal to some that we are now firmly seated in the "Anthropocene", a human

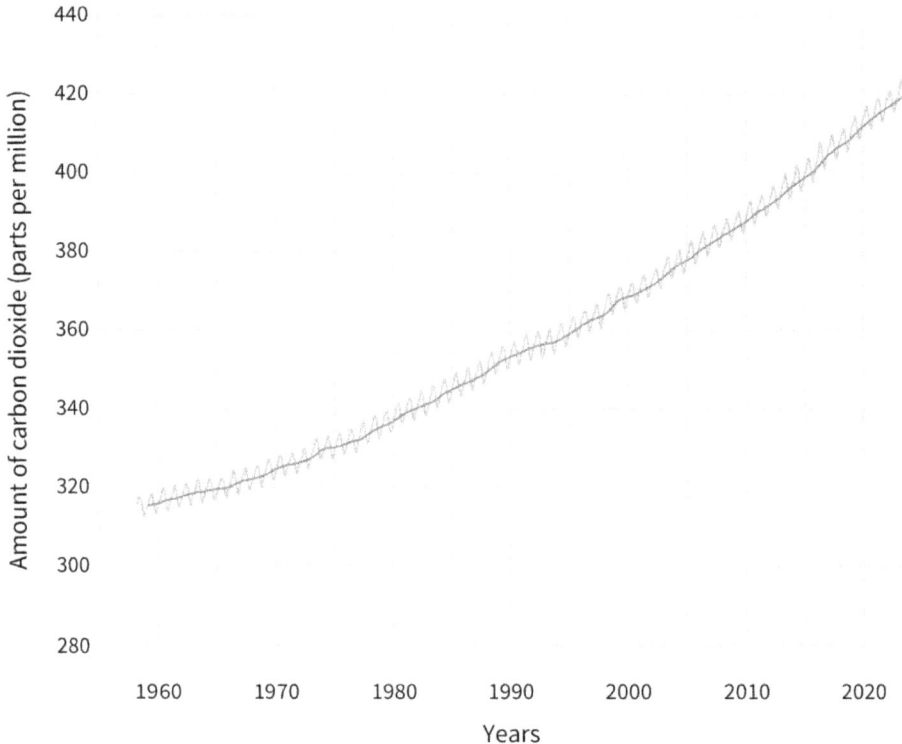

Figure 6.8 Atmospheric carbon dioxide levels
Source: Climate.gov (2023)

epoch where humans are having major and lasting impacts on the planet. This concept will be covered in Chapter 13.

Global temperatures

As noted in Section 6.7.2, air temperatures have been rising since the industrial revolution. The preponderance of evidence indicates human activities as being mostly responsible for making our planet warmer (NASA, 2023). According to the Goddard Institute for Space Studies at NASA, the average global temperature has increased by at least 1.1°C since 1880. The majority of this has occurred since 1975. The nine years leading up to and including 2022 were the warmest years for global temperatures on record and Figure 6.9 illustrates the temperature anomalies in 2022 versus the 1951–1980 average.

In the UK, 2022 was the warmest year since records began in 1884 and was the first year temperatures exceeded 40°C (40.3°C) and the belief of the Met Office (2023) is that this was made more likely by human-induced climate change.

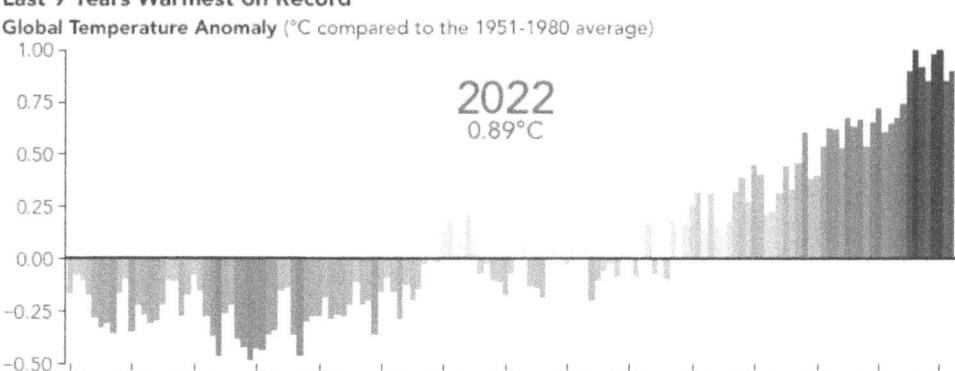

Figure 6.9 Temperature anomalies in 2022
Source: NASA (2023)

6.7.4 IPCC Sixth Assessment Report (AR6) on Climate Change

This 2023 report has provided the most comprehensive and best available scientific assessment of climate change (Boehm and Schumer, 2023). Some key findings (with *high confidence*) of relevance to the air transport industry, include:

- Human activities, principally through emissions of greenhouse gases, have unequivocally caused global warming, of 1.1°C.
- Widespread and rapid changes in the atmosphere, ocean cryosphere and biosphere have occurred and human-induced climate change is already affecting many weather and climate extremes in every region across the globe. Vulnerable communities who have historically contributed the least to current climate change are disproportionately affected.
- Observed climate change has caused adverse impacts on urban infrastructure, including transportation, water, sanitation and energy systems, which have been compromised by extreme and slow-onset events as well as an adverse impact on human health and livelihoods.
- Economic damages have been detected in climate-exposed sectors, such as agriculture, forestry, energy and tourism.

(IPCC, 2023)

There is much which can be done to positively influence these impacts, by reducing GHG emissions, scaling up carbon removal and building resilience. Chapter 14 will focus on the air transport industry's contributions to climate change and what the industry is doing to mitigate its impacts. The remainder of this chapter will focus on the broader impacts of climate change *on* the air transport industry.

6.8 The impacts of climate change on the air transport industry

There is considerable developing understanding that there is a two-way relationship between climate change and air transport (Gratton *et al.*, 2022). Air transport emissions contribute to climate change and conversely, as air transport is dependent on atmospheric conditions for its operations, any impacts here – such as changes in temperatures, wind speeds and extreme weather conditions – will affect the industry.

The impact of the air transport sector on climate change is widely referenced and acknowledged, however the impacts of climate change and the need for the industry to adapt has not been as well researched (Williams, 2016; Ryley, Baumeister and Coulter, 2020; Padhra and Kurnaz, 2023).

6.8.1 How climate change impacts air transport

Several organisations have undertaken surveys to establish the views of states in terms of readiness and risks related to air transport and climate change. ICAO (2019) surveyed its 193 member states and 65% of responding states believed climate change had already impacted air transport in their states and a further 15% believed they would be impacted by 2030. Eight climate impact categories were detailed:

1. Sea level rise
2. Increased intensity of storms
3. Temperature change
4. Changing precipitation
5. Changing icing conditions
6. Changing wind direction
7. Desertification
8. Changes to biodiversity.

Eurocontrol (2018) surveyed its constituent states and 86% of respondents believed that climate change adaptation would be essential for the air transport industry. Forty-eight percent considered themselves to be already impacted. Five major risk factors were identified:

1. Temperature increase
2. Changes to precipitation
3. Changes to storm patterns
4. Changes in wind patterns
5. Sea-level rise and storm surges

Gratton *et al.* (2022) stated that the effects of climate change on air transport included:

- increased take-off distances
- increased en-route flight times
- increased frequency and severity of encounters with clear air turbulence (CAT)
- changed patterns of wildlife activity
- shifting locations of flight safety hazards
- increased burdens upon airport and associated infrastructure.

Burbidge (2018) stated that extreme weather events at airports may cause:

- airfield flooding
- ground subsidence

Climatology and air transport 177

- inundation of underground infrastructure
- loss of utilities provision such as electrical power
- heat damage to the airport surface
- disruption to surface access.

Any infrastructure damage can delay flights, lead to cancellations and cause significant indirect costs, as well as costs to repair and protect the airport areas.

Ryley, Baumeister and Coulter (2020) researched literature on climate change and aviation, analysing significant issues affecting aviation in a changing climate and industry responses on climate change and adaptation. They stated that the adaptation concept for aviation could be categorised into three elements, which will provide the framework for the rest of this section:

1. Disruptions at airports
2. Disruptions in the air
3. Disruptions from changing patterns in passenger demand

6.8.2   Impact on airports

Airports face climate change challenges due to an increase in temperatures, precipitation, wind, rising sea levels and associated infrastructure issues. These are global impacts, as illustrated by the case study areas in this chapter (Figure 6.10), although there are many more and much research has focused on the USA and China.

Aircraft take-off performance

Higher temperatures result in lower air density, which reduces the ability of an aircraft to generate lift and impacts engine performance by reducing thrust force. During take-off, this results in a longer take-off distance on the runway (Padhra and Kurnaz, 2023).

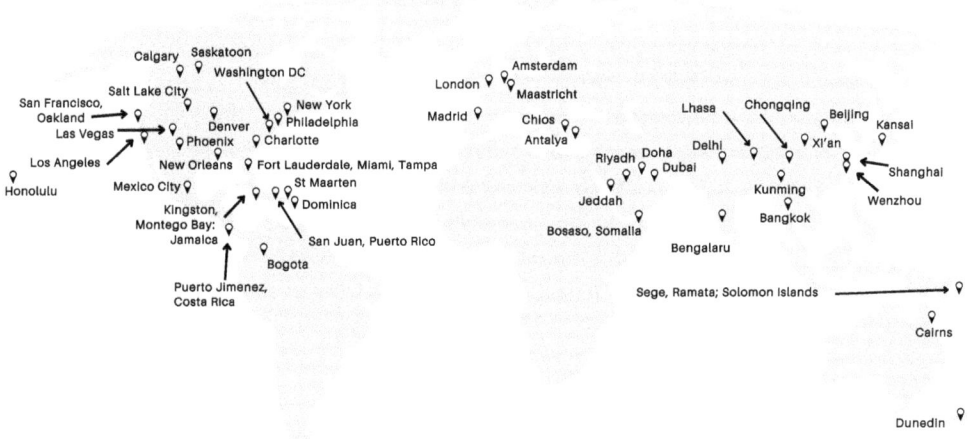

Figure 6.10 Global case study locations for airport climate change impacts
Source: author

As a result, any increase in temperatures could negatively impact aircraft take-off performance. At airports where there is ample runway length available, the main effect is of increased take-off distances; where runways are shorter, payload becomes limited (Gratton *et al.*, 2022). This could mean that certain aircraft may not be able to take off from specific runways or a reduced fuel load/payload may be required, such as fewer passengers or cargo (to reduce the aircraft weight) to take off.

Coffel, Thompson and Horton (2017) concluded that on average, 10–30% of annual flights departing at the time of daily maximum temperature may require some weight restriction below their maximum take-off weights (MTOWs), with mean restrictions ranging from 0.5 to 4% of total aircraft payload and fuel capacity, by mid–late century. Mid-sized (such as the Boeing 737–800) and large-sized (such as the Boeing 777-300) aircraft are both affected (with the largest experiencing the greatest impacts from weight restrictions) and airports with short runways and high temperatures will see the largest impacts. These will impose a non-trivial cost on airlines and impact aviation operations around the world.

T. Zhou *et al.* (2018) concluded that the impact of both 1.5° and 2° global warming temperature increase would mean a significant increase in weight-restriction days at Beijing, Shanghai and Lhasa Airports. They also stated that on 20 June, 2017, as temperatures at Phoenix Airport, USA, reached 119°F, airlines cancelled more than 40 flights.

Zhao and Sushama (2020) analysed aircraft take-off performance at Canadian airports and concluded that weight-restriction days would increase significantly by the end of the century for some of the studied airfields, especially at higher elevations such as Calgary and Saskatoon.

Y. Zhou *et al.* (2018) researched take-off distance and climb rate at 30 major international airports and concluded that, from the historical period (1976–2005) take-off distance would increase in summer by between 0.95–6.5% and climb rate would decrease by 0.68–3.4% in the 2021–2050 period. They also concluded that the Boeing 737–800, as a case study aircraft, would require between 3.5 and 168.7m in additional take-off distance in future summers.

ATG Case study 6.3

The impacts of climate change on Greek airports

Gratton et al. (2020) researched the impact of climate change on take-off performance for both an Airbus A320 and DHC-8–400 aircraft at ten Greek airports. For airports with longer runways, a steady but unimportant increase in take-off distances was discovered. However, airports with shorter runways experienced a steady reduction in available payload. The most extreme example was at the relatively short (1,511m) runway at Chios Airport, where the required reduction in payload would be equivalent to 38 passengers with their luggage, for the period between the A320 entry into service in 1988 and 2017. This reduction in revenue-generating potential negatively impacts the financial viability of some routes.

Sea-level rise

According to Eurocontrol (2018), *sea-level rise* will occur in the longer term and *storm surges* will happen in the nearer term, which increase the risk of airport flooding for coastal airports especially (Padhra and Kurnaz, 2023). When most of the world's major coastal airports were

Climatology and air transport 179

constructed, sea-level rise was not a consideration (Griggs, 2020). Now, there may be economic costs to protect airports or potential loss of airport capacity, as well as ground access being vulnerable. There may also be delay and perturbation from runway inundation as well as airline network disruption. Analysis by the E.U. Joint Research Centre identified 96 European airports at risk of inundation from one metre of sea-level rise, requiring adaptation such as dike building.

Yesudian and Dawson (2021) assessed more than 14,000 airports and concluded that a 2°C rise in global temperatures would place 100 airports below their mean sea level by 2100, while 1,238 airports are in the Low Elevation Coastal Zone (where 10% of the world's population lives). Many airports were deemed to be at risk in Europe, North America and Oceania, but risks were highest in Southeast and East Asia, such as at Suvarnabhumi Airport in Bangkok (BKK), Shanghai Hongqiao (SHA) and Wenzhou Longwan (WNZ) in China. Airports in the Solomon Islands such as Sege (EGM) and Ramata (RBV) and Bosaso (BSA) in Somalia were also deemed to be at high risk. Puerto Jimenez in Costa Rica and Sangster International, Montego Bay, Jamaica, in the LAC region, as well as Dunedin in New Zealand and Cairns in Australia were included as at risk. Amsterdam (AMS) in the Netherlands would have been deemed at even higher risk but for its flood protection measures.

In the USA, an FAA report (cited in Griggs, 2020), listed 13 of the country's largest coastal airports having at least one runway with an elevation within the reach of moderate to high storm surge, such that future sea-level rise will pose an increasing risk. These airports were:

- San Francisco
- Oakland
- Honolulu
- New Orleans
- Tampa
- Miami
- Isla Grande, San Juan
- Fort Lauderdale
- Washington Ronald Reagan
- Philadelphia
- New York Newark and JFK

American Airlines (2022) listed 12 airports within their network at risk of climate change-related flooding (both coastal and river), including Miami (MIA), New York JFK and LGA, Los Angeles (LAX), London Heathrow (LHR) and Charlotte (CLT).

ATG Case study 6.4

ECAC region and flood risk

Eurocontrol (2021a) concluded that 178 of the 273 coastal and low-lying airports in the European Civil Aviation Conference (ECAC) region were forecast to be at risk of marine flooding up to 2090. A one-day closure of a large airport due to full or partial/severe flooding could impact around 800–900 movements and up to 100,000 passengers per day.

On 4 September, 2018, Typhoon Jebi heavily damaged Kansai International Airport (KIX) in Japan. At the peak of the storm, 5-metre-high waves spilled over the seawalls resulting in severe flooding. The airport was closed for three days, causing significant physical, operational and economic damages. The IPCC predicts sea-level rise beyond which was taken into account at the design stage for KIX (ICAO, 2019).

6.8.3 Disruption in the air

As the core of air transport is the movement of people and freight by air, then any change in atmospheric processes will have an impact on the air transport industry. An increase in weather, such as thunderstorms, could cause turnaround procedures on the ground to be slowed or halted, refuelling to be stopped and rerouting to occur to avoid the worst of the weather, negatively impacting fuel burn and flight durations.

Eurocontrol (2018) analysed storm impacts in the Maastricht upper area airspace, predicting increases in spring and summer storms through 2050. As Gratton *et al.* (2022) point out, this can force longer routes, steeper climbs/descents and less efficient airspeeds, impacting flight times, emissions and costs, as well as potential safety issues.

Németh, Švec and Kandráč (2018) studied the influence of climate change on selected aircraft types and discovered that changes in temperatures and flight altitudes led to longer flight times and increased emissions.

Jet streams

As explained in Section 6.5, the presence of jet streams can have a significant impact on aircraft, both in terms of flight routings/durations and safety. Williams (2016) showed that if CO_2 were to double, the subsequent strengthening of the jet stream winds across the Atlantic would likely cause a shortening of eastbound flight times and lengthening of westbound flights. These would not cancel each other out and the net impact would be an increased flight time of more than 90 seconds. Whilst this may not sound a lot, Padhra and Kurnaz (2023) explain that based on 300 round trips per day, the additional flight time would be more than 2,000 hours and would add $22 million to airline fuel costs.

Turbulence

Prosser *et al.* (2023) stated that clear air turbulence (CAT) is projected to intensify in response to future climate change. They found evidence of large increases in CAT around the midlatitudes at aircraft cruising altitudes, with an average increase over the North Atlantic of light-or-greater CAT of 17% between 1979 and 2020 and severe-or-greater CAT increasing by 55% in the same time frame. Similar increases were found over the Continental USA.

Storer, Williams and Joshi (2017) concluded that by the period 2050–2080, there could be large relative increases in CAT, especially in the midlatitudes, with some regions experiencing several hundred percent more turbulence, highlighting the need to improve operational CAT forecasts.

Lv *et al.* (2021) analysed the Eurasian (at 250hPa) upper-level jet stream in winter and concluded that the jet stream will become consistently stronger for the period 2015–2100 in midlatitude regions between 40° and 60°N and the corresponding vertical wind shear in the same area will tend to increase.

Lee, Williams and Frame (2019) explained that the subpolar jet stream over the North Atlantic at flight cruising altitudes has become 15% more sheared since satellites began observing it in the late 1970s.

Kim *et al.* (2023) highlight the threat atmospheric turbulence poses to aircraft at their cruising altitudes and the impact of climate change on turbulence. They examined future frequencies of moderate-or-greater-intensity turbulence generated from sources such as CAT, mountain-wave and near-cloud turbulence and showed that turbulence from all three sources is intensified, with higher occurrences globally in a changed climate compared to the historical period.

6.8.4 Disruption to passenger demand

According to Eurocontrol (2021b), there are many aspects that inform tourist choices, including social, environmental and economic factors. A common consideration is climate, which can influence both the choice of destination and the time of year to travel. Both choices are associated with a perceived favourability towards a certain climate and acceptable comfort level by the consumer.

Their research utilised the Tourism Climate Index (TCI) to evaluate changes in destination favourability across the ECAC region, using five climate variables: temperature, humidity, wind speed, precipitation and sunshine. They concluded that across the ECAC region, countries are projected to have an increasing length of time in which the climate would be "good" or "ideal" for general low-level tourist activity; this would generally be evident in the autumn months of September, October and November, although countries in the south of the region (such as Southern Spain and Portugal) would see a small decrease in destination favourability in the summer months of June, July and August. In these three months, Northern and Central ECAC countries (such as the UK and the Baltic States) could see an increased TCI. The analysis indicates that many northern and central European countries, including France, Germany and Hungary, could have a similar destination favourability as that of Mediterranean countries such as Spain, Cyprus and Greece during the summer months, on the 2050 timescale. However, for countries around the Mediterranean coastline, in Spring, the tourist season may be extended.

Pentelow and Scott (2011) discuss how the world's small island states are estimated to contribute less than 1% of global GHG emissions but are projected to be some of the most impacted by climate change. The Caribbean is arguably the most tourism dependant region in the world, where in some nations, it exceeds 50% of GDP (Chapter 11). Many airports are only marginally above sea level, for example Sangster International Airport in Jamaica has an altitude of approximately 1m, therefore rising sea levels could create massive problems for much economic activity. Monioudi *et al.* (2018) projected that in 2030, both Sangster and Norman Manley Airports in Jamaica would on average need to deal with 65 days of disruption of air operations per year. Airports such as St Maarten and Dominica have also been significantly impacted by severe weather events in recent years.

ATG Case study 6.5

Airports within the top 100 exposed to high take-off weight-restriction risk

Pek and Caldecott (2020) analysed the world's top 100 airports by passenger traffic and found that of these, 19 were already exposed to high take-off weight-restriction risk (Table 6.3) due to at least one of three factors – high maximum daily temperatures, high elevation or short runways. All these airports are expected to experience an increase in the number of days when take-off restrictions are required, as well as an increase in the weight of required restrictions.

Table 6.3 Airports in world's top 100 by passenger traffic, currently exposed to high take-off weight-restriction risk

Bogota El Dorado	Dubai International
Mexico City Benito Juarez	Delhi Indira Gandhi
Kunming Changshui	Xi'an Xianyang
Denver International	Doha Hamad
Salt Lake City	Charlotte Douglas
New York LaGuardia	Madrid Barajas
Bengaluru Kempegowda	Chongqing Jiangbei
Riyadh King Khalid	Jeddah King Abdulaziz
Phoenix Sky Harbor	Antalya
Las Vegas McCarran	

Source: adapted from Pek and Caldecott (2020)

Linking to the disruption on the ground and take-off restrictions, any impacts here could have an impact on airlines – by reducing available payload – meaning fewer passengers and/or cargo carrying ability and perhaps smaller aircraft, therefore less revenue. This illustrates the inter-connected nature of various climatic factors within the broad air transport industry.

The measures which the air transport industry can take to try to reduce its climate impacts will be assessed in Chapter 14.

Chapter review questions

6.1 What are the principles involved in global atmospheric circulation and how does the three-cell circulation model influence global weather patterns?
6.2 What are the main global air masses and what weather conditions do they bring?
6.3 Explain the different types of weather fronts and the impacts these can have on aircraft. What are the major fronts present in your own areas and what are their seasonal variations?
6.4 What are jet streams and why are they important to air transport operations? For your local weather region, can you find out the upper wind speeds today and predict how they will impact airline routings?
6.5 What are tropical cyclones and what are the regions of the world they are from? What hazards do they bring? From this, research a case study airport of your choice from ONE of these regions and establish how air transport operations have been impacted by a tropical cyclone.
6.6 Explain the factors involved in climate change and their impacts and research the IPCC website to establish current thinking as to required government and industry responses to limit these impacts.
6.7 Analyse the impacts climate change has on the air transport industry. For a case study airport near you or in your country, research the possible impacts climate change may have. What impacts may climate change have on air transport demand in your country?

ATG trivia

Hurricane hunters

We have seen that hurricanes can cause huge problems to the air transport industry, both in flight and on the ground, however there are aircrews known as "hurricane hunters" who actively seek to fly into hurricanes to gather weather data. Pilots of the US NOAA fly into some of the world's worst weather to help forecasters make more accurate predictions during a hurricane and help hurricane researchers achieve a better understanding of storm processes, improving their forecast models.

Based in Lakeland, Florida, USA, the missions can last eight–ten hours. Lockheed WP-3D Orion four-engine turboprop aircraft are flown into the hurricane "eye" between 8,000 and 10,000 feet above sea level (ASL), avoiding more turbulent air below and icing above. A Gulfstream IV-SP is also used, which has a cruising altitude of 45,000ft, providing a detailed picture of weather systems in the upper atmosphere surrounding developing hurricanes.

Their three aircraft are nicknamed Miss Piggy, Kermit and Gonzo after the Muppet characters(!) (NOAA, n.d.).

References

American Airlines (2022) *Sustainability report 2022*. Available at: https://s202.q4cdn.com/986123435/files/images/esg/aa-sustainability-report-2022.pdf.

American Meteorological Society (2012) *Equatorial trough – glossary of meteorology*. Available at: https://glossary.ametsoc.org/wiki/Equatorial_trough.

Arrhenius, S. (1896) 'On the influence of carbonic acid in the air upon the temperature on the ground', *Philosophical Magazine*, 4, pp. 237–276.

Barry, R. and Chorley, R. (2010) *Atmosphere, weather and climate*. 9th edn. Abingdon: Routledge.

Boehm, S. and Schumer, C. (2023) *10 big findings from the 2023 IPCC Report on Climate Change*, *World Resources Institute*. Available at: www.wri.org/insights/2023-ipcc-ar6-synthesis-report-climate-change-findings.

Bryant, E. (2005) *Natural hazards*. 2nd edn. Cambridge: Cambridge University Press.

Buis, A. (2020) *Milankovitch (orbital) cycles and their role in earth's climate, climate change: vital signs of the planet*. Available at: https://climate.nasa.gov/news/2948/milankovitch-orbital-cycles-and-their-role-in-earths-climate.

Burbidge, R. (2018) 'Adapting aviation to a changing climate: key priorities for action', *Journal of Air Transport Management*, 71, pp. 167–174. Available at: https://doi.org/10.1016/j.jairtraman.2018.04.004.

Climate.gov (2023) *Climate change: atmospheric carbon dioxide*. Available at: www.climate.gov/news-features/understanding-climate/climate-change-atmospheric-carbon-dioxide.

Coffel, E., Thompson, T. and Horton, R. (2017) 'The impacts of rising temperatures on aircraft take-off performance', *Climatic Change*, 144, pp. 381–388. Available at: https://doi.org/DOI 10.1007/s10584-017-2018-9.

Eurocontrol (2018) *European Aviation in 2040; Annex 2 Adapting Aviation to a Changing Climate*. Available at: www.eurocontrol.int/sites/default/files/publication/files/challenges-of-growth-annex-2-01102018.pdf.

Eurocontrol (2021a) *Climate Change Risks for European Aviation 2021. Annex 3: Impact of Sea Level Rise on European Airport Operations*. Available at: www.eurocontrol.int/sites/default/files/2021-09/eurocontrol-study-climate-change-risk-european-aviation-annex-3.pdf.

Eurocontrol (2021b) *Climate Change Risks for European Aviation 2021: Annex 4: Impact of Climate Change on Tourism Demand*. Available at: www.eurocontrol.int/sites/default/files/2021-09/eurocontrol-study-climate-change-risk-european-aviation-annex-4.pdf.

FAA (2023) *Aviation weather handbook*. Newcastle, Washington: Aviation Supplies and Academics Inc.

Frey, R.P. (2017) *Whether the weather: aviation meteorology from A to Z*. Norderstedt, Germany: Books on Demand GmbH.

Goodwin, S. (2014) *Climatology and its applications*. Delhi: The English Press.

Gratton, G. et al. (2020) 'The impacts of climate change on Greek airports', *Climatic Change*, 160(2), pp. 219–231. Available at: https://doi.org/10.1007/s10584-019-02634-z.

Gratton, G.B. et al. (2022) 'Reviewing the impacts of climate change on air transport operations', *The Aeronautical Journal*, 126(1295), pp. 209–221. Available at: https://doi.org/10.1017/aer.2021.109.

Griggs, G. (2020) 'Coastal airports and rising sea levels', *Journal of Coastal Research*, 36(5), pp. 1079–1092.

Hansen, J. et al. (2006) 'Global temperature change', *Proceedings of the National Academy of Sciences of the United States of America*, 103(39), pp. 14288–14293. Available at: https://doi.org/10.1073/pnas.0606291103.

Hardiman, J. and Joshi, G. (2022) *How a British Airways 747 once flew 825mph, Simple Flying*. Available at: https://simpleflying.com/british-airways-747-825-mph/.

ICAO (2019) *2019 environmental report. Aviation and environment*. Available at: www.icao.int/environmental-protection/Documents/ICAO-ENV-Report2019-F1-WEB%20(1).pdf.

IPCC (2018) *Global warming of 1.5 °C —*. Available at: www.ipcc.ch/sr15/.

IPCC (2023) *Climate Change 2023 – Synthesis Report: Summary for Policymakers*. Intergovernmental Panel on Climate Change (IPCC). Available at: www.ipcc.ch/report/ar6/syr/downloads/report/IPCC_AR6_SYR_SPM.pdf.

Kearns, S.K. (2021) *Fundamentals of international aviation*. 2nd edn. Abingdon: Routledge (Aviation Fundamentals).

Kim, S.-H. et al. (2023) 'Global response of upper-level aviation turbulence from various sources to climate change', *npj Climate and Atmospheric Science*, 6(1), pp. 1–12. Available at: https://doi.org/10.1038/s41612-023-00421-3.

Lankford, T.T. (2001) *Aviation weather handbook*. New York: McGraw Hill.

Lee, S.H., Williams, P.D. and Frame, T.H.A. (2019) 'Increased shear in the North Atlantic upper-level jet stream over the past four decades', *Nature*, 572(7771), pp. 639–642. Available at: https://doi.org/10.1038/s41586-019-1465-z.

Lockwood, J.G. (2012) 'Atmospheric processes', in J. Holden (ed.) *An introduction to physical geography and the environment*. 3rd edn. Harlow: Pearson, pp. 77–116.

Lv, Y. et al. (2021) 'Increased turbulence in the Eurasian upper-level jet stream in winter: past and future', *Earth and Space Science*, 8(2), p. e2020EA001556. Available at: https://doi.org/10.1029/2020EA001556.

Maksel, R. (2018) 'Why was the discovery of the jet stream mostly ignored?', *Smithsonian Institute*. Available at: www.smithsonianmag.com/air-space-magazine/as-next-may-unbelievablebuttrue-180968355/.

McCabe, K. (2022) 'Jet stream and stormy weather', *RMetS*. Available at: www.rmets.org/metmatters/jet-stream-and-stormy-weather.

McClatchey, J. (2012) 'Global climate and weather', in *An introduction to physical geography and the environment*. 3rd edn. Harlow: Pearson, pp. 117–156.

Met Office (2023) *Record breaking 2022 indicative of future UK climate*. Available at: www.metoffice.gov.uk/about-us/press-office/news/weather-and-climate/2023/record-breaking-2022-indicative-of-future-uk-climate.

Met Office (n.d.a) *Development of tropical cyclones*. Available at: www.metoffice.gov.uk/weather/learn-about/weather/types-of-weather/hurricanes/development.

Met Office (n.d.b) *Global circulation patterns*. Available at: www.metoffice.gov.uk/weather/learn-about/weather/atmosphere/global-circulation-patterns.

Met Office (n.d.c) *Intertropical Convergence Zone (ITCZ)*. Available at: www.metoffice.gov.uk/weather/learn-about/weather/atmosphere/intertropical-convergence-zone.

Met Office (n.d.d) *What is the jet stream?* Available at: www.metoffice.gov.uk/weather/learn-about/weather/types-of-weather/wind/what-is-the-jet-stream.

MIT (2021) *Why did the IPCC choose 2°C as the goal for limiting global warming?, MIT Climate Portal*. Available at: https://climate.mit.edu/ask-mit/why-did-ipcc-choose-2deg-c-goal-limiting-global-warming.

Monioudi, I.N. et al. (2018) 'Climate change impacts on critical international transportation assets of Caribbean Small Island Developing States (SIDS): the case of Jamaica and Saint Lucia', *Regional Environmental Change*, 18(8), pp. 2211–2225. Available at: https://doi.org/10.1007/s10113-018-1360-4.

NASA (2005) *What's the difference between weather and climate?*. Brian Dunbar. Available at: www.nasa.gov/mission_pages/noaa-n/climate/climate_weather.html.

NASA (2006) *Weather and climate*. Available at: www.nasa.gov/centers/langley/pdf/245893main_MeteorologyTeacherRes-Ch2.r3.pdf.

NASA (2023) *World of change: global temperatures*. Available at: https://earthobservatory.nasa.gov/world-of-change/global-temperatures.

NASA (n.d.a) *Basics of space flight – solar system exploration: NASA Science, NASA Solar System Exploration*. Available at: https://solarsystem.nasa.gov/basics/chapter2-1/.

NASA (n.d.b) *Graphic: carbon dioxide hits new high, climate change: vital signs of the planet*. Available at: https://climate.nasa.gov/climate_resources/7/graphic-carbon-dioxide-hits-new-high.

NASA (n.d.c) *Tropical cyclones, Earthdata*. Earth Science Data Systems. Available at: www.earthdata.nasa.gov/topics/human-dimensions/natural-hazards/tropical-cyclones.

Németh, H., Švec, M. and Kandráč, P. (2018) 'The influence of global climate change on the European aviation', *International Journal on Engineering Applications (IREA)*, 6, p. 179. Available at: https://doi.org/10.15866/irea.v6i6.16679.

NOAA (2023a) *Air masses*. Available at: www.noaa.gov/jetstream/synoptic/air-masses.

NOAA (2023b) *Global atmospheric circulations*. Available at: www.noaa.gov/jetstream/global/global-atmospheric-circulations.

NOAA (2023c) *Inter-tropical convergence zone*. Available at: www.noaa.gov/jetstream/tropical/convergence-zone.

NOAA (2023d) *The jet stream*. Available at: www.noaa.gov/jetstream/global/jet-stream.

NOAA (2023e) *Tropical cyclone introduction*. Available at: www.noaa.gov/jetstream/tropical/tropical-cyclone-introduction.

NOAA (n.d.) *NOAA hurricane hunters | Office of Marine and Aviation Operations*. Available at: www.omao.noaa.gov/omao/noaa-hurricane-hunters (Accessed: 18 October 2023).

Nordhaus, W.D. (1977) 'Strategies for the control of carbon dioxide', in *Cowles Foundation Discussion Papers*. Cowles Foundation for Research in Economics, Yale University (443).

Padhra, A. and Kurnaz, S. (2023) 'Aviation and climate change: becoming a climate-neutral industry', in *Challenges and opportunities for aviation stakeholders in a post-pandemic world*. IGI Global, pp. 84–108.

Pek, S. and Caldecott, B. (2020) *Physical climate-related risks facing airports: an assessment of the world's largest 100 airports*. University of Oxford Sustainable Finance Programme.

Pentelow, L. and Scott, D.J. (2011) 'Aviation's inclusion in international climate policy regimes: implications for the Caribbean tourism industry', *Journal of Air Transport Management*, 17(3), pp. 199–205. Available at: https://doi.org/10.1016/j.jairtraman.2010.12.010.

Prosser, M.C. et al. (2023) 'Evidence for large increases in clear-air turbulence over the past four decades', *Geophysical Research Letters*, 50(11), p. e2023GL103814. Available at: https://doi.org/10.1029/2023GL103814.

Quantick, H.R. (2001) *Climatology for airline pilots*. Oxford: Blackwell Science.

Royal Meteorological Society (n.d.) *Global atmospheric circulation*. Available at: www.rmets.org/metmatters/global-atmospheric-circulation.

Ryley, T., Baumeister, S. and Coulter, L. (2020) 'Climate change influences on aviation: a literature review', *Transport Policy*, 92, pp. 55–64. Available at: https://doi.org/10.1016/j.tranpol.2020.04.010.

SKYbrary (n.d.) *Jet stream*. Available at: https://skybrary.aero/articles/jet-stream.

Storer, L., Williams, P. and Joshi, M. (2017) 'Global response of clear-air turbulence to climate change', *Geophysical Research Letters*, 44. Available at: https://doi.org/10.1002/2017gl074618.

Tampa Airport (2017) *National weather service names TPA first StormReady Airport in Florida*. Available at: www.tampaairport.com/national-weather-service-names-tpa-first-stormready-airport-florida.

UCAR (2023) *Weather fronts | Center for Science Education*. Available at: https://scied.ucar.edu/learning-zone/how-weather-works/weather-fronts (Accessed: 28 September 2023).

UNFCCC (2023) *Key aspects of the Paris Agreement*. Available at: https://unfccc.int/most-requested/key-aspects-of-the-paris-agreement.

Wickson, M. (2015) *Meteorology for pilots*. Marlborough, Wiltshire: Airlife. Available at: www.perlego.com/book/3157975/meteorology-for-pilots-pdf?queryID=7c7fff055d06d98c38f49c9e9bf14cbd&index=prod_BOOKS&gridPosition=6.

Williams, P.D. (2016) 'Transatlantic flight times and climate change', *Environmental Research Letters*, 11(2), p. 024008. Available at: https://doi.org/10.1088/1748-9326/11/2/024008.

World Meteorological Organization (1958) *Observational characteristics of the jet stream, a survey of the literature*. Available at: https://library.wmo.int.

World Meteorological Organisation (2020) *Tropical Cyclones*. Available at: https://public.wmo.int/en/our-mandate/focus-areas/natural-hazards-and-disaster-risk-reduction/tropical-cyclones.

World Meteorological Organization (2023) *Classification of tropical cyclones*. Available at: https://wmo.int/content/classification-of-tropical-cyclones.

Yesudian, A.N. and Dawson, R.J. (2021) 'Global analysis of sea level rise risk to airports', *Climate Risk Management*, 31, p. 100266. Available at: https://doi.org/10.1016/j.crm.2020.100266.

Zhao, J. and Sushama, L. (2020) 'Aircraft takeoff performance in a changing climate for Canadian Airports', *Atmosphere*, 11(4), p. 418.

Zhou, T. et al. (2018) 'Impact of 1.5 °C and 2.0 °C global warming on aircraft takeoff performance in China', *Science Bulletin*, 63(11), pp. 700–707. Available at: https://doi.org/10.1016/j.scib.2018.03.018.

Zhou, Y. et al. (2018) 'Decreased takeoff performance of aircraft due to climate change', *Climatic Change*, 151(3), pp. 463–472. Available at: https://doi.org/10.1007/s10584-018-2335-7.

7 Landforms, airports and geophysical hazards

Chapter outcomes

At the end of this chapter, you will be able to:

- Describe the spatial scale and classification of landforms.
- Understand the importance landforms play in airport planning and design.
- Explain the different types of surfaces available for runways and their potential advantages and disadvantages.
- Understand the role runway slopes play in aircraft operations.
- Explain the earthquake hazards faced by airports and potential mitigation strategies.
- Understand the hazards faced by airports by tsunamis and the different response strategies available to reduce potential impacts.
- Analyse the impacts of volcanic eruptions and attempts to mitigate these by air transport operators.

7.1 Introduction

Whilst the purpose of air transport – flying from a to b – takes place within the atmosphere, what goes up must come down to land. Or sea. Thus, the physical environment – and landforms – play a huge role in a safe and efficient air transport industry.

According to Chorley, Schumm and Sugden (2019), *geomorphology* is the scientific study of the geometric features of the earth's surface. These features, their locations and any changes will have an impact on where and how aircraft are able to land and take off, as well as the necessary infrastructure to facilitate the movement of passengers and freight – the *airport*.

According to ATAG in Kearns (2021), there were 41,788 airports in the world, used for military, airline and general aviation, and of these, 3,883 airports support scheduled commercial airline flights. The world leader in number of airports is the USA, with 13,513 in 2022 (CIA, 2023). All of these will have differing geomorphological processes influencing landforms and differing challenges in the planning and construction of the airport infrastructure.

The occurrence of a threatening condition from a natural phenomenon in a defined space or time can be considered a natural hazard, especially where there is human presence. Geophysical (or geological) hazards originate from internal earth processes, such as earthquakes, volcanic activity and tsunamis – although tsunamis essentially become a hydrological hazard in their

DOI: 10.4324/9781003405351-9

188 *Fundamentals of Global Air Transport Geography*

manifestation (UNDRR, 2007). These have massive implications for airport resilience and design principles.

This chapter will explore how landforms influence airport design and construction, as well as the impact on the air transport industry of geophysical hazards. The air transport industry is not only influenced by landforms but is also *responsible* for landform change – such as building airports on reclaimed land (e.g. Kansai in Japan) and this aspect will be covered in Chapter 13, on the impact of air transport *on* the physical environment.

7.2   An overview of geomorphology and landforms

> **ATG Did you know?**
>
> **Some definitions**
>
> *Topography* is the elevation and relief of the earth's surface. It is measured by differences in elevation.
>
> *Landforms* are the topographic features of the earth's surface. These can be as small as sand dunes or as large as mountain ranges.
>
> *Geomorphology* is the study of earth's surface processes and landforms. The word geomorphology originates from the Greek language: *Geo* – meaning "earth"; *Morph* – its "shape"; *Ology* – "the study of".
>
> Geomorphology is the broader scientific discipline providing the theoretical framework, landforms are the specific features on the earth's surface and topography is a representation of the earth's surface configuration.

As a science, geomorphology lies between the study of the natural environment and geology – the study of the solid earth. It will interact with other natural environmental sciences, such as soil science (pedology), climatology and hydrology, as well as geological subdisciplines such as tectonics and sedimentology – all of which have an influence on the construction and operation of aerodromes.

7.2.1   Spatial scales

According to Harvey (2012), there are four main spatial scales on earth (Figure 7.1):

a) **Global/continental** – the major features of the earth's surface, such as continental land masses (approximately 30% of earth's surface) and ocean basins (70%).
b) **Regional** – intermediate forms, such as individual mountain and hill ranges and river basins.
c) **Local** – rivers, hillslopes, beaches, glaciers.
d) **Micro** – details of the surface itself, such as weathering.

Airport development will be influenced by each of these scales. In some parts of the earth, such as Nepal and Tibet in the Himalayas, landforms are almost solely mountain systems, limiting the ability to build airports and runways. Individual mountains/hills will also create these issues throughout the world and building runways near rivers or by the coast may be more of an engineering challenge due to the land characteristics or subject the airport to flood risk. Weathering processes and soil erosion can have a negative impact on airport operations and the strength and safety of airport pavement areas.

Landforms, airports and geophysical hazards 189

a) Continental land masses and oceans

b) Runway in a mountain range (Lukla, Nepal)

c) Airport with runway on a beach (Barra, Scotland)

d) Soil erosion

Figure 7.1 Spatial scale landforms
Source: author from Canva.com.

Each of these four scales has resulted in landform formation over differing temporal periods. For example, the structure of the Himalayas has a much longer geological past than soil erosion at a particular airport.

7.2.2 Landform classification

Landforms can be grouped by the dominant set of geomorphic processes responsible for their formation. Some of these are given here (South Carolina Geological Survey DNR, 2005):

- tectonic landforms
- extrusive igneous landforms
- intrusive igneous landforms
- fluvial landforms
- karst landforms
- aeolian landforms
- coastal landforms

An in-depth study of these landforms is beyond the scope of this book, however each of these has an impact on airports and will be assessed later in this chapter when discussing natural hazards, in Chapter 8 on hydrological hazards and in Chapter 13 when discussing the impacts air transport has had on these landforms.

7.3 Physical geography of airports

In the early days of the 20th century, airports as we know them today did not exist. Indeed, the requirements of the Wright Brothers in 1903 were essentially related to an open area free of obstacles, landforms such as sand dunes and mowed meadows with good drainage and located where flight could occur into the wind (see Chapter 5). The "runway" was a 60-foot launching rail which the aircraft rode down on a small wheeled dolly and was referred to by them as the "Grand Junction Railroad" (Smithsonian Institute, n.d.). In more modern times, a runway is "a defined rectangular area on a land aerodrome prepared for the landing and take-off of aircraft" (ICAO, 2022).

7.3.1   The early airports

Locations for the earliest airports were chosen for convenience and a requirement for flat land, such as a beach or farmers' field. They didn't even have food courts or duty-free shops! The status as the first official airport is widely debated and, in many ways, depends on the interpretation of the word "airport".

Although the Wright Brothers had their first flight in North Carolina, their first "airport" was back home in Dayton, Ohio at the Huffman Prairie Field. This was in essence a cow pasture and was where they improved the design of their flying machine and achieved subsequent flying milestones, such as the first controlled turn.

If it is said that to be an airport, there must be two key things occurring – getting aeroplanes into and out the sky safely and second, to transfer goods and/or people from one mode of transportation to another – then Huffman Prairie Field may not fit these criteria. One airport which calls itself the oldest continually operating airport in the world is College Park Airport, in Maryland, USA, which was established in 1909 by the US Army Signal Corps. Table 7.1 lists many of the earliest airports.

Many airport runways began as grass and the size and weight of the early aircraft facilitated this. Indeed, many airfields today, operating smaller general aviation aircraft, still utilise grass runways. In 1928, Ford Field in Michigan became the first airport with a concrete runway in the USA and this trend continued largely due to the increased weight of aircraft.

Whilst airport physical requirements such as wind direction and flat land have not changed, much else has, from grass to asphalt runways, jet aircraft, runway length increases, the opening of passenger terminals, right through to the complex *airport cities* witnessed in many parts of the world today. However, due to the international nature of air transport, there must be standards and regulations which are global in nature and requirements for airport site selection are also generally very consistent and many of these relate specifically to landforms.

Table 7.1 Example earliest airports

Airport	Year of opening
College Park, USA	1909
Hamburg, Germany	1911
Shoreham, UK	1911
Bucharest, Romania	1912
Bremen, Germany	1913
Rome Ciampino, Italy	1916
Amsterdam, Netherlands	1916
Paris-Le Bourget, France	1919
Sydney, Australia	1920

Source: author

7.4 International airport regulations

Safety is the main objective in air transport and a way of achieving this can be consistent, standardised facilities, services and procedures.

7.4.1 ICAO Annex 14 – aerodromes

ICAO first adopted standards and recommended practices (SARPs) for aerodromes in 1951, being designated as Annex 14 to the convention. Since then, it has been one of the most updated annexes as technologies evolve, such as the introduction of the large Airbus A380, necessitating changes.

> **ATG Did you know?**
>
> **When does an aerodrome become an airport?**
>
> According to ICAO (2018), an aerodrome is a "defined area on land or water intended to be used either wholly or in part for the arrival, departure and surface movement of aircraft". It is a broad, generic term for any location used for take-offs and landings. In this sense, it includes not only airports, but also smaller airfields and heliports. Airports are generally larger and have more extensive facilities such as terminal buildings, immigration, etc. to support airline services. In other words, all airports are aerodromes but not all aerodromes are airports!

Each ICAO member state is obliged to issue a national set of SARPs regulating the relevant points for their **international** airports. To ensure regulatory compliance, an aerodrome to be used for international flights must be granted an aerodrome certificate by its relevant Civil Aviation Authority and submit a manual detailing how it meets the ICAO standards. This is not applicable to **domestic** airports as these fall outside the scope of international regulations, however if the principles are sound, then it makes sense, where possible, to take note of these at domestic airports.

As explained in Chapter 2, the uniform acceptance of *Standards* is mandatory and any state which cannot comply must notify ICAO. *Recommendations* are considered desirable, especially considering safety and consistency. Pilots landing at aerodromes around the world should know that procedures, markings, signage, etc., are as consistent as possible wherever they fly to.

> **ATG Did you know?**
>
> **Airport master plans**
>
> Airport master plans are documents which present the short-term (one–five years), intermediate term (six–ten years) and long-term (ten+ years) development goals of an airport. These are typically evaluated every five–ten years, and operate best as phased implementation plans, although they should be reviewed at least annually and adjusted as appropriate to reflect

192   *Fundamentals of Global Air Transport Geography*

current conditions. They recognise the need for a systematic and consultative process for airport developments (ICAO, 2022).

According to Kazda and Caves (2015), the types of planning during airport master planning include:

- policy planning
- economic planning (such as market forecasts)
- physical planning (such as the airfield configuration and land uses)
- environmental planning (such as an environmental impact assessment and neighbourhood impacts)
- financial planning (such as financing options)

7.4.2   *Aerodrome reference codes*

These are two-part categorisations of aircraft types, which simplify the process as to whether an aircraft can operate at a particular aerodrome. The first part is a numeric code based on the reference field length for the aircraft and the second letter is based on a combination of aircraft wingspan and outer main gear wheel span (Table 7.2).

Table 7.2 Aerodrome reference codes

Code number	Aeroplane reference field length	Typical aeroplane
1	<800m	Piper PA31
2	800m but <1,200m	Bombardier Dash 8 Q300
3	1,200m but <1,800m	Airbus A220
4	1,800m and over	Boeing 737-800

Code letter	Wingspan	Typical aeroplane
A	<15m	Piper PA31
B	15m but <24m	De Havilland DHC-6
C	24m but <36m	Airbus A320 / Boeing 737-800
D	36m but <52m	Boeing 767
E	52m but <65m	Boeing 777 / 787 / Airbus A330
F	65m but <80m	Boeing 747-800 / A380-800

Source: author

So, for example, the largest aircraft – A380–800 – is a code 4F, requiring >1,800m in runway length (it needs around 3,000m) and has a wingspan of 79.8m, requiring appropriate runway/taxiway/stand widths.

ATG Case study 7.1

Boeing 777–8 and 777–9 (B777X) – folding wingtips

Boeing's classic B777 aircraft (introduced in the 1990s) – the B777-200 and B777-300 – have wingspans of 64.8m and are classified as category 4E aeroplanes. The new updated B777X series is expected to enter service in 2026 and to increase fuel efficiency, and reduce fuel costs and emissions. To achieve the desired performance improvements, the aircraft wingspan was increased to 71.8m, which would have placed it into category 4F, which had implications at a number of airports where the widths, especially at the terminal stands, were not adequate. Introducing *folding wingtips* reduced the aeroplane to a category 4E (wingspan of 64.8m) meaning it can operate in the same way as the current B777. When the aircraft is on the ground, the wingtips can be folded up and then extended in-flight to achieve the aerodynamic benefits.

Table 7.3 Top five airports in the world by land area

Airport	Land area in square kilometres
King Fahd International, Saudi Arabia	776.0
Denver International, USA	135.7
Dallas/Fort-Worth International, USA	69.5
Orlando International, USA	53.8
Washington Dulles International, USA	48.6

Source: adapted from Statista (2023)

7.5 Runways

The amount of land required by airports can be significant. Table 7.3 illustrates the airports with the largest land area in the world.

ATG Case study 7.2

King Fahd International Airport, Saudi Arabia (DMM)

The airport with the largest land area in the world is at the King Fahd International Airport in Dammam, Saudi Arabia, which was opened in 1999. It has more than ten times the land area of the third-place airport in Table 7.3. It has two runways each of 4,000m in length and the entire land area is greater than the Kingdom of Bahrain and approximately eight times bigger than Paris!

Four of the top five airports for land area are in the USA. It is here where the number of runways per airport is the highest in the world. Denver has six, Dallas/Fort Worth has seven (Figure 7.2) and both Orlando and Washington Dulles have four. It is runways which generally take up the most land area within an airport's boundary, with large international airports commonly having runways of more than 3,000 metres and indeed Denver has the longest commercial runway in the USA, with one of its six being 4,877m (16,000ft) in length. The other five each measure 3,658m (12,000ft). Table 7.4 illustrates some of the longest commercial runways in the world. This takes up significant space and therefore the *landforms* must be suitable to allow safe aircraft operations or altered accordingly.

Figure 7.2 Dallas Fort Worth International Airport.

Source: OpenAerialMap (2023)

Table 7.4 Example longest commercial runways in the world

Airport	Region	Length (m)
Qamdo Bamda	Tibet Autonomous Region, China	5,500
Shigatse Peace	Tibet Autonomous Region, China	5,000
Ulyanovsk Vostochny	Russia	5,000
Embraer Unidade Gaviao Peixoto	Brazil	4,967
Upington	South Africa	4,900
Denver International	Colorado, USA	4,877
Hamad International	Doha, Qatar	4,850

Source: author

Landforms, airports and geophysical hazards 195

ATG Did you know?

Qamdo Bamda Airport

The Guinness World Records (2023a) states that the longest civil runway in the world is in Tibet – at Qamdo Bamda (Changdu Bangda) – BPX – with a length of 5,500m. It is also one of the highest runways in the world at an elevation of 4,334m. A runway at this elevation requires a much longer length than at sea level due to the much thinner air degrading aircraft performance. Qamdo Bamda has also been described as "the loneliest airport in the world" due to it being 136km away from the nearest town.

7.5.1   Runway surfaces

Runways are often referred to as "tarmac" but are rarely built from this material. They are constructed from a variety of different materials, depending on the weight of the aircraft, geological and atmospheric conditions. Broadly, they are characterised in two ways:

- **Natural surfaces** (grass, dirt, gravel, ice, sand). Often referred to as *unpaved*.
- **Human-made surfaces** (asphalt, concrete). Often referred to as *pavements*.

Take-offs and landings on water by **seaplanes** are called *waterways*.

Whatever the surface chosen, ICAO states that the surface should comply with four basic requirements:

- Provide good ride capability during the aircraft's movement.
- Provide good braking action, even on a wet surface.
- Have good drainage capability.
- The bearing strength must be appropriate to the operation of the aircraft.

7.5.2   Unpaved runway surfaces

Grass/turf runways

Before the 1930s, with the introduction of heavier aircraft such as the DC-3, aircraft could operate from grass surfaces, with take-off distances of less than 600m (2,000ft). With the weight of modern aircraft (the MTOW of an A380 is 575,000kg), landing on grass may not be safe, except for the smallest aircraft. However, there are a huge number of small airfields with grass strips around the world, often found at general aviation aerodromes – they are easier to set up and cheaper to maintain. Of course, the geomorphological processes will play a role in this! Unpaved surfaces are typically non-homogenous in character and may contain different soils. The California Bearing Ratio (CBR) is often used to measure the surface's ability to resist shearing due to the weight of an aircraft.

Soil and climate are key factors for selection of grass types suitable for turf runway usage. The US FAA (2022) advises the following:

- Use grasses for airport turf with deep, matted root systems producing a dense, smooth surface cover with a minimum of top growth.

196 Fundamentals of Global Air Transport Geography

- Select long-lived grasses that are durable and recover quickly from dormancy or heavy-use conditions.
- Refrain from using short-lived, shallow-rooted weak sod species.
- If seeding, time the planting to provide at least six weeks of favourable growing conditions to allow proper root development.

ATG Did you know?

Grass runways

Not all grass runways are short. Triple Tree Aerodrome in South Carolina, USA, has a 2,134m (7,000ft) grass runway. This is comparable with the paved runway at London Luton Airport – 2,162m (7,093ft) – which regularly accommodates the Airbus A321 and has on occasions hosted Boeing 747 aircraft.

Grass runways are more easily damaged and bad weather can deteriorate the surface. The underlying geomorphology, soil types, presence of rivers and subsurface drainage characteristics can all have an impact on the usability of grass runways. Pilots will need to pay particular attention, as the surfaces are more dynamic. According to Kazda and Caves (2015), a grass runway should ideally be located on flat land, on a natural layer of gravel covered with approximately 20cm of topsoil and the surface covered with a grass carpet. Even here, it may be hard to provide

Figure 7.3 Gravel kit example

Source: airforcefe from Richmond (2009)

regularity of operations throughout the year due to weather conditions. Accelerate-stop distances are typically 20% longer than for paved runways (FAA, 2022).

Gravel runways

Gravel is also often seen at smaller aerodromes. This may be suitable for lighter aircraft, but they lack versatility. There are many aerodromes with gravel runways in different parts of the world, particularly in Canada and Alaska. Alaska Airlines began flying the B737-200 combi in 1981, with the aircraft becoming known as "mud hens" as they could be operated on muddy gravel runways and it became the workhorse of intra-Alaska flying until 2007 (Singh, 2020).

For some aircraft to land on gravel, modifications are required, or considerations built into the designs. In 1969, Boeing introduced an "unpaved strip kit" to make existing B737-100/200 aircraft ready for gravel surfaces containing, for example, abrasion-resistant paint to the underside of the wings and fuselage, vortex dissipators fitted to the engine nacelles (Figure 7.3) and a nose-gear deflector to keep gravel off the underbelly (Brady, 2021).

ATG Case study 7.3

Air Inuit, Northen Canada

Air Inuit, operating primarily in the Nunavik region in Northern Quebec, also operates the B737-200 on routes serving gravel runways. To people living in Northern Canada, the B737-200 provides an important means of connecting with other parts of Canada, however it is a very old aircraft, with environmental issues, including larger emissions than newer aircraft. Air Inuit has ordered Newer NG B737–800s, however these are much more problematic in operating to gravel runways, with a lack of retrofit capabilities possibly requiring airstrip upgrades.

Sand runways

Sand runways are located directly on the beach and not on paved surfaces. There are very few in the world and one of the most notable is at Barra Airport, in Scotland.

ATG Case study 7.4

Barra Airport, Outer Hebrides, Scotland

The runway at Barra Airport is one of the only runways in the world where the beach is the runway (Figure 7.4). A small population (just over 1,000) and a shortage of land means the island does not have a conventional runway. Arrivals and departures are carefully timed to avoid the high tides, where the runway is covered in water and the runways are set out in a triangle facing the terminal, the use depending on winds. Only small aircraft can operate here (such as the 19-seat DHC-6 Twin Otter from Loganair) and it is one of the few airports in the world where a delay code note of "tides" can be used!

198 Fundamentals of Global Air Transport Geography

Figure 7.4 Barra Airport, Scotland
Source: Canva.com

Ice runways

These are airstrips built on ice or compacted snow to allow aircraft to operate in polar regions where paved runways are not practical. Many of these are temporary during specific seasons when aircraft operations are feasible and for purposes such as scientific research missions in Antarctica. Without these runways, most heavy materials would have to be brought by ship, then ferried inland, adding significant time.

Blue ice runways are constructed on blue ice and are generally denser and more compacted than white snow or ice. These are often preferred for aircraft operations as they can provide better support for heavier aircraft and reduce the amount of snow cover required to be cleared. Due to the long distances from Antarctica to the nearest country (the southern tip of Argentina is approximately 700 miles away) and the need for research station resupply, there is a requirement for larger aircraft to be able to land there.

Aircraft reverse thrust is often used to assist stopping, due to the low efficiency of surface friction, and hence the length requirements are longer like-for-like than paved surfaces.

ATG Case study 7.5

Ice runways, Antarctica

There are several ice runways now operating in Antarctica and Figure 7.5a provides five examples. The Union Glacier Blue-Ice runway provides expedition support and tours to the

Landforms, airports and geophysical hazards 199

Antarctic interior. Phoenix airfield opened in 2017, replacing Pegasus white ice runway and serves McMurdo Station. It is a compacted snow runway, being compacted by heavy rollers (weighing more than 160,000 pounds) into dense layers, almost as hard as concrete (USAP, n.d.). Williams Field is also a US Antarctic programme airfield, serving McMurdo station and New Zealand's Scott Base. Wolf's Fang runway is a 3,000m private runway located in Queen Maud Land, and was the first runway built to support large passenger jets, with an Airbus A340 landing in 2021. In November 2023, a Boeing B787 of Norse Atlantic Airways landed at the 3,000m Troll Airfield in Queen Maud Land transporting scientists and research equipment (Figure 7.5b).

Figure 7.5a Antarctic runways
Source: author

Figure 7.5b First Boeing 787 lands in Antarctica
Source: Norse Atlantic Airways (2023) via Norwegian Polar Institute

7.5.3   *Paved runways*

Although the first runways were on turf, there was a need for more suitable stronger surfaces due to progressively heavier aircraft being produced. In 1916, one of the world's first paved runways was built in Clermont-Ferrand in France, as part of the World War I effort. One of the first paved runways (concrete) to be constructed for commercial purposes was at Ford Field in Dearborn, Michigan, USA, in 1928. Since then, apart from the smallest of airfields, paved runway surfaces (**pavements**) are used at airports around the world. Aircraft with a mass greater than approximately 2,000kg in European climatic conditions require the use of hard surfaces (Kazda and Caves, 2015). These can be categorised into two types:

- flexible
- rigid

The choice of pavement type depends on the planned aircraft characteristics, geological conditions and landforms. ICAO Annex 14 – aerodromes – provides airports with specific guidance on airport pavement strength. The surface needs to be carefully designed to provide friction for the aircraft and the surface texture can be achieved through treatments, grooving or the use of aggregates. The main differences between the two types lie in the materials, structural composition, flexibility and requirements for maintenance.

Flexible paved runways

These are normally made of multiple layers of materials, with **asphalt** being the common choice. Asphalt is a liquid or semi-solid type of petroleum made from aggregate materials held together with a bitumen binder. They are designed to essentially deform during aircraft load, distributing this load over a broader area and are often constructed consisting of a surface layer, base course and sub-base layer. They often require more frequent maintenance including periodic resurfacing and crack sealing. However, they often have lower initial costs than rigid pavements.

Rigid paved runways

These are made of **concrete**, which is both rigid and inflexible and do not deform significantly under heavy loads. They usually consist of continuous, monolithic slabs, which provides strength and stability. The binder is cement-based and the use of expansion joints helps to control and manage cracks. Maintenance is generally less frequent than asphalt, but can often be more expensive, when required for slab replacement or joint repairs. They often have higher initial costs than asphalt but have longer-term durability. They may also be more environmentally friendly, due to asphalt's petroleum-based binder.

Asphalt is more vulnerable to damage from aviation fuel, therefore it is more common to have concrete in areas where fuel spillage may occur, such as aircraft stands.

Both types are used around the world and some airports even use both at the same airport. For example, New York JFK has three concrete runways and one asphalt (Hardiman, 2021). The use of either of these depends on a variety of factors, such as:

- airport design criteria
- traffic loads
- soil conditions
- climate
- budget

The geomorphology and landforms will play a key role in the materials used during runway construction, as well as the level of costs involved to prepare and stabilise the land.

ATG Case study 7.6

"Soil like toothpaste": the new runway at Brisbane Airport, Australia

Work began on Brisbane's new 3,300m runway in 2012 and was completed in 2020, at a cost of $1.1 billion. The project also included 12km of taxiways, high intensity approach lighting, a four-lane underpass, 300 hectares of landscaping and a 1.7km revetment wall.

Site preparation took five years and was situated on a reclaimed portion of the Brisbane River delta (a landform created by sediment carried by a river depositing at its mouth), which was extremely soft, up to 30m deep and comparable to the consistency of toothpaste! The soft, waterlogged soil needed to be loaded with sand to provide stability, and the dredging and reclamation works in Moreton Bay were a monumental challenge – 11 million m³ of sand was dredged. There were also 330,000 wick drains installed – these operate in a similar way to a straw, allowing a quicker exit of the water from the soil, to speed up the settling process, which then took three years before the construction was finalised (Brisbane Airport, 2021).

Both the runway and taxiways were made up of three layers (Figure 7.6) and the runway was topped with asphalt whereas the taxiways were topped with concrete. Soil and trees removed during the process were used for later landscaping. The runway is long and strong enough to accommodate the A380 aircraft.

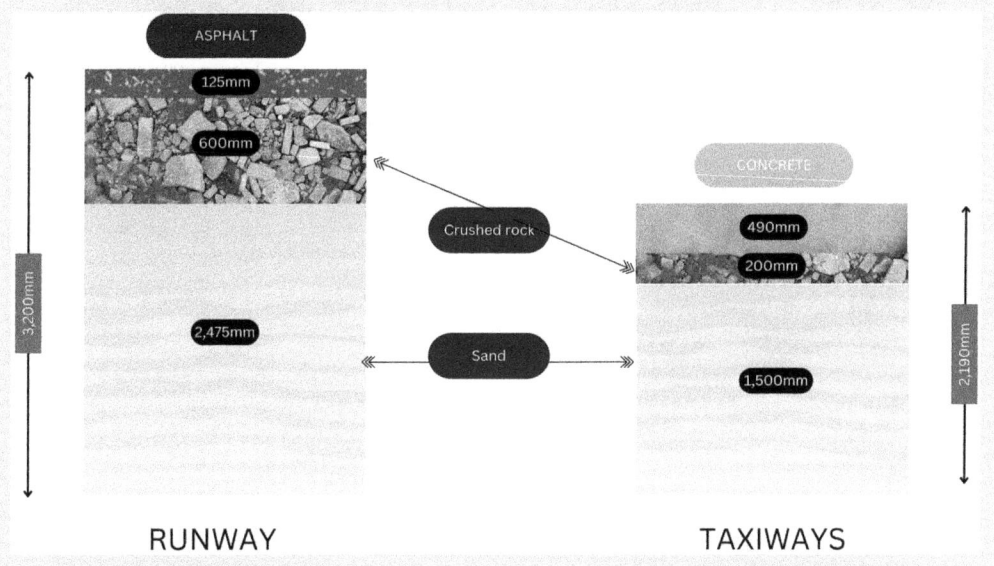

Figure 7.6 Pavement materials used for the new runway/taxiways at Brisbane Airport
Source: adapted from Brisbane Airport (2021)

7.5.4 Tabletop/mountain runways

Tabletop runways are located on the top of a plateau or hill with one or both ends adjacent to a steep precipice. In principle they are like mountain runways, however the difference lies in their construction and terrain.

202 *Fundamentals of Global Air Transport Geography*

Mountain runways are built on the slopes of mountains, adapting to the natural terrain, while tabletop runways are artificially built flat surfaces, raised above the surrounding landscape, to provide safe take-off and landing conditions in challenging terrains. Both these runway types can be challenging due to the surrounding terrain and they often leave little room for pilot error. Figure 7.7 illustrates two examples of tabletop and mountain runways.

Figure 7.7a Talcha Airport, Nepal
Source: Anuppanthi, 2016, Wikimedia Commons

Figure 7.7b Sedona Airport, Arizona
Source: Shane Torgerson, 2010, Wikimedia Commons

ATG Case study 7.7

Juancho E. Yrausquin Airport, Island of Saba, Caribbean Netherlands Antilles: the shortest commercial runway in the world

The "tabletop" runway at Saba is a mere 400m (1,312ft) long. It is near the ocean, not a mountain range, and is flanked on one side by high hills and cliffs which drop into the sea at both ends (Figure 7.8). It has been likened by some pilots as like landing "a bird on a postage stamp" (Peters, 2023). Aircraft using the airport include the small de Havilland DHC-6–300 Twin Otter.

Figure 7.8 Juancho E. Yrausquin Airport, Saba
Source: killians_red (2012) via Wikimedia Commons

7.5.5 Waterways

Before the increased development of land-based airports around the time of World War II, it was common for aircraft – or more correctly **seaplanes** – to land or take off on water. There are two main types of seaplanes – floatplanes and flying boats – the main distinction being how they are supported on water.

Floatplanes

Floatplanes have pontoons or floats attached to the fuselage or wings, which provide buoyancy and allow the aircraft to land or take off from water. These are limited by their inability to handle small wave heights. Examples include the de Havilland Canada DHC-2.

204 *Fundamentals of Global Air Transport Geography*

Flying boats

These have a boat-like hull which serves as the main structure to keep the aircraft afloat on the water. Examples include the Sikorsky S-42. These are generally larger and carry more payload.

ATG Did you know?

The St. Petersburg-Tampa Airboat Line

On 1 January, 1914, the world's first regularly scheduled heavier-than-air airline took off from the Municipal Pier in St-Petersburg and flew to the Hillsborough River in Tampa. The boat was known as Benoist Airboat Model XIV, no. 43, carried a solitary one passenger and took around 23 minutes (Figure 7.9).

Figure 7.9 Benoist first airline take-off

Source: airandspacemuseum.org (1914) via Wikimedia Commons

Seaplanes are still in use today for a few purposes (see the following list) and are invaluable in situations where "traditional" aircraft may face limitations, such as in regions with numerous lakes, islands or remote areas:

1. Passenger and cargo transport
2. Tourism and sightseeing
3. Search and rescue

4. Wildfire fighting
5. Aerial survey and surveillance
6. Military operations
7. Recreational flying

ATG Case study 7.8

Seaplanes in the Maldives

The Maldives, within the Indian Ocean, is 99% water with individual islands scattered across 90,000sq.km of which only 298sq.km is dry land. The highest island is only 3m above sea level. Passengers will generally fly into the regions' main airport in the capital Male (MLE) and then transfer, often by seaplane, to one of the 1,190 coral islands, grouped in a double chain of 27 atolls (High Commission of Maldives, n.d.).

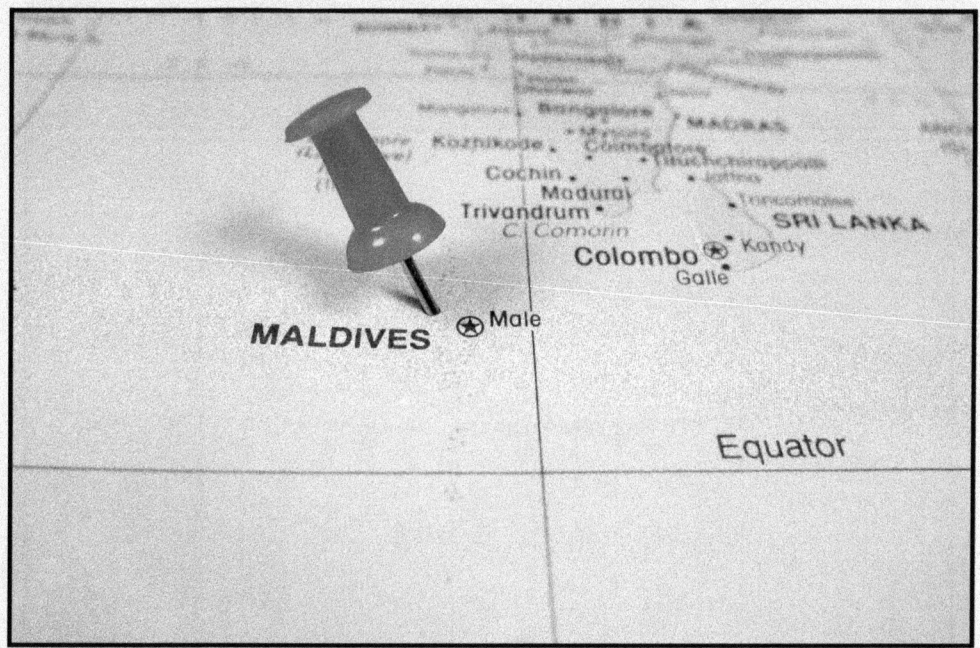

Figure 7.10a Map of the Maldives
Source: author adapted from Canva.com

Figure 7.10b Seaplanes in the Maldives
Source: author adapted from Canva.com

7.5.6 Runway slopes

Although one of the general considerations for siting runways is flat land, it is unlikely that the runway itself is *completely* flat, both in terms of its longitudinal and transverse profile.

Longitudinal slopes

Few runways are completely level throughout their longitudinal length. To make a runway completely flat would require considerable earthwork, but these can exist on artificial islands (Chapter 13). ICAO (2020) states that the allowable slope should be computed by dividing the difference between the maximum and minimum elevation along the runway centre line by the runway length and should not exceed:

- 1% where the runway code number is 3 or 4 (see Table 7.2).
- 2% where the code number is 1 or 2.

Along no portion of a runway should a longitudinal slope exceed:

- 1.25% where the code number is 4, except for the first and last quarter which should not exceed 0.8%.
- 1.5% where the code number is 3, except for the first and last quarter which should not exceed 0.8%.
- 2% where the runway code number is 1 or 2.

Any changes in longitudinal profile should be as gradual as practicable and abrupt changes or sudden reversals of slopes avoided. In special cases, a slope of up to 8% is permitted for one-way runways in airfields for agricultural activities, for special airfields for mountain rescue and also for commercial service in mountainous areas (Kazda and Caves, 2015). Beyond this there are some exceptional examples, such as Tenzing Hillary (Lukla) in Nepal, which has a slope of almost 12%.

ATG Case study 7.9

Courchevel Altiport, France

This altiport is located in the French Alps and is one of the highest (2,845m/9,334ft) paved runways in Europe at an extremely short 537m, and is also the steepest runway in the world at over 18.5% (Guinness World Records, 2023b). Pilots require extra certification and there is no go-around procedure due to the surrounding mountainous terrain (Figure 7.11).

208 *Fundamentals of Global Air Transport Geography*

Figure 7.11 Courchevel Altiport, France
Source: MartinPUTZ (2013) via Wikimedia Commons

Transverse slopes

ICAO (2020) recommends that to promote the most rapid drainage of water, the runway surface should, if practicable, be cambered except where a single crossfall from high to low in the direction of the wind, most frequently associated with rain, would ensure rapid drainage. The transverse slope should ideally be:

- 1.5% where the code letter is C, D, E or F.
- 2% where the code letter is A or B.

The transverse slope should not be any less than 1% except at runway or taxiway intersections where flatter slopes may be necessary and each side of the centre line should be symmetrical. Achieving these profiles may result in significant landform alterations.

7.6   Natural hazards

When a natural process poses a threat to human life or property, it is called a *natural hazard*. A hazard becomes a *natural disaster* when the event causes significant damage to life or property (Hyndman and Hyndman, 2017). Natural disasters affect millions of people every year and the number of people affected is not linear in any given year, as factors such as the type of hazard and its severity are clearly extremely important, however the location and concentration of human settlement also has huge importance.

As populations have grown, so has the need for transport infrastructure and hence a rise in the number and size of airports. Each year, airports and the air transport industry in general are impacted by natural hazards. Natural hazards can be grouped into five categories:

1. **Meteorological hazards** (cyclones, tornadoes, thunderstorms)
2. **Climatological hazards** (extreme temperatures)
3. **Geophysical hazards** (earthquakes, volcanoes and tsunamis)
4. **Hydrological hazards** (floods, landslides, mudslides)
5. **Biological hazards** (epidemics and pandemics)

Related to landforms, natural hazards could be grouped by:

1. **Geophysical hazards**
2. **Hydrological hazards**

Geophysical (geological) hazards are associated with the earth's geological features and processes caused by the earth's internal forces, such as tectonic plate movements, volcanic activity and seismic events, which can result in sudden and intense hazards with potentially widespread destruction. Earthquakes, volcanic eruptions and tsunamis are major hazards in this category.

Hydrological hazards are natural events or processes related to water that pose a threat to human life, property and the environment. These often operate on longer time scales, shaping landforms over extended periods. Landslides, mudslides, soil erosion and flooding are examples of hydrological hazards.

Meteorological and climatological hazards were covered in Chapters 5 and 6 and hydrological and biological hazards will be assessed in Chapter 8. The remainder of this chapter will focus on geophysical (geological) hazards and air transport.

7.7   Geophysical hazards – earthquakes

Earthquakes – the sudden shaking of the earth's surface caused by the movement of tectonic plates beneath the earth's crust – are natural phenomena that have shaped our landforms for millions of years. According to Davies, Korup and Clague (2021), 90% of all earthquakes occur along the subduction zones of the Pacific Ring of Fire, which form the longest continuous set of active plate boundaries on the planet, circumscribing the Pacific Ocean.

7.7.1   *Earthquake hazards and airports*

Earthquakes have the potential to cause widespread devastation, including significant loss of human life and infrastructure. Amongst the critical infrastructure vulnerable to seismic activity are airports, due to their role in transport connectivity and in any relief efforts. The impacts will depend on various factors, including the quake intensity, proximity of the epicentre to the airport and the airport's design and construction standards. A few impacts are possible:

1. **Structural damage.** This can be to any aspect of the airport, which may render the area unsuitable for use, either by significant direct damage or risks to structural integrity.
2. **Runway and taxiway integrity.** Ground displacement, cracks, subsidence or other damage may make these unsafe for use.
3. **Terminal and facilities.** Airport passenger and cargo terminals, ATC towers, fuel and other facilities may be impacted and areas rendered unusable.

4. **Utilities disruption.** Water, gas and power disruptions may all affect an airport's ability to continue operations.
5. **Communication and navigation systems.** ATC communications and on the ground operational communication damage may lead to delays and safety issues.
6. **Emergency response challenges.** The airport emergency personnel will have to be involved in rapid and coordinated responses, but their own infrastructure may be damaged and access may be impeded.

ATG Case study 7.10

Earthquake in Nepal, 2015

The epicentre of this magnitude 7.8 earthquake on Saturday 25 April, 2015, was around 85km (53mi) northwest of central Kathmandu and killed approximately 9,000 people (Reid, 2023) and also triggered an avalanche on Mount Everest, killing a number of climbers. It was the worst natural disaster to strike Nepal since 1934.

Shaking in Kathmandu was severe and weak structures suffered heavy damage. Numerous old buildings in cities and towns in the region collapsed, including a popular six-storey tourist hotel, killing at least 50 (Hyndman and Hyndman, 2017).

With many remote areas and villages, delivery of relief services was complicated due to their remoteness to the transportation network. Tribhuvan International Airport, serving Kathmandu, was closed immediately after the earthquake and sporadically afterwards due to aftershocks. It was also closed temporarily due to fear of runway damage and the airport facilities suffered damage. The airport runway was only designed for small and medium-size jets and when larger aeroplanes began flying in relief supplies, it began deteriorating under the weight of larger aircraft.

7.7.2   Airport earthquake mitigation

There are things that can be done by airports to try to prepare and to mitigate for some of the effects of earthquakes. These will vary due to locations, regulations and the airport's size and function, but the overall goal should be to create a comprehensive and resilient framework to minimise impacts.

1. **Building design.** Structural engineers can design buildings to withstand (or at least reduce the impacts of) seismic activity, using materials and construction techniques to absorb and dissipate energy.
2. **Seismic zoning and codes.** Airports should, where possible, adhere to national and international building codes which provide minimum building standards for construction in earthquake-prone areas and consider specific local factors, such as soil conditions.
3. **Emergency response planning.** Conducting regular training and emergency response drills, including all stakeholders, as well as maintaining appropriate medical supplies.
4. **Advanced warning systems.** For vulnerable airports, to consider seismic-warning systems, to at least provide more lead time for preparedness and response.

5. **Infrastructure resilience.** Building in redundancy in critical systems, such as power and communications, to help in remaining operational and providing back-up if required.

ATG Case study 7.11

Sabiha Gökçen Airport, Istanbul: seismic engineering

Upon opening in 2009, the 22 million passenger capacity terminal at Istanbul's Sabiha Gökçen International Airport was one of the world's most seismically isolated buildings. Principles of earthquake-resistant designs include structural rigidity and flexibility, and at Sabiha Gökçen this involved designing a base isolation system to ensure effective damping. The principle of isolation was used to separate the superstructure from the substructure. The terminal building sits on a platform that is essentially isolated from the ground below, with large spans and the terminal's 300 isolators enabled it to move in a controlled manner and absorb seismic energy in the event of an earthquake (Mara, 2010). It was designed to allow the airport to operate safely in an earthquake and facilitate relief operations. Testing suggested the building could withstand a force 7.5–8.0 earthquake on the Richter Scale.

7.8 Geophysical hazards – tsunamis

Tsunamis are water wave phenomena generated by the shock waves associated with seismic activity, explosive volcanism or submarine landslides. These shock waves can be transmitted through oceans, lakes or reservoirs (Bryant, 2005). Tsunamis can also be classed as hydrological hazards due to the sudden displacement of water, however according to NOAA (2023), the underlying cause is violent seafloor movement, most commonly from *geophysical activity*, such as earthquakes.

ATG Did you know?

Tsunami

The word tsunami (pronounced tsoo-nah'-mee) means **harbour wave** in Japanese (tsu = harbour / nami = wave). This refers to the fact that the waves rise highest where they are focused into bays or harbours.

7.8.1 *Tsunami hazards and airports*

Tsunamis can pose a significant threat to coastal areas, including coastal airports and these threats can be numerous:

1. **Flooding.** Airports in low-lying coastal areas could be subject to severe flooding, where runways, taxiways and other major infrastructure could be submerged by water.

212 *Fundamentals of Global Air Transport Geography*

2. **Infrastructure damage.** Airport buildings, control towers and other infrastructure may be damaged leading to operational repairs and delays.
3. **Debris.** The carriage of debris in the water can pose a significant threat to the safety of airport operations and saltwater may cause corrosion and equipment damage.
4. **Navigation systems.** Radars, ILS and other communications required for safe air transport may be damaged.
5. **Aircraft displacement.** Due to the huge power contained within tsunami waves, aircraft, as well as other airport vehicles, may be displaced and damaged by the storm surge.
6. **Emergency response.** Damage to emergency equipment may hinder possible relief efforts.

Hundreds of millions of people live in tsunami-prone zones. According to ICAO (2016), 25% of global aerodromes used for international air transport are located near the coast, at elevations lower than 10m and in Asia the figure is 45%, which is the most exposed region of all. There were 32 aerodromes identified as highly vulnerable based on their distance to tsunami-generating tectonic plates (Figure 7.12). Japan has fared the worst, for example Tokyo Haneda Airport has faced 29 tsunamis in the 20 years to 2023 (Hayward, 2023).

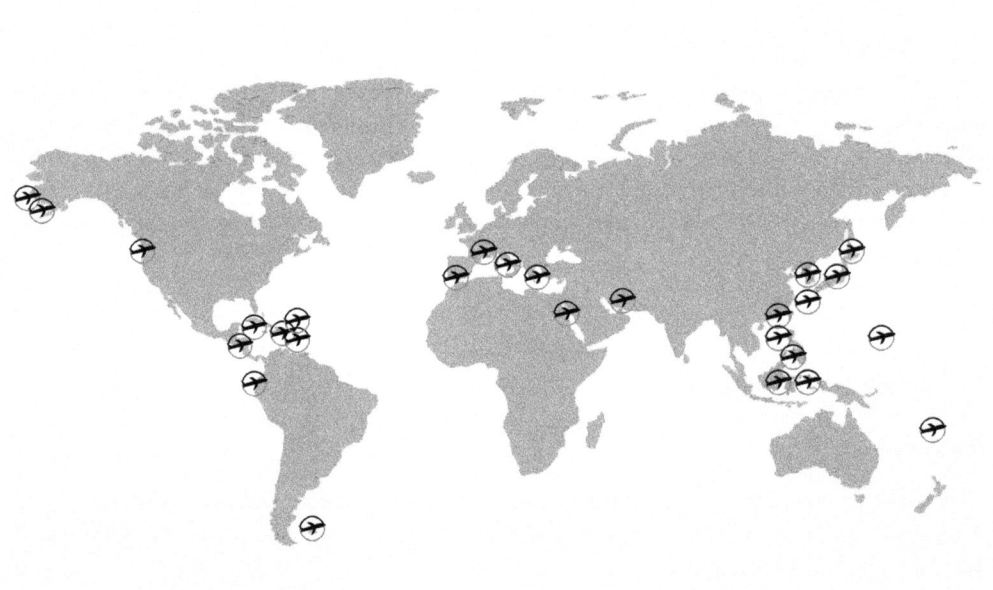

Figure 7.12 Tsunami-vulnerable airports
Source: adapted from ICAO (2016)

ATG Did you know?

Deadliest tsunami – Indian Ocean, 26 December, 2004

According to the NOAA (2023), the deadliest tsunami in history was from an earthquake of magnitude 9.1 off the coast of Sumatra in the Indian Ocean, generating a tsunami as high as 167 feet, causing flooding up to 3 miles inland. The impacts were observed in 17 countries, including Thailand and Indonesia, with approximately 230,000 deaths, 1.7 million people displaced and roughly $13 billion in 2017 economic dollar losses.

7.8.2 Airport tsunami mitigation

The most powerful tsunamis may result in widespread devastation, however there are a few measures airports could take to try to mitigate or at least reduce the impact of tsunamis and these would work best in combination:

1. **Land-use planning.** Limit airport development in high-risk tsunami zones.
2. **Elevated infrastructure.** Design critical infrastructure such as control towers and terminals to potentially be above inundation levels. This is clearly not possible in areas such as small islands, where the land barely rises above sea level.
3. **Early warning systems and evacuation plans.** Early detection of seismic activity to allow evacuation or movement of passengers and staff to higher ground and well communicated evacuation routes, as well as drills and training for airport staff.
4. **Protective barriers.** Erect physical barriers such as seawalls to reduce the tsunami force.
5. **Resilient design standards.** Apply relevant building regulations to reduce potential tsunami impacts.

ICAO Annex 14, Part 1 details aerodrome emergency planning for natural disasters and relevant SARPs, including for aerodromes close to water – establishment, testing and assessment of a predetermined response for specialist rescue services at intervals not exceeding two years.

ATG Case study 7.12

Tsunami near Sendai, Japan, 2011

On March 11, 2011, a magnitude 9.1 earthquake off the east coast of Japan generated a tsunami that caused huge devastation and was the strongest earthquake ever recorded in Japan. The tsunami reached 127 feet and travelled up to 5 miles inland, causing more than 18,000 deaths (mostly due to the tsunami) and approximately $243 billion in economic damage. This was felt as far away as California and Chile (NOAA, 2023).

214 *Fundamentals of Global Air Transport Geography*

Sendai Airport was the closest airport affected and the tsunami swept away cars, aeroplanes and ground aids as well as flooding buildings and radar systems (Figure 7.13). The mitigation measures in place prevented countless deaths and permitted the start of the evacuation of Sendai Airport 50 minutes before the tsunami hit and despite massive destruction and considerable financial losses, no human life was lost within the airport (ICAO, 2016). The airport was inundated with water, but flights resumed within three days and quickly became a base for disaster-relief efforts.

Figure 7.13 Sendai Airport, Japan, March 13, 2011
Source: U.S. Air Force photo/Staff Sgt Samuel Morse, 2011, Public Domain

7.9 Geophysical hazards – volcanoes

Volcanoes can provide some of the most spectacular displays of earth's dynamic processes. Of all the natural hazards, volcanoes can also be the most complex. There is a variety of volcanic forms and each event appears unique in the way it behaves and their physical and human consequences (Bryant, 2005). Volcanic hazards arise from the coupling of magmatic, hydrogeological and atmospheric processes and some can be tied to eruptions directly, whilst others may occur many years to millennia after a volcano has become extinct (Davies, Korup and Clague, 2021). "Tephra" is used as the catch-all term to describe erupted material regardless of size, but "ash" describes particles of less than 2mm in size (BGS, 2023).

Excluding the association of volcanoes with earthquakes and tsunamis, volcanic hazards can be grouped into six categories:

- lava flows
- ballistics and tephra clouds

- pyroclastic flows and base surges
- gases and acid rains
- lahars (mudflows)
- glacier bursts

7.9.1 Volcanic hazards and air transport

Volcanic eruptions pose a range of serious issues for both airports and aircraft in the air, impacting both safety, aircraft operations and travel infrastructure:

1. **Ash clouds.** Volcanic ash is composed of small pieces of pumice less than 2mm across, light enough to drift some distance on the wind. Ash can erupt into a column which can rise 6–20km in the air (Hyndman and Hyndman, 2017). Larger, heavier particles fall to the ground closer to the eruption whilst finer particles can be carried high into the atmosphere and may spread globally.

    Flying into a volcanic ash cloud should be avoided due to potential engine effects such as malfunction, flame-out and overheating, clogging of pitot-static probes, abrasion of aircraft components and skin surface corrosion. In addition, there is the matter of little or zero visibility for the pilots.
2. **Flight disruptions.** Volcanic eruptions can lead to airspace closures and disrupted flight schedules and cancellations in affected regions and perhaps precautionary closures in other adjacent areas. Prediction can be problematic due to possible wind speed and direction changes.
3. **Airport closures.** Airports may have to close due to ash fallout onto the terminal and airside areas.

ATG Case study 7.13

The 1991 Mount Pinatubo (Philippines) eruptions and their effects on aircraft operations

The explosive eruptions of Mount Pinatubo in June 1991 injected enormous clouds of volcanic ash and acid gases into the stratosphere to altitudes more than 100,000 feet. The largest ash cloud, from the 15 June eruption, was carried by upper-level winds to the west and circled the globe in 22 days. This covered a broad equatorial band from about 10°S. to 20°N. latitude and contaminated some of the world's busiest air traffic corridors. Sixteen damaging encounters were reported between jet aircraft and the drifting ash clouds. Three encounters occurred within 200 kilometres of the volcano, with ash clouds less than three hours old. Twelve encounters occurred over Southeast Asia at distances of 720 to 1,740 kilometres west from the volcano when the ash cloud was between 12 and 24 hours old. Encounters with the Pinatubo ash cloud caused in-flight loss of power to one engine on each of two different aircraft. A total of ten engines were damaged and replaced, including all four engines on a single jumbo jet.

Ashfall in the Philippines damaged aircraft on the ground and caused seven airports to close (Figure 7.14). Restoration of airport operations presented unique challenges, which were successfully met by officials at Manila International Airport (Casadevall, Reyes and Schneider, 1999).

Figure 7.14a Damage from volcanic ashfall at Clark Air Force Base, Philippines
Source: U.S. Geological Survey, 2016, Public Domain

Figure 7.14b A World Airways DC10 sitting on its tail due to the weight of wet volcanic ash
Source: U.S. Geological Survey, 2016, Public Domain

7.9.2 Volcanic mitigation

Attempting to mitigate the effects of volcanic activity is essential for air travel safety. Some key measures the industry has taken include:

1. **Volcanic ash advisory centres (VAACs).** These centres monitor volcanic ash clouds to provide timely information to the aviation industry, via satellite data, ground-based observations, aircraft in-flight and computer modelling.
2. **Airspace closures and redirection.** This stops aircraft flying into areas of potential eruptions and ash.
3. **Volcanic ash detection systems.** Aircraft are often equipped with on-board weather radar and detection systems with real-time information to help pilots identify and navigate around ash clouds. However, small particles in ash clouds may not always be detected.

ATG Case study 7.14

Eruptions of Eyjafjallajökull Volcano, Iceland, April 2010

These eruptions in Iceland were a series of volcanic activities which had a massive impact on the European air transport industry. The plumes released on 14 April covered almost the whole of Northern Europe (Figure 7.15), causing more than 20 countries to close their airspace resulting in essentially a six-day flight ban. Plumes reached 9km high and debris spread through most of Europe, with around 100,000 flights being cancelled and 10 million passenger journeys disrupted.

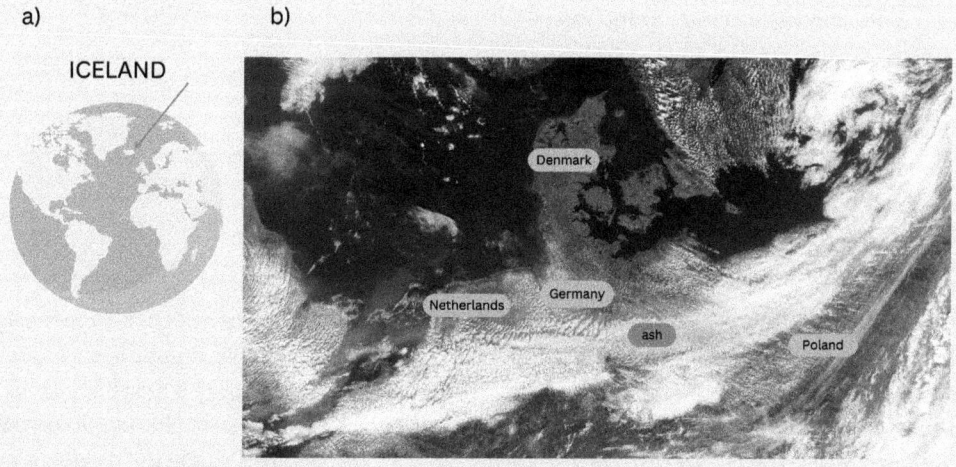

Figure 7.15a Iceland location
Source: author

Figure 7.15b Location of ash cloud on April 16, 2010
Source: NASA (2010) – Jeff Schmaltz

Chapter review questions

7.1 What are the different spatial scales related to landforms and how are landforms classified?
7.2 What are the key ICAO SARPs related to runway design? From the relevant sections in Annex 14, apply some of these key SARPs to an airport of your choice and assess their importance.
7.3 What are the different types of unpaved runway surfaces available for use? For any ONE of these, research an airport which uses this type and analyse the operational benefits and restrictions these can bring.
7.4 Why do airports sometimes use different materials for paved surfaces? Research your nearest major airport – can you find out which materials are used for both the runway and taxiway areas? Why?
7.5 Explain the regulations related to allowable runway slopes. Why might designs outside these limits sometimes be necessary and what could the possible safety risks be?
7.6 What are the potential impacts to airports of earthquakes? What are the potential mitigation strategies to reduce these impacts? Research any ONE case study where an airport has been affected by an earthquake – what did the airport do/should have done to mitigate some of the impacts?
7.7 What are the possible effects of tsunamis on airport infrastructure? Analyse the possible benefits of mitigation strategies.
7.8 Explain the different types of hazards faced by both airports and aircraft following volcanic eruptions. For any major volcanic eruption, research the impacts on the air transport industry and assess whether improvements have been made to reduce possible impacts of future eruptions.

ATG trivia

The Galunggung Glider

On 24 June, 1982, a British Airways Boeing 747-200 – BA009 – flying at FL370 from Kuala Lumpur to Perth at night encountered an ash cloud from a volcanic eruption on Mount Galunggung, just south of Jakarta. The pilots were unaware of this, as it did not appear on the weather radar. All four engines quickly failed and a 25,000-feet loss of altitude occurred, until four successful engine restarts were achieved at FL120, before a successful diversion to Jakarta on three engines, after one had to be shut down.

At the time, the engineless portion of the flight (around 13 minutes) entered the Guinness Book of Records as the longest glide in a non-purpose-built aircraft. In a classic example of understating events, Captain Eric Moody announced "Ladies and gentlemen, this is your captain speaking. We have a small problem. All four engines have stopped. We are doing our damnedest to get them going again. I trust you are not in too much distress" (Creedy, 2022).

References

airandspacemuseum.org (1914) *Benoist Type XIV first airline takeoff*. Available at: https://commons.wikimedia.org/wiki/File:Benoist_Type_XIV_first_airline_takeoff.jpg.

airforcefe from Richmond (2009) *737 gravel kit*. Available at: https://commons.wikimedia.org/wiki/File:737_Gravel_Kit_%284082334066%29.jpg.

Anuppanthi (2016) *Talcha airport in Nepal – the gateway to Rara Lake covered in heavy snow*. Available at: https://commons.wikimedia.org/wiki/File:Talcha_Airport.jpg.

BGS (2023) *Volcanic hazards*. Available at: www.bgs.ac.uk/discovering-geology/earth-hazards/volcanoes/volcanic-hazards/.

Brady, C. (2021) *Unpaved strip kit, the Boeing 737 technical site*. Available at: www.b737.org.uk/unpavedstripkit.htm.

Brisbane Airport (2021) *Brisbane's new runway, Brisbane Airport*. Available at: www.bne.com.au/corporate/projects/future-bne-projects/completed-projects/brisbanes-new-runway.

Bryant, E. (2005) *Natural hazards*. 2nd edn. Cambridge: Cambridge University Press.

Casadevall, T.J., Reyes, P.J.D. and Schneider, D.J. (1999) *The 1991 Pinatubo eruptions and their effects on aircraft operations*. USGS. Available at: https://pubs.usgs.gov/pinatubo/casa/.

Chorley, R.J., Schumm, S.A. and Sugden, D.E. (2019) *Geomorphology*. 1st edn. Routledge.

CIA (2023) *Airports*. Available at: www.cia.gov/the-world-factbook/field/airports/country-comparison/.

Creedy, S. (2022) 'Captain Moody's wisdom: "the aeroplane might break any minute" | Flight Safety Australia', 4 March. Available at: www.flightsafetyaustralia.com/2022/03/captain-moodys-wisdom-the-aeroplane-might-break-any-minute/.

Davies, T.R., Korup, O. and Clague, J.J. (2021) *Geomorphology and natural hazards: understanding landscape change for disaster mitigation*. American Geophysical Union.

FAA (2022) *Advisory Circular: Airport Design; 150/5300–13B*. Available at: www.faa.gov/documentLibrary/media/Advisory_Circular/draft-150-5300-13B-Airport-Design-chg1-ind-red.pdf.

Guinness World Records (2023a) *Longest runway*. Available at: www.guinnessworldrecords.com/world-records/63151-longest-runway.

Guinness World Records (2023b) *Steepest runway at an international airport*. Available at: www.guinnessworldrecords.com/world-records/100267-steepest-runway-at-an-international-airport (Accessed: 25 November 2023).

Hardiman, J. (2021) *What are the differences between concrete and asphalt runways?*, *Simple Flying*. Available at: https://simpleflying.com/concrete-asphalt-runway-differences/.

Harvey, A. (2012) *Introducing geomorphology. A guide to landforms and processes*. Edinburgh: Dunedin Academic Press.

Hayward, J. (2023) *How do tsunamis impact aviation?*, *Simple Flying*. Available at: https://simpleflying.com/how-tsunamis-impact-aviation/.

High Commission of Maldives (n.d.) 'About Maldives'. Available at: https://maldives.org.my/about-maldives.

Hyndman, D. and Hyndman, D. (2017) *Natural hazards and disasters*. 5th edn. Boston, MA: Cengage Learning.

ICAO (2016) 'When the sea affects the skies – tsunami risks in aviation as the world celebrates Tsunami Day', *Uniting Aviation*, 5 November. Available at: https://unitingaviation.com/amp/news/general-interest/when-the-sea-affects-the-air-tsunami-risks-in-aviaition/.

ICAO (2018) *Annex 14 to the Convention on International Civil Aviation, Volume 1: Aerodromes*. 8th edn. ICAO. Available at: https://elibrary.icao.int/reader/274803/&returnUrl%3DaHR0cHM6Ly9lbGlicmFyeeS5pY2FvLmludC9wcm9kdWN0LzI3NDgwMw%3D%3D?productType=ebook&themeName=Blue-Theme.

ICAO (2020) *Doc 9157, Aerodrome Design Manual Part 1 – Runways*. 4th edn. ICAO.

ICAO (2022) *Annex 14 — Aerodrome. Volume 1 Aerodrome Design and Operations*. 9th edn. Available at: https://elibrary.icao.int/reader/274803/&returnUrl%3DaHR0cHM6Ly9lbGlicmFyeeS5pY2FvLmludC9eS1saWJyYXJ5?productType=eBook&themeName=Blue-Theme.

Kazda, A. and Caves, R.E. (2015) *Airport design and operation*. 3rd edn. Bingley, UK: Emerald.

Kearns, S.K. (2021) *Fundamentals of international aviation*. 2nd edn. Abingdon: Routledge (Aviation Fundamentals).

killians_red (2012) *English: Saba Runway March 2012*. Available at: https://commons.wikimedia.org/wiki/File:Saba_Runway_march_2012.jpg.

Mara, F. (2010) 'Sabiha Gökçen Airport, Istanbul: seismic engineering', *The Architects' Journal*, 4 March. Available at: www.architectsjournal.co.uk/specification/sabiha-gokcen-airport-istanbul-seismic-engineering.

MartinPUTZ (2013) *English: the airport of Courchevel, called 'Altiport'*. Available at: https://commons.wikimedia.org/wiki/File:Altiport_Courchevel2.jpg.

NASA (2010) *Eruption of Eyjafjallajökull Volcano, Iceland*. NASA Earth Observatory. Available at: https://earthobservatory.nasa.gov/images/43684/eruption-of-eyjafjallajakull-volcano-iceland.

NOAA (2023) *Tsunamis*. Available at: www.noaa.gov/jetstream/tsunamis.

Norse Atlantic Airways (2023) *Media library – Fly Norse Corporate*. Available at: https://corporate.flynorse.com/en-gb/newsroom/media-library/.

OpenAerialMap (2023) *OpenAerialMap, CC-BY 4.0*. Available at: https://browser.openaerialmap.org/.

Peters, L. (2023) *Welcome to Saba: landing on the world's shortest commercial runway, Simple Flying*. Available at: https://simpleflying.com/welcome-to-saba-landing-on-the-worlds-shortest-commercial-runway/.

Reid, K. (2023) 'Nepal earthquakes: facts, FAQs, and how to help', *World Vision*, 4 November. Available at: www.worldvision.org/disaster-relief-news-stories/2015-nepal-earthquake-facts.

Shane Torgerson (2010) *Sedona Airport*. Available at: https://commons.wikimedia.org/wiki/File:SedonaAirport.JPG (Accessed: 20 November 2023).

Singh, S. (2020) *How some jet aircraft are able to land on gravel runways, Simple Flying*. Available at: https://simpleflying.com/jet-aircraft-gravel-runways/.

Smithsonian Institute (n.d.) *1903 Wright Flyer on Launching Rail*. Available at: https://airandspace.si.edu/multimedia-gallery/5820hjpg.

South Carolina Geological Survey DNR (2005) *Topography, landforms and geomorphology*. Available at: www.dnr.sc.gov/geology/pdfs/education/Landforms.pdf.

Statista (2023) *World's largest airports by land area 2022*. Available at: www.statista.com/statistics/485897/size-of-the-ten-largest-airports-worldwide-in-hectares/.

UNDRR (2007) *Hazard*. Available at: www.undrr.org/terminology/hazard.

U.S. Air Force photo/Staff Sgt Samuel Morse (2011) *Sendai Airport, March 13, 2011*. Available at: https://commons.wikimedia.org/wiki/File:SendaiAirportMarch16.jpg.

U.S. Geological Survey (2016) *Remembering Mount Pinatubo 25 Years Ago: mitigating a crisis*. Available at: www.usgs.gov/news/featured-story/remembering-mount-pinatubo-25-years-ago-mitigating-a-crisis.

USAP (n.d.) *United States Antarctic Program (USAP) Aircraft Landing Areas | NSF - National Science Foundation*. Available at: www.nsf.gov/geo/opp/support/landstrp.jsp.

8 Biogeography, hydrology and air transport

Chapter outcomes

At the end of this chapter, you will be able to:

- Explain the importance of biogeography to air transport.
- Assess the different types of wildlife management strategies employed at airports.
- Understand the types of wildlife hazards encountered at airports and potential mitigation strategies.
- Understand the importance of airports in preserving biodiversity.
- Analyse the impact of biological hazards such as COVID-19 to the air transport industry.
- Analyse the three key areas in which hydrology impacts the airport sector.
- Assess the risk of coastal and river flooding to airports.
- Examine the risk of aircraft de-icing as a potential airport hazard

8.1 Introduction

Chapter 7 discussed the importance of landforms to airport design and operations. It is not just the form of the land which impacts the air transport industry – it is the plants, animals and other organisms which live there. This, and their geographical location, plays a major role in where airports can be built and in any development and expansion plans.

Wildlife and habitat management is a crucial aspect of airport environmental planning and is often influenced by international, national and local policies. Their implementation will have impacts on species which inhabit those localities, to airport operations and to the safety of aircraft, because of possible bird and other wildlife strikes.

Some of the smallest of organisms can have the biggest impact and this was the case in 2020 when COVID-19 decimated the air transport industry, causing it to suffer its worst year in history. The air transport industry is not only impacted economically by these organisms and wider pandemics, but also has a role to play in restricting potential global transmission.

Water is essential for various aspects of airport operations, with uses such as passenger consumption, landscaping, firefighting, cleaning, maintenance and day-to-day operations. The air

DOI: 10.4324/9781003405351-10

transport industry can also have an impact on local water quality and airports require water management plans to limit the risk of operations having a detrimental impact on surrounding water sources. In addition, the risk of flooding is present in different global locations, especially at low-lying airports and various defences need to be implemented to reduce hazard risk – increasingly important in our warming planet.

The areas of biogeography and hydrology are therefore crucial to air transport and these two key aspects will be discussed in this chapter.

8.2 Biogeography and air transport

Biogeography describes the geography of the biological world. It involves the study of the distribution and patterns of life on earth and the underlying processes that result in these patterns (Holden, 2011). To understand, forecast and manage changes in biogeographical patterns, we need to examine the processes that create these distributions.

The *biosphere* is the surface of the earth, together with those parts above it and below that maintain life (Thomas, 2012). Many *ecosystems* are contained within the biosphere – these are geographical areas where plants, animals and other organisms, along with weather and landscapes, work together to form a "bubble of life" (National Geographic, n.d.).

ATG Did you know?

Biogeography and ecology

The links between biogeography and ecology are many. Both biogeography and ecology seek to understand the processes that determine natural patterns, but at different temporal and spatial scales. Ecology is the study of the interactions amongst organisms, whereas biogeography tends to focus more on distributions in time and space.

We are interested not only in the patterns of life on earth, but also how humans influence the biosphere through deliberate and inadvertent actions. The purpose of this chapter is not to analyse biogeography in detail, but rather to analyse the aspects of the air transport industry which are influenced by biogeography and consequently have a direct and indirect impact on it.

8.2.1 Habitat modification and fragmentation

With infrastructure developments across the world, landscapes and their inhabitants are increasingly being impacted. Habitat modification involves altering natural landscapes to accommodate human needs and the removal of vegetation and topographical alterations can have large implications for local ecosystems. Airports can account for a huge amount of land use, with runway(s), taxiways, passenger, cargo and general aviation terminals and associated infrastructure, and the need for sustainable planning and mitigation is vital.

8.3 Wildlife management at airports

Managing wildlife at airports is vital for several reasons, relating both to the impacts on wildlife itself and to air transport operations:

1. Aircraft safety
2. Minimising delays and disruptions
3. Protecting airport infrastructure
4. Protecting wildlife

The management of wildlife on or near airports is not usually an easily solved problem. This can be as diverse as habitat manipulation, to the use of predators to control wildlife. Many animals of all species can be affected by the construction of airports and their expansion, and this will be specific to the locale. Safety-related issues arise from both the local wildlife and migratory wildlife passing through.

8.3.1 Non-avian wildlife hazards

One reason for wildlife management at airports is to enhance aviation safety. Birds and other wildlife pose significant threats to aircraft during take-off and landing, causing potential damage to engines, wings, nose cone and other key components, which can compromise the structural integrity of the aircraft. According to the US FAA (2022), around 97% of safety issues involving wildlife were caused by birds, with mammals about 3% and reptiles less than 1%.

Due to the relative size of mammals (other than bats), any aircraft impact has the potential to be significant, especially with mammals such as bears or deer. These are very location-specific and one option for control is exclusion via perimeter fencing, although this is not universal at all worldwide aerodromes. According to SKYbrary (n.d.b), recorded strikes are greatest in North America and Europe, with over 40 deer strikes per year occurring regularly in North America.

ATG Case study 8.1

Aircraft accident at Kimberley, South Africa, 2010

On July 16, 2010, a South African Express Airways Bombardier DHC 8–300 collided with a wild aardvark shortly after touchdown on runway 20 at Kimberley aerodrome, at night. The impact was on the nose landing gear which caused the nose gear to collapse backwards, but no one was injured. The investigation found that the aerodrome fence had been constructed without a proper foundation, allowing animals to dig holes and trenches underneath the fence to gain access to the aerodrome property. The open areas between the runway and taxiways were found to consist mainly of savanna-type grassland (Figure 8.1), approximately 0.5m in height, which acted as good camouflage for several species of animals (South African CAA, 2010).

Figure 8.1 Kimberley Airport, South Africa
Source: Google Earth (n.d.)

8.3.2 Bird strikes

A bird strike is a collision between a bird and an aircraft, when it is in flight or on a take-off/landing roll. These can be extremely serious and pose a safety threat for passengers and crew, especially on take-off with possible maximum engine power and larger angle of ascent. Figure 8.2 illustrates the parts of an aeroplane most affected in bird strikes.

According to the FAA (2023), in 2022 in the USA alone, 17,190 wildlife strikes were reported and the loss to the US civil aviation industry was estimated to be 67,848 hours of aircraft downtime and $385 million in direct and other monetary losses.

For the period 1990–2022 in the USA:

- 272,016 wildlife strikes with civil aircraft were reported.
- Airports recorded significant increases in reported strikes per 100,000 movements.
- 54% of bird strikes occurred between July and October and around 70% were at or below 500 feet above ground level (AGL).
- 5% had a negative effect on flight, such as precautionary/emergency landings.
- For the 33 species of birds identified as being struck by civil aircraft, there was a strong correlation between body mass and the likelihood of the strike causing damage to aircraft – for every 100g increase in body mass, there was a 1.28% increase in the likelihood of damage.
- 81 strikes resulted in a destroyed aircraft.

Biogeography, hydrology and air transport 225

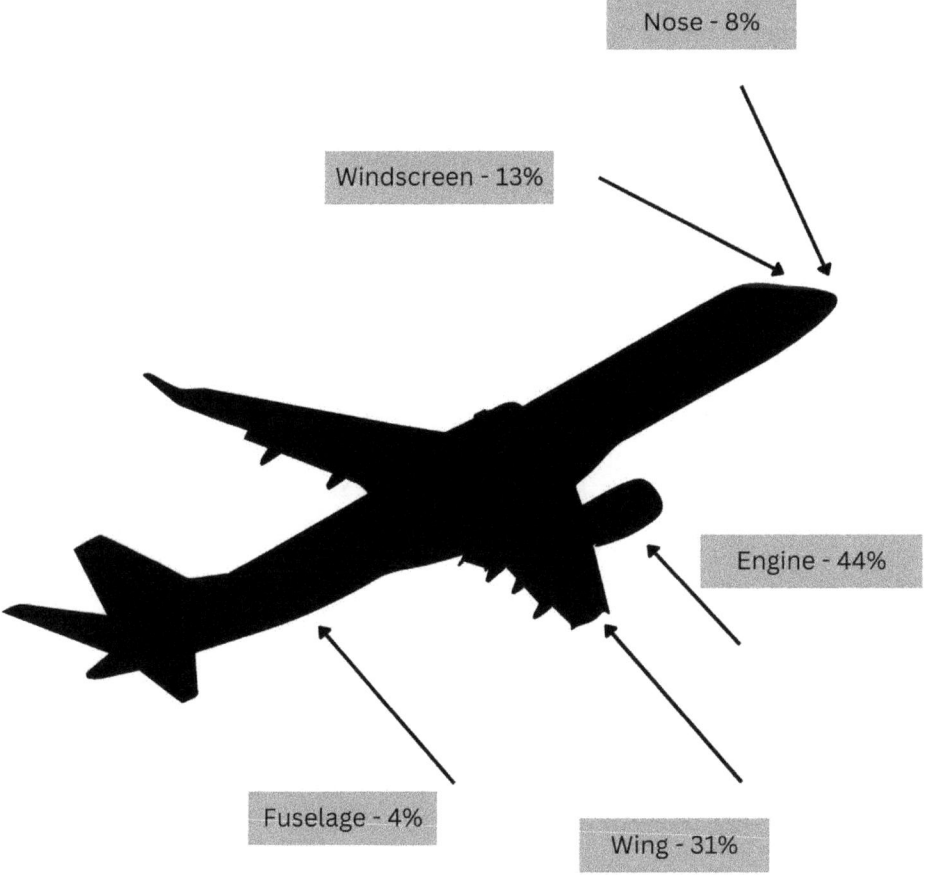

Figure 8.2 Locations of bird-strike damage
Source: author, adapted from Nicholson and Reed (2011)

ATG Did you know?

The first bird strike and bird strike fatality

The first recorded bird strike was by the Wright Brothers in 1905, in the Wright Flyer III, when their aircraft hit a flock of birds – one bird was killed – but there was no damage to the aircraft or its occupants (Aviation Safety Network, 2023).

In 1912, in Long Beach, California, Calbraith Perry Rodgers, the first man to make a transcontinental flight across the USA, became the first person to die as a result of a bird strike, flying into a flock of gulls that jammed the aircraft's controls, causing it to plunge out of control into the ocean (ICAO, 2020).

Smaller, propeller-driven aircraft are more likely to experience structural damage, such as windscreen damage or control surface damage, whereas larger jet-engine aircraft have a high risk with engine ingestion. Complete engine failure or serious power loss may be critical during the aircraft take-off phase and the risks are spread with flocks of birds across multiple engines.

ATG Case study 8.2

"Miracle on the Hudson River", New York City, US Airways 1549

Perhaps the most famous bird strike occurred on 15 January, 2009, when a US Airways Airbus A320 hit a flock of Canada geese (Branta canadensis) just after taking off from LaGuardia Airport, New York City. The ingestion of multiple geese into both engines resulted in a forced emergency landing on the Hudson River (ICAO, 2020). The impact was just under 3,000 feet and around 4.5 miles from the airport. The skilled piloting of Captain Chesley "Sully" Sullenberger and First Officer Jeffrey Skiles meant all 150 passengers and five crew members survived.

Scientists at the Smithsonian Institute concluded that the Canada geese were migratory geese that had probably come looking for food and open water in response to a cold snap and snow in their wintering grounds. These birds average eight pounds in size, much greater than the Airbus A320 engines had been designed to withstand ingesting (Zielinski, 2009).

Figure 8.3a US Airways Flight 1549 on Hudson River
Source: Greg (2009)

Figure 8.3b Canada geese
Source: Anderson (2017)

According to Kutbi (2014), contributory factors for bird strikes include:

- **Habitat features**, including open areas of grass and water, as well as shrubs and trees, which can provide food and roosting sites for birds.
- **Landfill and other waste disposal sites** can attract birds if not carefully managed, as can other agricultural activity.

- **Migrating birds** often follow well-defined flight paths in considerable numbers, which can create a problem if in the vicinity of an airport.
- Airports in **coastal locations** (and other water sources) may be at higher risk.
- Many airports contain areas of **grassland** which can be attractive to birds.

> **ATG Did you know?**
>
> **The highest bird strike**
>
> Although most bird strikes occur at low levels, the highest recorded bird strike was at 37,000 feet on a commercial flight over Abidjan, Ivory Coast, in 1973, where the airliner lost one engine, but landed safely. The bird was identified as a Ruppell's griffon vulture, one of the highest-flying birds in the world.

8.3.3 Wildlife hazard mitigation

ICAO Doc 9137 Airport Services Manual Part 3 – Wildlife Control and Reduction – informs aerodromes on action to be taken to decrease the risk to aircraft operations, by adopting measures to minimise the likelihood of collisions between wildlife and aircraft, often via Wildlife Hazard Management Programmes.

To manage wildlife hazards, *risk* should be assessed for each species present and the aerodrome area (and its vicinity) should be assessed, followed by strike assessments – both in terms of probability and severity. Knowledge of wildlife living in the vicinity is important, along with their movements and attractions.

ICAO (2020) suggests three risk levels be defined:

- Level 1 (green) – Acceptable. No further action is required.
- Level 2 (yellow) – Tolerable. Risk can be tolerated based on safety risk mitigation.
- Level 3 (red) – Intolerable. Take immediate action.

Habitat management is extremely important for airports, to eliminate or exclude food, water and shelter, which can limit the attractiveness of an aerodrome to wildlife, which can be the root cause of the hazards. Mitigation can involve the removal and alteration of attractive habitat features, such as adjusting the design of aerodrome buildings, preventing wildlife from accessing aerodrome property using fencing, reducing grass height, pruning or removal of trees and shrubs, as well as management of waste and removal of standing water such as ponds or puddles.

Other mitigation methods include:

- Management patrols.
- Noise to scare wildlife, such as sound generators, pistol shots and pyrotechnics as well as recorded distress or alarm calls. Variations in time and location may work best to avoid wildlife habituating.
- Visual repellents such as kites, balloons and scarecrows.
- Translocation to move individual animals to new locations.
- Culling, although airports need to be aware of relevant local and national regulations and be sensitive to public perception.

228 *Fundamentals of Global Air Transport Geography*

ATG Did you know?

The deadliest bird strike

Sixty-two people were killed in 1960 at Boston Logan International Airport, when a Lockheed Electra L188 struck a flock of European starlings (Sturnus vulgaris) just as it became airborne. The birds were ingested into three of the aircraft's four engines, causing the aircraft to lose power, stall and crash into the harbour (ICAO, 2020).

ATG Case study 8.3

Wildlife management at Singapore Changi Airport (SIN)

Wildlife management comes under four main headings at SIN:

1. **Understanding wildlife behaviour** involves conducting regular wildlife patrols, data analyses and hazard assessments. The main bird species include small birds such as swallows and larger birds such as the Egret.
2. **Wildlife control measures** involve improving dispersal capabilities, such as using aerolasers and long-range acoustic devices (LRADs), trapping, removal and management of habitat and food source, such as turf, water bodies and vegetation management.
3. **Appropriate staff training.**
4. **Education programmes** and increasing awareness, including to third-party organisations and the implementation of a Wildlife Management Committee.

8.4   Biodiversity at airports

Biodiversity is a term that describes the number and variety of species within an ecosystem (Holden, 2011). It refers to the diversity of life on earth at all scales, from genes to ecosystems, and we cannot have the healthy ecosystems we need without a diverse variety of animals, plants and microorganisms (ICAO, 2022). "Biodiversity matters because it supports the vital benefits humans get from the natural environment. It contributes to the economy, health and well-being and it enriches our lives" (JNCC, 2023).

Airports can impact biodiversity in several ways, including:

- Loss or degradation of habitats due to airport construction and expansion.
- Controlling wildlife for operational reasons.
- The impacts of noise and light pollution on certain species.
- Wildlife migration patterns.
- Propagation of invasive species.
- Airport infrastructure such as runways can cause changes in water runoff patterns, which may affect local ecosystems and water quality.

The biodiversity impacts of aviation should be addressed via airport planning applications and environmental assessments. In certain instances, some airport expansion has been delayed due to likely biodiversity impacts, such as at Lydd Airport, UK, in the 1990s while the impact of aircraft noise on the breeding success of birds at the adjacent internationally protected wetlands was investigated (Aviation Environment Foundation, 2023). As airports seek to expand and more land is being required all over the world, airports are facing a growing challenge in terms of the balance between growth and operations on one hand and the safeguarding of natural ecosystems and biodiversity conservation on the other. Airports are trying to find ways to manage these issues and incorporate these challenges into their business and sustainability strategies.

ATG Case study 8.4

Vancouver International Airport, Canada (YVR): a salmon-safe airport

YVR is located on 25sq.km of low-lying land at the mouth of the Fraser River. In 2016 it became the first airport in the world to achieve Salmon-Safe certification, which acknowledges the continuing efforts to transform land and water management practices to protect the water quality of the Fraser River and enhance the habitat, so Pacific salmon can continue to thrive.

To certify YVR, a team of independent Salmon-Safe experts conducted an assessment researching aspects such as wildlife and pest management, landscaping and construction practices, chemical containment and stormwater management. The airport has committed to several policies, including ensuring all development activities include zero sediment runoff from construction sites into waterways, operating a centralised and contained de-icing facility to protect water quality, developing a significant habitat restoration project on Sea Island and managing over 530 hectares of land airside with use of minimal herbicides and pesticides through a zoned approach. In 2021, the certification was renewed for another five years.

ATG Case study 8.5

Maun International Airport, Botswana (MUB): restoration and regeneration

The use of *exclosures* can assist with regenerating degraded and deforested areas, by excluding animals and humans to promote natural ecological rehabilitation. MUB was seeking to regenerate woody species through the use of exclosures and was able to enhance woody species richness, diversity and evenness as well as facilitating regeneration as part of the project (Kashe et al., 2023).

ATG Case study 8.6

Salvador Bahia Airport, Brazil (SSA): fauna management

SSA is located beside the Park of Dunes – an environmental reserve of 6 million sq.m – and preserves very rich flora and fauna, with 214 native animal species, including reptiles, mammals, amphibians and birds. To protect the local biodiversity, SSA implemented a Fauna Management Plan, involving biologists, engineers and veterinarians, which undertakes routine monitoring through sightings, collections and repair, as well as training for the airport community and non-harmful capture techniques to manage the challenge presented by simultaneously protecting airport operations and biodiversity. Since the bird capture actions began, bird strike damages were reduced by 80% (Tavares, 2019). In 2019, SSA was named Brazil's most sustainable airport, with other successes including zero waste to landfill, zero liquid discharge and 100% LED lighting.

ATG Case study 8.7

London Gatwick Airport, UK (LGW): Biodiversity Action Plan (BAP)

LGW has 75 hectares of woodlands, grasslands and wetlands within its airport boundary. In 2012, a Biodiversity Action Plan was launched, helping to protect over 2,400 different species of plants, animals and fungi. A range of habitats are located here, including scrub mosaic, wildlife ponds, ancient woodland, old hedgerows and floodplain meadows, providing grounds for a wide range and rare and threatened flora and fauna. The biodiversity monitoring programme has led to 35,570 biological records, from 2,490 different species, including 74 rare, declining or protected species, such as the nationally scarce long-horned bee.

In 2023, LGW won the Eco-Innovation Award at the ACI Europe Best Airport Awards and its BAP is aligned to the UK Wildlife Trust's Biodiversity Benchmark Standards (ICAO, 2022; Bicker, 2023).

8.5 Biological hazards

A biohazard is defined as a biological agent, substance or organism which poses a threat to human health, including bacteria and viruses. In the context of air transport, this refers to the potential risks associated with the transmission of diseases or pathogens via the movement of people, goods, animals and other species by air, posing a threat to air transport personnel, as well as the global community.

8.5.1 Disease transmission

Airports and aircraft can be environments where people from different countries and regions come into close contact, which can facilitate the potential transmission of infectious diseases. Aircraft cabins are critical in the spread (or not) of airborne disease transmission. Typically, aircraft have sophisticated filtration systems to minimise risks, such as high-efficiency particulate air (HEPA) filters to trap particles, including viruses and bacteria. Good ventilation systems and air exchange are also key in reducing airborne contaminants.

> **ATG Did you know?**
>
> **HEPA air filters**
>
> HEPA filters are used on most commercial aircraft, as they are very effective at trapping microscopic particles such as bacteria and viruses. According to the European air filter efficiency classification, HEPA air filters are any filters rated between 85 and 99.995% removal efficiency. For the production of aircraft with cabin air recirculation systems, manufacturers have generally chosen the higher efficiency filters, which are like those used in hospital operating theatres. Filtered, recirculated air provides higher cabin humidity levels and lower particulate levels than 100% outside air systems. Cabins are designed to operate most efficiently by providing around 50% outside air and 50% filtered, recirculated air. The total air supply is essentially sterile and particle-free (IATA, 2018).

8.5.2 Air transport and pandemics

A pandemic is a large-scale outbreak of infectious disease that can greatly increase morbidity and mortality over a wide geographic area and cause significant economic, social and political disruption (Madhav *et al.*, 2017). ICAO and the World Health Organization (WHO) collaborate to establish guidelines and standards for dealing with biological hazards in air transport and whilst each country will determine their own response, these guidelines should ensure a consistent and coordinated global response.

Due to the global nature of aviation and the risk in providing the means of transferring infectious diseases around the world in a short time frame, many controls are put in place during pandemics to slow or stop transmission by air, for example:

- Flight cancellations.
- Wearing of PPE (such as facemasks) on board.
- Health screening at airports, such as temperature checks and health questionnaires.
- Pandemic testing prior to flight travel.
- Quarantine protocols for arriving passengers.
- Airline and airport response plans, including coordinating with public health agencies, communication strategies and passenger management and education, such as awareness of good hygiene practices and adherence to health guidelines.

8.5.3 Severe acute respiratory syndrome

The epidemiological vulnerability of an inter-connected and extremely mobile society in many parts of the world was shown when the severe acute respiratory syndrome (SARS) virus rapidly spread from East Asia in 2003 (Budd, Bell and Brown, 2009). In fewer than four months, SARS spread from China to 25 countries and Taiwan, becoming the first new, easily transmissible disease of the 21st century.

According to Bowen and Laroe (2006), the story of SARS illustrates both the positives and negatives of globalisation. First, health professionals, using the latest information technology, were able to isolate and characterise the SARS-CoV rapidly, but it also spread rapidly due to global airline networks. The origin being very close to Hong Kong made it prone to diffusion by air travel, but according to Weiss and McLean (2004), had the origin been in Central Africa, SARS would have spread more slowly.

As part of the SARS outbreak, 32 international airports in 20 different countries applied surveillance measures to detect SARS cases, but there were uneven country-specific responses. The WHO issued advisories in stages from 15 March, 2003, and by 2 April, it issued what it called "the most stringent travel advisory issued in its 55-year history" as it recommended that all but emergency travel to areas of local transmission be postponed (Bowen and Laroe, 2006).

There were >8,000 global infections, with a mortality rate of approximately 10% and a devastating effect on global economies (LeDuc and Barry, 2004). Financial costs were estimated at US$40 billion for the Asia-Pacific region alone (Bowen and Laroe, 2006).

8.5.4   COVID-19

In an air transport industry characterised by sudden and severe "black swan" events, such as wars, terrorist acts and economic recessions, the impacts of the COVID-19 pandemic from early 2020 onwards were especially severe, in terms of both health and financial impacts.

> **ATG Did you know?**
>
> **A timeline of COVID-19**
>
> Cases of COVID-19 were first detected in China in December 2019, with the virus spreading rapidly to other countries across the world. The WHO declared COVID-19 to be a Public Health Emergency of International Concern (PHEIC) on 30 January, 2020, and characterised the outbreak as a pandemic on 11 March, 2020. More than three years later, on 5 May, 2023, the WHO emergency committee recommended that although the pandemic itself was not over, the global emergency it caused was (WHO, 2023).

World passenger traffic suffered an unprecedented decline because of global lockdowns and travel restrictions, with a decline in global passengers of 60% in 2020 versus 2019. By the end of 2022, there was still a reduction of 29% versus 2019 levels, thus the industry effects were being felt for a long period (Figure 8.4).

Global industry revenues were also severely impacted, with a drop of over 54% from 2020 versus 2019 (from $838 million to $384 million) and by 2022, although this had recovered somewhat, there was still a 12.2% drop versus 2019, to a revenue figure of $736 million. Passenger revenue experienced a massive drop, from $607 million in 2019, to only $189 million in 2020 and by 2022, this was still well down versus 2019, at a revenue of $436 million. Conversely, global cargo revenues performed well, with an increase from $100.8 million in 2019 to $140.4 million in 2020 and $210 million in 2021 (Figure 8.5).

In terms of the entire aviation value chain, freight forwarders and the air cargo sectors were the sole bright spots, with both sectors experiencing profits at the global level, with freight

Biogeography, hydrology and air transport 233

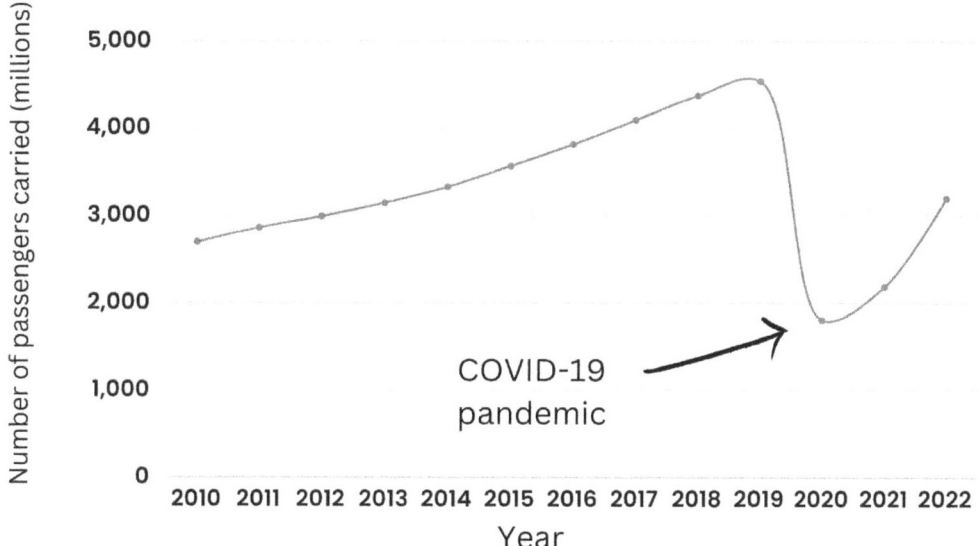

Figure 8.4 Total global passenger numbers, 2010–2022
Source: adapted from IATA (2024)

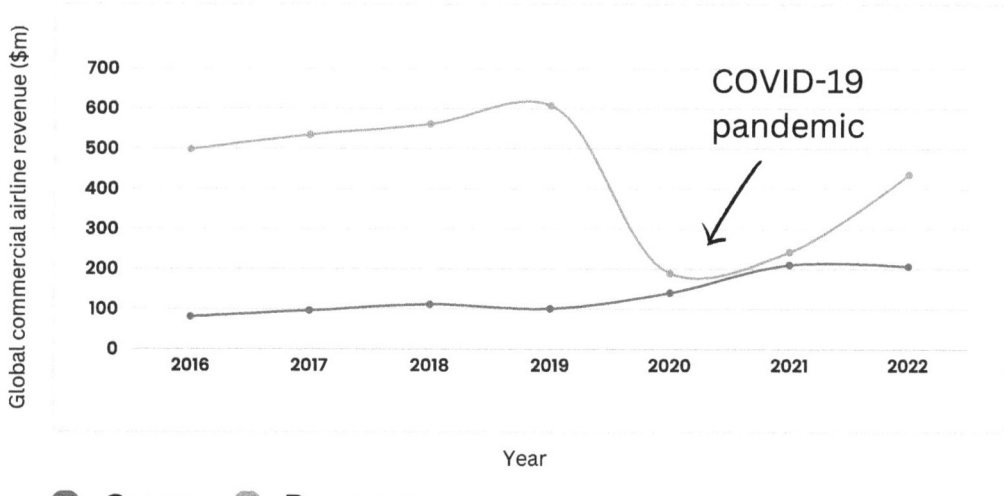

Figure 8.5 Global airline passenger and cargo revenues, 2016–2022
Source: adapted from IATA (2024)

forwarders experiencing economic profits of 4% and air cargo carriers 9% overall (McKinsey, 2022).

Initially, there was a sharp decrease in cargo capacity due to passenger flight cancellations and the resultant drop in "belly" hold capacity. During 2020, air carriers increased the number of dedicated freighters they used to partially offset this decline. Demand for air cargo was initially strong due to the need for protective personal equipment (PPE) and medications and then from the growth in e-commerce, with electronic products rising in share and level as consumers shifted their spending away from services and towards durable goods. This resulted in a steep increase in air freight rates versus 2019 (USITC, 2021).

Other aviation sectors – including manufacturers; aircraft lessors; air navigation service providers (ASNPs); catering; ground services; maintenance, repair and operations (MRO); and especially airlines – all lost money, with a combined loss of $230.1 million in 2020 (McKinsey, 2022).

The economic impacts of the pandemic were especially severe in emerging economies, where income losses caused by the pandemic revealed and worsened some pre-existing economic fragilities, as well as having a dramatic impact on global poverty and inequality (World Bank, 2023).

8.6   Hydrology

Hydrology comes from the Greek word for water and is the *science or study of water*. Contemporary hydrology is concerned with:
- The distribution of water on the surface of the earth.
- Water movement over and beneath the surface of the earth.
- Water movement through the atmosphere.

Thus, it is the study of fresh water that is the main concern, rather than the study of saline water, which is the remit of oceanography. The link with landforms and geomorphology (Chapter 7) are important, in terms of explaining the processes that lead to water moving around the earth (Davie and Quinn, 2019).

One of the most important concepts is the hydrologic (water) cycle, which is the process by which water is continuously transferred between the surface of the earth and the atmosphere, involving processes such as evaporation, transpiration and sublimation and this was analysed in Section 4.2.

ATG Did you know?

Where is earth's water?

Water covers approximately 75% of the earth's surface, is practically everywhere and is the only known substance that can exist as a liquid, a solid and a gas, within a small range of air temperatures and pressures found at the earth's surface. Liquid water is found in oceans, rivers, lakes and underground; solid ice is found in glaciers, snow and the north and south poles; whilst water vapour – a gas – is found in the atmosphere. The earth's water content is about 1.39 billion cubic kilometres and is found in the following proportions:

- 96.5% is in the oceans
- 1.7% is in lakes, rivers, streams and soil
- 1.7% is in polar ice caps, glaciers and permanent snow
- 0.001% is in water vapour in earth's atmosphere

Source: NASA Earth Observatory (2010)

Water resources are under increasing pressure due to a growing global population, urbanisation, poor water management and factors such as climate change. According to the World Resources Institute (Kuzma, Saccoccia and Chertock, 2023), demand for water has doubled since 1960 and 25 countries, housing one-quarter of the global population, face extremely high water stress (the ratio of water demand to renewable supply). The most water-stressed countries are Bahrain, Cyprus, Kuwait, Lebanon, Oman and Qatar. At least 50% (around 4 billion people) live under water-stressed conditions for at least one month of the year, jeopardising jobs, food and lives. Global water demand is projected to increase by 20–25% by 2050, whilst in sub-Saharan Africa alone, demand is expected to grow by 163% by 2050, mainly for irrigation and domestic water supply. Effective water management by both industry and individuals is therefore critical.

8.6.1 Regulations

Airports are often required to comply with local, national and international regulations regarding water management and environmental protection.

U.N. Sustainable Development Goals

The 2030 Agenda for Sustainable Development, which was adopted by all UN member states in 2015, provided a shared blueprint for peace and prosperity for people and the planet, now and into the future. At the core were 17 Sustainable Development Goals (SDGs) which are an urgent call for action by all countries in a global partnership (United Nations, 2023). Table 8.1 illustrates the broad range of these goals.

SDG number 6 is *Clean Water and Sanitation* and this was introduced to ensure the availability and sustainable management of water and sanitation for all. Whilst the SDGs are not legally binding, governments are expected to take ownership and establish national frameworks for achievement of the goals.

Table 8.1 UN Sustainable Development Goals

1. No poverty	5. Gender equality	9. Industry, innovation and infrastructure	13. Climate action	
2. Zero hunger	6. Clean water and sanitation	10. Reduced inequalities	14. Life below water	17. Partnerships for the goals
3. Good health and well-being	7. Affordable and clean energy	11. Sustainable cities and communities	15. Life on land	
4. Quality education	8. Decent work and economic growth	12. Responsible consumption and production	16. Peace, justice and strong institutions	

8.6.2 Hydrology and air transport

When studying the distribution and movement of water, the role of human interaction is important, both in terms of *water quantity and quality* and secure and reliable water resources are critical to airport operations. IATA (2023) believes that the demand for air travel is expected to double between 2023 and 2040, from a figure of around 4 billion to just over 8 billion origin–destination passengers. Water pressures at airports will only intensify and appropriate, sustainable water management strategies will be critical, not only for airports themselves, but other responsible public and private bodies.

In the context of airports, there are three key areas water needs to be considered (ICAO, 2021):

1. **Supply** – the water entering the facility and water demand management.
2. **Handling capacity** – managing issues such as flooding, erosion and drainage.
3. **Disposal** – ensuring water leaving the facility is clean and safe for the surrounding environment.

8.7 Water supply at airports

Water supply at airports is required for a multitude of reasons and any supply challenges may vary depending on the location, size and infrastructure and the range of passengers, aircraft and operational activities. Some key activities at airports requiring water, include:

- **Aircraft servicing**, such as potable (drinking) water for passengers and crew and water for sanitation systems and lavatories.
- **Firefighting systems** require a steady supply of water if required in an emergency.
- **Cooling systems** such as air conditioning units and refrigeration systems.
- **Potable water supply** for airport passengers and staff.
- **Terminal restrooms.**
- **Construction and expansion activities**, where water will be required for activities such as concrete mixing.
- **Cleaning and maintenance**, of both the terminal facilities as well as runways, taxiways and the aircraft themselves.
- **Landscaping of green spaces.**

Airports will rely on locally available water for their activities and therefore will also be dependent on the local and regional water supply, with airports in areas of water stress facing particular challenges. Airports can consider approaches such as the use of aerators, low-flush toilets, sink timer taps and rainwater capture systems.

> **ATG Case study 8.8**
>
> **Rajiv Gandhi International Airport, Hyderabad, India (HYD): water sustainability through efficient devices, recycling and replenishment**
>
> HYD has become self-sustainable in water management and has implemented several policies, via their "4R" strategy:
>
> - **Reduction**, via automation of the landscape irrigation system across 278 hectares, through a cloud-based central control system for the drip and sprinkler network as well as sensor-based water taps and aerators in the terminal buildings and offices.

- **Reuse** by recovering the 103 air handling units (AHUs) of the air conditioning system, condensate and feeding it to the cooling tower circuit, enhancing efficiency.
- **Recycling** of the airport wastewater through multi-stage treatment for WCs flushing and irrigation within the airport premises.
- **Replenishment** via rainwater harvesting, collecting runoff from paved areas, rooftops and open areas.

The total project cost was INR 338 million and savings over a three-year period were INR 361 million, with associated environmental benefits such as reduced stress on water resources, a 30.92% water use efficiency and improvement of the ground water table (ICAO, 2021).

ATG Case study 8.9

Aircraft "drywashing": Emirates Airline/Dubai International Airport (DXB)

Emirates Airline uses a "drywash" technique on its aircraft fleet in DXB, which saves millions of litres of water each year and improves the aerodynamic performance of the aircraft. During flights, aircraft accumulate dust and grime on their external surfaces, which make the aircraft heavier and burn more fuel. The name "drywash" is a clue to the technique and little or no water is used in cleaning the aircraft when thousands of litres would normally be used per wash, typically 11,500 litres for an Airbus A380. A small amount of cleaning agent is applied with cloths, then removed using clean microfibre fabrics, leaving the aircraft with a fine protective film. It takes a crew of 15 staff about 12 hours to clean an A380 (Aviation Benefits Beyond Borders, 2017; Emirates, 2023).

Other airports, such as Hong Kong International, also use "drywashing" (HKIA, 2020).

8.8 Water handling capacity at airports

Water handling capacity refers to an airport's ability to manage and handle water-related issues, such as rainfall, drainage and water runoff, and this is especially important during adverse weather conditions.

8.8.1 Airport drainage systems

Properly functioning drainage systems are essential for safe and efficient airport operations, to prevent water accumulation on runways, taxiways and other key areas. Runways and taxiways are graded to facilitate water runoff, often with a crown in the middle, to allow water to flow away from the centre line and edges of the pavement surfaces into drainage systems. Storm drains may be positioned strategically to collect and channel rainwater away from key areas and catch basins are often used to collect surface water and prevent debris from entering the drainage system. Underground pipes are used to transport water away from the airfield and into appropriate outlets and there may be retention ponds to temporarily store excess water during heavy rainfall.

There are times when the handling capacity of the airport is not enough to prevent hydrological hazards from occurring. Low-lying airports are especially at risk and there are three main hazards:

- coastal flooding
- river flooding
- landslides

Flooding is a natural phenomenon and should be expected, but each year flooding causes many deaths around the world and significant impacts on infrastructure. Floods commonly occur when rivers overtop their banks and can also occur through a very high tide at the coast, often made worse by a storm event. Flooding may also happen when very heavy rainfall cannot escape from an area (Holden, 2011).

Hazard risks are greatest in specific locations and as discussed in Chapter 6, with earth's warming climate, risks are increasing in many parts of the world.

8.8.2 Hydrological hazards – coastal flooding

Airports are at particular risk from coastal flooding due to the fact that many of the world's major cities are located along coastlines and their large international airports are typically built either very close to sea level or on filled shallow coastal waters (Griggs, 2020). Many of these are exposed to extreme flood events such as hurricanes and large storms, for example both New York JFK and Newark Airports suffered flood damage during Superstorm Sandy in 2012.

According to Yesudian and Dawson (2021), a modest amount of sea level rise, such as that associated with a global mean temperature rise of 2°C, would place 100 airports below sea level. Their analysis also concluded that 269 airports were at risk now and this could grow to 572 by 2100 and that the global cost of airport adaptation to maintain the risk in 2100 at present levels could cost up to $39 billion or even $57 billion, depending on sea-level rise scenarios.

The Maldives, with one of the lowest average land elevations above present-day mean sea level, is one of the world regions which could be worst impacted by mean sea-level rise (Amores et al., 2021).

Flood protection measures include constructing seawalls/dykes, levees, tidal gates, detention ponds, pumping stations, drainage equipment and possibly elevating runways. Singapore Changi Airport (SIN) has put in place initiatives such as building the airport on higher ground above mean sea level as well as increasing the capacity of the airports' drains, and sensors to monitor drain conditions (Changi Airport, 2021).

ATG Case study 8.10

Flood protection at Amsterdam Schiphol Airport, Netherlands (AMS)

AMS is located 4.5m below sea level (Schiphol, 2023) and like much of the Netherlands, where between 26% and 33% lies below sea level, is prone to flood risk. The airport was constructed on reclaimed land at the bottom of what was once Haarlemmermeer (lake), which was drained in 1852, with the first aircraft landing in 1916. Enabling the airport to survive its flood risk has involved engineering techniques, such as a 240km-long network of drainage structures and pumping stations, a system of dykes and other flood barriers as well as operational and management practices (Griggs, 2020). A system of retention, detention, conveyance, storage and discharge has been employed.

In 2015, AMS presented its Water Vision 2030 Strategy, from which a series of five key actions have been developed in their Climate Resilient Airports Framework, involving flood protection, dealing with weather extremes, achieving a good water quality, adaptive airport city planning and greening airport operations. New measures such as extra water retention, robust water connections, "sponges" (such as green roofs and rainwater collection for toilet flushing) and alternative water storage under parking lots and along runways have been developed (ICAO, 2022).

8.8.3 Hydrological hazards – river flooding

Rivers are found almost everywhere on earth's land surface. Understanding the dynamics of rivers requires an understanding of their behaviour in flood, as most changes in channel morphology occur when flow rates are high and sediment motion is rapid (Davies, Korup and Clague, 2021). Many airports are located on river floodplains all over the world and therefore understanding water processes is key to mitigate against possible floods. Airport infrastructure such as runways, taxiways, aprons and terminal buildings are at risk from flooding, especially in low-lying areas, and the possible risk of contaminants from leaving the airport environment because of floods has to be managed.

ATG Case study 8.11

Flooding at Don Mueang International Airport, Bangkok, Thailand (DMK)

In 2011, extreme rainfall caused Bangkok's main river to flood a large part of the city, including DMK, north of downtown Bangkok, causing diversions to Bangkok's other airport at Suvarnabhumi. The effects were so severe that it took a year to carry out the extensive repair work required and to replace electrical and IT systems (Sindhamani and Vorage, 2020).

At first glance, the watery rectangle in Figure 8.6 looks like a harbour, with structures extending into the water like docks, with small white dots like ships. However, the white dots are actually aeroplanes and the watery rectangle is the submerged runway complex.

240 *Fundamentals of Global Air Transport Geography*

Figure 8.6 Bangkok Don Mueang Airport, 31 October, 2011
Source: NASA Earth Observatory (2011)

ATG Case study 8.12

Flooding at La Vanguardia Airport, Colombia (VVC)

In May 2022, La Vanguardia Airport was closed after the terminal and runway were flooded by the nearby Guatiquia River, affecting one of the last operators of the DC-3 aircraft, Aliansa – Aeroloineas Andinas.

8.8.4 Landslides

Landslides are the downward movement of slope materials under the influence of gravity. Water is involved in many cases, at least to some degree, and slopes fail when shear forces acting on slope materials exceed the shear strength of the materials (Davies, Korup and Clague, 2021). Landslides can be caused by several factors, such as geophysical disasters (earthquakes and volcanic eruptions), removal of vegetative cover, wildfires, heavy rainfall and changes in groundwater levels.

As with any other human structure, airports can be impacted by landslides, resulting in significant damage, closures and flight cancellations. In 1979, Nice Airport (NCE) was hit by a tsunami, with waves reaching up to 3.5m in height. A section of fill slope failed, killing seven people. The landslide involved between 2 and 3 million cubic metres of fill and about 7 million cubic metres of underlying clay-silt (Petley, 2022).

8.9 Water disposal at airports

Water quality is essential to human health and economic development and polluted water has the potential to affect not only airport areas, but also surrounding areas, and depending on water and river flow, potentially some distances away. A key area of focus for airports is to avoid pollutants getting into water sources, as well as decreasing pollutant levels and making sure that polluted water is treated appropriately before being released into surrounding water areas.

> **ATG Did you know?**
>
> **Types of water**
>
> Water is often classified by its use:
>
> - **Wastewater** is water which has been subjected to a human use, either domestic or industrial.
> - **Grey water** is relatively clean wastewater from baths, sinks and laundry facilities. This can sometimes be captured and reused for other purposes, such as landscape irrigation.
> - **Black water** contains either sewage or water from toilets and this must be treated in proper facilities.

The risk of contaminants to water supplies is present from a few sources within airports:

- Ethylene or propylene glycols from de-icing/anti-icing of aircraft.
- Urea, acetates or formats from de-icing/anti-icing of runways, aprons and taxiways.
- Fuel from spills during aircraft refuelling and other vehicles.
- Fire suppressant chemicals and foams dispersed in firefighting exercises.
- Dust, dirt and hydrocarbons from paved surfaces.
- Herbicides and pesticides.

Source: ICAO (2021)

8.9.1 De-icing/anti-icing

De-icing is the process which removes ice, snow, slush or frost from aeroplane surfaces. Of particular concern at airports is aircraft *de-icing fluid*. If weather conditions are appropriate then icing conditions may be present, which can be extremely safety critical, as a loss of aircraft control is possible if these contaminants are not removed. The aerodynamic effectiveness of an aircraft requires it to be free from contamination from either frozen or semi-frozen moisture prior to take-off and airports which regularly experience snow and ice conditions incorporate de-icing processes into their operations. This may be done on the aircraft stand or in dedicated bays, usually on the way out to the runway, so the aircraft can be de-iced and then depart as soon as possible, to prevent re-freezing. Using de-icing fluid serves three main purposes:

- Removal of frozen moisture from critical aircraft external surfaces before flight.
- Surface protection between treatment and aircraft becoming airborne.
- Removal of any frozen moisture from engine intakes and fan blades prior to take-off.

(SKYbrary, n.d.a)

ICAO (2018) provides guidelines to air transport operators via Doc 9640 – Manual of Aircraft Ground De-Icing/Anti-Icing Operations. As de-icing fluids generally contain chemicals, care must be taken regarding discharge, to prevent them being released into groundwater. Of particular concern is the effect on aquatic life, which can be lethal to fish, plants and other amphibians. Eutrophication (an overabundance of nutrients such as nitrogen and phosphorus) can also be caused by the breakdown of chemicals in de-icing fluids.

8.9.2 Disposal management

In addition to de-icing fluids, other activities such as refuelling and gate servicing of aircraft have the potential to impact surface waters and any activity involving chemical agents should consider how to prevent these substances entering surface or ground water.

For de-icing, it is important to use the minimum amount necessary to de-ice safely. At the gate, butterfly valves (valves which isolate or regulate the flow of a fluid) can be used to contain any contamination on the apron. To the extent possible, de-icing fluids should be collected after use and segregated from stormwater systems – on a de-icing pad, a liner could be used underneath to protect groundwater. Spent glycol can be collected, separated from other water and recycled.

A spills response plan is recommended if spills occur. For example, contaminants should be kept away from catch basins and any storm drainage where possible and oil/water separators installed on aprons and high vehicular traffic areas.

ATG Case study 8.13

De-icing operations at Oslo Gardermoen Airport (OSL), Norway

At OSL, non-toxic de-icing fluid is used for aircraft and additive-free organic salt is used for the runways and taxiways. The de-icing fluid consists of a mixture of hot water and propylene glycol, plus several additives, and the fluid is not categorised as an environmental toxin, as the additives used make up less than 1% of the fluid and are not toxic to the environment around the airport in the quantities used.

De-icing takes place on three special platforms (Figure 8.7) and there are collection systems for the excess de-icing fluid, which is then treated. The airport surfaces, such as the runways, are de-iced using formates as the principal de-icing agent, which are organic salts and do not contain any harmful additives, with a collection system in place. The collected chemicals are taken to a purification plant, where they are removed and treated (Avinor, 2023).

Biogeography, hydrology and air transport 243

Figure 8.7 De-icing area at Oslo Gardermoen Airport
Source: Bjoertvedt (2013)

Chapter review questions

8.1 Why is the concept of biogeography so important to the air transport industry?
8.2 Assess why wildlife management strategies are so important to air transport.
8.3 Why are bird strikes such a risk to aircraft? What are the factors which influence their risk? Research any ONE accident/serious incident from the "Accident and Serious Incident Reports: Wildlife Strike" section of SKYbrary. What were the causes? Were the risks identified? What has the airport done to reduce future wildlife strike risks?
8.4 Why do airports have an important role in preserving biodiversity? For a major airport in your country, research the biodiversity at risk in this location and assess the airport's strategies to managing biodiversity.
8.5 Analyse the financial impact of the COVID-19 crisis to the air transport industry and assess if, to what extent and in what areas, the impacts of COVID-19 are still being felt now.
8.6 Identify the different reasons airports require an adequate quantity and quality of water supply. For any ONE airport in your country, what are the airport's strategies to manage its supply of water?
8.7 What are the key factors which influence flood risk at an airport, in terms of either coastal or river flooding. Choose any country you identify as having a risk of flooding – why is this and what strategies have been implemented to reduce the possible risks at any of the airports?
8.8 Why is aircraft de-icing a potential hazard at airports? What methods have been employed to reduce hazard risk?

244 Fundamentals of Global Air Transport Geography

ATG trivia

The "Rain Vortex" at the Jewel Changi Airport, Singapore

We have discussed the importance of water in the context of airports; however, one other innovative use of water occurs at the Jewel Changi Airport. The Jewel Changi is a ten-storey, nature-themed entertainment and retail complex surrounded by and linked to one of the passenger terminals at SIN, which opened in 2019. It is open to both passengers and non-passengers and includes attractions such as an indoor garden spanning five storeys – the Shiseido Forest Valley and the Canopy Park at the top.

The centrepiece of this complex is the "Rain Vortex", the tallest indoor waterfall in the world (Figure 8.8) at seven storeys and approximately 130 feet (40m) in height. Recirculating rainwater is pumped to the roof to free fall through a round hole at approximately 37,850 litres per minute, falling to a pool set in the basement. An even circular flow is achieved as the water is channelled to the ring in the middle of the oculus then distributed down the Rain Vortex (Jewelchangiairport, 2023).

Figure 8.8 The Rain Vortex at Jewel Changi
Source: Peel, 2023, via Wikimedia Commons. CC BY-SA 4.0

References

Amores, A. *et al.* (2021) 'Coastal flooding in the Maldives induced by mean sea-level rise and wind-waves: from global to local coastal modelling', *Frontiers in Marine Science*, 8. Available at: www.frontiersin.org/articles/10.3389/fmars.2021.665672.

Anderson, J. (2017) *Canada geese*. Available at: https://commons.wikimedia.org/wiki/File:Canada_Geese_%28199078783%29.jpeg.

Aviation Benefits Beyond Borders (2017) *Emirates showcases environment friendly aircraft cleaning technique*. Available at: https://aviationbenefits.org/newswire/2017/06/emirates-showcases-environment-friendly-aircraft-cleaning-technique/.

Aviation Environment Foundation (2023) *Biodiversity*. Available at: www.aef.org.uk/what-we-do/biodiversity/.

Aviation Safety Network (2023) *Incident Wright Flyer III*. Available at: https://aviation-safety.net/wikibase/222530.

Avinor (2023) *Water and soil – Avinor, avinor.no*. Available at: https://avinor.no/en/corporate/airport/oslo/community-and-environment/vann-og-grunn/avising.

Bicker, R. (2023) *London Gatwick's biodiversity action plan, International Airport Review*. Available at: www.internationalairportreview.com/article/189355/protecting-woodlands-wetlands-and-willows-under-london-gatwicks-biodiversity-action-plan/.

Bjoertvedt (2013) *Gardermoen deicing pit*. Available at: https://commons.wikimedia.org/wiki/File:Gardermoen_deicing_pit_IMG_5908.JPG.

Bowen, J.T. and Laroe, C. (2006) 'Airline networks and the international diffusion of severe acute respiratory syndrome (SARS)', *The Geographical Journal*, 172(2), pp. 130–144. Available at: https://doi.org/10.1111/j.1475-4959.2006.00196.x.

Budd, L., Bell, M. and Brown, T. (2009) 'Of plagues, planes and politics: controlling the global spread of infectious diseases by air', *Political Geography*, 28(7), pp. 426–435. Available at: https://doi.org/10.1016/j.polgeo.2009.10.006.

Changi Airport (2021) *"Google drains" – Changi Airport's new layer of defence against flash floods*. Available at: www.changiairport.com/corporate/media-centre/changijourneys/the-airport-never-sleeps/google-drain-map.html.

Davie, T. and Quinn, N.W. (2019) *Fundamentals of hydrology*. 3rd edn. Abingdon: Routledge.

Davies, T.R., Korup, O. and Clague, J.J. (2021) *Geomorphology and natural hazards: understanding landscape change for disaster mitigation*. American Geophysical Union, Wiley. Available at: www.perlego.com/book/2575349/geomorphology-and-natural-hazards-understanding-landscape-change-for-disaster-mitigation-pdf.

Emirates (2023) *Sustainability in operations | Our planet | About us | Emirates United Kingdom, United Kingdom*. Available at: www.emirates.com/uk/english/about-us/our-planet/sustainability-in-operations/.

FAA (2022) *Wildlife management*. Available at: www.faa.gov/airports/airport_safety/wildlife/management.

FAA (2023) *Wildlife strikes to civil aircraft in the United States 1990–2022*. 29. Available at: www.faa.gov/sites/faa.gov/files/Wildlife-Strike-Report-1990-2022.pdf.

Google Earth (n.d.). Available at: www.google.co.uk/earth/.

Greg, L. (2009) *US Airways Flight 1549 in the Hudson River New York, USA*. Available at: https://commons.wikimedia.org/wiki/File:US_Airways_Flight_1549_(N106US)_after_crashing_into_the_Hudson_River.jpg.

Griggs, G. (2020) 'Coastal airports and rising sea levels', *Journal of Coastal Research*, 36(5), pp. 1079–1092.

HKIA (2020) *Turning down the tap for aircraft washing*. Available at: www.hongkongairport.com/en/sustainability/environment/greenest-airport-updates/20-03-16.

Holden, J. (2011) *Physical geography: the basics*. Abingdon: Routledge.

IATA (2018) *IATA Briefing paper: cabin air quality – Risk of communicable diseases transmission*. Available at: www.iata.org/contentassets/f1163430bba94512a583eb6d6b24aa56/cabin-air-quality.pdf.

IATA (2023) *Global outlook for air transport – highly resilient, less robust*. Available at: www.iata.org/en/iata-repository/publications/economic-reports/global-outlook-for-air-transport----june-2023/.

IATA (2024) *Economics reports*. Available at: www.iata.org/en/publications/economics/economics-library/.

ICAO (2018) *Doc 9640. Manual of aircraft ground de-icing/anti-icing operations*. 3rd edn. Available at: https://elibrary.icao.int/home.

ICAO (2020) *ICAO Doc9137 airport services manual: part 3 – wildlife hazard management*. 5th edn. Available at: https://elibrary.icao.int/home.

ICAO (2021) *Water management at airports: eco airport toolkit*. Available at: www.icao.int/environmental-protection/Documents/Water%20management%20at%20airports.pdf.

ICAO (2022) *2022 environmental peport*. Available at: www.icao.int/environmental-protection/Pages/envrep2022.aspx.

Jewelchangiairport (2023) *Experience wonder at Jewel Changi Airport – Jewel Changi Airport*. Available at: www.jewelchangiairport.com/en.html.

JNCC (2023) *UK biodiversity indicators | JNCC – Adviser to Government on Nature Conservation*. Available at: https://jncc.gov.uk/our-work/uk-biodiversity-indicators/.

Kashe, K. *et al.* (2023) 'Restoration of diversity and regeneration of woody species through area exclosure: the case of Maun International Airport in northern Botswana', *Bothalia, African Biodiversity & Conservation*, 53(1). Available at: https://doi.org/10.38201/btha.abc.v53.i1.1.

Kutbi, N. (2014) *Bird strike risk reduction programme*. Available at: www.icao.int/MID/Documents/2014/Wildlife%20and%20FOD%20Workshop/Risk%20Reduction%20program%20-%20GACA.pdf.

Kuzma, S., Saccoccia, L. and Chertock, M. (2023) *25 countries, housing one-quarter of the population, face extremely high water stress*. Available at: www.wri.org/insights/highest-water-stressed-countries (Accessed: 21 December 2023).

LeDuc, J.W. and Barry, M.A. (2004) 'SARS, the first pandemic of the 21st century', *Emerging infectious diseases*, 10(11), p. e26. Available at: https://doi.org/10.3201/eid1011.040797_02.

Madhav, N. *et al.* (2017) 'Pandemics: risks, impacts, and mitigation', in *Disease control priorities: improving health and reducing poverty*. 3rd edn. Washington, D.C.: The International Bank for Reconstruction and Development/The World Bank. Available at: www.ncbi.nlm.nih.gov/books/NBK525302/.

McKinsey (2022) *COVID-19's impact on the global aviation sector*. Available at: www.mckinsey.com/industries/travel-logistics-and-infrastructure/our-insights/taking-stock-of-the-pandemics-impact-on-global-aviation.

NASA Earth Observatory (2010) *The water cycle*. Available at: https://earthobservatory.nasa.gov/features/Water.

NASA Earth Observatory (2011) *Flood waters inundate a Bangkok Airport*. Available at: https://earthobservatory.nasa.gov/images/76282/flood-waters-inundate-a-bangkok-airport.

National Geographic (n.d.) *Ecosystem*. Available at: https://education.nationalgeographic.org/resource/ecosystem.

Nicholson, R. and Reed, W.S. (2011) *Strategies for prevention of bird-strike events*. Available at: www.boeing.com/commercial/aeromagazine/articles/2011_q3/4/.

Peel, M. (2023) *The rain vortex, Jewel Changi, Singapore (www.mikepeel.net)*. Available at: https://commons.wikimedia.org/wiki/File:At_Jewel_Changi,_Singapore_2023_36.jpg.

Petley, D. (2022) 'The 1979 Nice Airport landslide and tsunami', *The Landslide Blog*, 4 February. Available at: https://blogs.agu.org/landslideblog/2022/02/04/nice-airport-1/.

Schiphol (2023) *Schiphol | Aanpassen aan klimaatverandering*. Available at: www.schiphol.nl/en/schiphol-group/blog/adapting-to-a-changing-climate/ (Accessed: 24 December 2023).

Sindhamani, V. and Vorage, P. (2020) *Flood warning! NACO*. Available at: https://airport-world.com/flood-warning/.

SKYbrary (n.d.a) *Aircraft ground de/anti-icing*. Available at: https://skybrary.aero/articles/aircraft-ground-deanti-icing.

SKYbrary (n.d.b) *Non avian wildlife hazards to aircraft*. Available at: www.skybrary.aero/articles/non-avian-wildlife-hazards-aircraft.

South African CAA (2010) *Aircraft Accident Report and Executive Summary: CA18/2/3/8805*. Available at: www.skybrary.aero/sites/default/files/bookshelf/1977.pdf.

Tavares, R. (2019) *Sustainability series: how does Salvador Airport protect the environment?*, *International Airport Review*. Available at: www.internationalairportreview.com/article/99311/sustainability-series-how-does-salvador-bahia-airport-protect-the-environment/.

Thomas, H.S.C. (2012) 'The biosphere', in *An introduction to physical geography and the environment*. 3rd edn. Harlow: Pearson, pp. 521–556.

United Nations (2023) *THE 17 GOALS | Sustainable Development*. Available at: https://sdgs.un.org/goals.

USITC (2021) *The impact of the COVID-19 pandemic on freight transportation services and U.S. merchandise imports*. Available at: www.usitc.gov/research_and_analysis/tradeshifts/2020/special_topic.html.

Weiss, R.A. and McLean, A.R. (2004) 'What have we learnt from SARS?', *Philosophical Transactions of the Royal Society B: Biological Sciences*, 359(1447), pp. 1137–1140. Available at: https://doi.org/10.1098/rstb.2004.1487.

WHO (2023) *Coronavirus disease (COVID-19) pandemic*. Available at: www.who.int/europe/emergencies/situations/covid-19.

World Bank (2023) *WDR 2022 Chapter 1. The economic impacts of the COVID-19 crisis*. Available at: www.worldbank.org/en/publication/wdr2022/brief/chapter-1-introduction-the-economic-impacts-of-the-covid-19-crisis.

Yesudian, A.N. and Dawson, R.J. (2021) 'Global analysis of sea level rise risk to airports', *Climate Risk Management*, 31, p. 100266. Available at: https://doi.org/10.1016/j.crm.2020.100266.

Zielinski, S. (2009) *Migratory Canada geese brought down Flight 1549, Smithsonian Magazine*. Available at: www.smithsonianmag.com/science-nature/migratory-canada-geese-brought-down-flight-1549-12575190/.

Part III
Human geography and air transport

9 Political geography and air transport

Chapter outcomes

At the end of this chapter, you will be able to:

- Explain the various ways in which governments can have an impact, both direct and indirect, on the air transport sector.
- Differentiate between multilateral, bilateral and national air transport regulations.
- Understand the importance of nationality in air transport regulation.
- Discuss why the Chicago Conference in 1944 was so important to the development of the air transport industry.
- Explain the key roles and responsibilities of ICAO.
- Describe the Nine Freedoms of the Air.
- Identify the key provisions in traditional airline bilateral agreements.
- Explain the importance and key features of the US Deregulation Act.
- Understand the term Open Skies and explain the key benefits to both airlines and consumers.
- Understand the extent to which the airline industry in each region of the world operates based on either more restrictive bilateral agreements or on a more liberal Open Skies basis.

9.1 Introduction

The importance of political factors within air transport cannot be overstated. Direct government involvement can include the ownership and the operation of key assets such as airlines, airports and air traffic control services as well as safety, economic and other types of regulation. Indirect involvement includes the setting of levels of taxation, employment policies, subsidies and investment. For an industry which is by its definition international in nature, multinational governmental agencies, such as ICAO, are essential for the safe and efficient operation of the air transport sector.

The need for a safe and efficient air transport sector is enshrined in the Chicago Convention of 1944 and the safety performance of the air transport sector is now amongst the best in the world. Aviation places its highest priority on safety and 2023 witnessed the lowest fatality risk and "all accident" rate on record, at 0.80 per million sectors, with jet operations seeing no hull losses or fatalities (IATA, 2024a). Safety can never be taken for granted though and it is the multinational nature of aviation, informed by ICAO Standards and Recommended Practices and implemented by respective governments, which is one of the key reasons for this excellent safety performance. This and what Syed (2019) refers to as "failure is the key to flying high". Learning from past mistakes is crucial to improving safety performance and the global dissemination and actioning of previous accident and incident findings. The importance of learning from mistakes can be illustrated by an exemplar in a different industry – vacuum cleaners – where Sir James Dyson made 5,127 prototypes of his vacuum cleaner before getting it right. Perhaps not as safety critical as aircraft, but the principle applies.

Two key characteristics in many parts of the contemporary world economy are globalisation and open market access. However, the air transport industry is one of the most highly regulated sectors in the world. The importance of this, via internationally agreed standards in the context of safety, is vital, however what about from an economic perspective? As Doganis (2019, p.20) put it, "airline managers are not free agents" and the industry is constrained by the many national and international regulations which exist today, some of which have their origins before the end of World War II.

Chapter 10 will focus on the economic geography of air transport, but before the economic performance of the industry can be assessed, it is important to understand the global regulatory framework which in many ways constrains the industry. To what extent do air transport organisations have commercial freedom and to what extent are they constrained by both multinational and national regulations? Two regulatory regimes essentially exist today. First, "*bilateralism*", where air service agreements are made between states regarding international flights to and from each territory – up to the late 1970s, these were essentially protectionist and very restrictive in nature. From the late 1970s in the USA and early 1990s in Europe, there was a move to more liberal "*Open Skies*" policies, reducing and often removing economic regulations, meaning airlines were subject to much fewer restrictions on where and how often they could fly.

Many governments have moved towards more liberal, open skies policies; however, this is far from global in nature and a number of nations still rely on heavily regulated bilateral agreements for their airline sector. The focus of this book being on the geography of air transport, this chapter will begin by analysing the historic trends and global regulations which underpin much of the current industry, before focusing on the regional aspect and which one of the "restrictive versus liberal" approaches regions have taken. However, the reality is that many regions and individual nations adopt elements of both regimes and even in the most liberal of nations, many restrictions, such as on foreign ownership, remain.

9.2   Nationality and sovereignty

Although multilateral regulation and close cooperation between nations is vital to the safe and efficient operations in air transport, one of the key areas underpinning international air transport regulations is that of *nationality* and has been for over a century. Citizens, airlines, aircraft – all have nationalities – and the concept is fundamental in what can and cannot be undertaken in the air transport industry.

The United Nations (2024) is currently made up of 193 nations, each of which has its own sovereign rights, with control over its land, territorial waters and airspace. The principle of *sovereignty* is the key which provides the foundation for other regulations in all aspects of life, not just air transport, and this was first enshrined for air transport after World War I, in a multinational treaty in 1919 by the Paris Convention, which stated that: "Each nation has absolute sovereignty over the airspace overlying its territories and waters. A nation, therefore, has the right to deny entry and regulate flights (both foreign and domestic) into and through its airspace" (ICAO, 2024b).

Political geography and air transport 253

ATG Did you know?

Multinational, bilateral and national regulations

International air law falls into three categories:

1. Multilateral agreements
2. Bilateral agreements
3. National regulations

Multilateral treaties (and other international agreements) are written agreements between three or more sovereign states (or between states and international organisations) governed by international law. They span the whole spectrum of international relations, from human rights, environmental matters, space, territorial boundaries and, of course, air transport. When these treaties are ratified by individual nations, they become part of national law. That said, the nation state still has its own sovereign powers in terms of rules and regulations and in the context of air transport, ICAO is the main supranational body.

Bilateral agreements are treaties between two countries which allows international commercial air transport to take place.

National regulations govern international air law, considering multilateral and bilateral agreements. The implementation of regulations is carried out by a nation state agent – usually a Civil Aviation Authority (CAA) – who is responsible for aspects such as legislation and licensing of aircraft and airports.

It is not just the issue of airspace where nationality is key, it is also the *ownership* of key components of the air transport industry, particularly airlines. Even in contemporary times, restrictions are placed on airlines in most countries of the world, in terms of who is permitted to own (and in what percentage) airlines registered in that country. This has important ramifications for airline performance and will be discussed in more detail later in the chapter and in Chapter 10.

ATG Case study 9.1

Gulf airspace blockade: Qatar and Saudi Arabia

An example of the importance of airspace sovereignty occurred in the Gulf region in 2017, when four countries – Bahrain, Egypt, Saudi Arabia and the UAE – cut ties with Qatar for political reasons, which also involved Saudi Arabia closing its airspace to overflights from Qatar national carrier, Qatar Airways. This meant lengthy and costly diversions for flights which would take off from Qatar and use Saudi airspace en-route to their destinations. Figure 9.1 illustrates this in the context of flights from Doha (DOH) Qatar to Khartoum (KRT) in Sudan, which would normally use Saudi airspace for a large part of the flight and take approximately 3hrs 35mins. After the airspace closure and to avoid Saudi airspace, the rerouted flight was almost 6 hours long, adding significant costs and extra emissions.

254 *Fundamentals of Global Air Transport Geography*

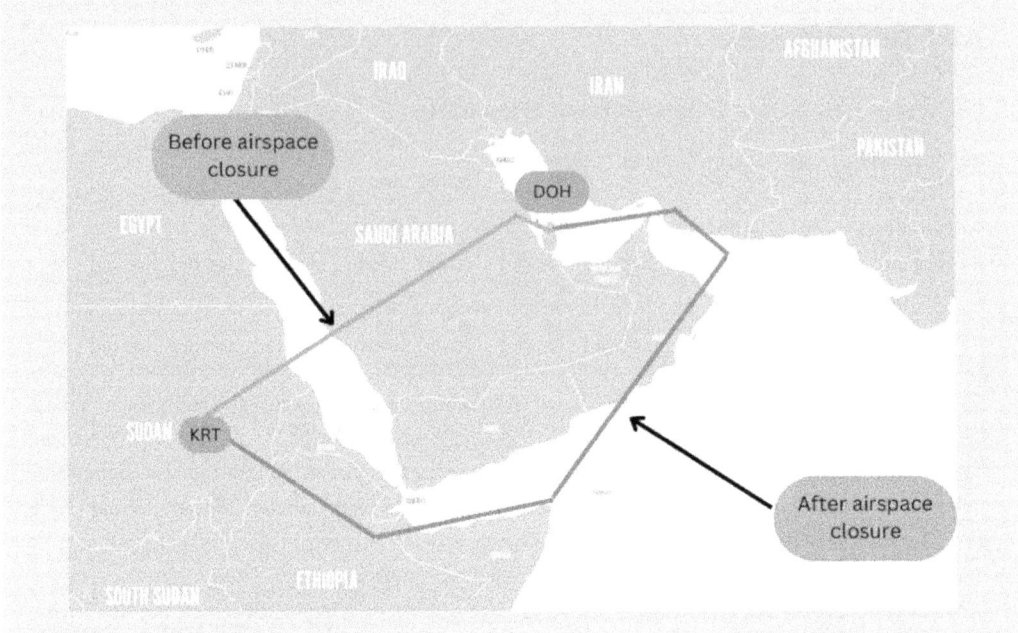

Figure 9.1 Approximate rerouting of Qatar Airways flight from Doha to Khartoum, June 2017
Source: author using data from Flightradar24 (2017)

9.3   Multinational regulations: Chicago Convention

Following the Paris Convention in 1919, there were two further Conventions – Madrid (1926) and Havana (1928) – however there was little in the way of the advancement of international law, and the rapid growth of international air transport between the two World Wars demonstrated the possibilities of civilian air transport. The single most important multilateral treaty in the history of air transport was drafted in Chicago in 1944 and ratified in 1947 – the *Convention on International Civil Aviation* (the Chicago Convention – explained in detail in Section 2.2). This is, as Meijer (2020, p.20) put it, the "constitution of the commercial aviation sector". Its primary objective was the development of international civil aviation "in a safe and orderly manner" such that international air transport services would be established on the basis of equality of opportunity and operated soundly and economically (ICAO, 2024b).

A key purpose was to obtain multilateral agreement in three key areas (Doganis, 2019):

1. The exchange of traffic rights – "freedoms of the air".
2. Control of fares and freight tariffs.
3. Control of flight frequencies.

9.3.1   ICAO

A key success from the Chicago Conference was the creation of the International Civil Aviation Organisation (ICAO), which came into being on 4 April, 1947, upon sufficient ratifications

Political geography and air transport 255

to the Chicago Convention being achieved. As a result, ICAO (a specialised agency of the United Nations) became the sole universal institution of international public aviation rights. For the first time in history, a single international organisation would standardise technical issues in aviation and harmonise practices between states. As outlined in Article 44, the purpose of ICAO was "to develop the principles and techniques of international air navigation and foster the planning and development of international air transport so as to insure the safe and orderly growth of international civil aviation throughout the world" (Convention on International Civil Aviation, 1944).

It is hard to overstate how important ICAO has been in terms of standardisation of key aspects in safety, security and in later times, environmental matters, via its standards and recommended practices (SARPs). Unfortunately, there was no general multilateral consensus on a regime for post-war *economic* regulation of international air transport (Humphreys, 2023). Final agreement on the rules governing air services became the remit of bilateral agreements between individual nation states.

9.3.2 Freedoms of the Air

The Chicago Convention reconfirmed the principle of sovereignty, stating that "each state has complete and exclusive sovereignty over the *airspace* above its territory … National sovereignty cannot be delegated". As a result, air transit and traffic rights required specific agreements, which were proposed multilaterally at Chicago. These were known as the **Freedoms of the air** – the freedom to cross the territory of another country and undertake commercial services to other countries (Figure 9.2).

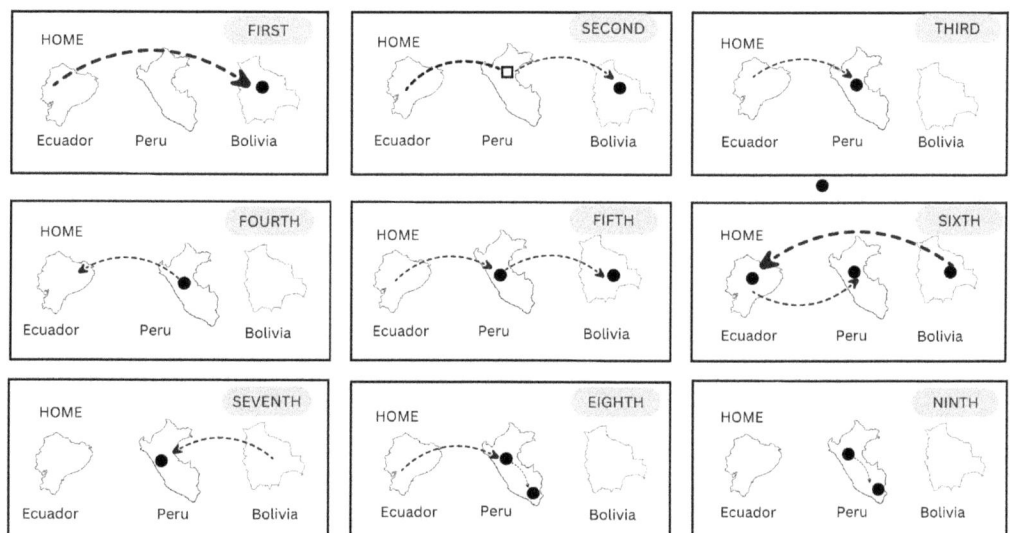

Figure 9.2 The Nine Freedoms of the Air

Source: author

256   *Fundamentals of Global Air Transport Geography*

There were two important instruments contained in the convention in this regard:

1. *The International Air Services Transit Agreement (or "Two Freedom" agreement)*, under which the aircraft of member states may:

   - Fly over each other's territory without landing (**First Freedom of the Air**). In Figure 9.2, with Ecuador as the home country, flying to Bolivia whilst overflying (but not stopping in) Peru.
   - Land in another country for non-traffic purposes (without picking up or dropping off passengers, cargo or mail), e.g. for refuelling or maintenance (**Second Freedom of the Air**). Flying from Ecuador to Bolivia and stopping in Peru to uplift fuel but not collect any revenue paying passengers or cargo.

As of 2024, 135 states had signed the Two Freedom Transit Agreement, the latest being Romania in 2021 and Brazil in 2022 (ICAO, 2024a). The agreement has in general been effective in simplifying overflight rights and assisting when diplomatic tensions arise. One exception to this was in the earlier Case Study 9.1 when, as Saudi Arabia had not ratified the Transit Agreement, they withdrew the privilege of overflight rights (CAPA, 2017).

ATG Case study 9.2

Russian airspace

Another non-signatory to the Transit Agreement – Russia – controls some of the largest airspace in the world and is very important for overflights between Europe and Asia. In 2011, the European Commission launched infringement procedures against Russia, as EU airlines were being obliged to pay Siberian overflight charges for routes to many Asian destinations, which it believed were in breach of Article 15 of the Chicago Convention, according to which "no charge shall be imposed by any Contracting State solely for the right of transit over or entry into or exit from its territory of any aircraft of a Contracting State or persons or property thereon" (European Commission, 2011).

In the Spring of 2022, following the Russian invasion of Ukraine, some countries placed sanctions on Russia and its airlines, including the EU closing its airspace to Russian traffic. Reciprocal measures by Russia meant airlines from many countries, such as from the EU, were not allowed to use Russian airspace. The closure of the Siberian flight corridor component had a severe impact on Europe–Far East traffic in terms of extra flight times, fuel burn and associated emissions. From Amsterdam Schiphol to Northern Asia, the most favourable route was often over Northern Russia and Siberia. Now, aircraft are having to fly the Southern route over Turkey and the Middle East or over the North Pole via Alaska (Schiphol, 2024).

For Japanese airlines, flights between Tokyo and London using Russian airspace used to take around 12 hours and now they can take over 14 hours, with a much more northerly routing via Alaska and Greenland (Figure 9.3).

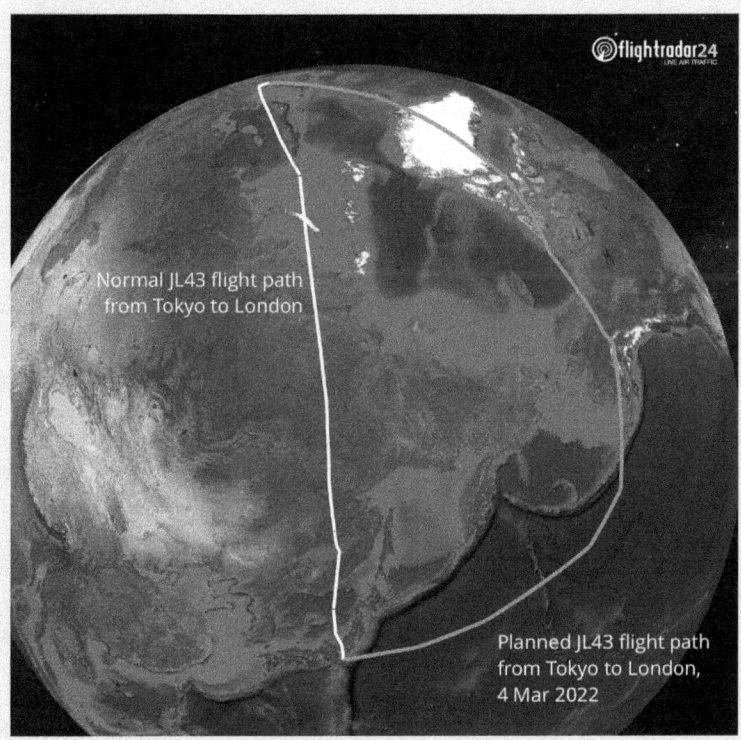

Figure 9.3 Normal route between Tokyo and London for Japan Airlines Flight 43 versus route avoiding Russian airspace, 4 March, 2022

Source: Flightradar24 (2022a)

2. *The International Air Transport Agreement (or "Five Freedoms" Agreement)*. In addition to the previous two freedoms, a further three concerning commercial transport were established, being the right to:

- Deliver paying passengers from a home country to a foreign country (**Third Freedom of the Air**). In Figure 9.2, carrying passengers from Ecuador to Peru.
- Deliver paying passengers from a foreign country to a home country (**Fourth Freedom of the Air**). For example, carrying passengers from Peru to Ecuador.
- Carry passengers from a home country to a foreign country, then drop off passengers, pick up new ones and carry them to a third, new country (**Fifth Freedom of the Air**). Carrying passengers from Ecuador to Peru, dropping off and then picking up new ones in Peru to fly to Bolivia.

Of the five Freedoms of the Air put forward at the Chicago Conference, only the first two were formally adopted. It was hoped by the delegates at Chicago that the third, fourth and fifth freedoms may be settled multilaterally but this was not agreed and therefore the only other way to secure these rights was via bilateral negotiations between individual states.

As an example of the impasse, there were nations who stated that a country's fifth freedom rights were another country's third and fourth freedom rights – allowing a foreign airline the right to operate direct international services from your territory. The USA was of the view that fifth freedom rights should be included, in order for their airlines to operate long-haul multi-stop services (Humphreys, 2023).

The "so-called" Freedoms of the Air

Since the Chicago Conference, there have been four more freedoms of the air added to the list, to make nine in total, although ICAO calls these the "so-called" freedoms, because only the first five have been officially recognised as such by international treaty:

- **Sixth Freedom of the Air**: the right or privilege, in respect of scheduled international air services, of transporting, via the home state of the carrier, traffic moving between two other states. It is the combining of third and fourth freedoms. In Figure 9.2 (with Ecuador as the home country) from Bolivia to Ecuador and Ecuador to Peru. This freedom is the basis for many of today's hub airlines such as Emirates at Dubai and Turkish Airlines in Istanbul.
- **Seventh Freedom of the Air**: the right or privilege, in respect of scheduled international air services, granted by one state to another state, of transporting traffic between the territory of the granting state and any third state with no requirement to include on such operation any point in the territory of the recipient state. For example, an Ecuadorian airline carrying passengers between Bolivia and Peru.
- **Eighth Freedom of the Air**: the right or privilege, in respect of scheduled international air services, of transporting cabotage traffic between two points in the territory of the granting state on a service which originates or terminates in the home country of the foreign carrier or (in connection with the so-called Seventh Freedom of the Air) outside the territory of the granting state (also known as "**consecutive cabotage**"). For example, an airline flying from Ecuador to Lima in Peru and then on to Cusco in Peru.
- **Ninth Freedom of the Air**: the right or privilege of transporting cabotage traffic of the granting state on a service performed entirely within the territory of the granting state (also known as "**stand alone**" *cabotage* or "*open skies*") (ICAO, n.d.). For example, an Ecuadorian airline flying passengers on a single leg flight between Lima and Cusco.

9.4 Bilateral air service agreements

As a result of the lack of agreement in terms of economic freedoms beyond the first two essentially transit "freedoms of the air", there then became requirements for inter-state negotiations to take place and for each state to negotiate relevant access for its carrier(s) with another state. Bilateral ASAs are agreements, often treaties, negotiated between countries that regulate the terms from which airlines from the respective countries can operate international services between their respective territories and sometimes beyond (fifth freedom).

One of the first bilaterals was an aerial navigation agreement between France and Germany in 1913. Ever since, bilateral ASAs have been the cornerstone of economic air transport regulation, with new ASAs regularly being negotiated, allowing existing and new airlines to expand their global footprint. According to the Australian Government (2024), who have negotiated 90 bilaterals themselves, there are currently over 3,000 international bilateral agreements.

9.4.1 Bilateral provisions

Bilateral ASAs contain a few key provisions:

1. **Traffic rights**: These are the routes which airlines can fly between the bilateral partners – for example, third and fourth freedoms – and sometimes beyond (fifth freedom, which requires agreement from the third country).
2. **Frequency and capacity**: How many flights or seats (often quoted weekly) can be offered between the two countries.
3. **Designation, ownership and control**: The number of airlines allowed to operate on the route. Following the Chicago Conference, this was often only one from each state and usually the national, or "flag", carrier, such as British Airways, Air France, Aeroflot, etc. Ownership and control restrictions were usually put in place, meaning that only nationals of the respective country were allowed to own (and essentially control) the designated airlines. Much of the ownership and control restrictions are still in force today.
4. **Tariffs (prices)**: Many early agreements required airlines to submit ticket prices to the relevant authorities for approval, generally based on operating costs plus a reasonable profit. Much of the tariff coordination was done through IATA for fare agreements and essentially fares were either set or approved by governments.
5. **Safety, security, customs and immigration provisions** were often also included.

Many of the original bilaterals between 1947 and the late 1970s (and a number still in force in parts of the world today) were very protectionist in nature, severely limiting competition and preventing airlines from operating routes where there may have been economic benefits, if they were not included in the bilateral.

ATG Case study 9.3

Bermuda I Agreement

At the Chicago Conference, the US was seeking more liberal agreements and a larger exchange of traffic rights, whilst many other countries, such as the UK, were more protectionist. However, in 1946, the US and UK met in Bermuda to discuss aviation policies and reached a compromise acceptable to both (although more so to the US side than the UK). The Bermuda Agreement established the framework for international air travel between the two countries, allowing for the regulation of air routes, capacity and fares and the exchange of landing rights and other privileges.

Bermuda I (as it came to be known) was a traditional bilateral, but slightly more liberal in the sense that fifth freedom rights were more widely available and there was no control of frequency or capacity on the routes between the two countries (Doganis, 2019).

The Bermuda I Agreement served as a prototype model for ASAs adopted by countries around the world over the next 30 years, during which time the USA entered into Bermuda I-type agreements with most of the 75 states with which it had aviation relations (Dempsey, 2016).

By the 1970s, the economic and political arguments for more liberal aviation policies began to seriously develop in certain parts of the world, such as in the USA, the Netherlands and Singapore. Arguments were being made that economic regulations were keeping airfares artificially high and that restrictive market access and restrictions on routes, frequency and capacity were stifling competition, to the detriment of the consumer. These were certainly not uniformly accepted arguments amongst the proponents of regulation, however, but by the late 1970s, regulatory changes were happening.

9.5   USA – airline deregulation

The year 1978 was famous for the movies *Grease* and *Superman*; it was also the year Liverpool FC won the European Cup and Argentina won the football World Cup; the year when the Dallas Cowboys won the Superbowl; and it was also one of the most momentous years for air transport – the year of the US Airline Deregulation Act. It did not involve Olivia Newton-John or John Travolta (who coincidentally is a qualified airline pilot, including on the Boeing 747!), but the impacts of this Act are still being felt, as the air transport industry evolves over 45 years later.

> **ATG Did you know?**
>
> **Deregulation**
>
> *Deregulation* (and the often-used term *liberalisation*) refers to the process of reducing or removing government regulations and restrictions on the airline industry to allow greater competition and efficiency. The process meant that air transport regions would become much more governed by market forces, and alongside technological developments it was responsible for much of the air transport industry we see today.

The US Airline Deregulation Act was signed into law by President Jimmy Carter on 24 October, 1978, and transformed the US airline industry. It removed federal control over fares, routes and market entry, leading to a more competitive market and the large growth of the US domestic airline market. Airlines could now fly where they wanted and charge what the market would withstand.

Even earlier, in 1977, was the deregulation of air freight services, which according to Humphreys (2023), was far less contentious than passenger services, even attracting the support of the established all-freight airlines.

Deregulation created dozens of new airlines such as PEOPLExpress and New York Air, with others such as Braniff, Eastern and Northwest extending their networks to compete with the largest carriers. Intrastate airlines, such as Southwest in Texas, were also able to expand (Smithsonian Institution, 2021). Southwest is an example of an airline which prospered from deregulation, originally a small carrier flying between Houston and Dallas, growing into what is now the fourth largest and one of the most consistently profitable airlines in the USA.

To increase their efficiency, some airlines adopted more of a *hub-and-spoke system* using a few major airports as central connecting points. This often created "fortress hubs" which effectively restricted market entry and had a massive impact on the geography of air transport within the USA. An exemplar of this is Delta and their Atlanta hub. The choice of routes became much greater for the consumers due to the connection possibilities, but to optimise connection times, flights were scheduled in waves which often led to congestion and delays and increased the airline's vulnerability to disruptive events (Ison and Budd, 2020).

The 1978 Airline Deregulation Act had significant impacts on the industry, with both positive and negative outcomes. Oster Jr. (1987) and Evans and Kessides (1993) point out some of the benefits of deregulation, including lower fares, better service and increased airline earnings. However, Martello (2018) and Card and Saunders (1998) discuss the negative consequences, such as major bankruptcies, unstable fares and labour disruptions.

9.5.1 US open skies

The US Deregulation Act was aimed at the *domestic* US airline market. Liberalising *international* markets is more complicated, as was shown at Chicago in 1944 and for many years afterwards, as two (or more) governments must agree. In the USA, the benefits of domestic deregulation meant that the pressures for this to be applied internationally were growing.

In 1978, the first revised bilateral which had a focus on increased competition was signed between the USA and the Netherlands (another champion of more liberalised air markets) including multiple airline designations, increasing the number of points in each country to be served with no frequency restrictions and no restrictions on sixth freedom rights. This was followed by revised bilaterals with countries such as Singapore and Thailand. However, airlines such as Delta felt that these did not go far enough and were pushing for a fully liberalised open skies environment, which the US Department of Transportation also concurred with in terms of consumer and airline benefits (Doganis, 2019).

ATG Did you know?

Open Skies

Open Skies agreements are bilateral or multilateral agreements between states that aim to liberalise international aviation markets. They have at their core a reduction in government restrictions on how airlines operate internationally between the respective states and are generally economic in nature. This does not include the key area of safety, where states maintain strict oversight of air transport operations.

Some key features of Open Skies agreements include:

1. **Open market access** to any airline (which meets the safety standards) to fly to any point in the respective countries without significant restrictions.
2. **Freedom of pricing** – airlines set their own fares.
3. **Capacity and frequency** – no limitations placed on the number of flights or number of seats offered.
4. **The ability to codeshare and operate other partnerships.**

Potential issues which need to be considered include:

1. Obtaining airport slots at congested airports such as Amsterdam Schiphol.
2. Operating to a level playing field. This is where the issue of state subsidies can become an issue.
3. Regulatory alignment, when dealing with different regimes.

In 1992, the USA launched its "Open Skies initiative" and the first essentially Open Skies bilateral was signed between the USA and the Netherlands, which allowed open route access between any two points in the two countries and unlimited fifth freedom access. The USA has 135 Open Skies partners, in all regions of the world, ranging from Canada to the UK, Colombia to Chile, Cameroon to Botswana and Malaysia to New Zealand. The most recent being with Mozambique in December 2023 (US Department of State, 2024a).

A key issue, which is witnessed not only in the USA but throughout most of the world, relates to how truly "open" the Open Skies agreements are.

First, there are very few countries which permit "Ninth freedom" (**cabotage**) rights. An example would be easyJet (a UK airline) flying between New York and Washington DC in the USA, which by law cannot happen. The US air cabotage law – 49 U.S.C. 41703 – "prohibits the transportation of persons, property, or mail for compensation or hire between points of the U.S. in a foreign civil aircraft" (Cornell Law School, n.d.).

Second, ownership and effective control restrictions are one of the few remaining parts of the traditional bilateral agreements still in place under Open Skies. US federal laws limit foreign ownership of US airlines to 25% and that the airlines are "effectively controlled" by US citizens, which limits investment potential and are some of the most restrictive in the world.

9.6   Europe

As in the US, there was an increasing desire in parts of Europe for increased liberalisation of air services and more open competition. A key difference in Europe was the multilateral approach taken with the various EU member states, rather than bilateral.

9.6.1   *The three packages*

The liberalisation of the EU internal aviation market was carried out gradually with the adoption of three packages of measures (European Parliament, 2019) and it was only in the third package that full "Open Skies" was established – a single air transport market in Europe.

- The First Package (1987)
  This limited the rights of governments to reject the introduction of new fares and provided some flexibility concerning seat-capacity-sharing.

- The Second Package (1990)
  This further opened the market, allowing greater flexibility over fares and capacity-sharing. It also allowed all EU carriers to carry an unlimited number of passengers or cargo between their home country and another EU country.

- The Third Package (1993)
  This completed the process by providing open market access and full airline freedom concerning fares and rates. Airlines from any EU member state could operate with full market access between any two points within the EU. This also included cabotage rights, enabling an airline from any EU member state to operate internally within any other EU country, regardless of airline nationality. So, for example, Ryanair (an Irish airline) could now operate between Frankfurt in Germany and Milan in Italy and on a cabotage basis between Milan and Rome, both in Italy.

One exception could be for routes with public service obligations (PSOs), but these were subject to conditions and for a limited period of time (European Parliament, 2023). Via treaties, these rights were extended to non-EU member states Iceland, Norway and Switzerland.

In many ways, it could be argued that EU Open Skies is even harder to achieve than in the USA, with currently 27 countries in the EU, each with their own laws, regulations, cultures and precedents. Many countries also had their own national "flag" carriers which had historically been protected through traditional bilaterals and powerful employee trade unions. However, national air carriers are now *"Community air carriers"*.

As in the USA, there are still restrictions on *foreign ownership*. Airlines registered in the EU must be "owned and effectively controlled by Member States and/or nationals of Member States and their principal place of business must be located in a Member State" (European Parliament, 2023). Essentially this means more than 50% must be in the hands of EU member states or nationals. However, the key here is that it is any EU member state. Therefore, a German national can own an airline registered in Spain, but a US national cannot.

9.6.2 *European Common Aviation Area (ECAA)*

The European Common Aviation Area (ECAA) was an agreement to progressively integrate the EU's neighbours in south-east Europe into the EU's internal aviation market, aiming to deliver economic benefits for air travellers … ensuring that ECAA airlines have open access to the enlarged European single market in aviation (EUR-Lex, 2020).

In addition to EU countries (plus Norway and Iceland), members include Albania, Bosnia and Herzegovina, Montenegro, North Macedonia, Serbia and Kosovo. It was signed in 2006 and entered into force in 2017.

"Neighbourhood agreements" were also reached by the EU with Morocco in 2006, Georgia and Jordan in 2010, Moldova in 2012 and Israel in 2013 (European Parliament, 2019).

9.6.3 *EU/non-EU Open Skies*

As a result of the US move towards Open Skies, one objective was to secure more Open Skies agreements with European countries. The first was with the Netherlands in 1992 and by the end of the 1990s, several such agreements had been concluded. As a result of the aviation single market, the European Commission decided to challenge individual bilateral agreements between EU and non-EU member states, with the aim of creating essentially multilateral agreements. A key issue here is that the bilateral agreements contained nationality clauses. So, in the Germany/US agreement, to take advantage of Open Skies, carriers had to be German or US. They could not be from another EU state, such as Spain, so a Spanish airline established in Germany could not benefit from the traffic rights.

The European Commission instigated legal action against the eight member states (Belgium, Denmark, Germany, Luxembourg, Austria, Finland, Sweden and the UK) which had signed Open Skies agreements with the USA. In 2002, the European Court of Justice (ECJ) ruled that the nationality clauses in the agreements were a violation of what was called the "Right of Establishment" enshrined in Article 43 of the EC Treaty. In other words, discrimination based on nationality was illegal within the Community and that carriers were "Community carriers", thus bilateral agreements between EU states and non-EU partners had to be brought in line with EU law – replacing "national designation" airline clauses with "EU designation". This was no easy task!

As stated by the European Parliament (2019), bringing agreements into line with EU law would involve two methods:
- Bilateral negotiations between each EU member state concerned and its partners, amending each bilateral air service agreement *separately*.
- Bilateral negotiation of single *horizontal agreements*, with the Commission acting on a mandate from the member states. The purpose of each horizontal agreement would be to amend the relevant provisions of all existing bilateral air service agreements in a single negotiation with a third country.

As of 2024:
- Separate bilateral negotiations: there have been changes with 73 partner states, representing 340 bilateral agreements corrected.
- Horizontal negotiations: there have been changes with 41 countries and one regional organisation with eight member states, representing an additional 670 bilateral agreements.

From an airline perspective, the major FSNC airlines began a period of consolidation following the 2002 judgement. Air France and KLM merged (although they still operated separately); the Lufthansa group now contains Swiss, Austrian Airlines and Brussels Airlines; and IAG was created, consisting of British Airways, Iberia, Vueling and Aer Lingus.

ATG Case study 9.4

EU–US Open Skies

Signed in 2007, the EU–US Open Skies agreement became effective in 2008. Following the ECJ ruling of 2002, the European Commission was given an authorisation to negotiate an air transport agreement with the USA that applied to the EU. At the time, 16 EU member states already had Open Skies agreements in place with the USA (EUR-Lex, 2022).

The agreement granted traffic rights and market access to any EU or US member state to any US or EU airline. Fifth freedom rights were also granted and there were no restrictions on frequency or tariffs. Codesharing possibilities were extended. EU and US airlines could now offer all-cargo flights under seventh freedom rights. The agreement also strengthened cooperation in security, safety and environmental matters.

What the agreement did not do was offer cabotage rights and thus, EU airlines are not able to operate domestically within the US or vice-versa. Nor did it alter the requirement for a maximum foreign investment by EU nationals in US airlines, of 25%.

There are now many more routes being offered and increased competition on many transatlantic routes, and the agreement has liberalised one of the world's largest international aviation markets.

9.7 Latin America and the Caribbean

In comparison to the USA and Europe, the process of liberalisation in the Latin American and the Caribbean (LAC) region remains generally fragmented. At one end of the spectrum, countries such as Chile, Uruguay, and Paraguay have chosen to offer foreign airlines an open and flexible framework for operations. At the other end, Argentina, Venezuela and Bolivia have historically opted for a more restrictive approach, often favouring the interests of national carriers and organised groups. Somewhere in the middle are Mexico, Brazil, Peru and Colombia, which are generally considered to have embraced a gradual approach to air liberalisation, with certain conditions on the granting of fifth freedom traffic rights (Echevarne, 2024). Changes in the political cycle within countries have often resulted in changes to aviation policies.

Several subregional agreements have been formalised, to pursue a more multilateral approach to liberalisation:

1. Latin American Civil Aviation Commission (LACAC)

The civil aviation organisation of Latin American States was founded on 14 December, 1973, and is formally called *Comisión Latinoamericana de Aviación Civil*. The primary objective of this Commission is to provide the civil aviation authorities of the member states with a suitable framework within which to discuss and plan all the necessary measures for cooperation and coordination of civil aviation activities. LACAC has played an important role in coordinating and establishing policies and common approaches covering a wide range of subjects pertaining to the development of civil aviation among member states (ICAO, 2024b).

In March 2008, the LACAC initiated a multilateral Open Skies agreement to unify all its 22 member states under a single aviation market. The Open Skies multilateral agreement for member states of the LACAC was finalised in 2010 and entered into force in 2019 upon ratification of three member states, namely: Brazil, Panama and Uruguay. The agreement allows for maximised market access up to cabotage rights with the flexibility to extend the granting of fifth and sixth freedom rights to third-party states (ICAO, 2022).

2. Caribbean Community (CARICOM)

CARICOM initially concluded a multilateral agreement concerning the operation of air transport services within the Caribbean Community in 1998. The agreement was revised in 2018 to implement the liberalised air transport services regime envisaged by the Revised Treaty of Chaguaramas. The revised multilateral air services agreement, which entered into force in 2020, aims to establish a single market for air transport services by removing barriers to airline ownership to allow for establishment of community carriers and liberalised market access up to cabotage rights (ICAO, 2022).

Table 9.1 The 22 member states of LACAC

Argentina	Aruba	Belize	Bolivia	Brazil	Chile
Colombia	Costa Rica	Cuba	Dominican Republic	Ecuador	El Salvador
Guatemala	Honduras	Jamaica	Mexico	Nicaragua	Panama
Paraguay	Peru	Uruguay	Venezuela		

Table 9.2 The 15 member and five associate member states of CARICOM

Member states			Associate members
Antigua and Barbuda	Bahamas	Barbados	Anguilla
Belize	Dominica	Grenada	Bermuda
Guyana	Haiti	Jamaica	British Virgin Islands
Montserrat	Saint Lucia	St Kitts and Nevis	Cayman Islands
St Vincent and the Grenadines	Suriname	Trinidad and Tobago	Turks and Caicos Islands

266 *Fundamentals of Global Air Transport Geography*

3. Association of Caribbean States (ACS)

The Air Transport Agreement between the member states and associate members of the Association of Caribbean States (ACS) was signed in 2004, which liberalised the market up to fifth freedom rights. The agreement entered into force in 2008 and was ratified by 12 states.

ATG Case study 9.5

Liberalisation in Argentina

A good example of a country which has been changing policies following swings in the political cycle is Argentina. It has historically adopted a more restrictive, protectionist approach to its airline industry. When Javier Milei became President in December 2023, he indicated a clear change in direction for liberalisation in Argentina, with plans to privatise Aerolineas Argentinas (the state-owned airline) and to introduce an Open Skies policy, which would allow any foreign entity to operate flights in Argentinian airspace. Airlines would have access to fly either domestic or international routes through its airspace.

In March 2024, a Memorandum of Understanding (MoU) was signed between Argentina and Brazil to establish an Open Skies policy between the two countries, which will eliminate the weekly limits on regular passenger flights and facilitate the authorisation of cargo flights, on a seventh freedom basis (Gov.br, 2024). By July 2024, Open Skies agreements had also been signed by Argentina with Chile, Peru, Panama, Ecuador, Uruguay and Canada, and carriers can now petition to operate as many routes as they want, subject to safety approval (Reuters, 2024).

The LATAC region has historically struggled to achieve profitability, with many regional differences and changing political climates, often causing changes in aviation policies. These can make it difficult for airlines to plan with some kind of certainty, and according to Echevarne (2024), the region lacks a supranational body like the EU.

When it comes to connectivity, this difference is illustrated by analysing intraregional international routes, with 462 due to operate in the LAC in Summer 2024, according to the Official Airline Guide (OAG). This compares to over 6,400 intraregional international routes in Europe (Aviation Week, 2024). There is also a lack of international routes from secondary destinations – for example, around 98% of international traffic in Peru – a country with 33 million people – operates via Lima.

Much of Latin America is also mountainous (see Chapter 11) and hence air transport is even more important for economic and social development, for which further liberalisation could provide a key to opening markets throughout the LAC region.

9.8 Middle East

The World Bank (2024a) states that the Middle East is comprised of Bahrain, Egypt, Iran, Iraq, Jordan, Kuwait, Lebanon, Libya, Oman, Qatar, Saudi Arabia, Syria, United Arab Emirates (UAE), West Bank and Gaza and Yemen. The main engines of growth in the region have come from the six Gulf Cooperation Council (GCC) members – Bahrain, Kuwait, Oman, Qatar, Saudi Arabia and the UAE. The GCC is a regional intergovernmental political and economic alliance.

The massive growth in air transport from, to and via the Middle East has reshaped global air transport geography and the key hubs such as Dubai in the UAE and Doha in Qatar, via their respective carriers Emirates and Qatar Airways, have been instrumental in these changes. Anyone who has ever visited Dubai International Airport at 0200 and witnessed it bustling like a downtown city (when airports such as London Heathrow are essentially asleep from a flying perspective) will testify to the level of activity here. According to ACI World (2024), Dubai is now the busiest airport in the world for international passenger numbers.

Almost all of Asia, Africa and Europe can be reached within eight hours of these hubs, making them ideal connecting points for traffic to/from Europe to Asia, all the way to Australia and New Zealand. Thousands of travellers arrive every hour from countries such as China, India, Australia and the UK, many of whom are using the geographic position of the Middle Eastern hubs to connect to other flights and thus do not even leave the airport. Sixth freedom rights are especially important here, for example passengers from London to Sydney, in transit at Dubai, are classed as sixth freedom travellers.

For many years, air traffic in the Middle East was heavily regulated, however there has been a gradual progression in liberalisation in the region over the last 20 years – in some countries more than others. Some states have extremely small, or no, domestic market – such as Dubai, Abu Dhabi and Qatar, whereas Egypt and Saudi Arabia have a much larger domestic market. Some international markets have been much more liberalised than the domestic markets, for example the USA has Open Skies agreements with Jordan (1996), UAE (1999), Bahrain (1999), Qatar (2001), Oman (2001), Kuwait (2006), Saudi Arabia (2011) and Yemen (2012).

Within the region, liberalisation allowed new airlines to commence operation. In the UAE, Air Arabia (2003) was the region's first low-cost carrier (LCC), operating out of Sharjah. Flydubai in Dubai and Wizz Air Abu Dhabi have continued this trend. To take advantage of Open Skies in Kuwait, Jazeera Airways was allowed to commence operations and compete with Kuwait Airways (O'Connell and Williams, 2010).

The Arab Air Carriers Organisation (AACO), which includes the major Middle Eastern carriers and has aero-political affairs as one of its focus areas, recognises the importance of liberalisation to trade, tourism, job creation and economic growth. However, although some states have adopted Open Skies policies, relations between some Arab states are still mainly bound by bilateral agreements mostly covering third and fourth freedoms, with a small number of fifth freedom rights granted.

The Damascus Convention, which was ratified in 2007 by Jordan, Lebanon, Morocco, Oman, Palestine, Syria, Yemen and the UAE, was established as the multilateral agreement for the liberalisation of air transport between Arab countries, covering aspects such as market access up to fifth freedom rights and fair competition (AACO, 2019). To date, it is the only multilateral agreement currently in force.

The benefits of more liberal aviation policies in the Middle East were assessed by Cristea, Hillberry and Mattoo (2015) who believed that plurilateral agreements (agreements between more than two countries, but not a great many) amongst Arab states could lead to a 30% increase in intraregional passenger traffic.

Potential issues associated with any international bilateral or Open Skies agreements do complicate matters. For example, American Airlines, United and Delta (dubbed the "US3") alleged in 2015 that the rapid growth of Emirates, Qatar and Etihad (the "ME3") was down to billions of dollars of unfair government subsidies, in violation of the US Open Skies policy.

ATG Did you know?

Liberalisation, globalisation and the lessons from different sectors

The trend towards liberalisation is not exclusively an air transport one, however it can be very beneficial overall for both consumers and producers, although the process is not easy. Evidence from the Retail Banking, Energy, Telecommunications and Media sectors showed the positive influences on competition, efficiency and costs (IATA, 2007). For example:

- **Energy markets.** In the EU electricity prices were 10–20% lower and gas prices 35% lower than without liberalisation. Increased competition improved capacity utilisation.
- **Retail Banking.** Relaxing ownership restrictions in the USA increased service quality and ease of access (although evidence for service quality improvements in the airline industry has been mixed). Improving profitability by lowering costs, improving efficiency and developing economies of scale.
- **Media.** Increased output and choice and helping to transfer managerial and technological best practice.
- **Telecommunications.** In Japan, productivity increases were experienced for both incumbents and new entrants. Increasing investment and lowering cost of capital, as firms have access to more efficient sources of capital.

As pointed out by Finger and Button (2017) in Humphreys (2023), airline liberalisation has been located amongst a broader movement which is at least partly the product of globalisation, as with the sectors just mentioned. The air transport industry has exerted increased pressure on governments for liberalisation.

9.9 Asia Pacific

Stretching west from Japan and Korea to Afghanistan and Pakistan, south and east from Mongolia to Australia, New Zealand and the Pacific Islands and with India and China in between, the Asia Pacific region has a vast array of differences in geographic size, populations, political systems, economic development and regulatory structures, creating an extremely complex air transport geography. In the air transport industry, it has been a key area of growth, especially in China and India, which according to IATA (2023) is expected to yield the highest annual passenger growth rate through to 2040 (4.5% versus a world average of 3.4%). When the two most populous countries in the world (India and China) are part of this region, accounting for the best part of 3 billion people, then a 4.5% growth rate could turn into a substantial actual number of passengers using air transport.

Singapore was an early adopter of liberalisation and was the first state in the Asia Pacific to make an agreement on Open Skies with the USA, first applied in 1997, just ahead of Taiwan. India liberalised domestic services and in the early 2000s, airlines such as SpiceJet and Indigo commenced operations. Australia was a key proponent of liberalisation, on both domestic and international routes, although compared to the over 60 Open Skies agreements signed by Singapore, Australia has seven – with the UK, USA, Singapore, New Zealand, India, China and Japan (UNSW, 2023).

Table 9.3 Air transport liberalisation arrangements within ASEAN countries

No.	Legal modalities	Scope	Signing year	Entry into force
1	ASEAN Multilateral Agreement on the Full Liberalisation of Air Freight Services	Unlimited Third, Fourth and Fifth Freedom Rights among all points with international airports in the ASEAN.	May 2009	November 2009
2	ASEAN Multilateral Agreement on Air Services (MAAS)	Unlimited Third, Fourth and Fifth Freedom Rights between ASEAN capital cities.	May 2009	December 2009
3	ASEAN Multilateral Agreement on Full Liberalisation of Passenger Air Services (MAFLPAS)	Unlimited Third, Fourth and Fifth Freedom Rights between any ASEAN cities.	November 2010	July 2011

Source: adapted from ICAO (2022)

Regional liberalisation in Asia was mainly centred around the Southeast Asian region via the Association of Southeast Asian Nations (ASEAN). In 2007, the 13th ASEAN Summit endorsed the establishment of the ASEAN Single Aviation Market (ASAM) by 2015 in support of the development of the ASEAN Economic Community. ASAM aims to liberalise the ASEAN market up to the seventh freedom right with cabotage rights and consider the introduction of the ASEAN Community Carrier via relaxation of ownership and control requirements. The development of ASAM was supported by the Roadmap for Integration of Air Travel Sector (RIATS), which, through this, a progressive liberalisation approach of the ASEAN market among its ten member states was established (Table 9.3).

Other dialogue partners have agreed liberalisation arrangements with ASEAN, including China in 2011 and the European Union in 2021. Further liberalisation beyond fifth freedom, including cabotage, are very much a work in progress.

Multilateral approaches in liberalising the air transport market are slowly finding their place within the Asia Pacific region despite challenges of geographical barriers, economic standing disparity, diverse technological developments and differing socio-political climates (ICAO, 2022). However, to date, Europe is the only region, in broad terms, where a multilateral regional Open Skies framework has been successful. The EU had a legal framework to implement Open Skies, although there was reluctance from many EU member states. No such legal framework exists across the Asia Pacific region.

9.10 Africa

Africa is the smallest world air traffic market, although in many ways this is surprising, considering it is the second largest in terms of geographic size, there are large distances between population centres and major urban areas and it has a relatively under-developed ground transport network across the continent. It is home to 12% of the world's people, but less than 1% of the global air service market (World Bank, 2024b). According to the World Bank, part of the reason for Africa's under-served status is that many African countries restrict their air services markets to protect the share held by state-owned air carriers.

9.10.1 *Yamoussoukro Decision (YD)*

In November 2000, a policy framework was adopted called the Yamoussoukro Decision (YD), to attempt to gradually liberalise market access in Africa. It was adopted from a recognition that the restrictive and protectionist bilateral ASAs in force were hampering air transport. The YD had its origins in 1988 when African Aviation Ministers met and agreed a New African Air Transport policy – the Yamoussoukro Decision. On paper, the aims of the YD were far-reaching:

- Three-phase process over eight years.
- Focus on internal market liberalisation and fair competition.
- Removing restrictions on traffic rights, capacity and frequency between city pairs.
- The YD having precedence over any other bilateral agreements.
- Compliance with established ICAO safety standards and recommended practices.

Its success was limited, however, as although 44 states signed it, a number did not implement it. Signatories felt free to ignore its provisions when the interests of their own airlines and economies were perceived to be at stake (Humphreys, 2023). There were also differences in macro-economic policies and strategies among states, different levels of development and deep concerns of individual states' interests, market position and the weak assurance of fair competition and a level playing field for both big and small airlines (AFCAC, 2023).

9.10.2 Single African Air Transport Market (SAATM)

Another liberalisation attempt was made in 2018, under the auspices of the African Union (AU) and their Agenda 2063, with a Declaration of the Establishment of a Single African Air Transport Market, with a focus on implementing the YD and to create an African Open Skies area like the EU. As the AFRAA (2019) stated, "towards a virtuous cycle of air transport development in Africa". By 2022, 35 AU member states, with a total population of over 800 million people and 61% of the African population, had joined SAATM. Some key objectives for SAATM included:

- Fully implementing the YD.
- Lifting market access restrictions on airlines.
- Liberalising frequency and capacity limitations for both passenger and cargo operations.
- Removing restrictions on ownership.

Time will tell as to the success of SAATM and the AFRAA (2019) recognises the following challenges:

- Delays on implementing and reporting on Concrete Measures (essentially the standards) by signatory states.
- Complex local procedures.
- Protectionism of national carriers.
- Granting limited frequencies and restricted capacity.
- High cost of operations due to unconventional taxes.

If implemented by all states, the YD and the SAATM have the potential to significantly improve connectivity, promote intra-Africa trade and tourism and create large socio-economic benefits across the African region. IATA (2024b) believes that if just 12 key African countries opened their markets and increased connectivity, an extra 155,000 jobs and $1.3 billion in annual GDP would be created in those countries.

ATG Case study 9.6

Air service liberalisation in Nigeria

Nigeria started to reform its air transport industry in the late 1990s, deregulating its domestic market and privatising airline and handling companies (InterVistas, 2014). In 2000 they entered into an Open Skies Air Transport Agreement with the USA, which entered into force in 2024 (US Department of State, 2024b) including provisions for unrestricted capacity and frequency of services, open route rights, a liberal charter regime and open code-sharing opportunities.

An analysis by Ismaila, Warnock-Smith and Hubbard (2014), researching the impacts of liberalisation on Nigeria, discovered that traffic volumes were significantly higher on routes with more liberal bilaterals and that the traffic increased with the level of liberalisation. They estimated that there could be at least a 65% increase in traffic growth if liberalisation of market access to the Open Skies agreement level were implemented. Traffic volumes could increase by 117–137% if ownership restrictions were removed and market access liberalised together and changed to the YD model.

Chapter review questions

9.1 What are the direct and indirect ways in which governments can influence the air transport industry? For any ONE example, how does the government in your own country influence the air transport sector?

9.2 What are the key features of and differences between multilateral, bilateral and national regulations? For your own country, can you find ONE example of each and how it is applied?

9.3 What is the maximum percentage ownership allowed of airlines by foreign nationals in both the USA and Europe? If you live outside these regions, can you find out what restrictions are in place in your own country? What are the pros and cons of having such nationality ownership restrictions?

9.4 Discuss the roles and responsibilities of ICAO. Why was the Chicago Conference in 1944 so important to the development of the air transport industry and which key aspects are still in force today?

9.5 Explain the principles in each of the nine Freedoms of the Air. Using flight tracking sources such as Flightradar24, flight search engines or airlines' own websites, can you find an example of each of the nine freedoms on flights in operation today?

9.6 Explain the key features of traditional airline bilateral agreements. What are the benefits and drawbacks of some of these features, for both airlines and consumers?

9.7 What were the key principles of the US Deregulation Act? Why has this been so important to air transport globally and why do you think the USA was the first country to proceed down this route?

9.8 Explain the key principles of Open Skies. In Europe, how has this transformed the air transport landscape? For your own country, list THREE benefits and three drawbacks of having an Open Skies policy.

9.9 For any ONE country outside the USA and Europe, can you discover how restrictive or liberal the air transport agreements are? Explain how these could benefit or hinder the air transport industry in this country, in terms of both airlines and consumers.

ATG trivia

Time travel: GMT, UTC and local times

Without possessing a Tardis like Doctor Who, how is it possible to fly thousands of miles and arrive before you left? Aside from the great speeds of air travel, the answer lies in the differences in time zones. One of the most famous examples was flying from London or Paris to New York on Concorde. Flight BA001 was due to leave London Heathrow (LHR) at 1030hrs local time and arrive in New York JFK around 0930 local time. The block time of the flight being around four hours and New York being five hours behind London meant the aircraft arrived before it departed.

More recently, a flight arrived in the previous year. United Airlines 858, a Boeing 777-300 travelling from Seoul to San Francisco, departed from Seoul at 0029 local time on 1 January, 2023, and landed in San Francisco at 1701 local time on 31 December, 2022 (Flightradar24, 2022b). Seoul is 16 hours ahead of San Francisco and the flight time was over nine hours. Happy New Year!

Inside the air transport industry, however, the concepts of Greenwich Mean Time (GMT) or Coordinated Universal Time (UTC) are used. GMT is a time zone and UTC is a time standard. Essentially, they refer to the same times and the purpose is to provide a standard global time usage. UTC (sometimes referred to as "Zulu time") is a coordinated time scale which is used to keep time synchronised across the world. This helps the industry to monitor flight operations without having to wonder how many hours ahead or behind one country is to another.

So, in the example of LHR–JFK, in the UK in winter, the departure time of BA001 would be 1030UTC (1030 local time, as UTC = local in the UK in winter) and the expected arrival in JFK would be 1430UTC, for a four-hour block time. As New York is five hours behind London, the local time would be 0930. So, travelling back in time. Or not?

References

AACO (2019) *Liberalization*. Available at: www.aaco.org/policy/liberalization.

ACI World (2024) *Top 10 busiest airports in the world shift with the rise of international air travel demand*. Available at: https://aci.aero/2024/04/14/top-10-busiest-airports-in-the-world-shift-with-the-rise-of-international-air-travel-demand/.

AFCAC (2023) 'Yamoussoukro Decision (YD) – AFCAC'. Available at: www.afcac.org/yamoussoukro_decision/.

AFRAA (2019) *The Single African Air Transport Market (SAATM). Towards a virtuous cycle of air transport development in Africa*. Available at: https://afraa.org/wp-content/uploads/2019/11/%E2%80%A2Single-African-Air-Transport-2019.pdf.

Australian Government (2024) *Bilateral air services system*. Available at: www.infrastructure.gov.au/infrastructure-transport-vehicles/aviation/international-aviation/bilateral-air-services-system.

Aviation Week (2024) *Bureaucracy continues to stifle Latin America's air transport market*. Available at: https://aviationweek.com/air-transport/airlines-lessors/bureaucracy-continues-stifle-latin-americas-air-transport-market.

CAPA (2017) *Qatar Airways Middle East landing & airspace restrictions; wider ramifications for global aviation*. Available at: https://centreforaviation.com/analysis/reports/qatar-airways-middle-eastlanding—airspace-restrictions-wider-ramifications-for-global-aviation-348493.

Card, D. and Saunders, L. (1998) 'Deregulation and labor earnings in the airline industry', in J. Peoples (ed.) *Regulatory reform and labor markets*. Dordrecht: Springer Netherlands, pp. 183–247. Available at: https://doi.org/10.1007/978-94-011-4856-6_5.

Cornell Law School (n.d.) *19 CFR § 122.165 – Air cabotage., LII / Legal Information Institute*. Available at: www.law.cornell.edu/cfr/text/19/122.165.

Cristea, A.D., Hillberry, R. and Mattoo, A. (2015) 'Open Skies over the Middle East', *The World Economy*, 38(11), pp. 1650–1681. Available at: https://doi.org/10.1111/twec.12314.

Dempsey, P.S. (2016) *The evolution of air transport agreements*. Rochester, NY. Available at: https://papers.ssrn.com/abstract=3295341.

Doganis, R. (2019) *Flying off course: airline economics and marketing*. 5th edn. Abingdon: Routledge.

Echevarne, R. (2024) 'Interview. ACI Latin America-Caribbean Director General'. Interview by author via MS Teams on 21 June, 2024.

EUR-Lex (2020) *European Common Aviation Area (ECAA)*. Available at: https://eur-lex.europa.eu/EN/legal-content/summary/european-common-aviation-area-ecaa.html.

EUR-Lex (2022) *European Union–United States aviation agreements*. Available at: https://eur-lex.europa.eu/EN/legal-content/summary/european-union-united-states-aviation-agreements.html.

European Commission (2011) *Air transport: infringements concerning bilateral aviation agreements with Russia*. Available at: https://ec.europa.eu/commission/presscorner/detail/el/MEMO_11_167.

European Parliament (2019) *EU external aviation policy*. Available at: www.europarl.europa.eu/RegData/etudes/BRIE/2019/642221/EPRS_BRI(2019)642221_EN.pdf.

European Parliament (2023) *Fact sheets on the European Union. Air transport: market rules*. Available at: www.europarl.europa.eu/factsheets/en/sheet/131/air-transport-market-rules#:~:text=In%201992%2C%20the%20third%20package,airlines%20operating%20within%20the%20EU.

Evans, W.N. and Kessides, I. (1993) 'Structure, conduct, and performance in the deregulated airline industry', *Southern Economic Journal*, 59(3), pp. 450–467. Available at: https://doi.org/10.2307/1060284.

Finger, M. and Button, K. (2017) *Air transport liberalization*. Cheltenham: Edward Elgar. Available at: www.e-elgar.com/shop/gbp/air-transport-liberalization-9781786431851.html.

Flightradar24 (2017) *Live flight tracker – real-time flight tracker map*. Available at: www.flightradar24.com/.

Flightradar24 (2022a) *Comparing the standard routing for JL43 between Tokyo and London to today's routing*. Available at: https://x.com/flightradar24/status/1499592904901287939/photo/1.

Flightradar24 (2022b) 'Flight UA858 took off in 2023 but will land back in 2022.', *X*. Available at: https://x.com/flightradar24/status/1609214964312440833.

Gov.br (2024) *Open skies aviation policy between Brazil and Argentina, Ministério das Relações Exteriores*. Available at: www.gov.br/mre/en/contact-us/press-area/press-releases/open-skies-aviation-policy-between-brazil-and-argentina.

Humphreys, B. (2023) *The regulation of air transport: from protection to liberalisation, and back again*. 1st edn. Routledge.

IATA (2007) *Airline liberalisation. IATA Economics Briefing No 7*. Available at: www.iata.org/en/iata-repository/publications/economic-reports/airline-liberalization-summary-report/.

IATA (2023) *Global outlook for air transport – highly resilient, less robust*. Available at: www.iata.org/en/iata-repository/publications/economic-reports/global-outlook-for-air-transport----june-2023/.

IATA (2024a) *2023 safest year for flying by several parameters*. Available at: www.iata.org/en/pressroom/2024-releases/2024-02-28-01/.

IATA (2024b) *The Single African Air Transport Market (SAATM)*. Available at: www.iata.org/en/about/worldwide/ame/saatm/.

ICAO (2022) *Overview of regulatory developments in international air transport*. Available at: www.icao.int/Meetings/a41/Documents/Overview_of_Regulatory_and_Industry_Developments_in_International_Air_Transport.pdf.

ICAO (2024a) *International air services transit agreement*. Available at: www.icao.int/secretariat/legal/list%20of%20parties/transit_en.pdf.

ICAO (2024b) *The postal history of ICAO*. Available at: https://applications.icao.int/postalhistory/annex_16_environmental_protection.htm.

ICAO (n.d.) *Freedoms of the Air*. Available at: www.icao.int/Pages/freedomsAir.aspx.

InterVistas (2014) *Transforming Intra-African Air Connectivity: the economic benefits of implementing the Yamoussoukro Decision*. Prepared for IATA. Available at: www.iata.org/contentassets/44c1166a6e10411a982b2624047e118c/intervistas_africaliberalisation_finalreport_july2014.pdf.

Ismaila, D.A., Warnock-Smith, D. and Hubbard, N. (2014) 'The impact of Air Service Agreement liberalisation: the case of Nigeria', *Journal of Air Transport Management*, 37, pp. 69–75. Available at: https://doi.org/10.1016/j.jairtraman.2014.02.001.

Ison, S. and Budd, L. (2020) 'Airline regulation and deregulation', in *Air transport management: an international perspective*. 2nd edn. Abingdon: Routledge, pp. 47–57.

Martello, W.E. (2018) 'Airline deregulation', in *The SAGE encyclopedia of business ethics and society*. Thousand Oaks: SAGE Publications, Inc., pp. 80–84. Available at: https://doi.org/10.4135/9781483381503.

Meijer, G. (2020) *Fundamentals of aviation operations*. 1st edn. Routledge.

O'Connell, J. and Williams, G. (2010) 'Air transport development in the Middle East: a review of the process of liberalisation and its impact', *Journal of Air Transport Studies*, 1, pp. 1–19. Available at: https://doi.org/10.38008/jats.v1i1.112.

Oster Jr., C.V. (1987) 'The economic effects of airline deregulation, by Steven Morrison and Clifford Winston. Washington DC: The Brookings Institution.', *Journal of Policy Analysis and Management*, 6(3), pp. 469–473. Available at: https://doi.org/10.1002/pam.4050060326.

Reuters (2024) 'Argentina rolls out aviation reform in bid to bring in foreign airlines – Kylie Madry', 10 July. Available at: www.reuters.com/business/aerospace-defense/argentina-rolls-out-aviation-reform-bid-bring-foreign-airlines-2024-07-10/.

Schiphol (2024) *Diversion due to closure of Russian airspace*. Available at: www.schiphol.nl/en/blog/are-planes-flying-a-large-curve-around-russia/.

Smithsonian Institution (2021) *Airline deregulation: when everything changed*. Available at: https://airandspace.si.edu/stories/editorial/airline-deregulation-when-everything-changed.

Syed, M. (2019) *Why failure is the key to flying high*, Matthew Syed Consulting. Available at: www.matthewsyed.co.uk/why-failure-is-the-key-to-flying-high/.

United Nations (2024) *About us*. Available at: www.un.org/en/about-us.

UNSW (2023) *Under 'open skies', the market would decide how often airlines fly into Australia*. Available at: www.unsw.edu.au/newsroom/news/2023/09/under-_open-skies--the-market-would-decide-how-often-airlines-fl.

US Department of State (2024a) *Open Skies partners*. Available at: www.state.gov/open-skies-partners/.

US Department of State (2024b) *The United States-Nigeria Open Skies Air Transport Agreement Enters into Force*. Available at: www.state.gov/the-united-states-nigeria-open-skies-air-transport-agreement-enters-into-force/.

World Bank (2024a) *Middle East and North Africa*. Available at: www.worldbank.org/en/region/mena.

World Bank (2024b) *Open Skies for Africa – Implementing the Yamoussoukro Decision*. Available at: www.worldbank.org/en/topic/transport/publication/open-skies-for-africa.

10 Economic geography and air transport

Chapter outcomes

At the end of this chapter, you will be able to:

- Explain the factors influencing air transport growth.
- Describe the overall economic performance of the sub-sectors within the aviation value chain.
- Describe the three main characteristics in the context of the economic performance of the airline industry.
- Analyse the main trends influencing airline industry profitability.
- Describe and explain the different types of ancillary revenue products and services offered by airlines.
- Analyse the different costs incurred by airlines and cost reduction strategies.
- Describe the regional trends in airline financial performance.
- Explain the global economic benefits of air transport.
- Explain the benefits of air connectivity to economic development.
- Examine the key connectivity trends in Africa and assess the benefits of improved air connectivity.
- Interpret global air transport industry forecasts through to 2042.

10.1 Introduction

Q. What is the quickest way to become a millionaire in the airline business?

A. Start as a billionaire and buy an airline.

This is an old joke within the airline industry, however one that resonates to the present day. The industry is an extremely capital-intensive industry – one Boeing 777–9 alone will set you back $442.2 million at list price. Combine this with extensive competition, fluctuating demand, regulatory and ownership restrictions, high fuel costs, environmental responsibilities and many more and the financial struggles of many sectors of the air transport industry can be explained.

The airline industry has been one of the poorest performing sectors within the aviation value chain. It has been characterised over the last 30 years by high demand and growth, yet declining yields. By increasing ancillary revenues and declining real unit costs, yet marginal profitability. By high debts and poor return on invested capital (ROIC). Yet there are airlines that have been successful through many of the challenges and there are others that, even in the good times, struggle to turn a profit.

DOI: 10.4324/9781003405351-13

276 *Fundamentals of Global Air Transport Geography*

Air transport as an industry is also an important enabler for economic growth and development. It facilitates integration into the global economy and provides vital connectivity on a national, regional and international scale. It helps to generate trade, promote tourism and create employment opportunities (World Bank, 2024a).

The benefits of air transport, whilst global in nature, are not shared equally amongst the regions of the world and within those regions. Improving air connectivity can be a key strategy amongst governments and policymakers to drive economic improvements.

10.2 Factors influencing air transport growth

Air transport has experienced high growth rates since the 1950s and the beginning of the jet age, but this has not been consistent in its regional variations. This growth has contrasted with the marginal profitability of some of its key sectors – particularly the airline industry. Many factors influence air transport growth, and some will be discussed and expanded throughout the rest of the chapter. The consequences of environmental problems associated with climate change, air quality and noise have the potential to restrict air transport growth and these are discussed in Chapters 6 and 14.

10.2.1 Economic growth

The overall health of the economy (global, national and regional) significantly impacts air transport growth. During periods of economic expansion, businesses and individuals have more resources for travel. There are a few methods to measure the economic strength and growth of countries and the most used is gross domestic product (GDP). Figure 10.1 illustrates the countries with the largest GDP in 2022.

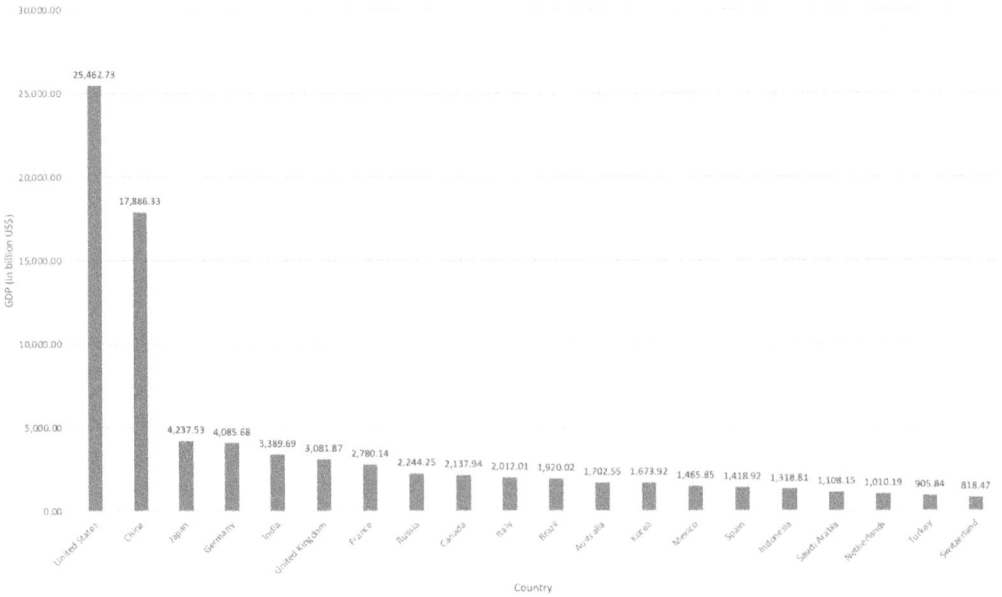

Figure 10.1 The 20 countries with the largest GDP in 2022 (US$ billion).
Source: adapted from IMF (2023)

The USA is in first place, with a GDP of around $25.5 trillion. According to the World Bank (2023), total global GDP in 2022 was just over $101 trillion, meaning the USA accounted for around one-quarter of global GDP. China was in second place on $17.9 trillion, followed by the established economies of Japan and Germany and the rapidly growing India, at $3.4 trillion.

ATG Did you know?

Gross domestic product (GDP)

The term GDP as a modern concept was developed in the USA in the 1930s, when Simon Kuznets, a Russian-born US economist, came up with an idea to assess the size of the economy by combining the financial value of individuals, companies and the government. It was adopted as the main measure of a country's economy at the Bretton Woods conference in 1944.

GDP measures the monetary value of final goods and services – that is those bought by the final user – produced in a country in a specified period. It counts all the output measured within the borders of a country (IMF, 2024).

An important factor in an economy is whether its total output of goods and services is growing or shrinking. To compare two time periods, *inflation* must be accounted for and "real" GDP calculated – has the value of output gone up because more is being produced or because the prices have gone up? GDP growth rate is used as an indicator of the general health of an economy. When real GDP growth is strong, employment is usually increasing. The average global GDP growth over a 40-year period between 1982 and 2022 was 3% (Figure 10.2). GDP growth

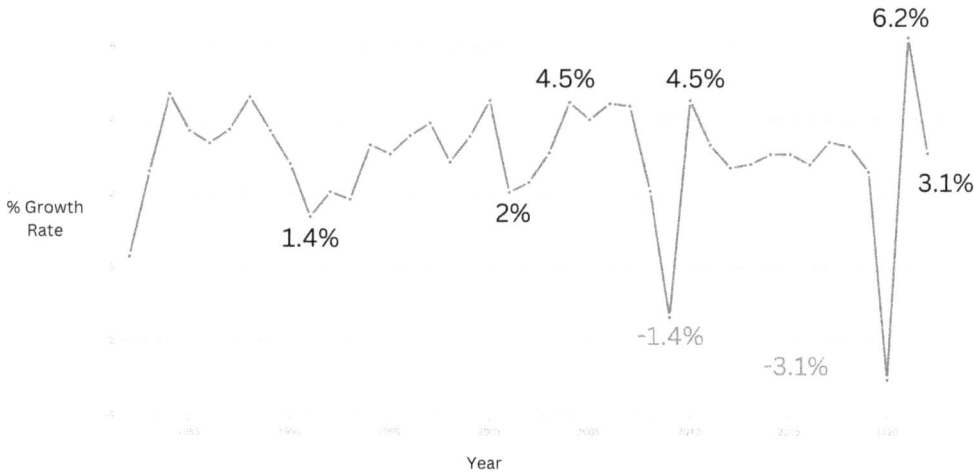

Figure 10.2 Annual world % GDP growth, 1982–2022
Source: adapted from World Bank (2024b), CC BY-4.0

Annual growth of GDP, 2022

Annual percent change in gross domestic product. This data is adjusted for inflation.

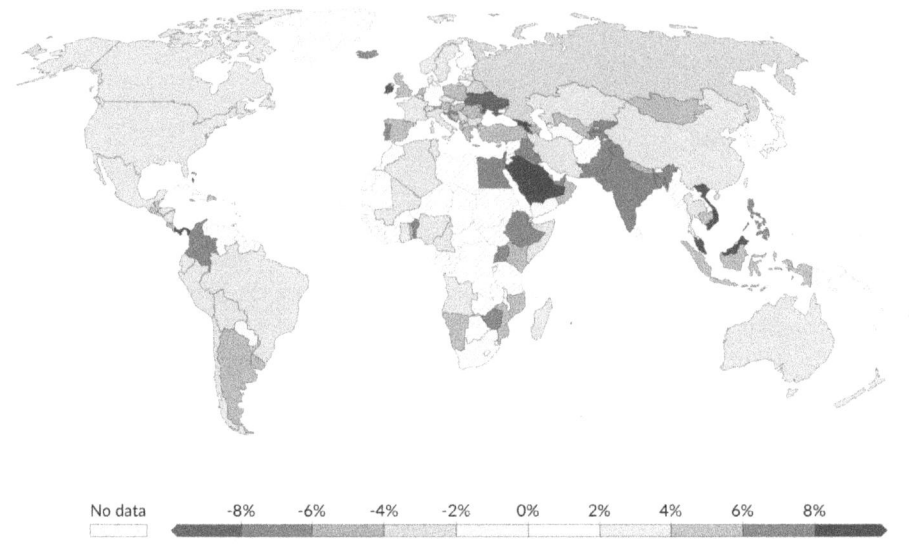

Figure 10.3 Annual GDP growth per country, 2022
Source: IMF via OurWorldInData (2023) – CC BY

moves in cycles over time and are sometimes in periods of boom (such as the mid-2000s), slow growth (such as in 2001 following the dot.com bust and 9/11 terrorist attacks) and periods of negative growth (such as following the economic recession in 2008 and COVID-19 in 2020).

There are also wide regional variations and these are not always consistent year on year. Figure 10.3 illustrates the annual GDP growth rates per country in 2022 and the three poorest performing countries were Ukraine (−29.07%), Belarus (−3.66%) and Russia (−2.07%) because of the Russian invasion of Ukraine. However, in 2021, Ukraine's GDP grew by 3.43% and Russia by 5.61%. There are countries experiencing consistently high GDP growth (although numbers for these periods can be skewed following COVID-19 recovery rates). Colombia grew by 11.02% in 2021 and 7.26% in 2022 and Saudi Arabia GDP grew by 3.92% in 2021 and 8.74% in 2022. Saudi Arabia is also experiencing rapid air transport growth, fueled in part by the Kingdom's Vision 2030 aspirations.

10.2.2 *Income growth*

IATA states that growth in incomes, often proxied by GDP, is the fundamental driver of air transport demand. An increase in economic prosperity often leads to employment creation and an increase in household income. It is important to note that a key factor here is *disposable income* – a proportional increase in taxation may not result in any more income to spend on discretionary activities such as leisure travel.

ATG Case study 10.1

India

The annual growth in GDP in India between 2010 and 2019 averaged a staggering 6.95% (IMF via OurWorldInData, 2023). Following a drop of 5.83% in 2020 due to COVID-19, growth between 2021 and 2023 averaged 7.54% – one of the highest GDP growth rates in the world. India currently sits as the fifth largest economy in the world in GDP and some analysts believe India's GDP will be in third place by 2027 and worth $10 trillion by 2030 (Zhan, 2024). According to the UN (2023), in April 2023, India surpassed China as the world's most populous country, in excess of 1.4 billion people.

Crucial to the air transport industry is the growth in the "middle class" which – according to the People Research on India's Consumer Economy – grew 6.3% per year between 1995 and 2021 and now represents 31% of India's population. It is projected to hit 38% by 2031 and 60% by 2047 (Skift, 2024b). The increase in disposable income amongst this demographic should signify an expanding market for travel and air transport. Alongside this, *population growth* in general will assist in stimulating air transport demand.

10.2.3 Globalisation

The interconnectedness of economies and businesses across the world increases the need for efficient and quick transportation of passengers and cargo. The growth of home working, video conferencing and environmental concerns may constrain demand, but the fundamental drivers of demand remain.

10.2.4 Liberalisation

As discussed in Chapter 9, the liberalisation of air transport markets – especially in the USA, Europe and parts of the Asia Pacific region – facilitated the growth of new airlines and significant new route development, with increasing competition creating more opportunities for air travel. Further developments in liberalisation in Africa and Latin America may result in an increased supply of air transport via new airlines, growth in existing airlines and route development opportunities in these regions.

10.2.5 Reduction in airfares

Allied to liberalisation and increased airline competition has been a reduction in airfares. Inflation-adjusted airfares in much of the world are now much less than in the past. Figure 10.4 illustrates this in the context of US domestic airfares, where average itinerary fares, based on constant 2023 inflation-adjusted dollars, have dropped from $600 in 2000 to $382 in 2023. The advent of LCCs – often known as the "Southwest effect" in the USA or the "Ryanair effect" in Europe – has led to much cheaper ticket prices, albeit the process of *unbundling* (charging separately for hold baggage, seat selection, etc.) increases the total trip cost, but still at much lower levels than historic averages.

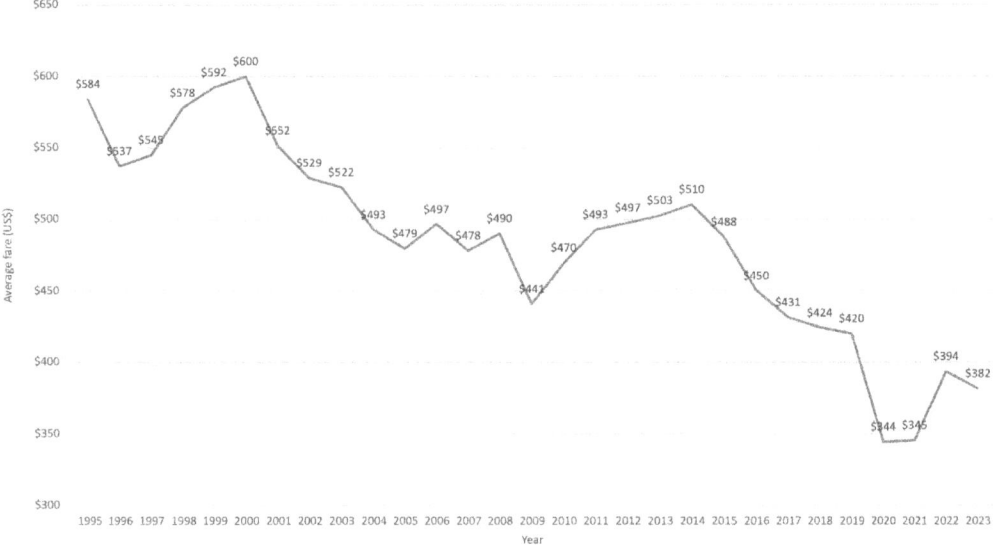

Figure 10.4 Annual US domestic average itinerary fare in constant 2023 US dollars
Source: adapted from Bureau of Transportation Statistics (2024)

10.2.6 Political stability and security

Stable political environments and enhanced security foster confidence and contribute to a positive climate for air travel growth. As discussed in Chapter 9, protectionist policies and regularly changing governments (as historically seen in many African countries) can cause weak traffic growth and undermine business activities, as well as leisure and tourism travel.

> **ATG Did you know?**
>
> **"Black Swans"**
>
> The Black Swan Theory refers to events which are difficult to predict in the normal course of business. They are considered "one-offs" due to the difficulty in predicting these events. Examples having a large impact on the air transport industry include:
>
> - 9/11 terrorist attacks in 2001.
> - Volcanic eruption in Iceland in 2010.
> - COVID-19 pandemic from 2020.

10.3 Aviation value chain

To undertake an analysis of the economic geography of air transport, a general overview of the key sectors and their overall performance is a good starting point. Chapter 2 introduced the concept of the aviation value chain (Figure 2.2), which is a diverse set of sectors in

Economic geography and air transport

terms of size, structure and performance – airlines, airports, aircraft manufacturers, freight forwarders, GDS companies, caterers, ground services providers, fuel producers, aircraft lessors, Air Navigation Service Providers (ANSPs) and maintenance, repair and operations (MROs). However, these are linked into one complex web, which ultimately is responsible for around 4.5 billion global airline passengers per year. The air transport industry generates positive social and economic externalities, both in terms of countries and cities with large aviation hubs and remote regions, which benefit from the connectivity provided and access to the global economy. The negative environmental externalities are assessed in Chapter 14.

In 2022, when the financial effects of the COVID-19 pandemic were still being felt, the aviation value chain still managed to generate revenues of around $1.45 trillion. Around half of this ($732 billion) was generated by the airline industry alone – the hub at the centre of the value chain (Figure 10.5). However, post-COVID, most sectors remained loss-making at the global level, generating an overall economic loss of $69 billion (IATA-McKinsey, 2024).

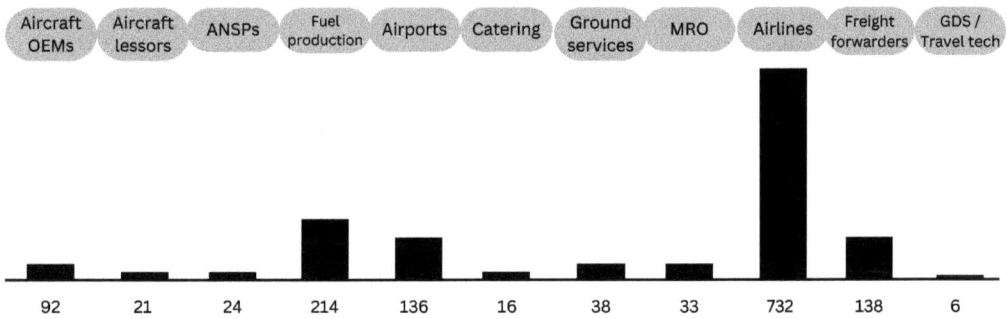

Figure 10.5 Aviation value chain revenue by subsector, 2022
Source: adapted from IATA-McKinsey (2024)

ATG Did you know?

Return on invested capital (ROIC) and weighted average cost of capital (WACC)

Both these metrics are extremely important in understanding economic benefits generated by companies/sectors.

ROIC measures the earnings available to pay debt and equity investors in relation to the capital invested. Essentially, it is how well a company is using its capital to generate profits.

WACC is the cost of capital. It can be viewed as an opportunity cost for an investor – what would the alternative return be if the capital were invested elsewhere?

The difference between the ROIC and WACC is the level of economic profitability. If ROIC is higher, then value is being created. If WACC is higher, then value is being lost.

282 *Fundamentals of Global Air Transport Geography*

Freight forwarders, fuel producers and ground services companies exhibit (on average) higher and more stable earnings. Airlines and airports exhibit some of the lowest average rates of returns, both below 5% over the 2012–2022 period. The airline industry is also very volatile, meaning the industry is more susceptible to external shocks such as economic recessions or other "black swan" events. In one of the ironies of the air transport industry, suppliers to airlines generally have higher returns and lower volatility than airlines themselves.

10.3.1 Airlines

Profitability was still elusive in 2022 for the airline industry globally, although a loss of $45.4 billion was much better than the collective loss of $137 billion in 2020. The airline industry is extremely sensitive to macro-economic shocks. Over the last 25 years, these have included the 9/11 terrorist attacks, 2008–2009 economic recession, geopolitical tensions in various parts of the world, natural disasters such as the Icelandic volcanic eruptions in 2010 and the COVID-19 pandemic. That said, when the economy grows, demand for air transport usually increases.

The truly global nature of the airline industry is evidenced in the positive correlation between ROIC and real GDP growth rates and that all global shocks had negative impacts on ROIC.

A key metric to illustrate the poor financial performance of the airline industry globally is ROIC versus WACC. Since 1996, ROIC has remained below WACC, with the closest it came in 2015 and the worst being in 2020 at the depths of the pandemic. The gap has closed again but it will be a huge effort to close the gap completely in the next few years. Section 10.4 will assess the financial performance of the airline industry in greater detail.

10.3.2 Aviation fuel production

Fuel producers were one of the most successful aviation subsegments in 2022, with an ROIC of 17% – versus 2% for airlines (IATA, 2024). Jet fuel is a component of refining crude oil (around 8% of output in 2022) and the prices usually have a correlation. As crude oil and jet fuel prices rise, normally so would the ROIC of fuel producers. Conversely, this places downward pressure

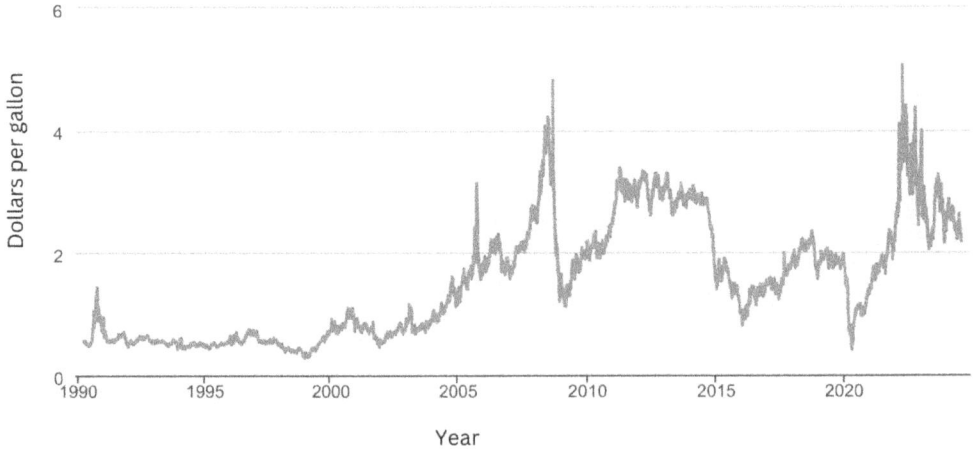

Figure 10.6 US Gulf Coast kerosene-type jet fuel spot price FOB, 1990–2024
Source: US Energy Information Administration (2024)

Economic geography and air transport 283

on airline profits. It is not just the overall price which creates issues for airlines. Figure 10.6 illustrates the average jet fuel spot price between 1990 and 2024 and it is apparent that there are vast differences over time resulting in one thing airlines do not like – *volatility*.

As an example, prior to the economic recession in the late 2000s – in July 2008 – the spot price reached a historic high of $4.17 per gallon. Nine months later, in March 2009 at the height of the recession, prices fell to $1.11 per gallon. It is very difficult for airlines to plan their business around such volatile fuel prices and whatever happens, it is imperative to reduce not only what is spent per unit, but also what is consumed. Prices can be *hedged* to protect against volatility, but airlines will also look to reduce fuel consumption, using tactics such as:

- Newer aircraft and engines.
- Improved aerodynamics such as wing sharklets.
- Reducing the weight of the aircraft, by removing or reducing IFE, in-flight magazines, catering items or using lighter weight trolleys and seats.
- Operational improvements such as using more optimal flight routings.
- Single-engine on-the-ground taxiing.

ATG Did you know?

Fuel hedging

Fuel hedging essentially involves buying an amount of fuel at a fixed price for delivery later. Fuel may be hedged in any percentage requirements, but it is normal to reduce the percentage hedged, the further into the future, due to uncertainty. This does not bring airlines any guarantees in saving money if the fuel prices drop after hedging has taken place, however it does create a level of certainty to better able airline planning and fare setting.

In their full year results report for 2022–2023, Ryanair stated that their fuel had been hedged over 80% at $64bbl (barrel; much less than the average spot price) saving the group over €1.4 billion (Ryanair, 2023).

10.3.3 Manufacturers

ROIC was positive for this subsector in 2022, for the first time since 2018. However, it was still not enough to cover WACC, posting a net economic loss of $1.7 billion. In 2023, the two main manufacturers achieved mixed results. Boeing lost $2.2 billion – improved from a loss of $5.1 billion in 2022 – largely down to losses at the commercial aircraft business ($1.6 billion), much of which was as a result of issues with the 737 MAX (Flight Global, 2024b). Airbus's commercial aircraft sector on the other hand achieved an adjusted earnings of €4.6 billion in 2023 (Flight Global, 2024a).

With a huge backlog of aircraft, as discussed in Chapter 2, if the 737 MAX issues and the Pratt and Whitney GTF engine issues on the A320neo can be resolved, there could be a bright future for both manufacturers, as airline demand is there.

10.4 Economic performance of the airline industry

> A recession is when you have to tighten your belt. A depression is when you have no belt to tighten. When you've lost your trousers – you're in the airline business!
>
> (Sir Adam Thomson, founder and Chairman of British Caledonian, *The Illustrated London News*, 1983, quoted in University of Alberta, 2017)

The economic performance of the airline industry over the last 40 years has been marginal at best. This contrasts dramatically with the high growth rates in most of the years since liberalisation, in many parts of the world. The first half of 2024 alone witnessed an average of almost 100,000 flights daily (OAG, 2024). Section 10.2 examined the factors influencing air transport growth broadly – this section will now place these in the context of the airline industry. Section 10.5 will break these global trends down into regional variations.

10.4.1 Characteristics of airline operations

According to Doganis (2019), the airline industry has three key characteristics:

1. The demand for air services is a *derived demand*

The air journey, for all but a handful of the 4.5 billion annual passenger journeys, is a means to an end. In other words, the journey is a means of travelling from a to b, but the purpose of the journey is getting to b – whether that be for reasons such as for business, holiday, visiting friends and relatives or sport and music events. The demand is therefore to travel to a destination for an activity of sorts – to travel from London to Singapore on business, from Norway to Tenerife for a holiday, from Portugal to Brazil to visit family, or to watch a Champions League tie between Bayern Munich and Real Madrid in Munich. In the last instance, had Paris St Germain been drawn against Real Madrid instead, demand would have been to Paris rather than Munich.

Competition is not just air versus air, it is also versus other transport modes, such as bus, car and train. This will be much higher the shorter the journey and/or the quicker and cheaper the alternative transport modes are. London to Singapore is unlikely to witness much competition other than air, however Brussels to Amsterdam most certainly will, due to high-speed rail links.

2. The airline product is *homogenous*

Essentially one airline seat is very much like another. This is especially true in economy class and even between low-cost carrier (LCC) and full-service network carrier (FSNC), their short haul seats are, in essence, very similar. There is more variation and indeed marketing and product development spend in the front cabins – in business and first class. Also, one aircraft is very like another and most airlines have similar aircraft for specific journeys, for example the A320 or B737 series for short–mid-haul services – there is not the variety that you will see on a car retailer's forecourt for example. The same applies for cargo – I am not sure cargo packages are aware which airline they are flying on! As a result, much of flying revolves around price, resulting in the three key factors influencing the choice of an airline for short-haul leisure flights, commonly quoted as being price, price and price. The reality is that factors such as frequency, airport proximity, catchment area and perceived service quality will also play a role, however the homogenous nature of the airline product is significant.

3. The airline product is *perishable*

This is also the case with services such as hotels, but essentially when the "expiry date" (the departure time) has passed, the product cannot be sold. This understandably means airlines wish to fill seats, however they also must fill them at appropriate fare levels and it is this balance which is at the core of airline *revenue management* – often described as not selling a ticket today, if you believe you can get more for it tomorrow (assuming it is not beyond its "expiry date").

As a result of these characteristics and other factors influencing air transport growth, different business models have been developed and these were analysed in Chapter 2. LCC or FSNC airlines? In North America or Asia Pacific? In liberalised or highly regulated environments? These factors can influence airline performance as much as individual management acumen and the rest of this chapter will examine the key global airline economic trends.

ATG Did you know?

Key economic terminology

Table 10.1 Key economic terms

Term	Description
Available seat kilometres (ASKs)	Measures an airline's supply – its passenger-carrying capacity. Calculated by multiplying the number of seats by the distance flown.
Available tonne kilometres (ATKs)	Measures an airline's supply – its transportation capacity. Calculated by multiplying the tonnes available for the transport of a revenue load (passengers and cargo) by the distance flown.
Revenue passenger kilometres (RPKs)	Measures an airline's demand – its traffic. Calculated by multiplying the number of revenue passengers by the distance flown.
Passenger yield	The average revenue generated per passenger kilometre flown. Some airlines use this metric to display average ticket prices paid.
Load factor (LF)	Measures the % of available seating capacity which is filled with passengers. Thus (RPK/ASK)*100.
Revenue per available seat kilometre (RASK)	The revenue generated by each seat kilometre. Calculated by dividing total operating revenue by ASK.
Cost per available seat kilometre (CASK)	The cost incurred to fly each seat one kilometre. Calculated by dividing total operating costs (expenses) by ASK. Can be used alongside RASK to assess profit per available seat kilometre.
Ancillary revenues	Additional products and services offered by airlines beyond the ticket price, such as baggage fees, seat selection, in-flight drinks, car hire and hotel bookings.
Operating profit margin (OPM)	Operating profit is calculated by subtracting operating costs (such as aircraft, aircraft fuel, salaries and maintenance) from operating revenues. OPM divides the operating profit by the revenue – expressed as a %. It is a key measure of business efficiency, in terms of generating profit through core operations.
Net profit margin (NPM)	Net profit is the operating profit minus interest and taxes – the "bottom line"; what is left after all costs have been paid. NPM divides the net profit by the revenue – expressed as a %. It is the % of profit left from the total revenue generated. Thus, an NPM of 5% means that for every £100 in revenue, £5 is kept after all expenses have been paid.

Source: author

Note: Kilometres have been used as the most common distance unit globally, however a number of airlines also use miles, especially in North America (thus ASK will be ASM).

286 *Fundamentals of Global Air Transport Geography*

10.4.2 A high growth industry

Since the 1950s and 1960s, there has been a significant growth in air travel. Whilst growth rates have generally slowed through the decades and there have been a few ups and downs following major shocks such as wars, terrorist attacks, economic recessions and pandemics, the overall long-term growth rates have been high. What can also be said is that geographically there have been large variations and these will be discussed in Section 10.5.

In the 1970s, annual passenger growth was close to 10% and around 6% in the 1980s, with an average of 5.2% in the 1990s (Doganis, 2019). Despite the dot.com bust period and 9/11 in the early 2000s and the 1% reduction in revenue passenger kilometres (RPKs) in 2008 following the economic recession, the industry still grew at an average of close on 4% in the 2000s. Between 2010 and 2019, RPKs grew at a higher rate of 6.5% thanks to the – relatively speaking – boom years of 2014–2018, before COVID-19 saw the industry largely grounded (except for cargo) for much of 2020 and into 2021 in many regions, and by 2023 RPKs had only recovered to around 95% of their 2019 peaks. Prior to COVID-19, for the 15-year period between 2004 and 2019, RPKs had grown by 141% overall, from 3.6 billion to almost 8.7 billion (Figure 10.7).

Factors (in a geographically uneven way) which have stimulated this growth include a rise in household income (and disposable income), an increase in liberalisation resulting in more airlines and more open markets, a rise in LCCs (in many ways because of liberalisation) and, as will be discussed in the next section, declining yields – good for passengers, not so for airline performance.

10.4.3 Declining yields

One of the key characteristics of the economic performance of the airline industry is a long-term decline in yields (alongside a large reduction in unit costs), with a near six times reduction in real yields since 1970 (IATA, 2023b). Figure 10.4 illustrates the 36% drop in US domestic

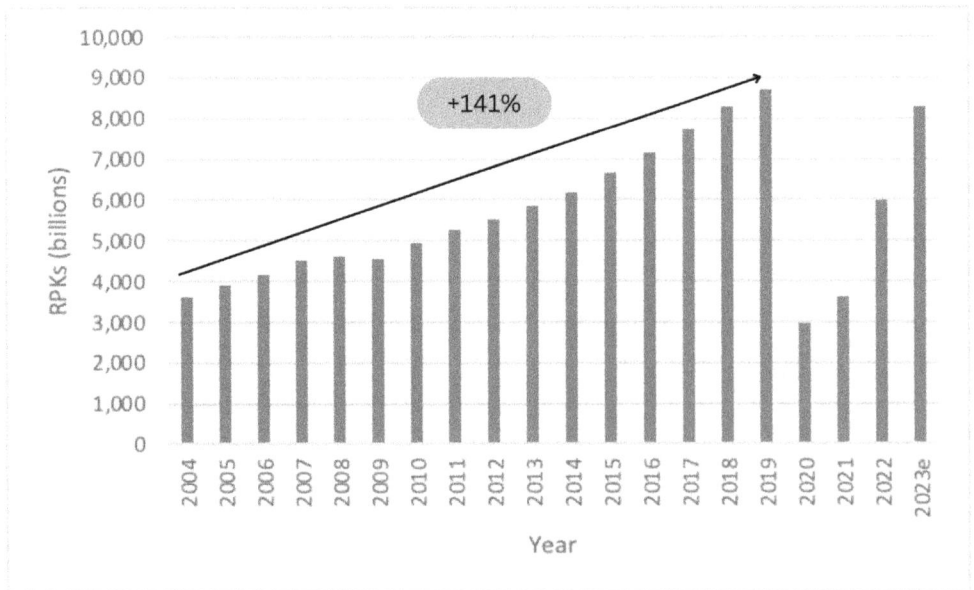

Figure 10.7 Airline RPKs, 2004–2023

Source: adapted from IATA (2023a) and Statista (2024c)

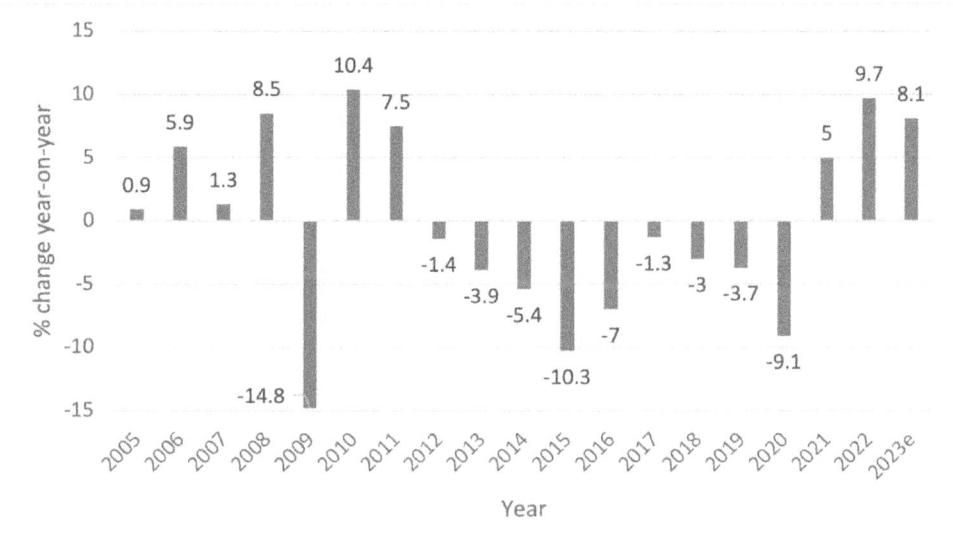

Figure 10.8 % annual change in airline yield: 2005–2023
Source: adapted from IATA (2023c); Statista (2024a)

inflation-adjusted airfares between 2000 and 2023. The increase in **competition**, largely created by liberalisation, has witnessed the start (and in many cases, failure) of new airlines in many parts of the world, especially in the USA since 1978 and Europe since the mid-1990s. Illustrating this is the fact that by 1986, the average passenger yield in the USA was 25% lower than in Europe, which was still regulated. Key to this has been the advent of LCCs, whose core *modus operandi* is to reduce their costs to offer reduced fares which stimulates demand, and to take passengers from higher-fare airlines.

Another key historic attribute of the airline industry is *overcapacity* – too many seats chasing too few passengers – which has resulted in fare decreases to fill seats. Often, when the economy is booming, airlines place orders for new aircraft with the manufacturers. Due to the often-long delays between ordering and actual delivery, these deliveries sometimes coincide with the end of a boom cycle or into a downturn, when there is less demand, exacerbating the problems.

Figure 10.8 illustrates the yield trend since 2005. The first few years of the period saw an increase in yield, apart from the effects of the economic recession in 2009, however from 2012 to 2019, every single year experienced an overall average yield reduction, culminating in a 9.1% reduction during the COVID-19 year of 2020. Many of these reductions were in part driven by declining fuel and other costs. The last three years have witnessed average yield increases, albeit from a low COVID baseline.

10.4.4 Growth in revenue and ancillaries

Despite the reduction in yields, the growth in RPKs and overall passenger numbers, airline revenue grew to record levels in 2019 and subsequently 2023, post-COVID. As Figure 10.9 illustrates, revenue grew by 121% between 2004 and 2019, from $379 billion to $838 billion. Following a massive contraction due to COVID-19, the estimate for 2023 shows an increase of almost 7% from the pre-COVID period of 2019. The reduction in yield is also evident here, as the growth in RPKs (Figure 10.7) between 2004 and 2019 was 141% – higher than total revenue growth.

288 Fundamentals of Global Air Transport Geography

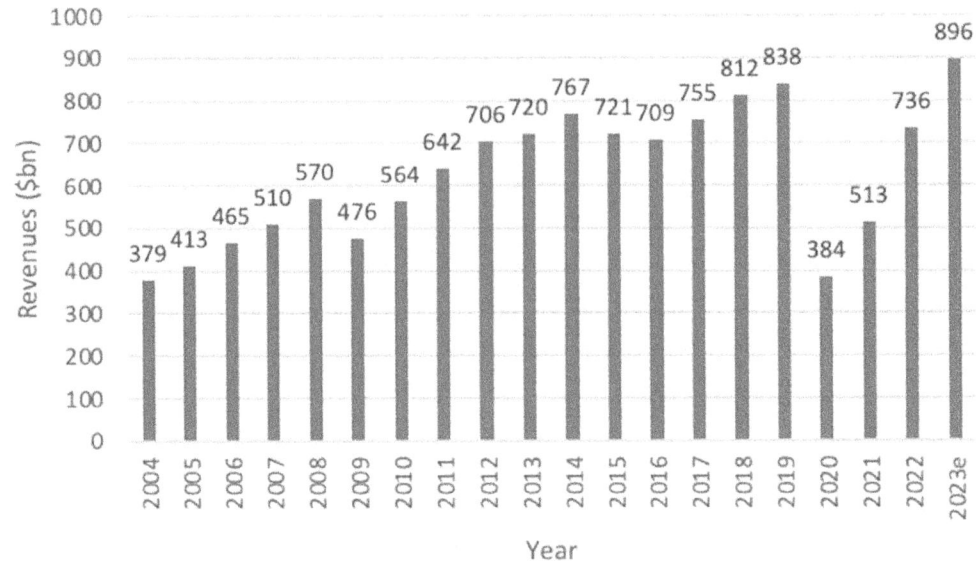

Figure 10.9 Total revenue of airlines worldwide, 2004–2023
Source: adapted from IATA (2023b); Statista (2024b)

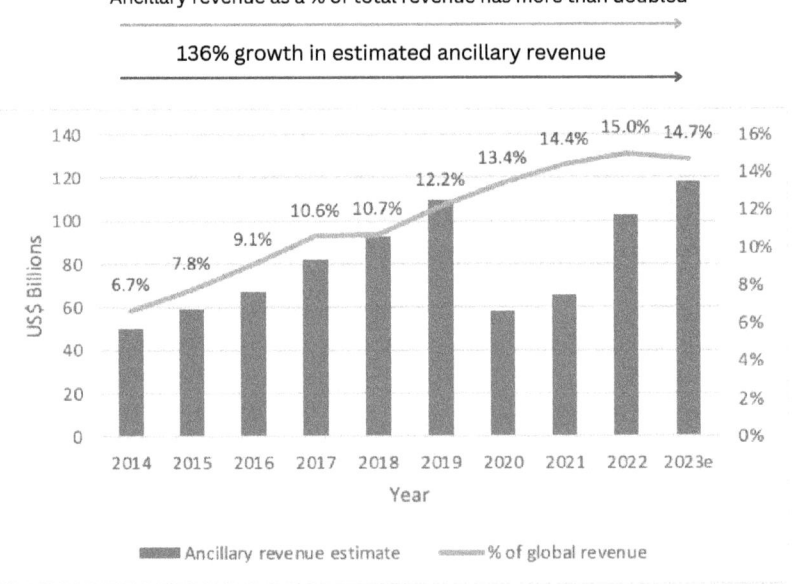

Figure 10.10 CarTrawler Worldwide estimate of ancillary revenue, 2014–2023
Source: adapted from IdeaWorksCompany (2023a)

Ancillary revenues

A major change during this period of revenue growth was the increasing importance of ancillary revenues. As Figure 10.10 shows, total estimated ancillary revenue has grown from $49.9 billion in 2014 to $117.9 billion in 2023. Ancillaries only accounted for 6.7% of total revenues in 2014, but by 2023 this figure had risen globally to around 15%.

ATG Did you know?

Airline ancillary revenues

The definition of ancillary revenue – offered by Jay Sorensen of IdeaWorksCompany in 2008 – has been widely adopted, as: "Revenue beyond the sale of tickets that is generated by direct sales to passengers, or indirectly as a part of the travel experience" (IdeaWorksCompany, 2023b, p.19). IdeaWorks also defines ancillary revenue using the five categories in Table 10.2.

Table 10.2 Airline ancillary revenue categories and examples.

Category	Description	Examples
A La carte features	Items consumers can add to their air travel experience	Checked and hand baggage fees, in-flight food, drink and gifts, priority check-in, seat selection, WiFi
Commission-based products	Commissions earned by airlines on sale of third-party products and services	Hotels, car hire, airport transfers, travel insurance, car parking
Frequent flyer activities	Sale of miles or points to programme partners who then sell these as part of their package to their customers	Co-branded credit cards, hotel chains, communication services and other retailers
Advertising sold by the airline	Advertising linked to passenger travel	Advertising in the in-flight magazine, aircraft fuselage, overhead lockers, seatbacks, tray tables and consumable items such as plastic glasses
Fare or product bundles	The bundling of different elements such as fares and specific ancillary items	Extra services may include seat selection, checked baggage and speedy boarding

Source: adapted from IdeaWorksCompany (2023b).

One of the biggest changes to the passenger experience from ancillaries is the concept of *unbundling*, largely because of the growth in LCCs, such as Ryanair and Air Asia. This is where products and services which were previously offered for free (as part of the ticket price) are now being sold separately, in addition to the ticket. Core unbundled products include:

- checked baggage
- seat selection
- in-flight food and drinks

As explained in Chapter 2, many FSNCs such as British Airways have adopted this strategy, especially on shorter flights, to reduce costs to be able offer lower ticket prices and attempt to compete with the LCCs more on price. Ultra low-cost carriers (ULCCs) such as Spirit Airlines have taken this a step further, charging for more items such as hand baggage (if it does not fit under the seat in front). Revenues generated from ancillary items on a per passenger basis have more than doubled between 2014 and 2022, from $19.28 to $42.11 (IdeaWorksCompany, 2023a).

The proportional importance of ancillary revenue is generally higher amongst the LCCs due to their business model of lowering ticket prices and selling ancillary products. Airlines such as

Spirit Airlines, Frontier, Allegiant, Ryanair, GOL and Volaris all have a high percentage (over 30%) of their total revenue as ancillaries.

However, in terms of total revenues, US FSNCs such as Delta, United and American generate the highest ancillary sums – over $7 billion each estimated in 2022 (IdeaWorksCompany, 2023b) – largely from the sale of mileage to organisations such as banks. The power of loyalty!

ATG Case study 10.2

Wizz Air, Hungary

Wizz Air is an ULCC, headquartered in Hungary and founded in 2004, operating mostly short-haul flights within Europe and to surrounding regions in North Africa and the Middle East. In 2024, they had over 200 A320 family aircraft and operated over 1,100 routes (Wizz Air, 2024). In 2021, it was estimated that Wizz Air had the highest percentage of ancillary revenue as a percentage of total revenue – at 56% – in the world (IdeaWorksCompany, 2023b).

They offer a broad range of a la carte products and services, such as checked and hand baggage fees, seat selection fees and in-flight food and drink for retail. A Discount Club provides added benefits for an annual fee. In addition, they also generate ancillary revenue through commission-based items such as car hire, airport parking and Wizz Experiences. Dynamic pricing is used for some ancillaries. CEO Josef Varadi stated that ticket revenue tends to fluctuate a lot more than ancillaries, due to market conditions and macroeconomic factors and ancillaries are easier to predict (Aerotime.aero, 2023).

Total revenue and RASK by top airlines

When comparing the revenue performance of individual airlines, it is easy to see the top performers – the "biggest" airlines in terms of revenue generated. Figure 10.11 illustrates the top 20 airlines in terms of total revenue for 2023.

This list is dominated by the major US carriers – Delta, United and American – each of whom generated more than $50 billion in total revenue, followed by the major three European legacy carrier groups – Lufthansa, Air-France-KLM and the British Airways and Iberia parent company: IAG. The Dubai-based carrier – Emirates – is sandwiched in between these. The top three airlines are also the largest in terms of fleet size. A key point to note is that of the top 20, only two are LCCs (Southwest and Ryanair) and the rest are FSNCs. This may not come as a huge surprise due to the size and scale of the mainly legacy, global carriers.

What is less easy to see, however, is who is the most *efficient*. A useful metric to analyse airline revenue efficiency is **unit revenue** – measured in the airline industry by RASK – which removes size and scale. RASK calculates the revenue generated for every seat kilometre flown and thus lends itself to comparative metrics between any airlines and is particularly useful for comparisons between those with similar business models. Figure 10.11, for 2023, shows that Delta is also one of the best performing airlines for RASK, at 13.26¢, with only Qantas (12th in revenue) at $15.10¢ and Cathay Pacific (18th in revenue) at 14.15¢ ahead of them. The Chinese carriers perform particularly poorly here in comparison to the other major airlines on the list.

Economic geography and air transport

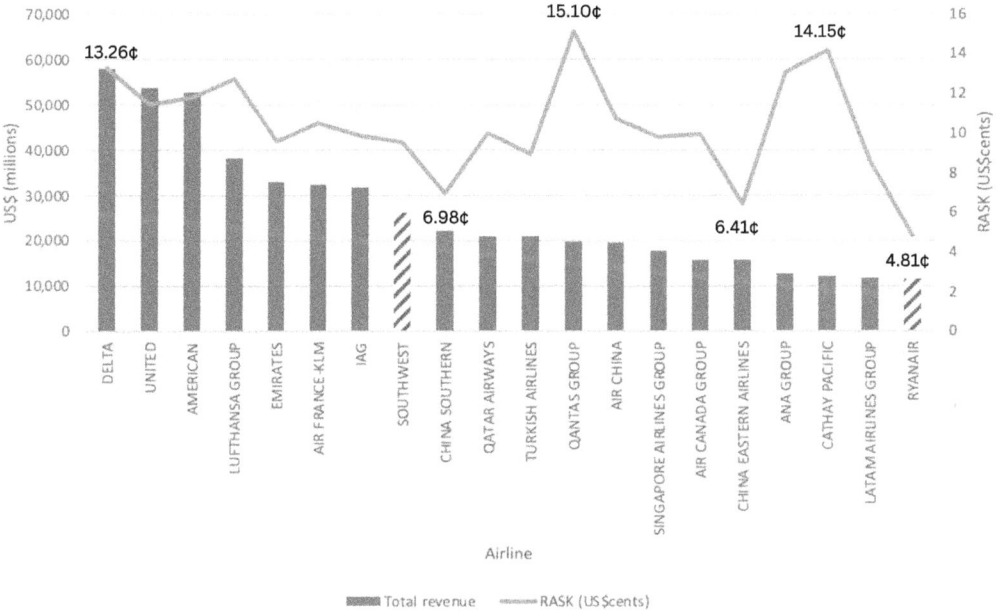

Note: Data taken from latest financial information available at time of writing. FedEx and TUI Group excluded from list due to no comparative ASK data.

Figure 10.11 Top 20 airlines for revenue and comparative RASK, 2023

Source: author from airline annual reports, Airline Business (2023) and CAPA (2024)

Load factor

In addition to RASK and *yield* (revenue received per passenger-km), another key factor used is that of *occupancy* – or load factor. How full are the planes? There is not much point in generating high revenue per seat if there are only a handful of passengers on a 300-seat plane, or conversely filling up the 300 seats at very little *yield*. The key (and one of the hardest parts for revenue managers) is to fill the planes at the highest possible yield and not to sell a ticket for a fare today, when it could have been sold for more tomorrow.

Average global load factors had been trending upwards in the 15 years prior to the pandemic, from 75.2% in 2005 to 82.6% in 2019 (Figure 10.12). The COVID-19 pandemic decimated the passenger airline business, but the estimate for 2023 is close to the 2019 peak, at 82%.

These figures are only averages and there are some who have regularly been achieving much higher LF – especially many LCCs and none more so than Ryanair. Ryanair adopts what it calls a "load factor active/yield passive" strategy, which essentially involves minimising ticket prices to fill the planes as much as possible. The lower the fares charged, the higher the load factor required to breakeven and then make money. An added benefit of this strategy is that for many LCCs, the ticket price is a starting point – a "way in" to then cross sell many other ancillary products and drive average revenue higher. Figure 10.12 shows that in 2019, when industry LF was 82.6%, Ryanair was achieving 96% and in 2023 as the recovery was almost complete, Ryanair LF was already at 93%.

Staying with Ryanair, it was the carrier with the lowest RASK on the top 20 list and the one with the lowest revenue, but a very high LF. This is a great example to illustrate that revenue

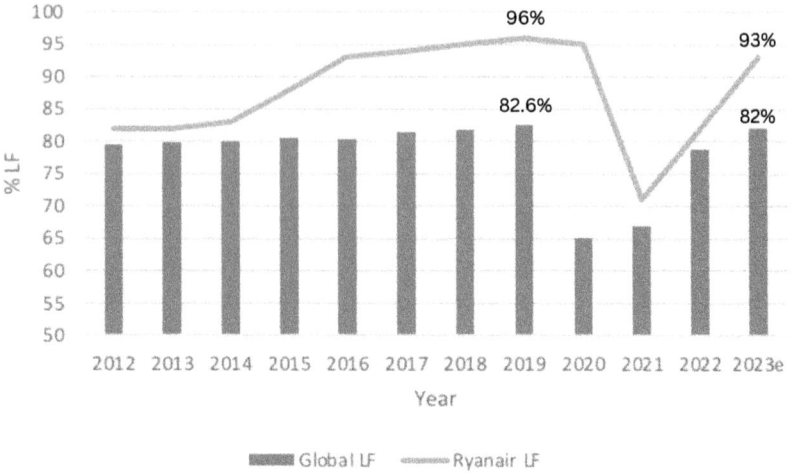

Figure 10.12 Average passenger load factor of commercial airlines, 2012–2023, with Ryanair as a comparison
Source: adapted from IATA (2023b); Statista (2023)

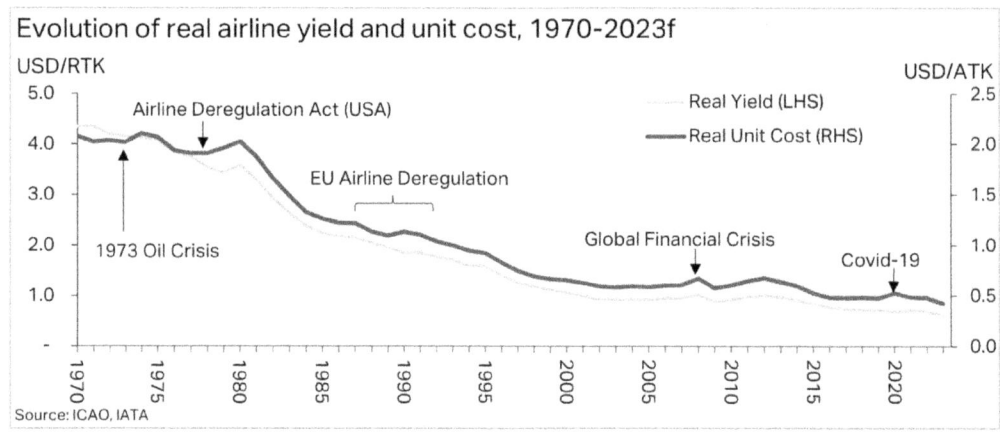

Figure 10.13 Evolution of real airline yield and unit cost, 1970–2023
IATA (2023b)

(and RASK) is only one side of a two-sided coin. If revenue was all that mattered, then Ryanair could be identified as having the poorest economic performance of the top 20. The fact that practically the opposite is true is testament to the fact that the other side of the coin has huge importance and must be considered in conjunction with revenue – that of costs.

10.4.5 Airline costs

As explained in Section 10.4.3, two key long-term trends in the airline industry have been declining real yields and declining real unit costs (Figure 10.13). This decline in unit costs has been remarkable, with a real unit cost reduction of over four times since 1970.

Economic geography and air transport 293

A key factor in this has been technological progress in terms of aircraft and engine efficiency, resulting in lighter aircraft, meaning greater fuel efficiency (historically the highest component of airline costs). Widespread industry competition since deregulation has also played a large part. This is despite the numerous global macroeconomic shocks the industry has experienced throughout this time.

Key costs for airlines include:

- fuel
- labour
- aircraft purchase (depreciation) and leasing
- maintenance
- airport and ATC charges
- sales and marketing

Factors such as fuel and airport charges are largely externally determined, but there are approaches airlines can take to influence the more internally controlled costs. Fuel can be hedged (a price agreed for a specific volume in advance) to ensure price stability, although this carries its own risk if hedged at a level the spot price drops below. Airlines can try to fly more optimum routings (subject to externally controlled airspace) and reduce the weight of their aircraft by using lighter seats, lighter in-flight trolleys and even individual knives and forks, which when grossed up over a flight and two or three meal services, can make a difference. Newer, more

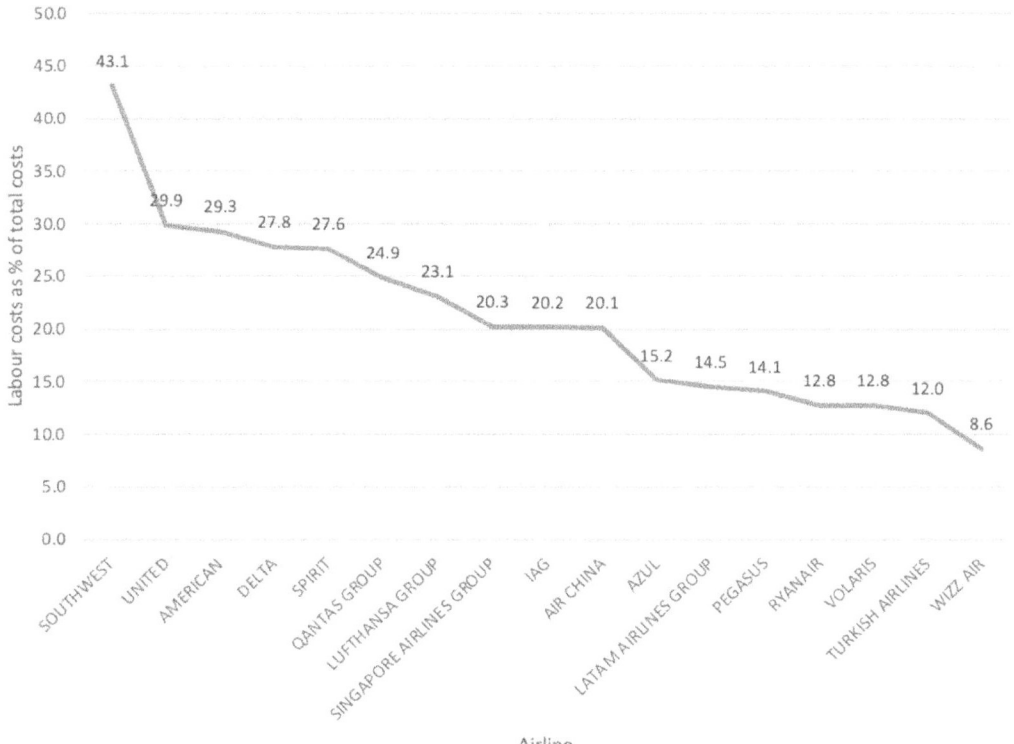

Figure 10.14 Labour costs as % of total costs, 2023

Source: author from latest financial statements

efficient aircraft, with more composite, lighter weight structures, more efficient engines and improved aerodynamics such as sharklets are also approaches which can save the amount of fuel burnt.

That said, the price of fuel is largely fixed and whilst there are global variations depending on where fuel is uplifted, the market sets the rates. A big issue here is not just the potential cost, but also the volatility (Figure 10.6). When prices are stable, it is much easier for airlines to plan the required revenues to stay (or become) profitable. Volatile prices of your largest cost base make airline management particularly tricky.

Historically, the second highest cost item has been labour. Much of this is influenced by macroeconomic variables such as the cost and standard of living in a particular country, as this will influence the prevailing wage rates. Having bases in countries with lower wage rates can reduce these costs, however this depends on the routes being operated. Figure 10.14 illustrates the labour costs as a % of total costs for a range of airlines. Labour costs are generally higher for airlines operating within higher income per capita jurisdictions such as the USA and Australia (the top five are all from the USA) and lower in others with lower wage rates, such as Pegasus and Turkish Airlines (Turkey), Volaris (Mexico) and Wizz Air (Hungary and several other bases – many in Eastern Europe).

A key measure of efficiency in analysing airline costs is the unit cost metric – CASK – which measures how much it costs an airline to fly one seat one kilometre. This allows a more direct comparison with the RASK metric to establish a profit per ASK (RASK–CASK). Figure 10.15 has the CASK metric overlaying the RASK to illustrate this for the top 20 airlines by revenue. The key is the RASK being higher than the CASK, otherwise the airline is likely

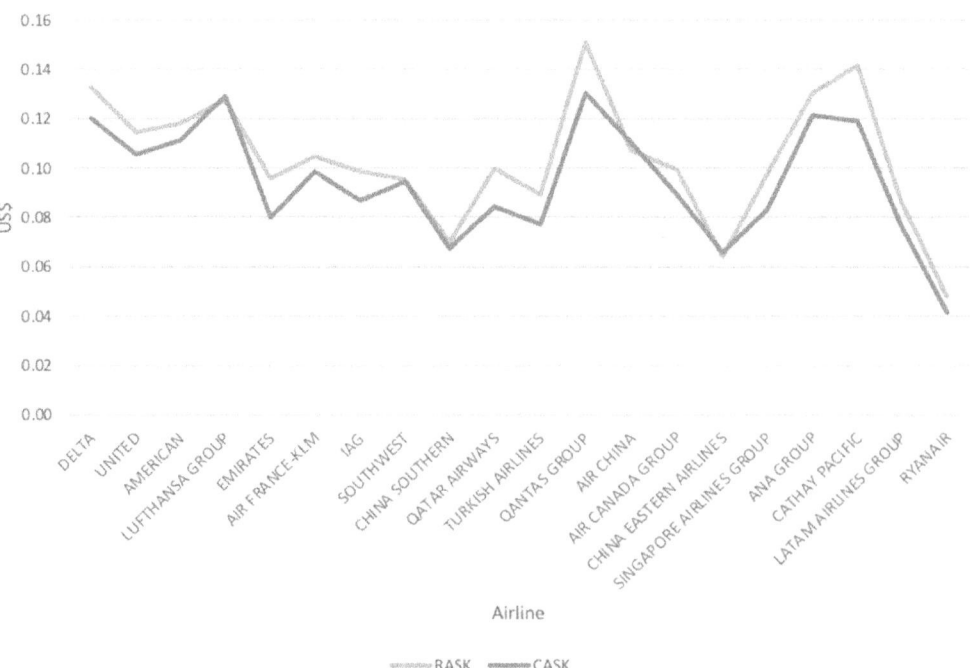

Figure 10.15 RASK and CASK for the top 20 airlines by revenue, 2023

Source: author from latest financial statements

Economic geography and air transport 295

loss-making – as with Air China and China Eastern, who were much later in the COVID-19 recovery phase due to Chinese restrictions.

A key point to note is that airlines can be "low cost" or "high cost". However, this may simply be a function of their business model. An FSNC such as Singapore Airlines should always have higher unit costs than an LCC such as Air Asia, due to the added services provided to the customer – such as meals, drinks, IFE and products in business and first-class cabins. That does not mean they wish to have any higher costs than they must, but these products and services cost money. If airlines wish to compete mainly on price, however, then a strict focus on costs is key, with the Ryanair mantra of "lowest-cost provider wins" having more than a grain of truth. Overall, though, whatever the cost base, what matters is that revenues exceed costs and the more the better, whether an airline has a "low" or "high" cost base. The example of Ryanair illustrates this well. Ryanair has the lowest RASK on the list, but it also has the lowest CASK and achieves a very healthy financial performance.

As stated earlier, it is often wise to use the RASK and CASK metrics to compare airlines within the same business model, to provide a more like-for-like comparison. Figure 10.16 displays the CASK for a selection of key low-cost airlines. The lower the costs in a price-dominated, elastic market, the easier it is to keep fares low and achieve financial success – providing the revenue side of the coin outperforms costs.

It is therefore easier for airlines such as Ryanair and Air Asia, with some of the lowest CASK figures, to achieve profitability, which they regularly do. Airlines such as Southwest must achieve far higher RASK to achieve financial success.

The CASK metric does have its limitations. Analysts often calculate CASK adjusting for the length of the flight, to standardise calculations based on a *common stage length*. Seat

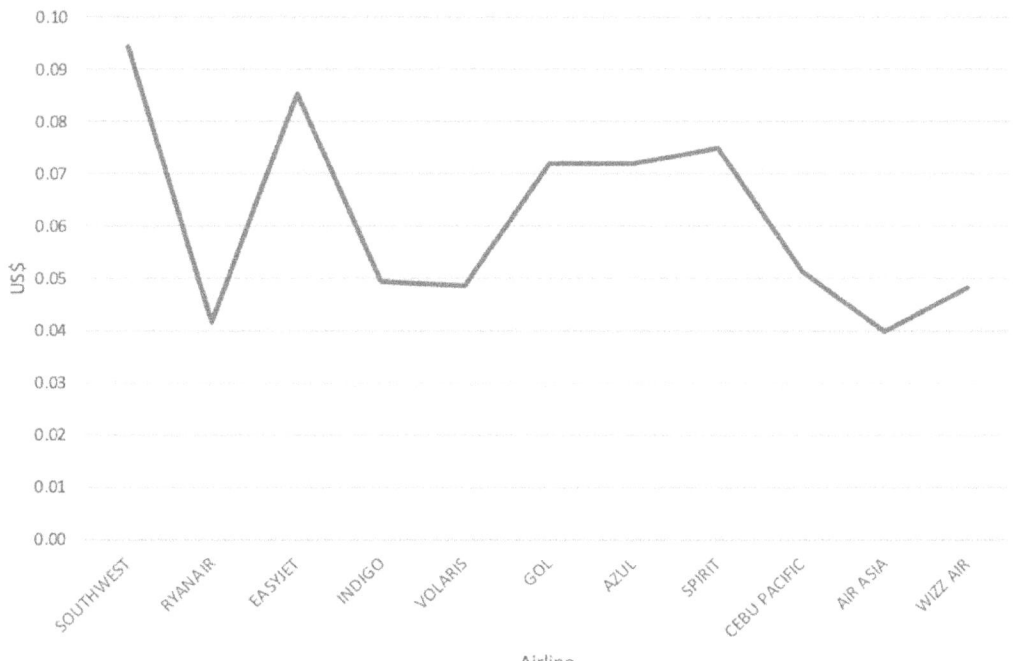

Figure 10.16 CASK for select LCCs, 2023

Source: author from latest financial statements

configurations also play an important role in the CASK denominator and some airlines have the same aircraft types in their fleet, but in different configurations. Different accounting practices can also distort the figures somewhat. However, at a broad level, CASK does give a good indication of an airline's unit cost performance.

10.4.6 Airline profitability

Having analysed revenue and cost, the final area for focus is on profitability. This chapter began with a quote relating to the difficulties in achieving profitability in the airline industry and the marginal profit regularly achieved by the industry over time (Doganis, 2019). This is despite the high growth rates and demand witnessed for much of the last 50 years.

Two commonly used metrics are operating profit (revenue minus cost) and net profit (operating profit minus tax and interest – the "bottom line"). Operating profit is a key metric to measure efficiency and allows a more like-for-like comparison between airlines, whereas net profit shows ultimately what is left for the business and different countries' tax regimes and interest rates can distort global comparisons. Figure 10.17 illustrates the global airline industry operating profit between 2010 and 2023. The best performance in this period was in the 2015–2017 period, with a record profit of $62 billion in 2015. The nadir was reached during the depths of the COVID-19 pandemic and an operating loss of $110.8 billion globally in 2020. When examining the net level, the loss was even worse at $137.7 billion.

When examining the years of profitability, it is hard to know how "good" these are, unless they are viewed in the context of revenues generated and that is where the operating profit margin is very useful. This measures the % of revenue retained as operating profit. Figure 10.17

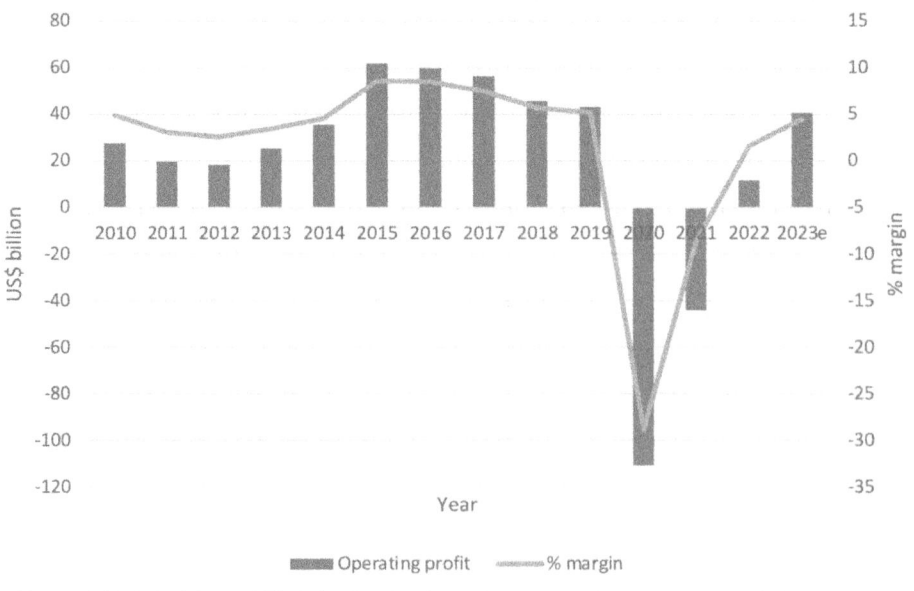

Figure 10.17 Operating profit and operating profit margin for airlines worldwide, 2010–2023
Source: IATA (2023c)

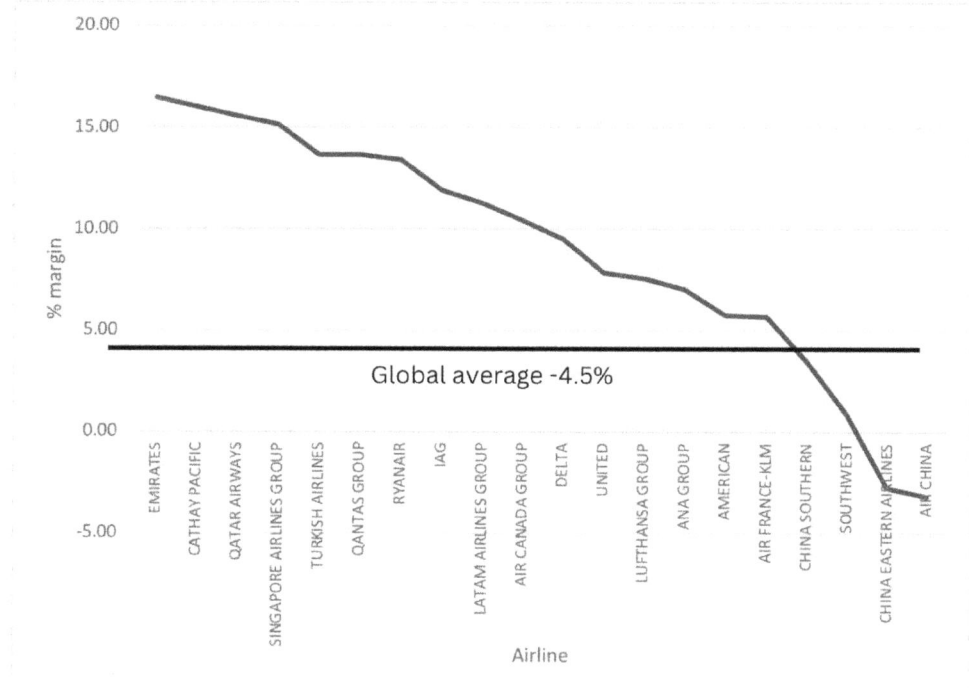

Figure 10.18 Operating margins for the top 20 airlines for revenue, 2023
Source: author from financial statements; Airline Business (2023); CAPA (2024)

shows that the best performing years for operating margin were also 2015–2017, with a peak of 8.6% in 2015 and a disastrous −35.8% in 2020. Even in the best years, less than $10 was kept before tax, for every $100 taken in revenue.

In both the good times and the bad times, there are those airlines that outperform and underperform. For example, in 2015 Alitalia still made losses. Figure 10.18 illustrates the operating profit margins of the top 20 airlines for revenue in 2023. The operating margin globally is estimated at 4.5% and 16 of the 20 exceeded this, with three of those who didn't from China – again due to the late re-opening following COVID-19. Several airlines are achieving double digit margins, such as the regularly outperforming Ryanair, and if these were to continue then at least for some airlines in the industry, regularly covering their cost of capital may be achievable.

10.4.7 *Regional economic performance*

Whatever the industry trends in revenue, profitability or growth, one thing that has held true is that these are inconsistent amongst the regions of the world (Table 10.3).

North America has traditionally been the largest region and hence the performance here has often been the driver in terms of global performance. In the pre-COVID years of 2014–2019, the region accounted for the largest percentage of global net profit and the highest operating profit margin in each of the years. That said, the decade between 2000 and 2009 saw the US industry alone lose around $50 billion. Post-COVID, the region has been the main driver of global net profitability improvements.

Table 10.3 Net profit, operating margin and RPK growth by world region, 2014–2023

	2014	2015	2016	2017	2018	2019	2020	2021	2022	2023e
Global										
Net post-tax profit (US$ billion)	13.8	36	34.2	37.6	27.3	26.4	−137.7	−41	−3.8	23.3
Operating profit margin (%)	4.6	8.6	8.5	7.5	5.7	5.2	−29	−8.6	1.6	4.5
RPK growth (%)	6	7.4	7.4	8.1	7.4	4.2	−65.8	21.8	64.9	38.4
Africa										
Net post-tax profit (US$ billion)	−0.9	−0.8	−0.4	−0.2	−0.1	−0.3	−1.8	−1.1	−0.8	−0.5
Operating profit margin (%)	−2.7	−2.1	1.1	0.8	1.5	1	−16.9	−6.8	−3.3	−0.3
RPK growth (%)	0.6	3.4	7.3	5.5	6.1	4.5	−68.2	17	84	40
Asia-Pacific										
Net post-tax profit (US$ billion)	0.5	7.5	7.4	10.5	6.1	4.9	−45	−13.7	−13.6	−0.1
Operating profit margin (%)	2.4	6.9	7.4	6.3	4.7	3.7	−34.3	−11.9	−8.9	−0.3
RPK growth (%)	7.8	9.6	11.1	10.8	9.3	4.8	−62	−13	32	98
Europe										
Net post-tax profit (US$ billion)	1.9	7.1	8.5	8.9	9.1	6.5	−34.5	−12.1	4.1	7.7
Operating profit margin (%)	3.1	5.5	6.1	7.9	6.2	4.8	−27.1	−8.9	4	6.5
RPK growth (%)	6.5	5.8	5.3	9.1	7.5	4.3	−69.5	28	102	22
Latin America										
Net post-tax profit (US$ billion)	0.1	−1.6	0.4	0.5	−0.8	−0.7	−11.9	−7	−3.9	−0.6
Operating profit margin (%)	5	5	5.6	6.2	2.7	2.9	−28.5	−9	−4.1	−0.2
RPK growth (%)	6.4	6.7	4.5	7.3	7.4	4.1	−62.5	40	63	16
Middle East										
Net post-tax profit (US$ billion)	1.1	2.1	1.3	0.1	−1.5	−1.5	−9.4	−4.9	1.4	2.6
Operating profit margin (%)	3	6.3	2.2	−3	−4.6	−5.2	−24.3	−12	5.1	6.1
RPK growth (%)	11.9	9.9	11.4	6.8	5	2.3	−72.1	9	145	35
North America										
Net post-tax profit (US$ billion)	11.1	21.7	17	17.8	14.5	17.4	−35.1	−2.3	9.1	14.3
Operating profit margin (%)	9.1	14.4	13.7	11.2	9.1	9.6	−27.3	−5.9	6	7
RPK growth (%)	3	4.5	4.3	4	5.3	3.9	−65.1	75	46	16

Source: adapted from IATA (2020b, 2023a)

Alongside the Middle East, the **Asia Pacific** region witnessed the largest RPK growth rates pre-COVID as economies expanded, liberalisation was extended and competition (especially via LCCs) and the middle classes both grew. China was a key performer here as well as India, both of which witnessed extraordinary growth rates. Post-COVID it lagged other regions in terms of recovery, largely due to slower re-openings following pandemic restrictions, especially in China. It is expected that the Asia Pacific region will be the main driver of growth over the next 20 years.

The **Middle East** has also seen large growth rates. This region more than any other illustrates the geographical fluidity of the global air transport industry. Most of the world's population is within an eight-hour flight of the region and the main carriers, especially Emirates, Etihad and Qatar, have taken full advantage of this via their hub-and-spoke networks in Dubai, Abu Dhabi and Doha respectively. Their economies have developed and the growth in air transport has had a reciprocal and symbiotic relationship, further assisting the growth of economies in the UAE and Qatar. Saudi Arabia, with its Vison 2030, has huge plans for air transport as a key component in the Kingdom's aspirations.

ATG Case study 10.3

Emirates and Dubai

Emirates began operations in October 1985, with $10 million in seed funding, with a Boeing 737 and A300 B4 wet-leased from Pakistan International Airlines (Emirates, 2024). In 2024, they operate an exclusively wide-bodied fleet of 269 aircraft, including both A380s (where they operated around half the total A380s ever produced) and B777s, with A350s on order. They are the largest international airline in the world by RPKs and Dubai International Airport (DXB) has the largest volume of international passengers globally. In 2024, it was announced that Dubai Al Maktoum (World Central/DWC) Airport, which lies to the south of Dubai, will replace operations at DXB by 2034 and be able to accommodate a capacity of 260 million passengers (2023 saw 86.9 million at DXB), with five passenger terminal buildings, five parallel runways and 70sq.km total area (Skift, 2024a).

Europe, as one of the most mature regions (alongside North America), has also been a key driver of global profitability post-COVID. Alongside North America, it is the region with the most developed LCC sector, thanks largely to EU Open Skies. The growth in air transport is driven by many LCCs such as Ryanair, Wizz and easyJet and these compete with, amongst others, the three legacy FSNCs in Lufthansa, IAG and Air France-KLM. Supply-side issues such as airport infrastructure and lack of capacity, especially at major hubs such as London Heathrow, has constrained growth and with environmental pressures and government policies, especially in countries such as France and the Netherlands, to increase the proportion of passengers travelling by train, alongside airspace infrastructure capacity constraints, could hinder future growth.

Latin America as a region was consistently achieving positive operating margins pre-COVID, but collectively losing money generally at the net level and it has struggled to achieve profitability since. However, the region was one of the first to recover from the pandemic with passenger numbers consistently topping 2019 levels in 2023. According to IATA Director General Willie Walsh (in Dunn, 2023), the market potential is huge but the region has cost headwinds greater than any other region, including higher average fuel costs, taxes and charges. Alongside infrastructure problems and currency volatility, it increases the issues in achieving regional profitability.

Africa has consistently performed the poorest of the world regions and has not been able to turn a net profit at the regional level in the last ten years. It is often considered the last frontier for air transport development but faces various economic, infrastructure, protectionist and connectivity challenges and these all impact performance. However, there is robust demand in many areas for air travel and the potential is massive (although the challenges are large). Indeed, Africa was one of the quickest regions to recover post-COVID. LCC market share has grown in recent years, to around one in five seats by October 2023 (CAPA, 2023), however this is concentrated largely in North Africa and flying to Europe; many markets have virtually no LCC presence. The launch of the SAATM in 2018 saw some progress but the true growth potential can only be realised when industry and government are able to work together to drive air transport growth forward. The performance of Ethiopian Airlines in recent years in terms of growth, profitability and becoming a truly international player suggests what can be possible by having a supportive home government and populous country, although its geography also helps, with the centrality of capital Addis Ababa, assisting in its network and traffic flows.

10.5 Air transport and economic development

So far, we have been focusing on the economic performance of the air transport industry, however what about the impact of air transport *on* global economic development and geography?

10.5.1 Economic impacts of air transport

Air transport provides a rapid worldwide transport network unlike any other and has a vital role in economic development via business activity and tourism. According to ATAG (2020), prior to the pandemic, air transport supported 87.7 million jobs worldwide, with an economic impact of $3.5 trillion globally and 4.1% of global GDP supported by air transport. The pandemic hammered the industry and much of the global economy, which by 2024 is now recovering to levels close to, and in some places exceeding, these numbers. The economic impacts can be characterised in four ways:

Direct impacts

These are economic activities which would not happen without air transport, such as local salaries, airport landing fees, airport retail outlet revenues and manufacturing activities (such as the Airbus final assembly line in Tianjin, China). According to ATAG (2020), "normal world" (pre-COVID) direct impacts of air transport amount to almost $1 trillion.

Indirect impacts

These are activities which are linked with the air transport industry, such as hotels, travel agencies, rental car companies, fuel suppliers and service firms such as IT systems and legal services. These originate from demand created (at least in part) from air transport such as hotels, which develop around airports, and train and buses serving airports. Indirect economic impacts are estimated around $816 billion.

Induced impacts

These are derived from a "multiplier effect" involving secondary spending from direct and indirect employees such as a new airline employee who buys a house in the local area and the builder will then spend money from the sale on their own goods and service locally (Vasigh and Pearce, 2023). Estimated induced impacts are around $700 billion.

Catalytic tourism impacts

This is spend from tourists who arrive by air, such as on entertainment, museums, attractions and retail spend, and amounts to around $1.7 trillion in a "normal" year.

When focusing on the even broader benefits of air transport, these are linked to other sectors and businesses in the economy such as investment, family ties, education, healthcare, labour supply and global trade, and these economic impacts are substantial.

10.5.2 The importance of air connectivity

City-pair connections are vital in supporting the flows of economic activities, especially internationally. Whilst economic growth is a key driver of air connectivity, the symbiotic relationship means that air connectivity generates and develops economic growth (Figure 10.19). Access to air transport links is fundamental to a city (or region, or country) for trade and to allow the mobility of people and goods, and is therefore a key factor in economic development.

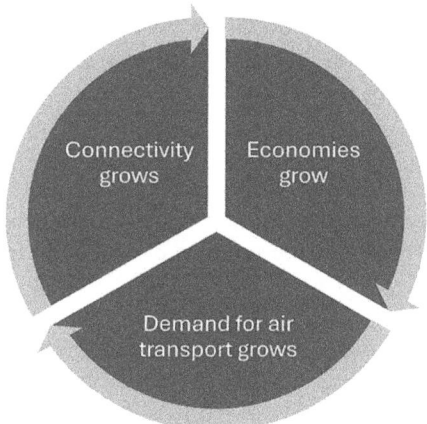

Figure 10.19 Circular and symbiotic relationship between economic growth and air connectivity
Source: author

ATG Did you know?

Air connectivity

According to IATA (2020a), air connectivity reflects how well a country is connected to cities around the globe and to what extent air transport connections support a country's economic development.

ACI Europe (n.d.) defines airport connectivity as the extent to which an airport is connected to the rest of the world, via the sum of both *direct connectivity* (the number of direct flights between two airports and the number of frequencies) such as LHR to ORD and *indirect connectivity* (the number of places people can fly to from connecting flights at airports flown from a particular airport) – for example, flights to destinations from ORD following the flight from LHR.

ICAO (n.d.) states that connectivity is based on the concept of the movement of passengers, mail and cargo which makes the trip as short as possible, with optimal user satisfaction at the minimum price possible.

From a passenger or cargo perspective, it could be as simple as the ability to seamlessly travel between a and b in the shortest time possible.

In 1919, it took around 28 days to fly from the UK to Australia, which reduced to around ten days before World War II. By the late 1940s it was still around 58 hours of travel time and by the mid-1960s it was around 30 hours. By the 1990s it was 23–24 hours with one stop, and by 2018 passengers could fly direct to Perth in under 17 hours. The advent of the jet engine and subsequent technological developments have enabled connectivity between opposite ends of the world to be achieved in times which could only have been dreamt of in the early days of aviation. Passengers and shippers can now access markets around the world in much reduced times and with greater links, thanks to improvements in aircraft technology and connectivity.

In 2019, the number of unique city pairs was 21,736, which was more than double the number in 1996, when there were less than 10,000. COVID-19 brought a huge decline of around 6,000 city pairs but by 2023, this had recovered to around 2019 levels, albeit with likely lower frequencies due to lower capacity (IATA, 2023a).

10.5.3 Benefits of air connectivity

There are many benefits of air connectivity to economies and people all around the world and the greater the connectivity, the higher the catalytic effects of air transport:

1. Trade and commerce

    Air connectivity provides rapid passenger and goods transport across long distances at a reduced cost and time, meaning businesses can access global markets more efficiently. Air transport accounts for less than 1% of world trade by volume but around 35% of global freight by value (IATA, 2020a).

2. Investment

    Regions with better connectivity are more attractive to businesses and investors, encouraging foreign direct investment (FDI) and expansion. Multinational corporations may establish regional headquarters and offices in cities with good air connectivity, for example along the "M4 corridor" between London Heathrow and Central London.

3. Tourism

    Air transport is the primary mode of transport for international tourists and improved connectivity makes it easier for tourists to visit different destinations, providing local economic benefits in sectors such as hotels, retail and entertainment and associated catalytic benefits. Of international tourists, 58% travel by air (ATAG, 2020).

4. Regional development

    Airports can act as hubs for regional development by connecting remote or underserved areas to global markets. For example, London Heathrow has public service obligation (PSO) routes (routes which may not otherwise by commercially viable) to Derry in Northern Ireland, Dundee in Scotland and Newquay in England, whilst the Scottish government provides support for flights from the largest city of Glasgow to the Highlands and Islands of Scotland (Aviation Week, 2023). Air transport and remote geography will be discussed in more detail in Chapter 11.

5. Employment

    The industry itself creates millions of jobs and the connectivity benefits spill over into jobs within related sectors such as tourism, retail and hospitality. There are an estimated

Economic geography and air transport 303

11.3 million direct jobs (airports, on-airport companies, airlines, civil aerospace and ANSPs) and a further 76 million indirect, induced and tourism catalytic jobs provided by air transport (ATAG, 2020)

6. Cultural exchange
    Connectivity promotes cultural exchange by facilitating interactions between people from different parts of the world. This will be discussed in Chapter 12.

7. Improved productivity
    Connectivity gives countries access to a wider market for production inputs, encouraging knowledge and technology transfers via both exports and imports. In addition, there are benefits such as agglomeration economies and improved productivity due to increased competition (IATA, 2020a).

10.5.4   Global and regional air connectivity

The extent of connectivity in a country will in many respects depend on the size of its economy and population – the larger the economy and population, in theory the greater connectivity. Direct connectivity is generally perceived as more valuable than indirect due to the reduction in total travel times; however, hub connectivity does allow many more routes to be commercially viable than with a pure direct, point-to-point model.

IATA (2020a) has developed a connectivity indicator to measure the degree of integration of a country into the global air transport network, based on factors including the number and economic importance of destinations and number of onward connections available and is calculated using the number of available annual seats to each destination and relevant weightings. Figure 10.20 shows that pre-COVID, in 2019, the most connected region in the world was Asia, in both 2014 and 2019. Asia has a booming domestic market, with large populations in countries such as China and India and alongside the Middle East (with its large hub airlines and airports in Dubai, Abu Dhabi and Doha), had the highest growth rates between 2014 and 2019.

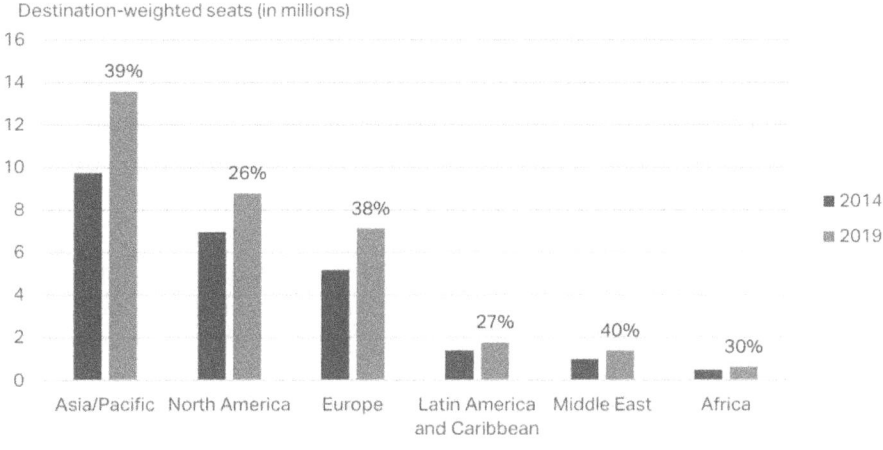

Figure 10.20 Air connectivity and growth rates by region, 2019 versus 2014

Source: adapted from IATA (2020a)

304 *Fundamentals of Global Air Transport Geography*

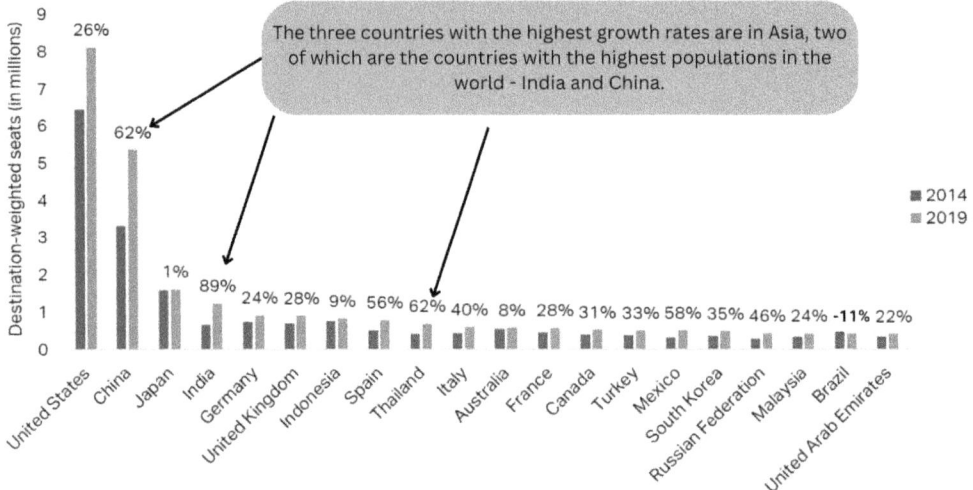

Figure 10.21 Top 20 most connected countries in the world and growth rates, 2019 versus 2014
Source: adapted from IATA (2020a)

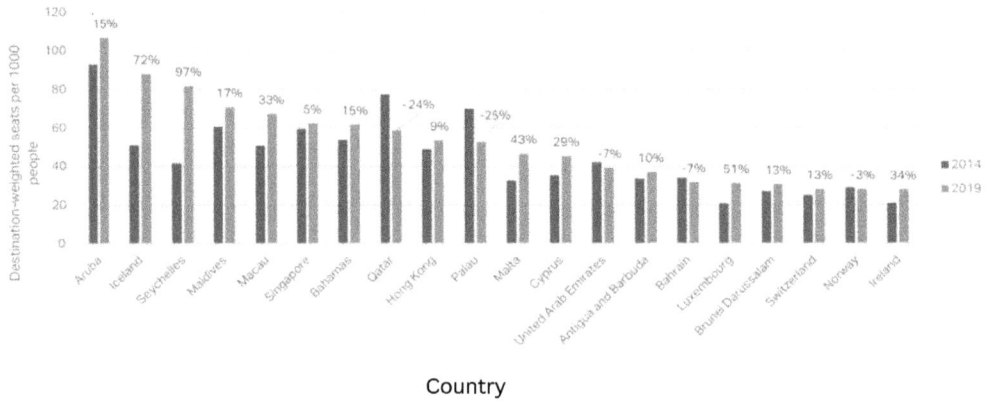

Figure 10.22 Top 20 most connected countries for air connectivity relative to population size and growth rates, 2019 versus 2014
Source: adapted from IATA (2020a)

The most connected country in the world in 2014 and 2019 was the USA, followed by China (Figure 10.21). There is then a huge gap to the next countries on the list – Japan and India – although India has the highest growth rate of the top 20 and the highest three growth rates were all in Asia – China, India and Thailand.

Another useful way to analyse connectivity is air connectivity relative to population size (Figure 10.22). This will introduce countries and city states that essentially punch above their weight based on their population size and generally have strong economies, with Macau, Singapore, Switzerland and Qatar in the top 20.

It is also useful to analyse connectivity at a city level, as there can be large regional disparities within countries and an emphasis on one or two major cities. For example, Figure 10.23

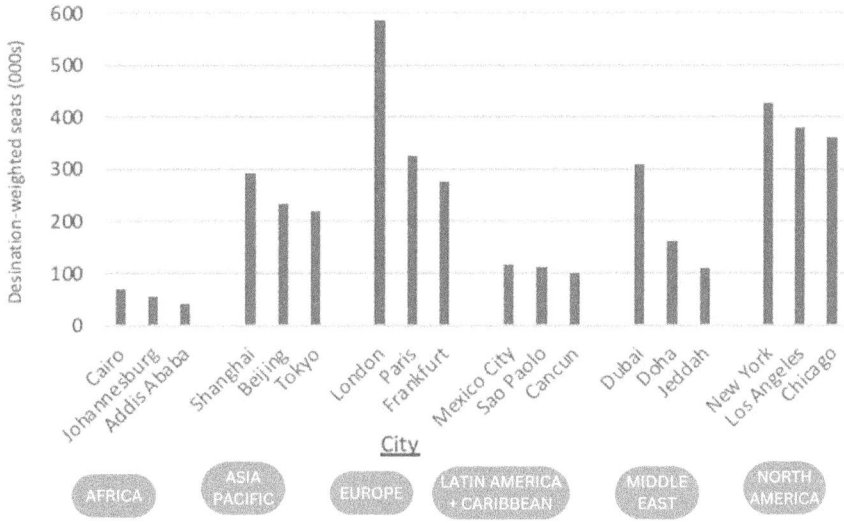

Figure 10.23 Top three cities for connectivity per world region, 2019
Source: adapted from IATA (2020a)

demonstrates that London was the most connected city in the world in 2019, however cities in the north of England, such as Manchester (which is only 20th in Europe), and cities in other countries in the United Kingdom – Scotland, Northern Ireland and Wales – all score much lower and in many ways rely on London for much of their connectivity, particularly via the hub airport of London Heathrow (alternatively, hubs in neighbouring countries such as Amsterdam and Paris can be used).

In addition to London, a few major global cities are at the top of the list, such as New York, Los Angeles, Chicago, Dubai, Paris, Shanghai and Frankfurt. Africa overall has the weakest connectivity; however, many cities here have experienced significant growth over the period 2014–2019, none more so than Addis Ababa, with a growth in connectivity of 83%. Latin America also had some fast-growing cities in terms of connectivity, such as Mexico City (although infrastructure issues are restricting growth here), Cancun (also in Mexico) and Santiago (Chile). Middle Eastern hubs such as Dubai and Doha continue to be key in connectivity in the Middle East, but Saudi Arabian cities such as Jeddah and Riyadh also experienced significant connectivity growth and Saudi Arabia has huge plans for further growth.

ATG Case study 10.4

Air connectivity in Africa

As of December 2023, there were 45 economies designated by the United Nations as the least developed countries (LDCs), of which 33 were in (sub-Saharan) Africa (UNCTAD, 2024). Air transport has the potential to provide bridges within countries, the continent as a whole and to international markets. The continent itself is the second largest and second most populous, but the least well-connected world region for air transport. According to the

African Development Bank (2023), only 53% of roads on the continent are paved and only 43% of Africa's main population has access to an all-season road, therefore air transport has the potential to connect cities, enabling economic flows and mobility.

ATAG (2020) estimated that around 7.7 million jobs were supported in Africa by air transport and $63 billion in GDP. This was 2.2% of all employment and 2.7% of all GDP in African countries in 2018. Of economic activity, $6 was supported elsewhere in Africa for every $1 of gross value added, directly created by the air transport sector.

In 2019, South Africa, Egypt and Morocco had the best air connectivity, with Ethiopia growing strongly in fourth place, almost tripling in the decade to 2019. Ethiopia is a leader in terms of airline performance and boasts Ethiopian Airlines, which has a solid international network, both within the continent and outside, in the USA and Asia especially, and the country also has a young and growing population. It was the fastest growing economy in Africa in 2019 and continuing connectivity improvements should help to foster further economic developments.

The importance of **liberalisation** in growing air connectivity and economic development cannot be overstated and the overall benefits, including new routes, frequencies, lower fares, increased tourism and trade, are displayed in Figure 10.24. A study by InterVistas (2014) found that:

- A more liberal air market between South Africa and Kenya in the early 2000s led to a 69% rise in passenger traffic.
- The operation of an LCC service between South Africa and Zambia resulted in a 38% reduction in discount fares and 38% increase in passenger traffic.
- The 2006 Morocco–EU Open Skies deal led to a 160% rise in traffic and the number of routes increasing from 83 in 2005 to 309 in 2013.

Network density and connectivity have been found to be essential components of business models for delivering profits, in an African context (Heinz and O'Connell, 2013). Abate (2016) concluded that there was a 40% increase in departure frequency in routes in Africa that experienced some form of liberalisation, versus restrictive bilateral agreements. Ismaila, Warnock-Smith and Hubbard (2014) stated that, with air service liberalisation in Nigeria, liberalisation of market access could increase passenger traffic by at least 65% and liberalisation of ownership and control could increase passenger traffic by around 34%.

The Kenyan tourism industry would benefit from increased connectivity and all household groups could experience an improvement in their welfare, if coordinated with rural development initiatives (Eric, Semeyutin and Hubbard, 2020). A study by Tolcha et al. (2021) on 52 African countries concluded that aviation liberalisation promoted economic freedom and improved air connectivity and that the economic contribution of air transport liberalisation varied across African countries relative to their economic freedom and air connectivity.

The importance of connectivity is further illustrated by the fact that Africa is projected to be the second fastest growing region in the world in terms of passengers over the next 15 years (IATA, 2020a). Liberalisation will be key, but reducing operating costs, such as in fuel, charges and taxes, improving infrastructure in airports and ATC systems and managing the environmental impacts of air transport, will also be necessary if Africa is to be able to fulfil its growth potential.

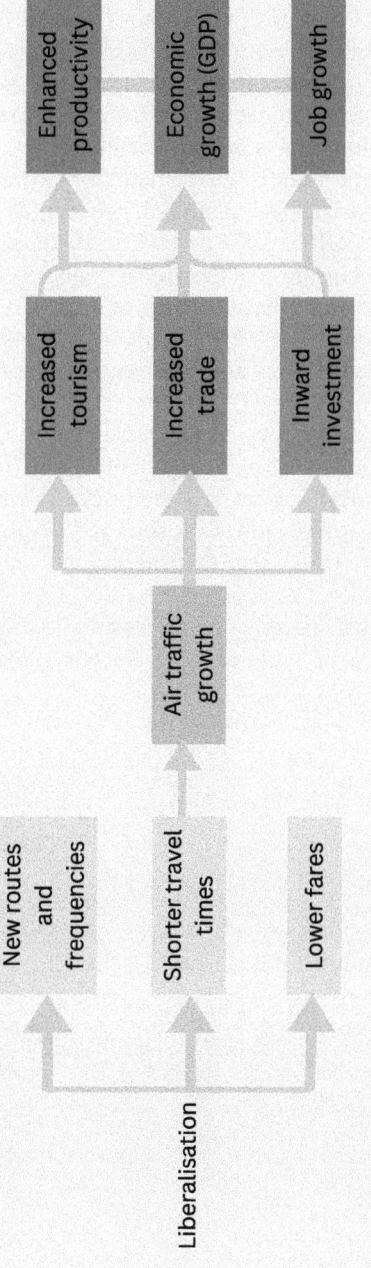

Figure 10.24 Air transport liberalisation benefits
Source: Adapted from InterVistas (2014)

308 *Fundamentals of Global Air Transport Geography*

10.6 The economic outlook for the air transport industry

According to IATA (2023a), global passenger traffic looks set to double by 2040. One thing for sure is that any growth will not be completely linear and there will be peaks and troughs along the way. Macro factors such as conditions in the global economy, terrorist threats, natural disasters, geopolitical instability, regulatory obstacles and the need to ensure the industry grows in a sustainable way, will all have an influence on the industry's growth rates over time and how these affect different regions of the world. In addition, supply-side factors such as airport congestion and slots, airspace congestion and fragmentation and the timely availability of new aircraft, can also leave demand unfulfilled. However, history has taught us that air transport has been very resilient and adaptable to changes in market conditions.

Another thing which does seem certain is that demand for air transport is there and is growing. Airbus (2023) forecasts that the underlying outlook between 2019 and 2042, at a compound annual growth rate (CAGR), for GDP is 2.5% and world trade 2.9%, with an increase in the world population of 1.5 billion and increase in the global "middle class" of 1.9 billion. The median growth in air transport passenger traffic is forecast at 3.6% CAGR which would be around 1.4 times that of GDP and freight traffic of 3.2% CAGR. Growth is expected to be a little more modest in mature markets such as the USA and Western Europe and stronger in Asia and the Middle East, led by India and China. Boeing (2023) forecasts that the world economy will grow from $92 trillion in 2022 to $155 trillion by 2042. Passenger traffic is forecast to grow from 6 trillion RPKs in 2022 to 20 trillion RPKs and cargo RTKs from 260 to 630 billion by 2042.

A good metric to use when analysing growth is the forecast demand for new passenger and freighter aircraft. Figure 10.25 illustrates both the Airbus and Boeing new aircraft demand forecast through to 2042 and both are estimating that the number of aircraft will approximately double in the 20-year period to 2042.

Both manufacturers forecast that narrowbody aircraft will dominate demand, with around 80%, and widebody around 20% of total demand. Increasingly, longer-range narrowbody

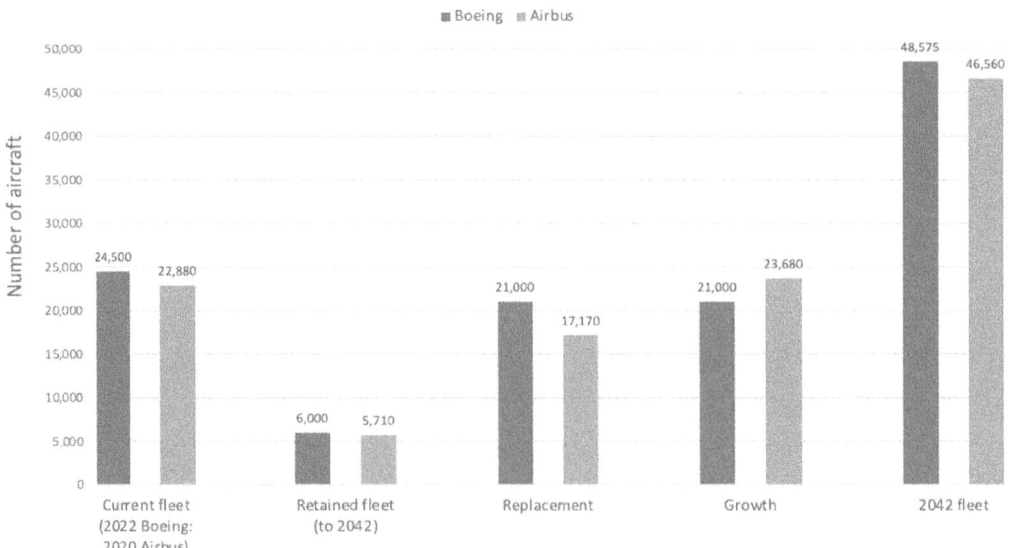

Figure 10.25 New aircraft demand to 2042

Source: adapted from Airbus (2023); Boeing (2023)

aircraft such as the A321LR are providing airlines with the opportunity to operate longer range flights, which would have originally required widebody aircraft, on more efficient narrowbody aircraft. These aircraft are making thinner routes more viable with their lower seating configurations and lower unit costs, potentially helping to develop secondary airports at the expense of slot-restricted hubs.

One of the key issues facing air transport is sustainability and working towards net zero goals (Chapter 14) will result in many changes, including operational improvements and more sustainable aviation fuel (SAF) usage. One of the key technologies being researched is hydrogen-powered aircraft and if these are successful, then aircraft demand may be on aircraft which look different to those flying today.

Chapter review questions

10.1 What are the key factors influencing air transport growth? For any chosen country, can you understand how these factors may have helped, or hindered, air transport growth?

10.2 What is the overall financial performance of the sectors within the aviation value chain? For any (non-airline) sector, can you explain why and what do you think the forecast for future performance will be and the factors which will influence these?

10.3 What are the three main factors underpinning airline industry financial performance and how have these changed over time?

10.4 Explain the main trends influencing airline profitability and how these have impacted specific countries or specific airlines.

10.5 What are the different categories of airline ancillary revenue? For any ONE airline, research which of these are utilised by the airline and the importance of each. Has there been any recent changes you can find in terms of ancillary strategy?

10.6 Describe the main costs incurred by airlines. For any ONE cost, evaluate the potential for reducing the total costs and the benefits and risks of any cost reduction strategies.

10.7 What are the differences between world regions in terms of airline financial performance? For the region you live in, analyse the factors contributing to its financial performance.

10.8 What are the main economic impacts experienced globally because of the air transport industry?

10.9 How does air connectivity contribute to economic development? For your nearest city and/or main airport, how would you describe its level of air connectivity in terms of number of destinations and its connection to different countries and world regions? How important do you think these are for the economic development of the city and surrounding area?

10.10 Using Africa as a case study, what are the key connectivity trends and issues found within different parts of the continent? For any ONE country, research the connectivity from its major city and assess how important you think these are for its economic development? How do you feel it could improve its connectivity?

10.11 Using either Boeing or Airbus information from their Commercial Market Outlook, what are the major trends forecast for the air transport industry through to 2042?

ATG trivia

The missing minute

Have you ever booked a flight which had a departure time of 2359hrs? Or 0001hrs? Or 11.59pm? Or 12.01am? It is common practice in the airline industry to use the 24-hour clock in airline and airport operations, although A.M and P.M is common in passenger communications.

You will rarely see a departure time of 0000, however. The reason is potential confusion. If a flight is scheduled to depart at 0000 on 14 November, is this the beginning of the day or the end of the day? Imagine turning up 24 hours late for your flight – although 24 hours early may pose fewer problems, aside from embarrassment! Technically 0000 on 14 November is the start of the day, but to avoid potential confusion, many airlines will schedule at either 2359 on 13 November or 0001 on 14 November to avoid any missed holidays, commute back home or a whole day spent in an airport terminal waiting for your scheduled on-time flight!

References

Abate, M. (2016) 'Economic effects of air transport market liberalization in Africa', *Transportation Research Part A: Policy and Practice*, 92, pp. 326–337. Available at: https://doi.org/10.1016/j.tra.2016.06.014.

ACI Europe (n.d.) *Air connectivity*. Available at: www.aci-europe.org/air-connectivity.html.

Aerotime.aero (2023) '"We are an airline of the future." Jozsef Váradi, Wizz Air – AeroTime', 10 April. Available at: www.aerotime.aero/articles/executive-spotlight-we-are-an-airline-of-the-future-wizz-air-ceo-jozsef-varadi.

African Development Bank (2023) *Africa must tackle huge infrastructure gap to unlock opportunities for transformation – Report*, African Development Bank Group. Available at: www.afdb.org/en/news-and-events/press-releases/africa-must-tackle-huge-infrastructure-gap-unlock-opportunities-transformation-report-65734.

Airbus (2023) *Global market forecast 2023*. Toulouse. Available at: www.airbus.com/en/products-services/commercial-aircraft/market/global-market-forecast.

Airline Business (2023) *Airline Business, Flight Global*. Available at: www.flightglobal.com/airline-business.

ATAG (2020) *Supporting economic & social development*. Available at: https://atag.org/industry-topics/supporting-economic-social-development.

Aviation Week (2023) *UK reforms PSO routes, extends support beyond London*. Available at: https://aviationweek.com/air-transport/airports-networks/uk-reforms-pso-routes-extends-support-beyond-london.

Boeing (2023) *Commercial market outlook*. Available at: www.boeing.com/commercial/market/commercial-market-outlook/index.page.

Bureau of Transportation Statistics (2024) *Annual U.S. domestic average itinerary fare in current and constant dollars*. Available at: www.bts.gov/content/annual-us-domestic-average-itinerary-fare-current-and-constant-dollars.

CAPA (2023) *Out of Africa – will things ever change?* Available at: https://centreforaviation.com/analysis/reports/out-of-africa--will-things-ever-change-664099.

CAPA (2024) *Aviation news*. Available at: https://centreforaviation.com/news.

Doganis, R. (2019) *Flying off course: airline economics and marketing*. 5th edn. Abingdon: Routledge.

Dunn, G. (2023) *Why Latin American airline sector struggles to make money, Flight Global*. Available at: www.flightglobal.com/analysis/why-latin-american-airline-sector-struggles-to-make-money/155516. article.

Emirates (2024) *History timeline | About us*. Available at: www.emirates.com/uk/english/about-us/timeline/.

Eric, T.N., Semeyutin, A. and Hubbard, N. (2020) 'Effects of enhanced air connectivity on the Kenyan tourism industry and their likely welfare implications', *Tourism Management*, 78, p. 104033. Available at: https://doi.org/10.1016/j.tourman.2019.104033.

Flight Global (2024a) *Airbus intends to deliver 720 commercial aircraft in 2023*. Available at: www.flightglobal.com/air-transport/airbus-intends-to-deliver-720-commercial-aircraft-in-2023/152090.article.

Flight Global (2024b) *Boeing lost $2.2 billion in 2023 amid fresh 737 Max scrutiny*. Available at: www.flightglobal.com/airframers/boeing-lost-22-billion-in-2023-amid-fresh-737-max-scrutiny/156725.article.

Heinz, S. and O'Connell, J.F. (2013) 'Air transport in Africa: toward sustainable business models for African airlines', *Journal of Transport Geography*, 31, pp. 72–83. Available at: https://doi.org/10.1016/j.jtrangeo.2013.05.004.

IATA (2020a) *Air connectivity. Measuring the connections that drive economic growth*. Available at: www.iata.org/en/iata-repository/publications/economic-reports/air-connectivity-measuring-the-connections-that-drive-economic-growth/.

IATA (2020b) *Industry statistics fact sheet: June 2020*. Available at: www.iata.org/en/iata-repository/publications/economic-reports/airline-industry-economic-performance-june-2020-data-tables/.

IATA (2023a) *Global outlook for air transport – highly resilient, less robust*. Available at: www.iata.org/en/iata-repository/publications/economic-reports/global-outlook-for-air-transport----june-2023/.

IATA (2023b) *IATA economics chart of the week: declines in unit cost and yield expected to continue in 2023*. Available at: www.iata.org/en/iata-repository/publications/economic-reports/declines-in-unit-cost-and-yield-expected-to-continue-in-2023/. All rights reserved.

IATA (2023c) *Industry statistics fact sheet December 2023*. Available at: www.iata.org/en/iata-repository/publications/economic-reports/industry-statistics-fact-sheet-december-2023/.

IATA (2024) *Aviation value chain*. Available at: www.iata.org/en/iata-repository/publications/economic-reports/aviation-value-chain/.

IATA-McKinsey (2024) *Aviation value chain*. Available at: www.iata.org/en/iata-repository/publications/economic-reports/aviation-value-chain/#:~:text=The%20aviation%20value%20chain%20consists,and%20efficiently%20across%20the%20globe.

ICAO (n.d.) *Connectivity*. Available at: www.icao.int/sustainability/Pages/Connectivity.aspx (Accessed: 29 May 2024).

IdeaWorksCompany (2023a) *Airline ancillary revenue reaches record $117.9 billion worldwide for 2023 – press release*. Available at: https://ideaworkscompany.com/airline-ancillary-revenue-reaches-record-117-9-billion-worldwide-for-2023-press-release/.

IdeaWorksCompany (2023b) *CarTrawler yearbook of ancillary revenue – 2023*. Available at: https://ideaworkscompany.com/2023-cartrawler-yearbook-of-ancillary-revenue-report/.

IMF (2023) *Countries with the largest gross domestic product (GDP) 2022. Statista*. Available at: www.statista.com/statistics/268173/countries-with-the-largest-gross-domestic-product-gdp/.

IMF (2024) *Gross domestic product: an economy's all*. Available at: www.imf.org/en/Publications/fandd/issues/Series/Back-to-Basics/gross-domestic-product-GDP.

IMF via OurWorldInData (2023) *Annual growth of GDP, our world in data*. Available at: https://ourworldindata.org/grapher/real-gdp-growth.

InterVistas (2014) *Transforming intra-African air connectivity: the economic benefits of implementing the Yamoussoukro Decision*. Prepared for IATA. Available at: www.iata.org/contentassets/44c1166a6e10411a982b2624047e118c/intervistas_africaliberalisation_finalreport_july2014.pdf.

Ismaila, D.A., Warnock-Smith, D. and Hubbard, N. (2014) 'The impact of air service agreement liberalisation: the case of Nigeria', *Journal of Air Transport Management*, 37, pp. 69–75. Available at: https://doi.org/10.1016/j.jairtraman.2014.02.001.

OAG (2024) *Airline frequency and capacity statistics | Aviation Data*. Available at: www.oag.com/airline-frequency-and-capacity-statistics.

Ryanair (2023) *Ryanair FY23 results*. Available at: https://investor.ryanair.com/wp-content/uploads/2023/05/FY23-Ryanair-Results.pdf.

Skift (2024a) *Dubai begins work on 'world's largest airport' project*. Available at: https://skift.com/2024/04/28/dubai-begins-work-on-worlds-largest-airport-project/.

Skift (2024b) *Is the world ready to meet the Indian middle class?* Available at: https://skift.com/2024/01/09/is-the-world-ready-to-meet-the-indian-middle-class-megatrends-2024/.

Statista (2023) *Commercial airlines: passenger load factor worldwide 2023*. Available at: www.statista.com/statistics/658830/passenger-load-factor-of-commercial-airlines-worldwide/.

Statista (2024a) *Passenger yield in worldwide air traffic 2023*. Available at: www.statista.com/statistics/655381/passenger-yield-of-commercial-airlines-worldwide/.

Statista (2024b) *Revenue of airlines worldwide 2023*. Available at: www.statista.com/statistics/278372/revenue-of-commercial-airlines-worldwide/.

Statista (2024c) *RPKs in air passenger traffic 2000–2020*. Available at: www.statista.com/statistics/1261233/air-revenue-passenger-kilometers-worldwide/.

Tolcha, T. et al. (2021) 'Effects of African aviation liberalisation on economic freedom, air connectivity and related economic consequences', *Transport Policy*, 110, pp. 204–214. Available at: https://doi.org/10.1016/j.tranpol.2021.06.002.

UN (2023) *UN DESA Policy Brief No. 153: India overtakes China as the world's most populous country | Department of Economic and Social Affairs*. Available at: www.un.org/development/desa/dpad/publication/un-desa-policy-brief-no-153-india-overtakes-china-as-the-worlds-most-populous-country/.

UNCTAD (2024) *UN list of least developed countries*. Available at: https://unctad.org/topic/least-developed-countries/list.

University of Alberta (2017) *Legal eagles – four aviation law experts talk legal careers in aviation*. Available at: www.ualberta.ca/en/law/about/news/2017/4/legal-eagles.html.

US Energy Information Administration (2024) *U.S. Gulf Coast kerosene-type jet fuel spot price FOB (dollars per gallon)*. Available at: www.eia.gov/dnav/pet/hist/eer_epjk_pf4_rgc_dpgD.htm.

Vasigh, B. and Pearce, B. (2023) *Air transport economics: from theory to applications*. 4th edn. London: Routledge. Available at: https://doi.org/10.4324/9781003388135.

Wizz Air (2024) *WIZZ – dream more. Live more. Be more*. Available at: https://wizzair.com.

World Bank (2023) *Gross domestic product 2022*. World Development Indicators Database. Available at: https://databankfiles.worldbank.org/public/ddpext_download/GDP.pdf.

World Bank (2024a) *Air transport*. Available at: www.worldbank.org/en/topic/transport/brief/airtransport.

World Bank (2024b) *World Bank open data*. Available at: https://data.worldbank.org.

Zhan, Y. (2024) *India's stock market will soar to $10 trillion by 2030 and the country's growth is impossible for investors to ignore, Jefferies says, Markets Insider*. Available at: https://markets.businessinsider.com/news/stocks/stock-market-outlook-india-10-trillion-worlds-third-largest-economy-2024-2.

11 Urban and rural geography and air transport

Chapter outcomes

At the end of this chapter, you will be able to:

- Understand the processes of urbanisation and globalisation and the factors which have influenced their development.
- Describe the global spatial trends in urbanisation.
- Understand the ways in which urban geography and air transport are interlinked.
- Explain the ways in which urban geography can influence air transport.
- Explain the ways in which air transport can have an impact on urban geography.
- Understand and analyse the terms "airport city" and "aerotropolis".
- Describe the types of air transport services operating to rural and remote areas.
- Assess the impact air transport has on economic development and tourism in rural and remote areas.
- Understand the importance of government subsidies in the viability of air transport services in rural areas.
- Explain the importance of air transport in disaster response and healthcare in remote areas.

11.1 Introduction

Airports occupy large areas of land. The largest commercial airport in the USA in terms of area – Denver International Airport – occupies over 135sq.km. Runways, taxiways, passenger and cargo terminals, fuel facilities, catering, hotels and offices are just some of the facilities found at airports. Some of these airports are found in downtown urban areas, others in the suburban areas and some newer airports are outside the city boundaries where there may be more available land. Wherever the airports are located, they can have a significant impact on the urban geography of the town or city. The associated surface transport infrastructure also adds to the impacts.

The airport city and aerotropolis are concepts which have been popularised to illustrate the airport's increasingly influential impact on the urban form. These concepts signify a focus on the airport at the centre of urban areas and subsequent developments, which are much broader than the traditional functions, including aspects such as logistics parks and entertainment corridors.

314 *Fundamentals of Global Air Transport Geography*

Although urban geography has traditionally been thought of as a sub-field of human geography, studying cities or urban areas is becoming more concerned with the inter-relationship between cities and biophysical processes, such as sustainability, which also brings in physical geography and the study of airports fits neatly into the overlap between the human and physical disciplines.

Airports are not only located in cities. Rural, remote and island communities are equally reliant, sometimes more so, on airports in terms of links with towns, cities and the rest of the world. Government-subsidised commercial services (where normal commercial operations are not financially viable) are common in remote areas, providing an economic and social lifeline for remote communities. General and emergency medical services, cargo flights and "fly-in, fly-out" operations, for industries such as mining, are other areas where air transport provides invaluable access to and for remote communities.

Whatever their location, airports can be significantly influenced by urban or rural geography, as well as having a significant impact on urban or rural areas.

11.2   Urbanisation and urban geography

11.2.1   Urbanisation

Today, around 56% of the world's population – 4.4 billion people – live in cities and the World Bank (2024) estimates that this will reach almost 70% by 2050. What is most incredible is the rate of change. At the beginning of the 20th century, only 14% of the global population lived in cities (Kaplan and Holloway, 2014). Despite the huge growth in urban areas, these still only cover 3% of earth's land surface, excluding Antarctica (European Commission, 2019), and this can be vividly seen through the lens of NASA's Earth Observatory (Figure 11.1), which utilises satellite images to provide a view of "earth at night". The key concentration of urban areas can be seen to be along the eastern coast of North America, parts of Europe, India, China and east Asia.

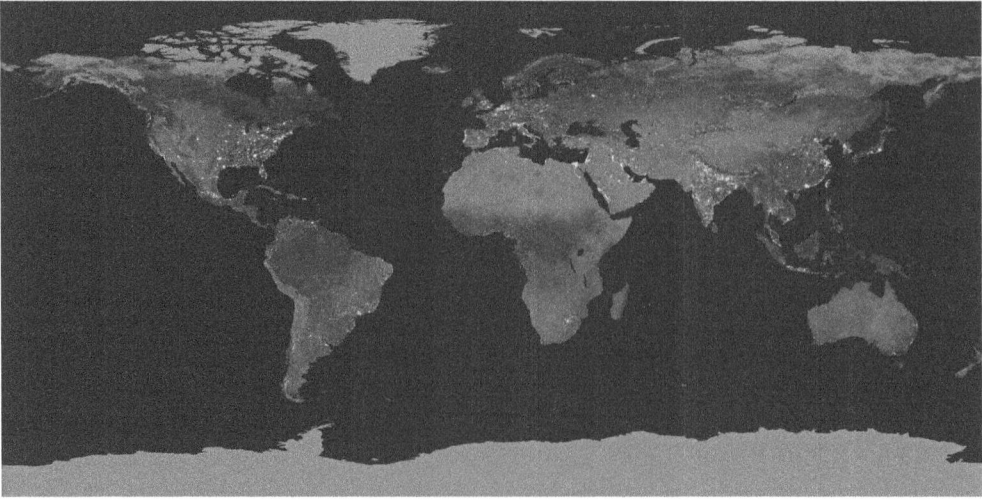

Figure 11.1 The earth at night
Source: NASA Earth Observatory (2016)

Urban and rural geography and air transport 315

The United Nations (2024a) forecasts that future increases in the urban population will be highly concentrated in just a few countries and together, India, China and Nigeria will account for 35% of the projected growth in the global urban population through to 2050, with India alone adding 416 million urban inhabitants. Considering that the global urban population in 1950 was only approximately 751 million, these growth figures are significant. Asia alone plays host to around 54% of the world's population. The most urbanised global region is North America (82%) with the least urbanised being Africa (43%).

ATG Did you know?

World's largest cities

The world's largest city is Tokyo, with an agglomeration of over 37 million people, followed by Delhi with over 33 million, Shanghai with almost 30 million, Dhaka in Bangladesh with almost 24 million and Sao Paolo, Cairo, Mexico City and Beijing, all with over 22 million inhabitants (World Population Review, 2024a).

More than 80% of global GDP is generated in cities and hence urbanisation is a significant contributor to the global economy. However, future sustainability will be a challenge, as the speed and scale of urbanisation has been significant in many areas. The demand for services such as (affordable) housing, education, medical care and employment creates a range of challenges for governments, particularly as there are nearly 1 billion urban poor who live in informal settlements to be near opportunities (World Bank, 2024). In 2022, only half the world's urban population had convenient access to public transportation and between 2010 and 2020, global cities expanded physically faster than their population growth rates, with average annual land consumption rates of 1.5% compared to population growth rates of 1.2% (United Nations, 2023).

Once a city is built, its physical form and land use patterns can be locked in for generations, leading to unsustainable sprawl and therefore understanding the key trends in urbanisation is crucial. In working towards Goal number 11 of the United Nations Sustainable Development Goals (SDGs) on "Sustainable Cities and Communities", transport – including air transport – has a key role to play.

11.2.2   Urban geography

Urban geography looks to explain town and city distribution and the socio-spatial similarities and contrasts between and within them. The character of urban environments is the result of outcomes of interactions between environmental, economic, technological, social, demographic, cultural and political factors, at a variety of global and local scales (Pacione, 2009).

According to Kaplan and Holloway (2014), urban geography studies urban areas using two approaches – the *inter*metropolitan and *intra*metropolitan approaches.

1. **Intermetropolitan** (urban systems) approach – stresses relationships *amongst* a system or groups of cities at the local, national or global level.

From an air transport perspective, this can be researched by focusing firstly on connectivity. Does the urban area have an airport? If so, how many? For example, London has Heathrow, Gatwick, City, Stansted, Luton and Southend within its (albeit loose!) boundaries. An assessment of route connectivity and destinations needs a focus on all. These routes can be direct, point-to-point services or transfer "hub" flights. Including hub flights may increase connectivity by multiples. For example, flying from London Heathrow to New York with British Airways is one route. Connecting on to partner airlines such as American Airlines at New York means potentially hundreds more US domestic and international flights are available for that one ticket from London Heathrow. This, along with the inbound flight connection possibilities, can create a substantially positive economic impact for the urban area, via business, tourism and associated economic activity.

2. **Intrametropolitan approach** – focuses on the internal locational arrangements *within* urban areas.

The location of the airport and its size has a significant bearing on its impact on and within the city (see Chapters 7 and 8). This can be positive but potentially negative – especially regarding environmental issues (see Chapters 13, 14 and 15). Land use and zoning around airports impacts the urban areas, as does the transportation infrastructure to and from the airport, in terms of both public transport and availability of an efficient (or inefficient) road network. Any expansion plans often involve consultation with local councils and populations, as witnessed with the proposed third runway at London Heathrow, which has met with local resistance.

In both these approaches, Kaplan and Holloway (2014) emphasise the importance of the levels of interaction or linkages among places – both between and within cities – with the volume of the traffic flows being a measure of spatial interaction.

Air transport and urban geography are linked and each shapes and influences the other. As urban areas grow, the relationship between urban areas and air transport becomes more significant, affecting factors such as economic development, urban planning and sustainability. Airports are increasingly being recognised as general urban activity centres and key assets for cities and regions, as economic generators and catalysts of investment (Salewski, Boucsein and Gasco, 2014). The rest of the section will focus on this reciprocal relationship – the impacts urban geography has on air transport and the impacts air transport has on urban geography.

11.3   Impact of urban geography on air transport

Whilst the core function of an airport, in terms of facilitating the carriage of passengers and cargo to fly between two points, has not changed, airports have grown from simple airstrips to complex, multifunctional aviation centres, sometimes utilising huge areas of land. Chicago O'Hare has eight runways, New York JFK has five terminal buildings and King Fahd International Airport in Saudi Arabia covers 776sq.km. Urban geography shapes where airports can locate (or expand) and interlinking with urban development and planning can create both opportunities and challenges.

Some urban areas have one airport to service their needs, others have multiple and one example of the latter is the global city of London – which has six main airports, with a combined total of around 180 million annual passengers (Routes Online, 2024). As Figure 11.2 demonstrates, there are airports located north, south, east and west of the city, as well as London City within the central urban area. London Heathrow is one of the largest (79 million passengers in 2023)

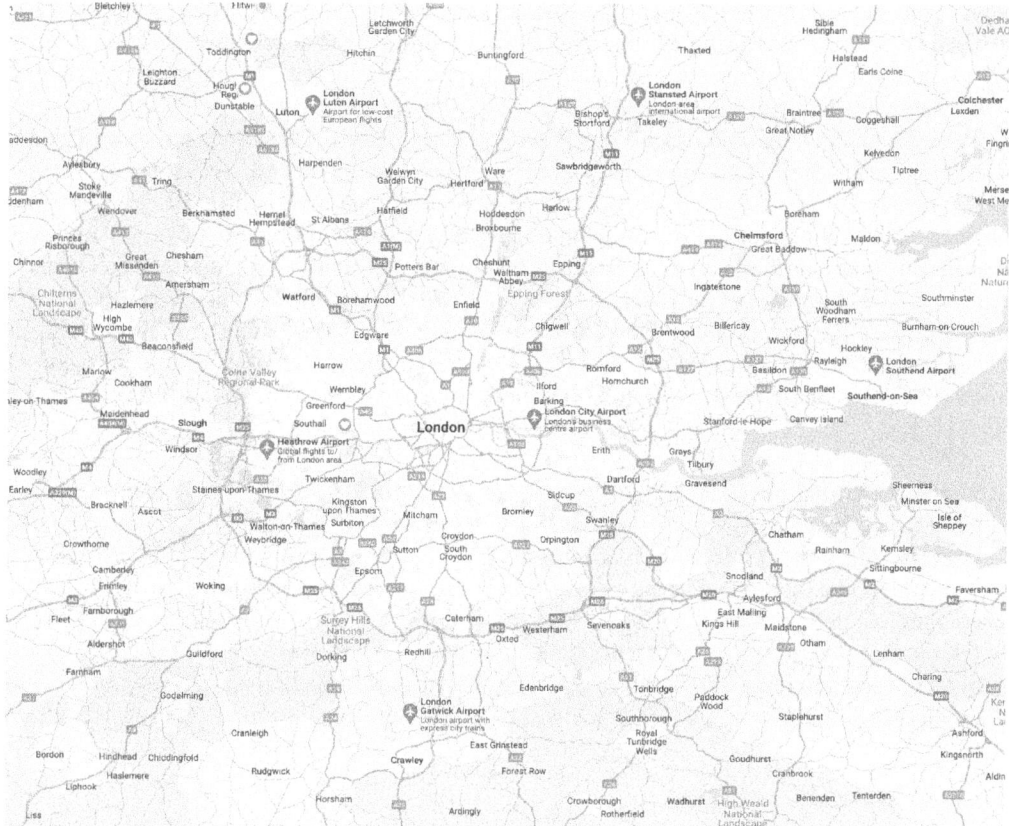

Figure 11.2 The airports of London
Source: Google Maps (n.d.)

and most connected airports in the world, serving much of the globe, whilst London Luton (16 million passengers in 2023) has short and medium routes within Europe at its core. Each of these airports requires appropriate transport infrastructure to access, as well as the area required for the airport itself and associated industries.

Most of these London airports have issues with expansion due to their location and the fact that the UK is a small country, with a shortage of available land in many areas. Heathrow has faced opposition to its proposed third runway for many years, largely on environmental grounds.

Two new global airports have been built in recent years in completely new areas where land has been available to accommodate demand – in Istanbul, Turkey and Beijing Daxing, China and one which was abandoned in 2018 while part way through construction – Mexico City Texcoco Airport.

11.3.1 Airport land use

The land use around airports has changed and developed over time as urban areas have grown, and Section 11.4 will discuss the influence air transport has had on urban geography. As

318 *Fundamentals of Global Air Transport Geography*

populations have grown – and with increasing urbanisation and globalisation – airports have also grown, and the increasing commercialisation of airports has accelerated this trend; the requirement for more land has been evident throughout the world, both for airport expansion and completely new airports (Chapter 7).

The growth in nonaeronautical revenues (such as from retail, food and beverage and car parking) has facilitated the requirement for more space. Indeed, according to Graham (2023), around 46% of overall airport revenues were nonaeronautical in 2019 and at some airports, such as Singapore, it was much higher (61%). Many airports are now retail and commercial destinations themselves, mirroring those outlets found within the central business district (CBD). Airports are now integral to most urban regions and in many areas, even where airports are on the outskirts, they have become core in terms of urban land requirements and the success of the urban area itself. As Goetz (2019) puts it, the largest airports have become "mini-cities" and land use development around airports could be conceptualised in three temporal stages:

1. **Simple extended outpost** with airport-related low-order low-density activities, such as car parking and rental cars. Airports are generally geographically peripheral.
2. **Diversifying outer node** with middle-order low-density activities, such as industrial parks and conference facilities. Urban growth is moving outward towards airports, which are becoming less peripheral.
3. **Globalised metropolitan centres** with high-order medium-density activities, such as corporate headquarters and high-end restaurants. The airport is playing an increasing role in the urban area.

11.3.2 *Categorisation of impacts*

According to Salewski, Boucsein and Gasco (2014), although the impacts of large airports on the built environment are widely acknowledged, there is no agreed-upon categorisation of airport-related urbanisation effects.

Each airport is different in terms of its topography, history and development, but some of the impacts of urban geography on airports can be categorised in the following ways:

Site selection and spatial planning

Airports are constructed for access to population centres – whether for global cities or small towns – and proximity to catchment areas is key. In and around densely populated areas, finding land for new airports, or for airport expansion, can be challenging. The presence of natural features such as hills, mountains and seas play a key role in airport development, as airports need essentially flat land (Chapter 7). Ideally, airports often prefer a coastal location where there are likely to be fewer obstacles, as was the case with San Francisco, which is one of the hilliest cities in the world, where the airport is located 14 miles south on the San Francisco Peninsula, next to the San Francisco Bay.

Several airports have been constructed by reclaiming land (Chek Lap Kok, Hong Kong) or even on completely artificial islands (Kansai, Japan), where suitable land has not been available (Chapter 13). Others have been closed and relocated, due to congestion and the lack of available land. Berlin Brandenburg Airport opened in 2020, replacing three airports – Tempelhof (which was closed much earlier) Tegel and Schönefeld (where Brandenburg was built).

ATG Case study 11.1

Mariscal Sucre Quito International Airport (UIO), Ecuador

Urban density can also be a key factor in limiting airport development. The old Mariscal Sucre Airport in Quito was in the middle of a mountainous city (at approximately 2,800m above mean sea level [AMSL]), which could no longer be expanded to accommodate larger aircraft or an increase in air traffic, due to the city essentially being built around the airport (Figure 11.3). The airport was moved in 2013 to a location 18km east of Quito near Tababela (approximately 2,400m AMSL), where there was available space and the new airport (on a 1,500ha site) is ten times larger than the previous 150ha site (Airport Technology, 2018). The runway at the new airport is 4,100m in length, as opposed to 3,120m at the old airport, which helps in terms of payload, especially due to the high altitude and potential aircraft performance restrictions.

Figure 11.3 Location of the old and new Mariscal Sucre Airports, Quito
Source: adapted from Google Maps (n.d.)

Surface access

Efficient surface access is essential for airport accessibility. Airports in well-planned urban areas can assist with connectivity, especially in relation to congestion. This can include road networks and public transport options, such as buses and trains. Many global airports have been trying to develop their public transport options due to environmental issues and attempting to restrict private car usage. For example, London Stansted has a £7 charge for cars entering the drop-off area.

Development of multimodal airports (integrated transport hubs offering connectivity for transferring passengers and cargo between air, rail, road and/or sea) is increasing at some airports. The success of this strategy is dependent on the range of options and the transfer efficiency between the different modalities, as well as the associated costs. Amsterdam Schiphol, for example, aims to facilitate connectivity on all scales, from widebody aircraft to cycle lanes and pedestrian zones, with their impacts on surface access and urban transport planning. Poland is planning the Centralny Port Komunikacyjny (CPK) to create a national transportation system that efficiently integrates air, rail and road transport, with a new airport to be built between Warsaw and Łódź and a high-speed rail network (CPK, 2024).

Environmental impacts

Airports can create significant noise pollution and therefore their location and those of the other urban areas is critical. Those airports which are in areas negatively affected by *noise* could struggle to accommodate future demand growth (such as London Heathrow, which is essentially full and has a night curfew on operations). Operational noise abatement strategies (Chapter 15) can reduce noise impacts, but this issue will continue to play a leading role in airport developments in the urban setting.

The same applies to local *air quality* and particularly in Europe, there are stringent controls on local emissions, therefore proximity of urban residential areas could significantly impact airport developments. Positive local community engagement and communications can be pivotal.

As we move well into the 21st century, the influence of urban geography on airports will continue to grow, especially from a sustainability perspective.

11.4   Impact of air transport on urban geography

Air transport can provide significant social and economic benefits, by facilitating trade, promoting tourism, cultural exchange and providing local, regional and global connectivity. Airports have become key components in urban landscapes, often influencing the success of the urban areas they serve. Airport development often acts as a catalyst by facilitating the business of other sectors of the economy and acting as an economic magnet for the region they serve (Graham, 2023).

Air transport can have a profound impact on urban geography in a myriad of ways, by influencing patterns of urban development, spatial organisation and economic growth and for broader urban development such as housing, transport infrastructure and business premises – both related and unrelated to air transport directly.

The proximity of airports to cities throughout the world has led to the creation of development corridors between the airport facilities and the core city and beyond, to facilitate movements of people, goods and services between the two places. The M4 corridor (a key trading area within the UK) between London and South Wales, with Heathrow Airport as an important node, is a good example. The links from airports to other land uses create other centres and activity areas along development lines in a city, hence creating different levels of interdependencies and inter-relationships that affect the activities and settlements in the city. The development axis to the core is a crucial link in connecting people to opportunity at and beyond the local level; and for integration of societies (ICAO, 2016).

Airport hubs create nodal hubs in a city region, being connected to outlying areas and the city itself by transport corridors to facilitate the movement of people and goods and, therefore, transport links are required from areas such as the CBD of a city to the airport. Airports are crucial planning tools that can have an impact on the development character of the city and other development options, which can enhance economic activity.

11.4.1 The aerotropolis and airport city

Since the turn of the century, there has been an increasing focus on airports sometimes being centre stage in metropolitan development. The premise is that airports are essentially transforming into urban centres, not just serving them. A key proponent of this idea is University of North Carolina Professor John Kasarda, who stated that airports are shaping urban space in the 21st century much as highways did in the 20th century, railroads in the 19th century and seaports in the 18th century (Aerotropolis, 2024). Kasarda has suggested that airports are engines for local and regional economic development and attract aviation-linked activities, as well as indirectly aviation-orientated land uses via accessibility, agglomeration and prestige economies in the immediate airport area, as well as along the transport corridors (Freestone and Baker, 2011).

As a result, many airports may essentially be changing in terms of functional and spatial evolution, from *city airports*, where airports undertake their traditional functions, to more all-encompassing purposes – as *airport cities*. Airport cities and the associated larger aerotropolis concept have the potential to have a large impact on urban geography.

> **ATG Did you know?**
>
> **The aerotropolis and the airport city**
>
> **Airport city** – An airport city is a city built around runways. It is the core of the aerotropolis, consisting of the terminals and on-airport facilities such as cargo, offices, commercial and retail facilities. It is grounded in the fact that in addition to their core aeronautical infrastructure and services, airports have developed significant nonaeronautical infrastructure, services and revenue streams and are extending their commercial reach and economic impact beyond airport boundaries (Kasarda, 2008).
>
> **Aerotropolis** – As an increasing number of aviation-oriented businesses are being drawn to airport cities and along transport corridors radiating from them, a new urban form is emerging – the aerotropolis – extending up to 30km from some airports. According to Kasarda (Aerotropolis, 2024), these are shaped like traditional urban areas with their central city and rings of commuter linked suburbs. The aerotropolis consists of an airport city and outlying corridors and clusters of aviation-linked businesses and residential development which feed off each other and their accessibility to the airport.

Essentially, the airport provides speedy market connectivity to businesses in the aerotropolis, which in turn provides passengers and cargo for the airports, with reciprocal benefits. Facilities such as distribution centres, logistics parks, research technology parks and just-in-time manufacturing can be located close to the airport. Airports such as Amsterdam Schiphol, Hong Kong Chek Lap Kok and Hyderabad conform to the aerotropolis concept (Kasarda, 2008).

Land use planning is much easier when developing an aerotropolis around an existing airport, where the new airport was built on mostly reclaimed land (as per Hong Kong). In many countries, the potential community nuisance, where there are close residential areas to airports, may make aerotropolis development more problematic. The principle though is to attract goods and services to the region to create jobs and stimulate local economic development.

ATG Case study 11.2

Incheon Airport, Seoul, South Korea (ICN)

ICN is the largest airport in South Korea and one of the busiest in the world, ranking seventh in international passenger traffic in 2023 (OAG, 2024). It is built on artificially reclaimed land between two islands, with work beginning in 1992 and the airport opening in 2001.

Kasarda and Chen (2021) call the aerotropolis around ICN "an exemplar of 21st century airport-centric development". They believe that it leads the world in the combined magnitude, range and quality of commercial investment in both its airport city core and peripheral business and urban clusters.

There are dual CBDs – Air-City, which incorporates offices, convention centres, hospitality facilities, shopping, leisure and logistics complexes – essentially the features of a modern metropolitan centre and Songdo International Business District, which is a $35 billion aviation-oriented smart city built on land reclaimed from the sea, which cornerstones the broader Incheon Free Economic Zone. ICN and Air-City are linked by a 21.4km bridge.

The airport is currently undergoing a further transformation to evolve the airport city and aerotropolis concepts, with a three-phase International Business Complex (IBC) project. In IBC-I, there are facilities including hotels, conference centres, commercial facilities and a golf course. In IBC-II there are more commercial, office and hotel facilities as well as tourist facilities and in IBC-III there is a planned resort complex and entertainment hub. These are aimed at leveraging the international accessibility of the airport, with ambitions to become a thriving hub for tourism and business in Northeast Asia (Incheon Airport, 2024).

According to Kasarda, a key challenge for the aerotropolis concept lies in planning. If there is not appropriate planning (urban, regional, airport and business-site), airport-area development will be spontaneous and ultimately unsustainable (Aerotropolis, 2024). Kasarda (2022) has since introduced the concept of the 4.0 Aerotropolis model, with at least eight broader functional zones: a knowledge zone, ICT zone, modern industries zone, express zones, logistics park and free trade zone, meetings incentives conventions and exhibitions (MICE) zone, bright lights zone, and an office zone that may evolve into a more complete airport edge city. Lifestyle destination structures, such as the Jewel Changi at Singapore Airport, are also being constructed.

Myroniuk (2020) points out the importance of considerations such as sustainability, tax-free zones, education and training, health centres, expo centres and events and entertainment being included in the conversations and development of airport cities. For example, Airport City Stockholm has advanced plans in place regarding developing the cycling infrastructure and all buildings to be within walking distance of a bus stop. DriveLAB Stockholm is a gathering place for training and development in traffic and traffic safety and is in the middle of Airport City Stockholm. Hotels, conference facilities, shops, restaurants and creative workspaces are all being developed – with a plan for Arlanda not just to be a place to fly to or from, but "to meet at and experience" (Swedavia, 2024).

Whilst the aerotropolis concept helps to provide a theoretical framework for airport and urban developments in some areas, it cannot be applied in all instances. Salewski, Boucsein and Gasco (2014), in their research on Amsterdam, Zurich and Singapore, discovered that the effects of airports do not form the built environment in the concentric aerotropolis model nor into a linear city of the airport corridor, but the urbanisation patterns are specific to each city region – the airport being a "relevant force", but one of many.

Goetz (2019) states that aerotropolis projects have been criticised for issues such as remote locations and a lack of vibrant activity, also stressing the importance of the state as playing a leading, if not determinative, role in the aerotropolis development projects at Seoul Incheon, Dubai and Amsterdam, relying on their proximity to the airport and international accessibility to market their development potential.

Bednarek (2024) discusses the highly contested nature of airports and urban development and that whilst there is a general agreement on the positive relationship between airports and economic development (termed "boosterism"), it has varied in nature and over time and that some scholars have questioned the cause and effect of airports and economic development.

With the forecast doubling of air transport over the next 15–20 years, there will undoubtedly be a huge growth in airport developments and their impacts on urban geography. Perhaps the key factor is the extent to which airports can continue to improve their sustainability, however as Goetz (2019) points out, whilst airports will continue to be important in terms of metropolitan regions, they will be one of many activity centres, including the CBD, historic districts and suburban downtowns.

ATG Case study 11.3

Eastern Airport City Project and U-Tapao Rayong-Pattaya International Airport, Thailand

This is one of the strategic projects of the Eastern Economic Corridor (EEC) in Thailand. The U-Tapao Airport is a Vietnam-era airport and with Bangkok's two international airports – Suvarnabhumi and Don Mueang – operating beyond capacity, the government intends to turn U-Tapao into a third major airport (Figure 11.4). The airport upgrade includes an airport terminal and a new runway and taxiways and will be spread over 1,040ha, with the aim of handling 60 million annual passengers. It is expected to be opened by 2029.

A key part of the project is the construction of the Eastern Airport City, encompassing urban development within an approximately 30km radius around the airport, which is planned to be a 24-hour international leisure and entertainment destination. The proposal also features over 3 million sq.m. of mixed-use commercial areas, Thailand's largest meetings, incentives, conferences and exhibitions (MICE) facilities, and a special free-trade zone retail destination, as well as a high-speed rail link connecting the three airports (UTA, 2024).

324 *Fundamentals of Global Air Transport Geography*

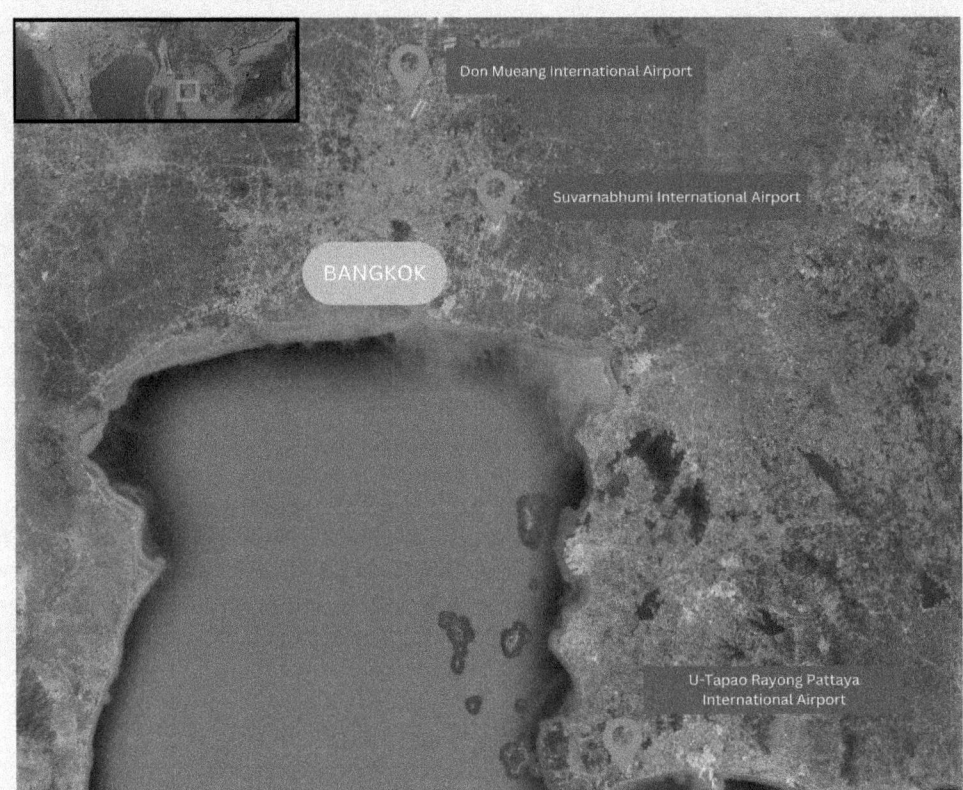

Figure 11.4 Don Mueang, Suvarnabhumi and U-Tapao Rayong-Pattaya International Airport, Thailand

Source: adapted from Google Maps (n.d.)

11.5   Rural geography and air transport

Approximately 44% of the world's population lives in what are called "rural" areas (World Bank, 2024). As recently as 1960, this figure was over 66%, illustrating how significant the global urbanisation trend has been over the last 60+ years. However, due to the increasing overall global population, the total rural population has still increased from around 2 billion in 1960 to over 3.4 billion in 2024, although it has been relatively static since around 2010 (Figure 11.5).

The first part of this chapter discussed the importance of air transport to urban geography and its importance to rural geography can be said to be equally, if not more, important in several ways, but what is rural geography?

Woods (2009) defined rural geography as the study of people, places and landscapes in rural areas and of the social and economic processes that shape these geographies. The definition of what is "rural" has been contested and is certainly open to interpretation. A definition of "non-urban" would seem to be too broad, however some kind of categorisation would be useful. It is not the purpose of this chapter to seek more adequate definitions, however using the broad term of "non-urban", we can ascertain three key spatial locations for which air transport has important impacts: **rural**, **remote** and **island**.

Urban and rural geography and air transport

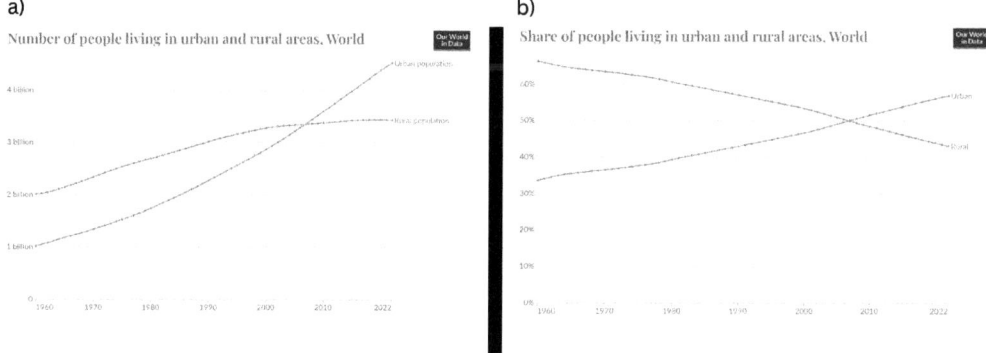

Figure 11.5a Number of people living in urban and rural areas

Figure 11.5b Share of people living in urban and rural areas

Source: Ritchie, Samborska and Roser (2024)

ATG Did you know?

Non-urban areas

Rural – rural areas can be defined as geographic locations outside towns and cities – sometimes classified as the "countryside".

Remote – remote areas could also be classed as rural and in many ways, it depends on just how rural. Therefore, remote could be better defined as places that are further out or even more secluded. In other words, a more extreme version of rural, marked perhaps by a lack of infrastructure such as roads and especially when the land topography is considered – such as in mountainous areas.

Island – in simple terms, an island is a piece of land surrounded by water and the key consideration here is the possibility of the island being detached or isolated. There are islands, such as Great Britian, which cannot exactly be described as isolated (other than in pure physical terms), however there are many islands which are and none more so than what the United Nations calls the Small Island Developing States (SIDS) – see Case Study 11.6.

There are a wide variety of air services operating to rural and remote areas. Merkert (2020) identified a range of market segments related to air transport and remote regions:

1. Passenger services not viable unless publicly subsidised.
2. Regular/scheduled air services which are commercially viable, to remote centres and tourist destinations.
3. Charter tourism and corporate flights to remote destinations, such as the fly-in, fly-out (FIFO) operations for mining and natural resources.
4. Cargo, especially in terms of mail, newspapers and perishable freight.
5. Government and military traffic, surveillance, border and maritime patrols.
6. Emergency and general medical services.
7. Supply chain resilience, disaster relief and humanitarian missions.

326 *Fundamentals of Global Air Transport Geography*

A key issue in the context of air transport is understanding the accessibility and connectivity of rural, remote or island locations to larger urban areas, especially where other forms of transportation are inadequate, inefficient or requiring significant travel times. They may not just be required to develop business, but act as a lifeline to the rest of the world. Remote locations are often characterised by geographical isolation and limited infrastructure, facing unique challenges which air transport can address.

ATG Case study 11.4

Colombia – Andean geography

Colombia has the second largest population in Latin America (World Population Review, 2024b) with over 52 million inhabitants in 2024, behind Brazil (with around 217 million). It is located in the northwestern part of the continent, bisected by the Andean mountains.

As Colombia is a mountainous country, travelling between locations can be challenging – an example of remote geography. The two largest cities by population are the capital Bogotá and Medellín, which are around 260 miles (420km) apart. Travelling between Bogotá and Medellín (Figure 11.6) is through winding, mountainous terrain with no main highways, which can take nine–ten hours, whereas the flight time is less than one hour and thus air links provide a key means of linking these two cities, in terms of the economic and social development of Colombia. According to IATA (2019), air transport is an extremely important economic facilitator in Colombia, contributing more than US$7.5 billion in GDP and generating 600,000 jobs, supporting tourism and facilitating trade.

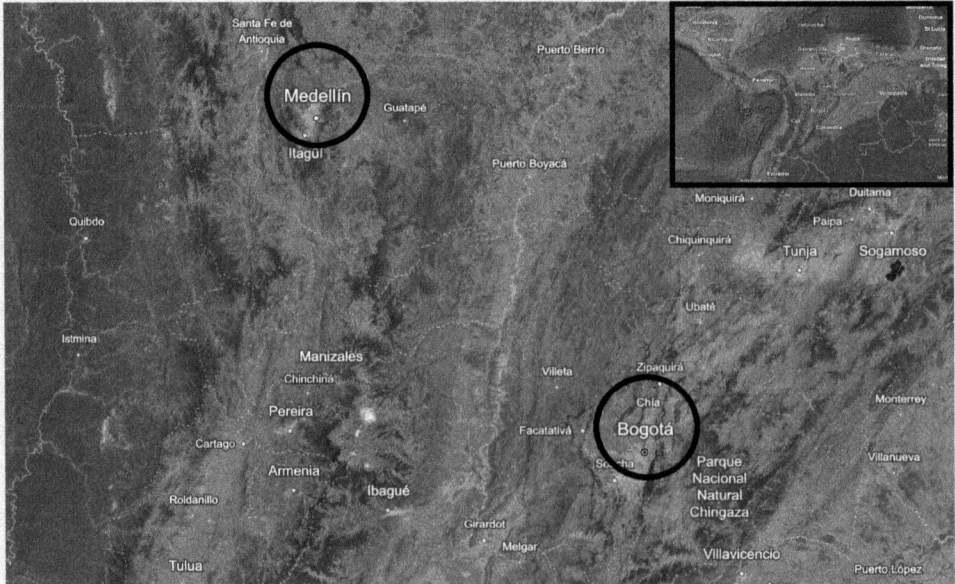

Figure 11.6 Location of Bogotá and Medellín, Colombia

Source: adapted from Google Earth (n.d.)

11.6 Economic development and air transport in remote areas

One of the greatest benefits air transport can bring to remote areas is in terms of economic development (as shown in Case Study 11.4).

11.6.1 Enhancing trade and market access

Enhancing trade and market access is one of the most significant air transport impacts in these regions. Air transport can provide the opportunity to access new markets, expand an industrial base and develop economic growth and diversification.

Remote regions often produce agricultural products or raw materials valued in more distant markets, and air transport allows these to reach market much more quickly. For example, Njoya *et al.* (2023) researched how air cargo could play a role in reducing poverty in Vietnam, demonstrating that air cargo growth stimulated the production of the sectors that employed semi-skilled and unskilled rural workers and that the effects of air cargo growth on poverty were more pronounced in rural households.

In Europe, hard to reach areas such as the Scottish Highlands and the Swiss Alpine regions rely on air transport for the timely and efficient flow of goods. Electronic components for the technology hubs in these regions are often sent by air. Just because areas may be remote, does not mean that consumers are any less sophisticated in their tastes for the latest electronic goods or fashion items, and air transport can facilitate this supply.

It is not just remote areas as such, *landlocked countries* (which cannot rely as much on sea freight) have a key requirement for air cargo transport. Switzerland is a watchmaking hub, requiring rare material imports such as mother-of-pearl and sapphire crystals for watch dials, and their delicate nature and high value often requires air transport. Automotive production in countries such as the Czech Republic and Slovakia relies on the timely delivery of specialised components such as turbochargers and the just-in-time manufacturing processes rely on air transport.

In general, air transport facilitates the import of goods which may not be locally available and the export of key goods, for which remote populations may need to sustain an economic livelihood.

ATG Case study 11.5

Darién National Park, Panama

The Darién National Park is in the southernmost part of Panama – around 200 miles (325km) from Panama City – and is one of the most important World Heritage Sites in Latin America. It is a natural bridge spanning the northern and southern American continents, covering 575,000 hectares, partially bordered by Colombia and the Pacific Ocean (Figure 11.7).

The "Darién Gap" refers to the roadless forest which is the "missing link" in the Pan-American Highway. The Pan-American Highway is a system of roads that run through North, Central and South America, covering around 30,000km, apart from a stretch of about 100km of marshland and mountains – the Darién Gap. As a result, air transport is a necessity, not a luxury, for economic and social development in and between Panama, Colombia and surrounding countries (Echevarne, 2024).

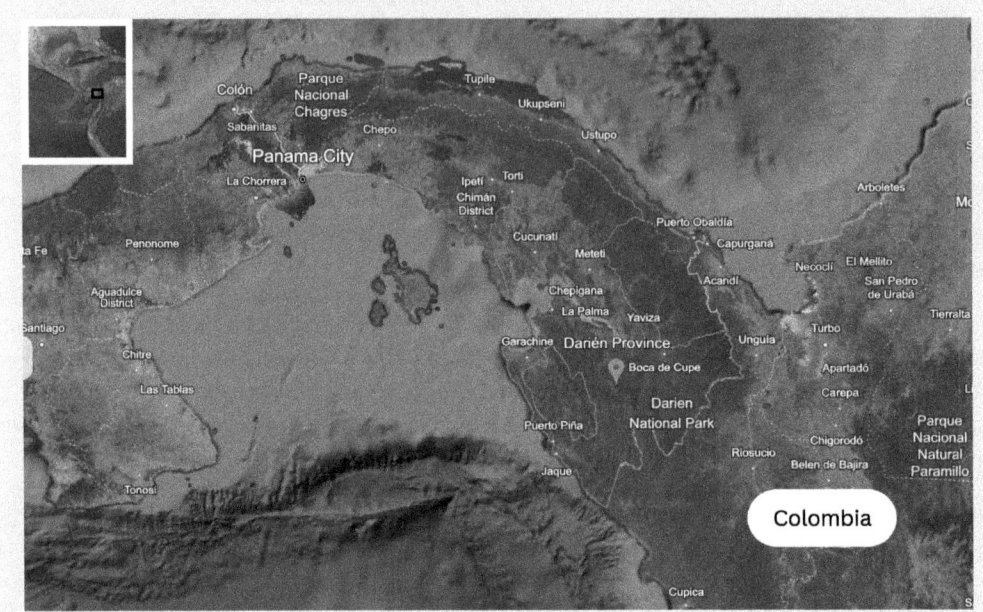

Figure 11.7 Darién National Park, Panama
Source: adapted from Google Earth (n.d.)

Merkert (2020) also identifies the importance of air transport in the context of charter tourism and corporate flights to remote destinations, such as for *mining and resource companies*, in countries as diverse as Australia, Russia, Brazil and Norway, as well as in Africa.

In many remote and rural locations, infrastructure is limited and the diverse topography can often mean that airport and *runway construction* is difficult. Runways are sometimes made from alternative surfaces such as gravel (in Alaska) and sand (in Barra, Scotland) – see Chapter 7.

11.6.2   Tourism

Tourism is a key sector for which rural, remote and island communities can benefit from air transport links. Remote areas can offer natural beauty, serenity, unique wildlife and "off the beaten track" vacations. Air transport is the key industry in making these destinations attractive for regional and global visitors. Tourism has a high potential to stimulate economic growth and social change in rural areas because it contributes to local economies, supports other product value chains, distributes benefits on both seasonal and geographic grounds, and promotes the conservation of cultural and natural heritage (UNWTO, 2024b).

There are distinctions between rural and island tourism. For example, UN Tourism (2024) defines rural tourism as "a type of tourism activity in which the visitor's experience is related to a wide range of products generally linked to nature-based activities, agriculture, rural lifestyle / culture, angling and sightseeing". Not all visitors to the party areas on the island of Tenerife, for example, are necessarily visiting for some of these reasons. However, the majority can fall under the heading of "non-urban" tourism and the potential benefits can apply.

The European Committee of the Regions (2024) studied the impact and potential for tourism to foster rural development across the European Union, where it emerges as a vehicle for economic diversification, job creation and the sustenance of local services, as well as the economic expansion of rural areas and preservation of cultural heritage. Air transport is key in this regard. However, if not managed carefully, there are several challenges, such as environmental impacts, economic disparities and changes in local dynamics.

Tourism dependant locations

The World Economic Forum (2020) stated that globally, 10% of jobs and GDP were in travel and tourism, contributing $8.9 trillion to the global economy. The importance of air transport is shown in Table 11.1, illustrating the countries which are extremely dependant on tourism for their economies and jobs, with a preponderance of island nations, mostly in the Caribbean, dependant on air travel for many of their visitors.

Table 11.1 Travel and tourism relative contribution to employment in 2019

Rank	Location	% of total employment
1	Antigua and Barbuda	90.7
2	Aruba	84.3
3	St Lucia	78.1
4	US Virgin Islands	68.8
5	British Virgin Islands	66.4
6	Macau	65.5
7	Maldives	59.6
8	St Kitts and Nevis	59.1
9	Bahamas	52.2
10	Anguilla	51.3

Source: World Economic Forum (2020)

ATG Case study 11.6

Small Island Developing States

According to the United Nations (2024b), Small Island Developing States (SIDS) are a distinct group of 39 states and 18 Associate Members of United Nations regional commissions – with an aggregate population of 65 million, less than 1% of the world's population. They are recognised by the UN as facing unique social, economic and environmental vulnerabilities and for many this is caused or exacerbated by their remote geography. They are in three global regions (Figure 11.8):

1. Caribbean.
2. Pacific.
3. Atlantic, Indian Ocean and South China Sea (AIS).

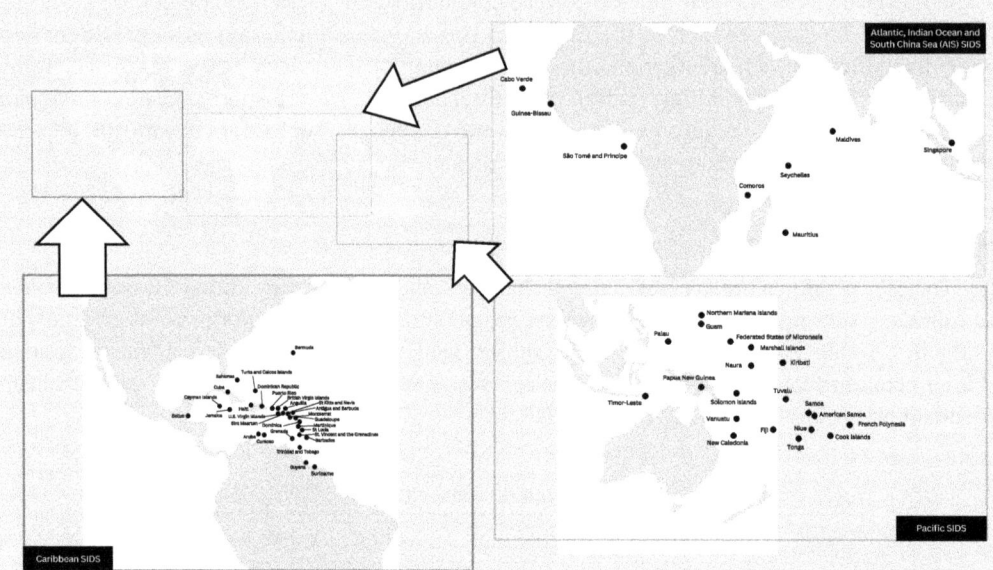

Figure 11.8 Small Island Developing States and Associate Members of United Nations Regional Commissions

Source: author, from United Nations (2024b)

Industries such as tourism and fisheries can constitute over half the GDP of small island communities and many have a reliance on air transport for their visitors and the import and export of goods. The geographical situation of the SIDS, and their natural and cultural heritage richness, make them unique for visitors. In 2023, around 38% of export revenues in SIDS (excluding Singapore) came from international tourism, reaching up to 85% in some destinations (UNWTO, 2024a). However, SIDS are particularly vulnerable to exogenous economic shocks and fragile land and marine ecosystems make them particularly vulnerable to biodiversity loss and climate change, as some may lack economic alternatives. Thus, *tourism sustainability* is key for these SIDS, such as implementing climate adaptation strategies and developing resilient tourism infrastructure.

11.7 Essential air service and public service obligation

Many air services to remote locations are not viable on a commercial basis (Fageda *et al.*, 2018; Merkert, 2020). Government tools such as the essential air service (EAS) and public service obligation (PSO) are essential to ensure that air services are available in many of these areas, where market forces alone would be unlikely to provide services.

In many parts of the world, air services to rural and remote communities are essential for economic growth and development, as well as providing lifelines for local communities. Government support is provided in many parts of the world and two key regions with schemes are:

USA: essential air service (EAS) – The Airline Deregulation Act of 1978 gave air carriers almost total freedom to determine which markets to serve domestically and what fares to charge for that service (Chapter 9). The EAS programme was put in place to guarantee that small communities that were served by certificated air carriers before airline deregulation maintained a minimal level of scheduled air service. The US DoT is mandated to provide eligible EAS communities with air services, generally accomplished by subsidising two round trips a day with 30- to 50-seat aircraft, or additional frequencies with aircraft with nine seats or fewer. These are usually to a large- or medium-hub airport and currently serve approximately 60 communities in Alaska and 115 communities in the lower 48 contiguous states that otherwise would not receive any scheduled air service (US Department of Transportation, 2023).

Europe: public service obligation (PSO) – As with the US EAS, in Europe the PSO framework is used to maintain appropriate scheduled air services on routes which are vital for the economic development of the region they serve and not feasible to be delivered commercially. In Scotland, these are often called "lifeline" services.

ATG Case study 11.7

Wick John O'Groats Airport (WIC) to Aberdeen International Airport (ABZ), Scotland – PSO

In the case of a PSO in Scotland, a public body (such as the Highlands Council with support from the Scottish government) signs an agreement with an airline to provide services to a set schedule and specification. The two parties work out how much it will cost and likely fare levels, with the government providing funding to make up the shortfall for the services to be operated.

One example is the PSO route between Wick and Aberdeen, which is designed to support the community of the Northern Highlands (around Wick) by providing air access to Aberdeen (a major urban centre) and beyond, to:

- Provide better access to business, leisure, healthcare and social opportunities.
- Support the business community by providing access from WIC to ABZ, where connections can then be made from the region to the rest of the UK.
- Benefit tourism businesses in the region by providing better air connectivity.
- Encourage inward migration and investment.
  (Highland Council, 2024)

Air services between WIC and ABZ were operated by Eastern Airways for a three-year period from April 2022, funded by Transport Scotland (£1 million per year) and Highland Council (£300,000 per year).

Some key areas of focus for these policy instruments, include:

- **accessibility** – in terms of flights, schedule and frequency
- **affordability** – pricing to be within the reach of the average local consumer
- **quality and reliability** – such as mail delivery timeframes

ATG Case study 11.8

Rural air services in East Malaysia

Rural air services (RAS) are air services carried out in the interior of East Malaysia (Figure 11.9) where air transport is an essential mode to connect people in rural areas with other parts of the country. The RAS programme is run by the Federal Government and airlines that operate the RAS routes may be entitled to receive subsidies, as they are defined as non-economic air services.

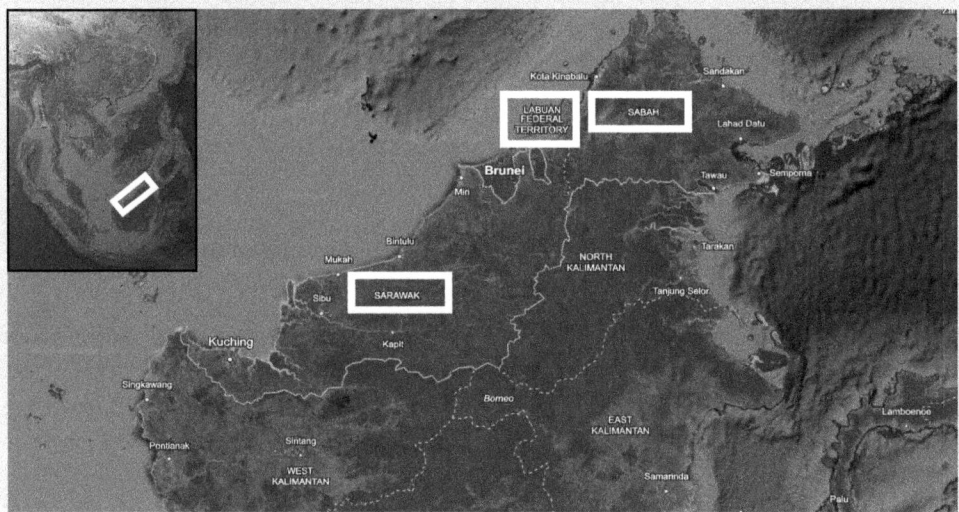

Figure 11.9 East Malaysia
Source: adapted from Google Earth (n.d.)

East Malaysia (sometimes called Malaysian Borneo) consists of Sabah, Sarawak and the Federal Territory of Labuan, where much of the land is rugged and it is difficult to use surface transport due to the size and condition of the roads. In many areas, transport is only possible using air or boat and therefore air transport is seen by the Malaysian government as essential for the rural residents living in East Malaysia (Table 11.2). There are services on 40 routes (Ministry of Transport Malaysia, 2024).

Table 11.2 Travel time for selected destinations in East Malaysia

From	To	Duration (by river/land)	Duration (by flight)
Kuching	Sibu	6hrs 30min by road	45mins
Marudi	Bario	4 days (by logging trail)	40mins
Kota Kinabalu	Sandakan	6hrs 30min	45mins
Kota Kinabalu	Lahad Datu	7hrs 30min	1 hour

Source: MAVCOM (2024)

Air transport is a necessary way of life for these communities, to obtain basic provisions in addition to medical aid and better education.

11.8 Disaster response and healthcare

Air transport not only provides lifelines for communities and plays a key role in the economic development of rural and remote areas, it also provides several other very important functions.

11.8.1 Disaster response

Remote areas are often vulnerable to natural disasters, such as floods, wildfires and earthquakes, where essential services can be damaged and disrupted and communities can become isolated. Air transport plays a key role in the immediate disaster response, by enabling rapid deployment of emergency personnel, supplies and equipment and the evacuation of injured or displaced people. It can also provide long-term resilience, by facilitating personnel and equipment for the rebuilding of infrastructure and providing supplies to the local communities (see Case Study 7.10 on the 2015 earthquake in Nepal). This is especially the case where other transport infrastructure, such as road or rail, may have been affected.

Air transport infrastructure can also be impacted by these natural hazards. In 2010, following an earthquake in Haiti, the local airport had to move from an average of 35 flights a day to more than 100. With a damaged ATC tower, the US FAA brought in a temporary ATC facility to support the response effort (ATAG, n.d.).

The United Nations Humanitarian Air Service (UNHAS), managed by the World Food Programme (WFP), responds to the need for access to the world's most remote and challenging locations, often under difficult conditions of security. They provide passenger and light cargo transport for the humanitarian community, where no safe surface transport or viable commercial aviation options are available. In areas of conflict, air transport may be the only way for life-saving projects to be implemented and monitored. In 2022, 390,000 passengers flew UNHAS and there were 320 regular destinations offered to humanitarians (World Food Programme, 2024). In 2018, relief support was provided for an Ebola outbreak in the Democratic Republic of Congo, earthquakes in Papua New Guinea, a cyclone in the Socotra Island in Yemen and floods in Somalia.

11.8.2 Healthcare access

People in remote communities may not have timely access to key medical facilities, or specialist support following accident or illness. *Air ambulances* – both helicopters and fixed-wing aircraft – are used to transport patients from remote or inaccessible areas to medical facilities, especially important in life-threatening situations. *Organ transplants* are often facilitated by air transport where time is of the essence. *Medical specialists* can fly to remote areas to provide consultations or perform surgeries. *Vaccines and medications* can be supplied rapidly via air transport, such as during the COVID-19 pandemic, especially where there are limited road transport options. *Bloods and biological samples* can be sent via air, to ensure timely diagnosis and treatment.

ATG Case study 11.9

Royal Flying Doctor Service – Australia

Perhaps the most famous example of using air transport for remote access to healthcare is the "Flying Doctor" service in rural and remote Australia, which is one of the largest aeromedical organisations in the world. Australia is a huge country – 7.69 million sq.km. – and two-thirds of the population lives around the regional coastlines. Around 7 million people live across rural and remote areas – with sometimes more than a six-hour drive to the closest township.

The Royal Flying Doctor Service (RFDS) brings emergency medical and primary healthcare to families and communities living in the "Bush", using a fleet of 81 aircraft, including the Pilatus PC-12 and PC-24, the King Air B350C and B200C, the new Beechcraft King Air 360CHW turboprop aircraft and the aeromedical helicopter service in Western Australia. In 2022/23, 36,951 people were flown by the RFDS (Royal Flying Doctor Service, n.d.).

Chapter review questions

11.1 What are the factors influencing the development of urbanisation and globalisation?
11.2 What have been the spatial trends in urbanisation? Which countries – and cities – have seen the greatest urban developments in recent years?
11.3 Explain the variety of ways in which urban geography and air transport are linked.
11.4 How can urban geography influence air transport? For your local urban airport, either undertake desktop research or visit, if safe to do so, to ascertain the extent to which the airport is constrained by its surrounding land uses.
11.5 How can air transport influence urban geography? For your local urban airport, to what extent does it follow the airport city concept? What are the different industries and facilities which have developed in the areas surrounding the airport?
11.6 Explain what is meant by the term aerotropolis. To what extent do you feel that airports have influenced urban geography using this framework?
11.7 What are the various types of air transport services operating to remote or rural areas?
11.8 How has air transport assisted in the economic development of rural areas? What are the closest locations to you, which would not have been able to fully develop their tourism industry without air transport?
11.9 How important are government subsidy schemes like the EAS and PSO to rural and remote areas?
11.10 Research case studies where air transport has been important in responding to disasters and providing healthcare in remote areas.

ATG trivia

Flying vehicles

Have you ever seen the film *Back to the Future Part II*, the 1989 science fiction film starring Michael J. Fox, where the lead characters travel to 2015, which is a year full of flying cars? Sadly, these had not materialised by 2015, however urban air travel is expected to change soon, thanks to urban air mobility (UAM).

According to EASA (2024), UAM is a new safe, secure and more sustainable air transportation system for passengers and cargo in urban environments, enabled by new technologies and being integrated into multimodal transportation systems. This is performed by electric aircraft, taking off and landing vertically, either remotely piloted or with a pilot onboard (Figure 11.10). EASA believes that by 2030, 340 million people will live in EU cities and experience UAM, either for passenger transport or the delivery of goods.

Figure 11.10 Urban air mobility concepts
Source: NASA (2021)

These may not be hoverboards or DeLorean cars (or will they?), but they may revolutionise urban transport.

References

Aerotropolis (2024) *Aerotropolis*. Available at: https://aerotropolis.com/airportcity/index.php/about/.
Airport Technology (2018) *Mariscal Sucre International Airport, Quito*. Available at: www.airport-technology.com/projects/quitoairport/.
ATAG (n.d.) *Rapid disaster response*. Available at: https://aviationbenefits.org/social-development/rapid-disaster-response/.
Bednarek, J. (2024) *Airports and urban development, Oxford bibliographies in urban studies*. Available at: www.oxfordbibliographies.com/display/document/obo-9780190922481/obo-9780190922481-0002.xml.
CPK (2024) *Centralny Port Komunikacyjny – CPK*. Available at: www.cpk.pl/en/.
EASA (2024) *What is UAM*. Available at: www.easa.europa.eu/en/what-is-uam.
Echevarne, R. (2024) 'Interview. ACI Latin America-Caribbean Director General'.
European Commission (2019) *WAD | World Atlas of Desertification*. Available at: https://wad.jrc.ec.europa.eu/urbanplanet.

European Committee of the Regions (2024) *Tourism and rural development*. LU: Publications Office. Available at: https://data.europa.eu/doi/10.2863/099682.

Fageda, X. et al. (2018) 'Air connectivity in remote regions: a comprehensive review of existing transport policies worldwide', *Journal of Air Transport Management*, 66, pp. 65–75. Available at: https://doi.org/10.1016/j.jairtraman.2017.10.008.

Freestone, R. and Baker, D. (2011) 'Spatial planning models of airport-driven urban development', *Journal of Planning Literature*, 26(3), p. 263.

Goetz, A.R. (2019) 'The airport as an attraction; the airport city and aerotropolis concept', in A. Graham and F. Dobruszkes (eds) in *Air transport: a tourism perspective*. Oxford: Elsevier, pp. 217–232.

Google Earth (n.d.). Available at: www.google.co.uk/earth/.

Google Maps (n.d.) Available at: www.google.com/maps.

Graham, A. (2023) *Managing airports: an international perspective*. 6th edn. Abingdon: Routledge.

Highland Council (2024) *Wick public service obligation*. Available at: www.highland.gov.uk/info/1523/transport_and_streets/1048/wick_public_service_obligation.

IATA (2019) *The value of air transport in Colombia*. Available at: www.iata.org/contentassets/bbff04f2b67140638dffb0b5cc5fc8e1/the-value-of-air-transport-in-colombia.pdf.

ICAO (2016) *Synergy between airports and urban development for sustainable development. Concept note*. Available at: www.icao.int/ESAF/Documents/meetings/2016/UN-Habitat-ICAO%20Experts%20Meeting%202016/Documents/Concept%20Note_Compressed%20-%2004%20August%202016%20publishing%20version.pdf.

Incheon Airport (2024) *About airport city*. Available at: www.airport.kr/co_cnt/en/majbus/airport/airove/airove.do.

Kaplan, D.H. and Holloway, S. (2014) *Urban geography*. 3rd edn. Hoboken, NJ: Wiley.

Kasarda, J. (2008) *Airport cities: the evolution*. London: Insight Media.

Kasarda, J. and Chen, M. (2021) *The Incheon aerotropolis: an exemplar of 21st-century airport-centric development*. Rochester, NY. Available at: https://doi.org/10.2139/ssrn.3806194.

Kasarda, J.D. (2022) 'Aerotropolis 4.0 – airport world'. Available at: https://airport-world.com/aerotropolis-4-0/.

MAVCOM (2024) 'How Rural Air Services (RAS) helps', *Malaysian Aviation Commission (MAVCOM)*. Available at: www.mavcom.my/en/industry/public-service-obligations/how-rural-air-services-ras-helps/.

Merkert, R. (2020) 'Air transport in regional, rural and remote areas', in L. Budd and S. Ison (eds) *Air transport management. An international perspective*. 2nd edn. Abingdon: Routledge, pp. 357–372.

Ministry of Transport Malaysia (2024) *Rural Air Services (RAS)*. Available at: www.mot.gov.my/en/aviation/operators/RAS.

Myroniuk, V. (2020) 'Airport cities: more than just offices, hotels and restaurants', *International Airport Review*. Available at: www.internationalairportreview.com/article/126351/airport-cities-sustainability-expo-health-education/.

NASA (2021) 'Autonomous systems – NASA', 7 December. Available at: www.nasa.gov/centers-and-facilities/armstrong/autonomous-systems/.

NASA Earth Observatory (2016) *Collection – earth at night*. NASA Earth Observatory. Available at: https://earthobservatory.nasa.gov/collection/1595/earth-at-night.

Njoya, E.T. et al. (2023) 'Examining the impact of air cargo growth on poor Vietnamese rural and urban households', *Transport Economics and Management*, 1, pp. 112–125. Available at: https://doi.org/10.1016/j.team.2023.08.001.

OAG (2024) *Busiest airports in the world 2023*. Available at: www.oag.com/busiest-airports-world-2023.

Pacione, M. (2009) *Urban geography. A global perspective*. 3rd edn. Abingdon: Routledge.

Ritchie, H., Samborska, V. and Roser, M. (2024) 'Urbanization', *Our world in data* [Preprint]. Available at: https://ourworldindata.org/urbanization.

Routes Online (2024) *Overview | London Southend Airport | Routes*. Available at: www.routesonline.com/airports/7308/london-southend-airport/.

Royal Flying Doctor Service (n.d.) *Health access*. Available at: www.flyingdoctor.org.au/what-we-do/health-access/.
Salewski, C., Boucsein, B. and Gasco, A. (2014) 'Towards an effect-based model for airports and cities', in S. Conventz and A. Thierstein (eds) *Airports, cities and regions*. London: Routledge, pp. 257–281. Available at: https://doi.org/10.4324/9780203798829.
Swedavia (2024) *Airport city*. Available at: www.swedavia.com/future-airports/stockholm-arlanda-airport/Airport-city/.
UN Tourism (2024) *Rural tourism*. Available at: www.unwto.org/rural-tourism.
United Nations (2023) *THE 17 GOALS | Sustainable Development*. Available at: https://sdgs.un.org/goals.
United Nations (2024a) *68% of the world population projected to live in urban areas by 2050, says UN*. Available at: www.un.org/uk/desa/68-world-population-projected-live-urban-areas-2050-says-un.
United Nations (2024b) *About Small Island Developing States | Office of the High Representative for the Least Developed Countries, Landlocked Developing Countries and Small Island Developing States*. Available at: www.un.org/ohrlls/content/about-small-island-developing-states.
UNWTO (2024a) *Small Islands Developing States (SIDS)*. Available at: www.unwto.org/sustainable-development/small-islands-developing-states.
UNWTO (2024b) 'Tourism and rural development: a policy perspective | World Tourism Organization', *Books* [Preprint]. Available at: www.e-unwto.org/doi/epdf/10.18111/9789284424306.
US Department of Transportation (2023) *Essential air service*. Available at: www.transportation.gov/policy/aviation-policy/small-community-rural-air-service/essential-air-service.
UTA (2024) *Eastern Airport City*. Available at: www.uta.co.th/airport-airportcity/the-city.
Woods, M. (2009) 'Rural geography', in *International encyclopedia of human geography*, pp. 429–441. Available at: https://doi.org/10.1016/B978-008044910-4.00900-7.
World Bank (2024) *Urban development*. Available at: www.worldbank.org/en/topic/urbandevelopment/overview.
World Economic Forum (2020) *10 destinations that depend on tourism for jobs*. Available at: www.weforum.org/agenda/2020/08/destinations-rely-most-on-tourism-travel/.
World Food Programme (2024) *UN humanitarian air service*. Available at: www.wfp.org/unhas.
World Population Review (2024a) *Largest cities by population 2024*. Available at: https://worldpopulationreview.com/world-cities.
World Population Review (2024b) *South America population 2024*. Available at: https://worldpopulationreview.com/continents/south-america.

12 Population geography and air transport

Chapter outcomes

At the end of this chapter, you will be able to:

- Explain the ways demand generation influences population geography and air transport.
- Explain the importance of global cities to population geography and air transport.
- Understand the importance of economic impact and development in population geography.
- Explain the key role of tourism in population geography and air transport.
- Understand the key factors influencing migration and air transport.
- List the world's busiest route networks.
- Explain the concept of mobilities and how its various components are influenced by, and have an influence on, air transport.

12.1 Introduction

World population growth over the last 75 years has been staggering, especially in more recent times. On 15 November, 2022, the global human population reached 8 billion. In 1950, it was an estimated 2.5 billion and there have been 1 billion added since 2010 and 2 billion since 1998. The United Nations (2024) also predicts that this could rise to around 9.7 billion by 2050, although this will be dependent on future fertility rates.

These trends have been driven largely by an increasing survival rate to reproductive age, increasing urban growth, accelerating migration and a gradual human lifespan increase. The United Nations states that countries with the highest fertility rates generally are those with the lowest income per capita, and therefore population growth has become more focused on the poorest countries in the world, such as in sub-Saharan Africa. These trends have had far-reaching implications and will do for many years to come.

Barcus and Halfacree (2017, p.1) talk about life being "lived across space", with population geography being the study of such experiences. Hazen, Alberts and Zaniewski (2023, p.4) state that population geography can best be described as the "application of geographic approaches and methods to topics related to human populations". Population geography will therefore involve areas such as fertility, mortality, migration and the concept of mobilities, as well as urbanisation, globalisation and spatial inequalities.

DOI: 10.4324/9781003405351-15

Population geography and air transport 339

The connection between population geography and air transport is multifaceted and interconnected, both influencing the other in different ways. These can be categorised under three general, often interlinked, headings, which will be developed throughout the chapter:

1. **Demand generation** – population growth, density, distribution and demographics.
2. **Economic impact and development** – tourism, business travel, trade and route networks.
3. **Social and cultural exchange** – globalisation, migration and mobility.

12.2 Demand generation

For air transport to grow and develop, there must be demand. Demand comes from people (even for cargo, it is driven by people) and therefore where these people are located, the density of the population, how the population is growing and expected to grow and the demographics of the population, are all key factors which influence air transport. The concept of *urbanisation* is also important, and this was covered in Chapter 11.

12.2.1 Population density and distribution

Population density relates to the population per land unit area, in a particular locale. Distribution refers to the spatial arrangement. These play a pivotal role in shaping air transport services, in terms of destinations served, frequencies, total number of passenger and cargo movements, as well as the location of the airports themselves and associated surface access links and related industry.

In areas with higher population density, demand for air transport is usually higher, although demographic factors such as disposable income will play an important role. These densely populated regions often serve as economic and cultural hubs, for both domestic and international travel, creating business, tourism and personal travel demands. These are key factors for airlines to consider setting up new bases or routes and may also stimulate competition, which could assist in lowering fares for the local population.

The growth of low-cost carriers (LCCs) has stimulated demand across much of the world, allowing people to fly who had never flown before and stimulating multiple annual trips, perhaps rather than just a two-week summer holiday, to an increasing number of trips, including for activities such as weekend breaks.

Global cities

The prime examples of areas with the highest population densities are the so-called "global cities", whose role and influence often extends far beyond their national boundaries, acting as key global commerce nodes.

> **ATG Did you know?**
>
> **Global cities**
>
> Global (or world) cities are explained by Derudder (2020) to be the command-and-control centres of the global economy, functioning as the key organising centres that sustain contemporary globalisation. They are urban centres which enjoy significant competitive advantages as hubs of global economic systems.

The role of air transport in the creation and subsequent development of global cities is extremely important, to facilitate the rapid, global movement of people and cargo. From an air transport perspective, these are generally characterised as having mega-airports (often with several airports serving the metropolitan area, such as London with six), with significant aircraft movements, passenger and cargo volumes, servicing business, tourism, visiting friends and relatives and other mobilities. They are often at what may be termed the "geographical crossroads" for major air routes, such as Dubai – linking Europe, Asia and Africa and also Hong Kong – acting as a critical node to link East and Southeast Asia with the rest of the world.

Oxford Economics (2024) produces a Global Cities Index, ranking the top cities in the world based on five categories: economics, human capital, quality of life, environment and governance and the overall top ten are displayed in Table 12.1. In 2023, the 1,000 largest cities in the world accounted for 60% of global GDP and over 30% of the world's population.

Most of these cities are global air transport hubs, with significant passenger volumes. One exception is San Jose, in California, which plays host to major tech and internet companies and is one of the wealthiest cities in the world, which has witnessed increasing air services, although San Francisco is also a major airport in the area.

Table 12.1 Top ten global cities by overall score

City	Score
New York, USA	100.0
London, UK	99.4
San Jose, USA	98.5
Tokyo, Japan	97.8
Paris, France	96.3
Seattle, USA	95.8
Los Angeles, USA	95.4
San Francisco, USA	94.7
Melbourne, Australia	94.6
Zurich, Switzerland	94.2

Source: Oxford Economics (2024)

ATG Case study 12.1

New York airport system

The number one ranked global city, as per Table 12.1, was New York. From an air transport perspective, the New York region is an example of a **diversified airport system**. As shown in Table 2.12, the busiest airport in the world in 2023 for passenger numbers was Atlanta Hartsfield-Jackson, with over 104 million passengers. New York did not appear in the list of top ten airports, but this is because the passengers are spread over four commercial airports.

However, the total annual passengers flying through the New York airport system in 2023 was over 144 million (Table 12.2), second globally only to London, with over 168 million passengers (CAA, 2024).

Table 12.2 Total passengers at commercial New York Airports, 2023

Airport	2023 passenger numbers
John F. Kennedy International Airport (JFK)	62,464,331
Newark Liberty International Airport (EWR)	49,084,774
La Guardia Airport (LGA)	32,382,960
New York Stewart International Airport (SWF)	253,865
Total New York system	144,187, 930

Source: Port Authority of New York and New Jersey (2024)

JFK airport alone contributed around $45.4 billion in economic activity to the New York–New Jersey metropolitan region, supporting nearly 238,000 total jobs.

Population distribution

The spatial arrangement of people across a region will also impact the demand for air transport. In large countries with vast geographic expanses such as Australia or Canada, air transport provides essential links for rural and remote regions with urban centres (Chapter 11). It may be the only transport option in some countries, where the physical landscape creates barriers to surface transport, such as in Brazil and between Panama and Colombia. Whatever the reasons, air transport often provides a lifeline to remote populations.

There are also economic reasons for this distribution, such as better employment prospects. In Mumbai, India, thousands of people move from rural to urban areas each year due to pull factors, such as the perception of more job opportunities, higher wages, better housing, healthcare and living conditions. However, the reality does not always meet the perception and there are many challenges, which in Mumbai may be poorer housing conditions, slum dwelling, lower wages and unemployment.

ATG Case study 12.2

Air transport in Indonesia

According to the United Nations, Indonesia's population was around 279 million in 2024 – 41% of the Southeast Asian total. Over 58% of the population lived in urban areas in 2023, up from 52% in 2013.

The geography and infrastructural developments have led to a wide variation in the Indonesian population structure. It is, however, the largest economy in Southeast Asia, charting impressive economic growth since the late 1990s (World Bank, 2024b).

The air transport sector employs 115,000 people in Indonesia, however when employment for local suppliers, tourism and related spending is considered, 4.2 million jobs are supported by air transport and tourists arriving by air, alongside 2.6% of Indonesian GDP ($24 billion). Of arrivals, 97.2% are from the Asia Pacific region – 118 million passengers. The air transport market is forecast to grow by 219% by 2037, supporting around $77 billion of GDP and 6.9 million jobs (IATA, 2019).

Due to Indonesia being an archipelago, with more than 17,000 islands, air travel is necessary for population mobility and trade. It includes some famous islands, including Sumatra, Borneo and Bali, however over 56% of its population lives on Java, the most populous island, with a density of over 140 people per sq.km (World Population Review, 2024).

Airport improvements have taken place to improve regional connectivity, for example the construction of a new airport in West Java – Kertajati International Airport (KJT) and the expansion of Ngurah Rai International Airport (DPS), serving the Denpasar metropolitan area and the tourist island of Bali. These can assist in the balancing of Indonesia's population distribution.

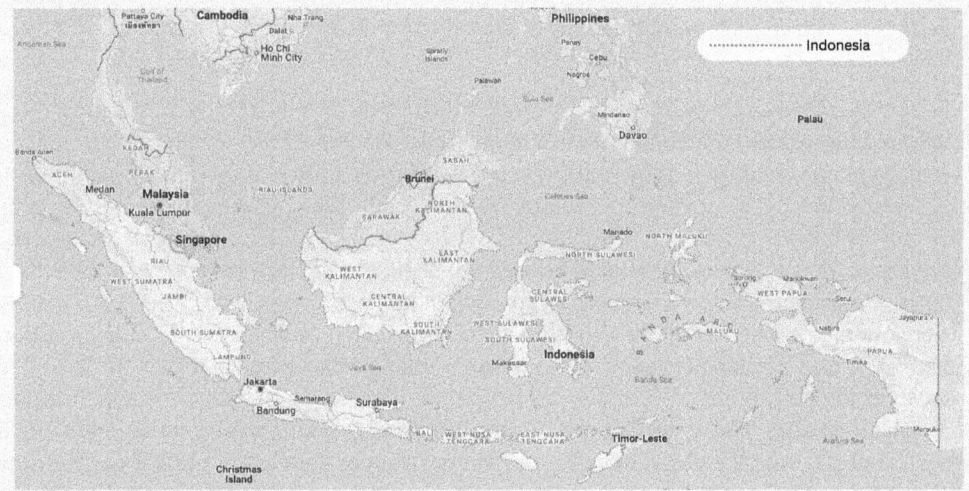

Figure 12.1 Map of Indonesia
Source: adapted from Google Maps (n.d.)

12.2.2 Population growth and demographics

As populations grow, the demand for air transport will also rise. Larger population centres will generate more passengers and the requirement for more goods and cargo due to increased economic activity. This is especially true in urban areas with the growth in urbanisation (Chapter 11), where economic concentration often leads to more demand for business and leisure travel.

Demographics are also key. Higher income levels are more likely to travel by air and with the advent of LCCs, this has opened the affordability of air travel to many more people, although "affordability" is a relative term. The rise in the "middle classes", as seen in countries such as India and China, has resulted in large air transport demand and growth, for both domestic and international travel, and these trends are especially prevalent in developing countries.

ATG Case study 12.3

Growth in China

The growth of the urban population in China has been nothing short of astonishing. In 1960, just 16% of the population lived in urban areas but by 2023, this figure had risen to 65%. This was a rise in urban population numbers from around 108 million in 1960 to over 910 million in 2023 (World Bank, 2024a). During the same time frame, the total population grew by 110%, from over 660 million in 1960 to over 1.4 billion in 2019. In 2000, roughly 3% of the country's population was classed as "middle class", but by 2018 this had grown to more than 50% – at 707 million people (Business Insider, 2021).

This growth has led to a large increase in demand for air travel, for which China has invested massively in terms of airlines, airports and associated infrastructure. Between 2000 and 2019, the number of annual airline passengers in China has grown by 965% – from 61.9 million to 659.6 million – a growth of 965%. As a comparison with a mature air travel market, over the same time frame, the USA had grown by only 39% (World Bank, 2024c).

The number of airports in China has grown from 139 in 2000 to 254 by 2022 (Statista, 2024). Available seat kilometres (ASKs), as a measure of airline supply system capacity, more than doubled from 759 billion in 2012 to 1.57 trillion in 2019, prior to the pandemic (CAPA, 2024). Boeing (2023) forecasts that China's commercial airliner fleet will double to nearly 9,600 jets by 2042, which will include new jet deliveries of around 8,560. China is expected to account for 20% of the world's airplane demand through 2042.

Working age population

It is the "working age" group (typically defined as 15–64 years old) which is a key driver of air transport demand, due to their involvement in economic activities, business and leisure travel.

Business travel often includes travel for meetings, conferences, client visits and other professional purposes. The working-age group is the primary segment driving business travel, which is a significant source of revenue for airlines and indeed the business class cabins of many full-service network carriers (FSNCs) are critical for profitability, due to their much higher average fares than in the economy cabin. Following the economic recession in 2008–2009, it was the lack of travel in the business class cabin – either business passengers not travelling or downgrading to economy class – which was one of the key factors contributing to airline losses in this period.

Business passengers also generally book their tickets later than leisure passengers, and with airline revenue management generally working on the principle of "the closer to departure, the higher the fare", higher yields can be a result. The global cities in particular, such as New York, London and Singapore, experience significant business travel. A key factor for business passengers is *frequency* and having the option of multiple daily flights is a key "pull" factor for airlines in attracting business passengers.

As an example, from published schedules for Summer 2024, for flights on the key business route between London Heathrow and New York JFK, there were 23 daily departures. Virgin (7) and its joint venture partner Delta (2) had 9 between them, British Airways (8) and transatlantic joint venture partner American Airlines (4) had 12 between them, plus JetBlue with 2 (Pearson,

2023). Adding in four flights from London Gatwick to JFK plus flights from Heathrow to New York Newark with United Airlines and British Airways, these offer passengers significant frequency travelling on the London–New York route.

The working-age group also drives a significant portion of leisure travel. For those with enough disposable income and holiday time, and access to LCC networks especially, this group often frequently flies for family visits, short breaks, longer holidays and other recreational purposes.

The demographic composition of the world is changing, with many advanced economies witnessing a growth in the "ageing" population, whilst many emerging markets have seen growth particularly in the working-age group. A good example is the emerging market of Vietnam. The expanding middle class and working-age population in Vietnam has led to a growth in air travel and growth in LCCs. According to IATA, it had the fastest-recovering domestic aviation sector after the pandemic and was one of the leading markets in the Asia Pacific in terms of air passenger growth.

According to Airbus (2024), GDP, trade and population are the main drivers of air traffic growth. GDP is expected to grow by a compound annual growth rate (CAGR) of 2.6% between 2023–2043, the world population to grow by 1.3 billion in the same period, the urban population to grow more (by 1.5 billion) and the middle class by even more (1.7 billion). Subject to successfully managing the serious consequences posed by environmental issues, population-related factors should be key in driving future air transport growth.

12.3 Population geography, tourism and route networks

The economic benefits of air transport in the geographical context were analysed in Chapter 10, however an important economic factor in the context of population geography and air transport is tourism.

The movement of people for tourism purposes has the potential to reshape population patterns as well as air transport route networks. Tourism influences where people live, work and travel, and airlines facilitate the growth and spread of tourism, with impacts being economic, social and environmental – positive and negative.

The concept of urbanisation can be influenced by population movements due to tourism. As destinations develop, infrastructure is needed to accommodate these visitors, via hotels, restaurants and relevant tourist attractions. This brings employment in terms of construction and then in terms of jobs in the tourist resort, attracting workers.

Cities such as Las Vegas, Nevada, have developed from small towns into massive urban centres, largely down to a growth in tourism and entertainment. The Las Vegas population has grown from 164,000 in 1980 to over 660,000 by 2023, much facilitated by air transport networks.

Tourism can lead to uneven population distributions, as significant tourist destinations – in cities, coastal areas, islands and culturally significant areas – create a flow of people to these destinations. In Thailand, for example, although the country is rapidly urbanising, only seven out of 76 provincial cities were classed as "growing cities" with the others either "stagnant" or "shrinking". This is an even bigger concern as the United Nations (2020) projects that Thailand will soon have an ageing and declining population. As a result, the working-age population will witness a downward trend, from 43.2 million in 2020 to 36.5 million in 2040.

12.3.1 Seasonality

Tourism in many destinations creates seasonal population flows. Areas such as the Greek islands see significantly larger tourist arrivals in the summer than the winter, which creates more seasonal employment opportunities. At peak times, this can strain local resources and services and create challenges to manage seasonal population growth. Often, local people rely on earning enough income in the summer to sustain themselves for the remainder of the year.

Dobruszkes, Decroly and Suau-Sanchez (2022) discovered that 36% of airports worldwide experienced a significant degree of seasonality, although most large airports were not significantly affected by seasonality patterns, due to the wide variety of travel purposes throughout the year. There are also other resorts which have good weather throughout the year, such as the Canary Islands, which see much less seasonality than others with more seasonal weather patterns.

From an air transport management perspective, a key decision is which level of demand to plan for. Planning for 100% demand in summer could leave infrastructure significantly underused in anything other than the peak days, coming at a potentially significant cost. IATA discusses possibly using the second busiest day of an average week in the peak month as a design peak day (DPD). Geneva Airport, for example, had a peak in the months of January through to March due to the ski season and constructed an alternate terminal, which only opened during the winter ski season.

From the perspective of population geography, more consistent flight schedules should lead to less seasonal employment opportunities, making these destinations less transient and more attractive for permanent residence.

12.3.2 Environmental and social challenges

The growth in tourism due to air transport has many positive effects. However, this can also lead to negative environmental and social challenges, which require managing. Higher emissions resulting in air pollution and climate change (Chapter 14) affects not just the tourist resort, but a much broader area. Noise (Chapter 15) can become more of a nuisance as the volume of flights grow, especially the closer the airport and the runway departure and approach paths are to the local population. Surface access issues such as transport congestion can also be a negative for the local population (and the visitors) if not correctly managed.

Also, the concept of *overtourism* is a potential issue. This is where the number of tourists exceeds the capacity of a destination to accommodate them comfortably. This can be due to a lack of local investment, a lack of land in which to construct appropriate infrastructure or just simply not enough carrying capacity. The problem does not have to be calculated using statistics to define overtourism – perception is also key – and when local residents struggle to afford rents due to holiday rentals, narrow roads become jammed with tourist vehicles and fragile environments become degraded, then overtourism can become a serious issue.

Barcelona, for example, has seen a large growth in air transport arrivals this century, with passenger numbers at Barcelona-El Prat Airport increasing from 19 million in 2000 to around 50 million in 2023 (AENA, 2024). Although many visitors to Barcelona are day-trippers from the cruise ships which dock in port, the growth in air transport has compounded the issue of overtourism. Barcelona has increased tourist taxes and is planning to ban short-term holiday rentals as part of its plan to reduce the negative impacts.

ATG Case study 12.4

Overtourism in Venice

Venice is a good example of how tourism has shaped population geography and air transport. Venice is one of the world's top tourist destinations, with around 20 million

346 *Fundamentals of Global Air Transport Geography*

visitors every year (Responsible Travel, n.d.). However, the influx of tourists has led to a reduction in residents, with a population in the centre of only around 50,000, a significant drop over the previous 50 years, due to the high cost of living and impact of tourism on daily life. Although day-trippers from cruise ships have played a significant part in the tourist numbers, arrivals via Venice Marco Polo Airport have grown significantly, from around 4 million passengers in 2001 to over 11 million in 2019 (Venice Airport, 2024).

Concerns have been raised about the sustainability of the city and as a result, cruise ships have been banned in the centre since 2021, and a tourist tax – called an access fee – of €5 per person per day levied on specific days.

12.3.3   The world's busiest route networks

Whether travelling for business or leisure purposes, there are airports which consistently dominate the global rankings. Atlanta Hartsfield-Jackson was the busiest in the world in terms of passenger numbers in 2023 (largely due to it being the main hub for Delta) and Dubai the busiest for international passengers (again being a major hub – this time for Emirates).

However, from a population perspective, individual routes are important to analyse as this can be key to understand the importance of certain airports and cities, and also passenger flows, whether it be for business or leisure or even the broader term of "mobility", which will be analysed in Section 12.4.

According to OAG (2024), nine of the top ten busiest global domestic routes in 2023 were in the Asia Pacific region (Table 12.3). Of these, most are between the two biggest cities in the country (such as Hanoi–Ho Chi Minh City in Vietnam and Melbourne–Sydney in Australia) as business, leisure and visiting-friends-and-relatives traffic will all play a role.

From an international route perspective, eight out of the top ten were in the Asia Pacific region (Table 12.4), with the only route not in this region or the Middle East being the long-haul route between New York JFK and London Heathrow. Many of these are classed as global cities.

Table 12.3 Busiest domestic flight routes in the world by seat capacity, 2023

Rank	Route	Country	Airline domestic seat capacity
1	Jeju–Seoul Gimpo	The Republic of Korea	13,728,786
2	Sapporo Chitose–Tokyo Haneda	Japan	11,936,302
3	Fukuoka–Tokyo Haneda	Japan	11,264,229
4	Hanoi–Ho Chi Minh City	Vietnam	10.883,555
5	Melbourne-Sydney	Australia	9,342,312
6	Beijing Capital-Shanghai Hongqiao	China	8,355,225
7	Tokyo Haneda-Okinawa Naha	Japan	7,982,218
8	Jeddah-Riyadh	Saudi Arabia	7,902,142
9	Mumbai-Delhi	India	7,276,430
10	Jakarta-Denpasar Bali	Indonesia	7,190,961

Source: OAG (2024)

Table 12.4 Busiest international flight routes in the world by seat capacity 2023

Rank	Route	From–to	Airline international seat capacity
1	Kuala Lumpur–Singapore Changi	Malaysia–Singapore	4,891,952
2	Cairo–Jeddah	Egypt–Saudi Arabia	4,795,712
3	Hong Kong–Taipei	Hong Kong–Taiwan	4,568,280
4	Seoul Incheon–Osaka Kansai	Republic of Korea–Japan	4,218,484
5	Seoul Incheon–Tokyo Narita	Republic of Korea–Japan	4,155,418
6	Dubai–Riyadh	UAE–Saudi Arabia	3,990,076
7	Jakarta–Singapore Changi	Indonesia–Singapore	3,910,502
8	New York JFK–London Heathrow	USA–UK	3,878,590
9	Bangkok–Singapore Changi	Thailand–Singapore	3,478,474
10	Bangkok–Seoul Incheon	Thailand–Republic of Korea	3,362,968

Source: OAG (2024)

12.4 Migration and mobilities

12.4.1 Migration

Migration is as old as time itself. Since the earliest of times, humanity has been on the move. Sometimes in search of work or other economic opportunities, to study, to create a new life or to join family. Others move to escape conflict, persecution or human rights violations or to escape the adverse effects of environmental factors or natural disasters.

Cohen (1997, in Boyle, 2015) uses the term "diaspora" in providing a typology for these movements:

- **Victim diasporas** (forced exile – such as the African diaspora and the Atlantic slave trade).
- **Labour diasporas** (mass migration in search of work – such as the Indian and Turkish diasporas).
- **Trade diasporas** (seeking to open trade routes and links – such as the Chinese diasporas).
- **Imperial diasporas** (serving and maintaining empires – such as the British and French diasporas).
- **Cultural diasporas** (movement through a process of chain migration – such as the Caribbean diasporas).

The scale of international migration has substantially increased in recent years, becoming a global phenomenon (Figure 12.2), with an estimated 281 million international migrants by mid-2020 – 3.5% of the global population, compared to 2.8% in 2000 and 2.3% in 1980. Of migrants, 15% are below 20 years of age and 73% of "working age" (United Nations, n.d.). In terms of migrant numbers, the USA is the top migrant destination and India the top origin for emigrants (Table 12.5).

In 2024, the top international country-to-country migration corridors (International Organisation for Migration, 2024c) were:

- Mexico–USA
- Syrian Arab Republic–Turkey
- Ukraine–Russian Federation
- India–United Arab Emirates

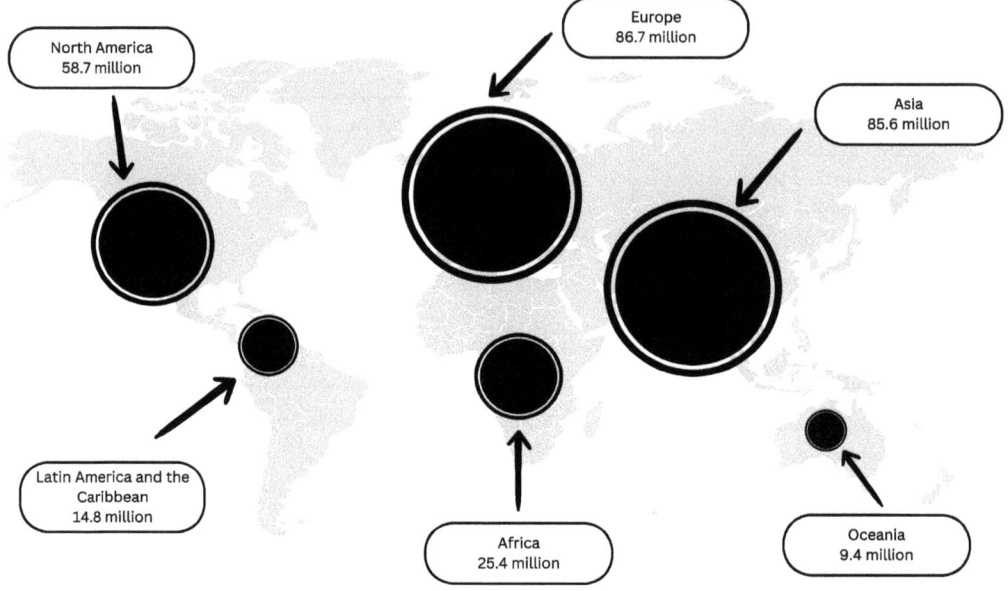

Figure 12.2 International migrants at mid-year 2020

Source: author, using data from United Nations (n.d.)

Table 12.5 Top five countries of migrant origin and destination in 2020, by number and proportion of total population

Origin			Destination		
Country	Emigrants (million)	% of population	Country	Immigrants (million)	% of population
India	17.79	1.3	USA	43.43	13.1
Mexico	11.07	7.9	Germany	14.22	17.0
Russian Federation	10.65	6.8	Saudi Arabia	13.00	37.3
China	9.80	0.7	Russian Federation	11.58	7.9
Bangladesh	7.34	4.3	UK	8.92	13.1

Source: International Organisation for Migration (2024b)

ATG Did you know?

Defining migration and mobilities

The International Organisation for Migration (2024a) defines a *migrant* as "a person who moves away from his or her place of usual residence, whether within a country or across an international border, temporarily or permanently, and for a variety of reasons". This is regardless of:

- The person's legal status.
- Whether the movement is voluntary or involuntary.
- What the causes for the movement are.
- What the length of the stay is.

Migration and mobility are related but distinct concepts in the context of the movement of people. *Mobility* is a broader term referring to the ability or capacity of people to move from one place to another. According to Hannam, Sheller and Urry (2006, p.1):

> the concept of mobilities encompasses both the large-scale movements of people, objects, capital and information across the world, as well as the more local processes of daily transportation, movement through public space and the travel of material things within everyday life.

Issues of mobility are centre stage in much of everyday life and this led to the development of a "new mobilities" paradigm (Sheller and Urry, 2006).

Thus, migration is a form of mobility, but not all mobility is migration. The influence of air transport in these concepts has been significant in contemporary times, in many parts of the world.

12.4.2 Mobilities

The mobilities literature makes the argument that the movement of people is complex in the contemporary world of low-cost airline travel and social media. People can move and communicate much more rapidly and more frequently than before, involving multiple homes and multiple mobilities which may be for work, leisure, tourism, retirement, finding a school for their children or combinations of all of these. As O'Reilly (2003, p.301) wrote: "As flexible forms of migration undermine the distinction between tourism and migration, analyses of human mobility demand new conceptualisations and characterisations."

Urry (2007, p.18) argued that the,

> mobilities paradigm is not just substantively different, in that it remedies the neglect and omissions of various movement of people, ideas and so on. ... It enables the "social world" to be theorised as a wide array of economic, social and political practices, infrastructures and ideologies that all involve, entail or curtail various kinds of movement of people, or ideas, or information or objects.

One illustration of the mobility concept being broader than migration was from a survey in the European Union by Salamooska and Recchi (2016), which concluded that while only 12.7% of the participants had lived in another EU state for more than three months, the majority had several experiences of mobility, such as communicating regularly with friends and family abroad (53.8%), watching TV in another language (52.7%) and visiting another EU country in the 24 months before the survey (52.4%). Mobility can be frequent and multifaceted and air transport can play an important role.

Kraemer *et al.* (2020) conducted a study aggregating mobile phone location data from over 300 million smartphone users, covering 65% of the earth's populated surface and discovered that people from lower-income countries tended to travel shorter distances than those countries with higher incomes, and smaller countries experienced greater cross-border mobility.

Button and Vega (2008) stated that air transport has facilitated an increasingly spatially dynamic global labour market and makes short-term, long-distance migration viable and allows migrants to maintain contact with their home country, by limiting the social costs. In addition, in larger countries such as the USA, "weekday" migration is more viable, with spouses working in different parts of the country, returning at weekends.

ATG Case study 12.5

EU enlargement in 2004

Since the late 1990s, the growth of LCCs in Europe facilitated mobility, both in terms of the number of route connections and reductions in real airfares. The year 2004 was key in Europe, as it involved the greatest enlargement in the history of the EU, with ten new accession member states (A10) – Czechia, Cyprus, Estonia, Hungary, Latvia, Lithuania, Malta, Poland, Slovakia and Slovenia – all joining the initial 15 member states (EU15), which included Germany, France and the United Kingdom (pre-Brexit). With Open Skies (Chapter 9) being present in the EU, this facilitated the growth and development of airlines such as Ryanair, easyJet and Wizz and the overall EU air network.

The period since EU enlargement has been marked by a continued convergence between the initial EU15 and new A10, in terms of air travel demand.

- **Air connectivity** has increased by 85% since accession for the A10, compared to 32% for the EU15. This has boosted the economic performance of the A10, as each 10% increase in direct connectivity of a country yields a 0.5% increase in GDP per capita (ACI-Europe, 2024).
- **Air passengers** grew by 215% in the A10 compared to 77% in the EU15, due to market maturity. This facilitated significant increases in mobility.
- The **ratio of population to air passengers** in the A10 decreased from 2.31 to 0.59 (0.26 in the EU15) indicating that many more EU citizens now had access to air travel, increasing mobilities.

In 2007, Bulgaria and Romania also became members of the EU, increasing air transport connectivity, passenger numbers and mobilities in both these countries.

Hannam, Sheller and Urry (2006) also emphasise that migrants will bring parts of their culture with them, such as souvenirs and foods, and that they often travel back home "to visit friends and relatives while ostensibly 'on holiday' in their country of origin" (p.10).

There are three key areas which underpin the concept of mobilities:

- labour mobilities
- lifestyle mobilities
- leisure mobilities

Labour mobilities

According to Long and Ferrie (2006), labour mobility involves "changes in the location of workers both across physical space (geographic mobility) and across a set of jobs (occupational mobility)" (p.1). There are considerable economic benefits of labour mobility. They also state that "the reallocation of workers across regions permits the exploitation of complementary resources as they are discovered in new places, while reallocation across sectors makes possible the use of new technologies and the growth of new industries" (p.1). People's economic welfare could be improved by mobility, enabling the best job location for their skill sets and motivations.

Demographic changes due to labour mobility in the EU15 were discussed by Larsson, Sforza and Turmann (2004), in the context of the ageing European population, arguing that the mobility of labour was a fundamental element within competitive economies. The shortage of skilled labour could be reduced by the dissemination of knowledge by mobile employees. The on-going change in organisations in terms of economic productivity could be facilitated by labour mobility.

In the context of air transport, labour mobilities can be observed across a variety of industries and situations, such as:

- **Seasonal tourism workers**, such as in ski resorts in the Alps in winter.
- **Executives and professionals** travelling for business, such as meetings and conferences – very important for international organisations.
- **Healthcare and aid** – medical professionals and specialists, such as responding to crises.
- **Construction and infrastructure** projects, such as employment on developments in Dubai and Saudi Arabia.

Lifestyle mobilities

Lifestyle mobilities can be seen as an entrepreneurial effort to sustain a mobile way of life, either through working temporarily in an industry or by being self-employed, with the aim of financing a different lifestyle (Benson, 2013; Cohen, Duncan and Thulemark, 2015).

"Lifestyle mobility differs from temporary mobility in that it is sustained as an ongoing fluid process, carrying on as everyday practice over time" (Cohen, Duncan and Thulemark, 2015, p.158). "A return to any identified 'origin' cannot be presumed. ... Through lifestyle mobility, there is no 'one' place to which to return, and through time, there may be multiple 'homes' that one can return to and/or revisit" (p.159).

Migrants can have many motivations and travel to and from a number of different places – some returning every year, others moving on continually – thus taking place in different time and space. O'Reilly and Benson (2016) purport that the social construction of places and how they are perceived frequently influence the choice of desired destinations. A key factor in the decision to migrate is the role of imagination. Lifestyle mobilities involve "escaping" – either from or to something or somewhere.

In the context of air transport, lifestyle mobilities can be observed across a variety of industries and situations, such as:

- **Digital nomads**, working remotely while travelling.
- **Retirees looking for seasonal living**, such as moving to warmer climates in the winter.
- **Long-term travellers and ex-pats**, living abroad for extended periods, while maintaining connections to their home countries.
- **Second-home owners.** The growth in LCCs has facilitated this throughout Europe especially.
- **Wealthy, private jet owners**, who may be even more mobile.

Leisure mobilities

Leisure and tourism mobilities are not a process that occurs once or one-way, there are many threads that intertwine, interflow and overlap. The concept sees humans as "inhabiting a mobile world in which we not only travel physically but also virtually, communicatively and imaginatively" (Lean, 2012, p.153). From a mobilities perspective, tourism studies no longer must be seen as just an ephemeral practice of social life outside of home life. Rather they should be "viewed as being bound up with both everyday and mundane journeys as well as with the more exotic encounters" (Hannam, Butler and Paris, 2014, p.172).

Hannam and Knox (2010) claim that at the heart of mobilities is the feeling of movement, which also applies to all social life. Leisure and tourism mobilities are related not just with the movement of people but also the social implications of these movements. Sheller and Urry (2004, p.1) further point out that, "many different mobilities inform tourism, shape the places where tourism is performed, and drive the making and unmaking of tourist destinations."

ATG Case study 12.6

Poland and LCCs – post-2004

Economic migration created a demand for fast and cheap travel after Poland joined the EU in 2004, especially to the UK, Germany and Ireland, which was largely fulfilled by the growing LCCs. There were no low-cost airline routes from Polish airports to the UK in Winter 2003/04, but this had increased to 65 by Winter 2008/09 (Olipra, Pancer-Cybulska and Szostak, 2011). Prior to 2004, most emigrating Polish workers used very long land transport options, but when flight routes were established and fares more affordable for many, migration increased, as did mobility, as the increase in flights facilitated an increased number of return trips home for holidays or vacation.

There was enormous growth and changes in the geographic distribution of new low-cost routes as they appeared following EU enlargement, "creating new mobility infrastructures between east and west" (Badcock and Burrell, 2022).

Wizz Air, along with Ryanair and other airlines, were key to these new mobilities. Wizz Air, headquartered in Budapest, Hungary, had its first flight in May 2004, from Katowice in Poland to London Luton in the UK; 20 years later, on 31 March, 2024, it had 924 routes across Europe and the Middle East (Table 12.6), including 173 from Poland, second only in number to Romania (176), providing huge opportunities for mobility. In the 2023/24 fiscal year, Wizz Air carried 62 million passengers.

Table 12.6 Most routes operated by Wizz Air as of 31 March, 2024 – selected countries

Country	Number of routes operated
Romania	176
Poland	173
Italy	135
Hungary	71
United Kingdom	62
Albania	53
Bulgaria	46

Source: adapted from Wizz Air (2024)

Chapter review questions

12.1 Analyse the ways in which demand generation influences population geography and air transport.
12.2 Assess the importance of global cities in terms of air transport. What is your nearest global city? What makes it so? Research the destinations and potential routings from this city.
12.3 What are the various population factors which influence economic development? What are the major industries in your local area? How important was air transport in the development of these and how important is it now?
12.4 Consider the different ways tourism has an influence on population geography and air transport. For your local tourist areas, where are the visitors from? How important is tourism for local development and employment?
12.5 Explain the key types of migration and assess their importance in the context of air transport. For your country, how important has air transport been in terms of its demographic composition?
12.6 What are the busiest route networks in the world? For any ONE of these, how many reasons can you find for why this route is so popular?
12.7 Explain the concept of mobilities and what is meant by labour, lifestyle and leisure mobilities.

ATG trivia

Extreme time zones

Thanks to a concept known as Coordinated Universal Time (UTC) offset – which is how many hours and minutes between UTC and the time at a particular location – there will always be a place which is furthest ahead, and indeed furthest behind, UTC.

These "extreme" honours fall to Kiribati, which is +14 hours ahead of UTC and both the Baker and Howland Islands, which are −12 hours behind UTC (Figure 12.3). Therefore, the maximum possible distance between times on earth is 26 hours. This is due to a quirk of the international date line, which roughly follows the 180th meridian north to south through the middle of the Pacific Ocean. Kiribati is located almost directly south of Hawaii but celebrates New Year a full day earlier.

The last inhabited islands to celebrate New Year are Niue and American Samoa, however the Baker and Howland Islands are further behind, but uninhabited. It should be noted that political factors can play a role. For example, Samoa (not American Samoa) used to be one of the locations furthest behind UTC, until it decided to change time zones in 2011 to align with its trading partners, Australia and New Zealand, and is now one of the places furthest ahead of UTC (National Geographic, 2022).

354 *Fundamentals of Global Air Transport Geography*

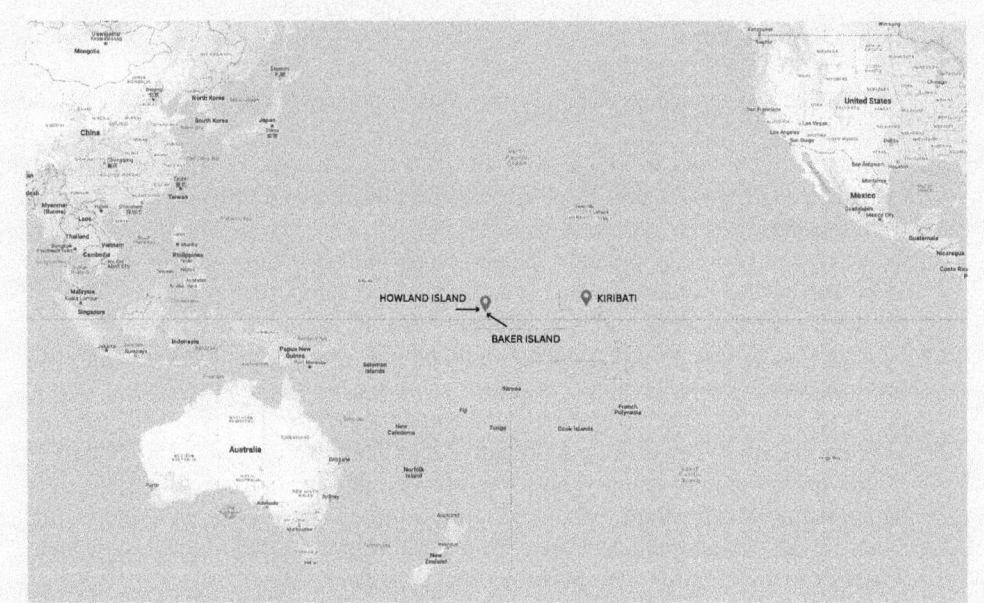

Figure 12.3 Location of Kiribati and Baker and Howland Islands
Source: adapted from Google Maps

What about airports? Kiribati has a number of airports, including Bonriki International Airport (TRW) and Cassidy International Airport (CXI). As both Baker and Howland Islands are uninhabited, not surprisingly these do not have airports. However, that is not the whole story! An airstrip was constructed on Howland Island to serve as a refuelling stop on Amelia Earhart and Fred Noonan's round-the-world-flight in 1937, but they sadly did not arrive, leaving one of aviation's greatest unsolved mysteries (Smithsonian Institute, 2021).

References

ACI-Europe (2024) *European airports celebrate 20th anniversary of EU enlargement showcasing key aviation benefits*. Available at: www.aci-europe.org/downloads/mediaroom/24-05-01%20European%20airports%20celebrate%2020th%20anniversary%20of%20EU%20enlargement%20showcasing%20key%20aviation%20benefits%20PRESS%20RELEASE.pdf.
AENA (2024) *Network of airports in Spain*. Available at: www.aena.es/en/passengers/our-airports.html.
Airbus (2024) *Global market forecast*. Available at: www.airbus.com/en/products-services/commercial-aircraft/market/global-market-forecast.
Badcock, M. and Burrell, K. (2022) 'Staying connected: low cost airlines in the lives of Polish migrants', in W. Lin and J.-B. Frétigny (eds) *Low-cost aviation*. Elsevier (Contemporary Issues in Air Transport), pp. 121–133. Available at: https://doi.org/10.1016/B978-0-12-820131-2.00009-6.
Barcus, H.R. and Halfacree, K. (2017) *An introduction to population geographies: lives across space*. 1st edn. Abgindon: Routledge.
Benson, M. (2013) 'Postcoloniality and privilege in new lifestyle flows: the case of North Americans in Panama', *Mobilities*, 8(3), pp. 313–330.
Boeing (2023) *Commercial market outlook*. Available at: www.boeing.com/commercial/market/commercial-market-outlook/index.page.

Boyle, M. (2015) *Human geography: a concise introduction*. Chichester: John Wiley and Sons.
Business Insider (2021) *China's middle class is starting to look a lot like America's, and that's not a good thing*. Available at: www.businessinsider.com/china-middle-class-starting-to-look-like-americas-2021-12.
Button, K.J. and Vega, H. (2008) 'The effects of air transportation on the movement of labor', *GeoJournal*, 71(1), pp. 67–81. Available at: https://doi.org/10.1007/s10708-008-9116-1.
CAA (2024) *Annual airport data 2023*. Available at: www.caa.co.uk/data-and-analysis/uk-aviation-market/airports/uk-airport-data/uk-airport-data-2023/annual-2023/.
CAPA (2024) *Schedules analysis for China*. Available at: https://centreforaviation.com/data/profiles/countries/china/schedules-analysis.
Cohen, S., Duncan, T. and Thulemark, M. (2015) 'Lifestyle mobilities: the crossroads of travel, leisure and migration', *Mobilities*, 10(1), pp. 155–172.
Derudder, B. (2020) 'World/global cities', in *International encyclopedia of human geography*. 2nd edn. Oxford: Elsevier, pp. 291–296. Available at: https://doi.org/10.1016/B978-0-08-102295-5.10358-0.
Dobruszkes, F., Decroly, J.-M. and Suau-Sanchez, P. (2022) 'The monthly rhythms of aviation: a global analysis of passenger air service seasonality', *Transportation Research Interdisciplinary Perspectives*, 14, p. 100582. Available at: https://doi.org/10.1016/j.trip.2022.100582.
Google Maps (n.d.) *Google Maps*. Available at: www.google.com/maps.
Hannam, K., Butler, G. and Paris, C.M. (2014) 'Developments and key issues in tourism mobilities', *Annals of Tourism Research*, 44, pp. 171–185. Available at: https://doi.org/10.1016/j.annals.2013.09.010.
Hannam, K. and Knox, D. (2010) *Understanding tourism: a critical introduction*. London: Sage.
Hannam, K., Sheller, M. and Urry, J. (2006) 'Editorial: mobilities, immobilities and moorings', *Mobilities*, 1(1), pp. 1–22. Available at: https://doi.org/10.1080/17450100500489189.
Hazen, H.D., Alberts, H.C. and Zaniewski, K.J. (2023) *Population geography: social justice for a sustainable world*. 1st edn. London: Routledge. Available at: https://doi.org/10.4324/9781003143253.
IATA (2019) *The importance of air transport to Indonesia*. Available at: www.iata.org/en/iata-repository/publications/economic-reports/indonesia--value-of-aviation/.
International Organisation for Migration (2024a) *About migration*. Available at: www.iom.int/about-migration.
International Organisation for Migration (2024b) *Who migrates internationally and where do they go? International*, World Migration Report. Available at: https://worldmigrationreport.iom.int/what-we-do/world-migration-report-2024-chapter-4/who-migrates-internationally-and-where-do-they-go-international-migration-globally-between-1995-2020.
International Organisation for Migration (2024c) *Top international country-to-country migration corridors*. Available at: https://worldmigrationreport.iom.int/resources/chapter-2-figure-1-top-international-country-country-migration-corridors-2024.
Kraemer, M.U.G. et al. (2020) 'Mapping global variation in human mobility', *Nature Human Behaviour*, 4(8), pp. 800–810. Available at: https://doi.org/10.1038/s41562-020-0875-0.
Larsson, A., Sforza, L. and Turmann, A. (2004) *A new European agenda for labour mobility: report of a CEPS-ECHR task force*. Brussels: Centre for European Policy Studies.
Lean, G.L. (2012) 'Transformative travel: a mobilities perspective', *Tourist Studies*, 12(2), pp. 151–172. Available at: https://doi.org/10.1177/1468797612454624.
Long, J. and Ferrie, J. (2006) *Labour mobility. Oxford encyclopedia of economic history*. Available at: https://faculty.wcas.northwestern.edu/fe2r/papers/Labour%20Mobility.pdf.
National Geographic (2022) *Who celebrates the new year first and last? It's complicated.*, Culture. Available at: www.nationalgeographic.com/culture/article/new-year-map-first-last-countries-international-date-line.
OAG (2024) *Busiest flight routes in the world 2023*. Available at: www.oag.com/busiest-routes-world-2023.
Olipra, L., Pancer-Cybulska, E. and Szostak, E. (2011) 'The impact of the migration processes on the low cost airlines' routs between EU countries and Poland after its accession to the EU, and on the territorial cohesion of Polish regions', *ERSA conference papers* [Preprint]. Available at: https://ideas.repec.org/p/wiw/wiwrsa/ersa11p1774.html.
O'Reilly, K. (2003) 'When is a tourist? The articulation of tourism and migration in Spain's Costa del Sol', *Tourist Studies*, 3(3), pp. 301–317. Available at: https://doi.org/10.1177/1468797603049661.

O'Reilly, K. and Benson, M. (2016) 'Lifestyle migration: escaping the good life', in *Lifestyle migration: expectations, aspirations and experiences*. Abingdon: Routledge, pp. 1–13.

Oxford Economics (2024) *Global cities index*. Available at: www.oxfordeconomics.com/global-cities-index/.

Pearson, J. (2023) *27 daily flights: inside the enormous London to New York JFK market next summer, simple flying*. Available at: https://simpleflying.com/london-new-york-jfk-summer-2024-market-analysis/.

Port Authority of New York and New Jersey (2024) *2023 airport traffic report*.

Responsible Travel (n.d.) *Overtourism in Venice – responsible travel*. Available at: www.responsibletravel.com/copy/overtourism-in-venice.

Salamooska, J. and Recchi, E. (2016) 'Europe between mobility and sedentarism: patterns of cross-border practices and their consequences for European identification', *SSRN Electronic Journal* [Preprint]. Available at: https://doi.org/10.2139/ssrn.2879338.

Sheller, M. and Urry, J. (2004) *Tourism mobilities: places to play, places in play*. Abingdon: Routledge. Available at: www.routledge.com/Tourism-Mobilities-Places-to-Play-Places-in-Play/Sheller-Urry/p/book/978041584570 (Accessed: 11 August 2024).

Sheller, M. and Urry, J. (2006) 'The new mobilities paradigm', *Environment and Planning A: Economy and Space*, 38(2), pp. 207–226. Available at: https://doi.org/10.1068/a37268.

Smithsonian Institute (2021) *Amelia Earhart*. Available at: https://airandspace.si.edu/explore/stories/amelia-earhart.

Statista (2024) *China: number of civil airports 2022*. Available at: www.statista.com/statistics/258207/number-of-civil-airports-in-china/.

United Nations (2020) *Thailand economic focus: demographic change in Thailand: how planners can prepare for the future | United Nations in Thailand*. Available at: https://thailand.un.org/en/96303-thailand-economic-focus-demographic-change-thailand-how-planners-can-prepare-future, https://thailand.un.org/en/96303-thailand-economic-focus-demographic-change-thailand-how-planners-can-prepare-future.

United Nations (2024) *Population*. United Nations. Available at: www.un.org/en/global-issues/population.

United Nations (n.d.) *Migrants*. Available at: www.un.org/en/fight-racism/vulnerable-groups/migrants.

Urry, J. (2007) *Mobilities*. Cambridge: Polity Press.

Venice Airport (2024) *Venice Airport passenger numbers (2001 to 2018) – historical traffic statistics*. Available at: www.veniceairport.net/passenger-statistics.shtml.

Wizz Air (2024) *WIZZ – Dream more. Live more. Be more*. Available at: https://wizzair.com.

World Bank (2024a) *China*. Available at: www.worldbank.org/en/country/china.

World Bank (2024b) *The World Bank in Indonesia*. Available at: www.worldbank.org/en/country/indonesia/overview.

World Bank (2024c) *World Bank open data*. Available at: https://data.worldbank.org.

World Population Review (2024) *Indonesia Population 2024 (Live)*. Available at: https://worldpopulationreview.com/countries/indonesia-population.

Part IV
Environmental geography and air transport

13 Impact of air transport on geomorphology and landform change

Chapter outcomes

At the end of this chapter, you will be able to:

- Understand the geomorphic and landform changes associated with air transport.
- Explain the types of direct and indirect anthropogenic processes created by air transport.
- Explain the rationale behind the creation of artificial islands and possible environmental impacts.
- Describe the process of land reclamation for creating land to build or extend airports.
- Explain the importance of land levelling to airport construction.
- Understand the risks of air transport development to coastlines.
- Explain the potential reasons for the construction of ridges, embankments and dykes in airport development.
- Understand how using recycled materials in airport construction can reduce overall material flows.

13.1 Introduction

Air transport is at the core of economic growth. It drives social and economic progress, creates employment, facilitates trade and enables tourism. Geographically broad, quick and efficient air connectivity is key for towns or cities to take part in their local, regional and global economy. It provides a lifeline for remote communities – in rural areas or island nations. It is a fundamental pillar in 21st-century global society.

Unfortunately, the growth of air transport has also brought associated issues which need to be reduced or resolved. At the global level, there is the issue of climate change because of emissions, predominantly from aircraft (Chapter 14). At the local level, there are air quality and aircraft noise problems (Chapter 15) caused by aircraft and associated airport infrastructure. The air transport industry has come a long way in reducing its impacts, in terms of both emissions and noise per aircraft movement, however, the growth of the industry over time has meant that the cumulative impacts continue to create problems. If the industry is to grow sustainably, it needs to develop timely, robust and effective measures to reduce and manage its environmental impacts further, in terms of noise, air quality and climate change, amongst other challenges such as biodiversity and indeed geomorphic changes.

A key component of the air transport industry, which has created significant impacts, is on the ground – in terms of geomorphology and landform change. The amount of land required for airports is substantial. The airport with the largest land footprint in the world – King Fahd International Airport in Dammam, Saudi Arabia – is roughly the size of the Kingdom of Bahrain and eight times bigger than Paris. To make land suitable for airports, alterations (sometimes significant) are required; from a requirement for flat land, to clearing approach paths through to building on completely reclaimed land, geomorphic and landform change is required. Whether building completely new airports, a new runway, new terminal building or even improving drainage, these may all result in some form of geomorphic change.

13.2   Geomorphology

Chapter 7 discussed the role geomorphology and landforms have on airport design and construction. However, the focus of this chapter is not on how the industry is influenced by these processes, rather the impact the air transport industry has had *on* geomorphic processes and landform change. In other words, how airport developments and related infrastructure such as surface transport links have changed the physical characteristics of the land.

> **ATG Did you know?**
>
> **Some definitions**
>
> *Topography* is the elevation and relief of the earth's surface. It is measured by differences in elevation.
>
> *Landforms* are the topographic features of the earth's surface. These can be as small as sand dunes or as large as mountain ranges.
>
> *Geomorphology* is the study of earth surface processes and landforms. The word geomorphology originates from the Greek language: *Geo* – meaning "earth"; *Morph* – its "shape"; *Ology* – "the study of".
>
> Geomorphology is the broader scientific discipline providing the theoretical framework, landforms are the specific features on the earth's surface and topography is a representation of the earth's surface configuration.

The impact of humans on geomorphology has a very long history dating back centuries. Fire, domestication, the adoption of agriculture and pastoralism were early changes in land cover and land use. Since then, there has been the adoption of irrigation, the Secondary Products Revolution, urbanisation, the exploitation of minerals, globalisation and the harnessing of energy (Goudie and Viles, 2016), all of which have had geomorphological consequences.

Since the turn of the century the term *anthropocene* has been used by a variety of scholars, coined by authors such as Crutzen (2002) and Steffen, Crutzen and McNeill (2007), who believe that this term referred to a new epoch in earth's history, where

human impacts rivalled or exceeded the forces of nature in terms of the earth system functioning. Goudie (2019) states that they have identified three stages in the anthropocene:

1. The "industrial era" – c.1800–1945.
2. The "great acceleration" – 1945–c.2015
3. Greater awareness of the extent of human impact and possible stewardship of the earth system – from c.2015.

Other authors have argued that the anthropocene began even earlier and others say it is not an era, but a sub-part of the Holocene era. Regardless, it is not the purpose of this chapter to discuss the possibilities of a new epoch, however anthropogenic effects and processes are those which have been derived from human activities as opposed to occurring in natural environments without human influences (European Environment Agency, 2024) and the impacts of anthropogenic processes in the context of air transport have been significant.

For a more detailed overview of geomorphology and human impacts on the natural environment, textbooks by Anderson and Anderson (2010); Goudie and Viles (2016); Chorley, Schumm and Sugden (2019) and Goudie (2019) are useful resources.

13.3 Areas of human impact in geomorphology

There are very few areas of human activity which do not create landforms and landform changes, even indirectly (Szabó, Dávid and Lóczy, 2010). *Anthropogenic geomorphology* can be thought of as the study of human involvement in creating landforms and modifying geomorphological processes.

Examples of anthropogenic landforms include:
- embankments
- reservoirs
- dykes
- canals
- banks alongside roads
- terracing
- plains
- artificial islands.

Some landforms are created because of direct human actions, usually deliberately, but there are also indirect human actions which are more difficult to spot (acceleration of existing processes rather than creating new ones) and often result in environmental changes inadvertently caused by humans (Table 13.1).

Table 13.1 Anthropogenic processes

Direct anthropogenic processes	
	Examples
1. Constructional	Tipping, land reclamation and creation
2. Excavational	Digging, mining, trawling of seabed
3. Hydrological interference	Flooding, damming, dredging, draining
Indirect anthropogenic processes	
	Examples
1. Acceleration of erosion and sedimentation	Agricultural activity and clearance of vegetation and engineering, such as road construction and urbanisation
2. Subsidence	Collapse, settling (such as from mining)
3. Slope failure	Landslides, flows
4. Earthquake generation	Fracking
5. Weathering	Acidification of precipitation

Source: adapted from Goudie (2019)

In 1950, there were an estimated 2.5 billion people in the global population and by mid-2022 this had increased to 8 billion, with a forecast of 9.7 billion in 2050 (United Nations, 2024). Boeing and Airbus forecast that the number of commercial aircraft in use will approximately double in the 20-year period through to 2042. As the global population increases and the world becomes more industrialised, human impact expands and the study and understanding of these forms and processes becomes ever more important.

The largest airports are usually in urban areas or in the periphery of towns and cities and the term *urban geomorphology* was introduced by Coates (1976) as understanding the impacts landforms and their process creation has on urban areas. The impacts here are greatest due to the large volumes of people and the sheer scale and size of some of the mega airports in the world (Chapter 7). For example, Chicago O'Hare has eight runways – more than any other commercial airport in the world. The surface area at some of the major US international airports illustrates this scale further – for example at Denver (135.7km²), Dallas/Fort Worth (69.6km²) and Orlando (53.8km²), much of which is the sheer number of runways constructed here. However, some other global airports are not far behind, such as Beijing Daxing (46km²) and Shanghai Pudong (40km²) in China and Bangkok Suvarnabhumi in Thailand (32.4km²).

Airports accommodating widebody aircraft usually require runways of at least 2,500m in length but often more than 3,000m to accommodate the largest aircraft such as the A380. Even those which accommodate the most popular narrowbody A320/B737 aircraft usually have at least 2,000m runways. The associated infrastructure such as taxiways, hangars, fuel facilities, catering units, as well as passenger and cargo terminals, means space requirements within the airport perimeter are and will continue to be substantial. As discussed in Chapter 11, the development of *airport cities* has meant increased development of facilities such as hotels, conference centres and retail outlets, as well as the associated transport infrastructure to service the airport facilities.

Some airports have been essentially surrounded by the urban environment and are unable to expand (such as Kai Tak in Hong Kong, necessitating airport construction in areas outside the city – Chek Lap Kok Airport, in the context of Hong Kong).

Airports also provide key links to rural, remote and island communities and geomorphic impacts can be as great, if not greater, in some of these regions, especially as they are often in more environmentally challenging areas for airport development; for example, in low-lying atolls and islands in the world, like the Maldives and Montego Bay and Kingston in Jamaica, through to mountainous terrain in areas such as Quito, Ecuador and La Paz, Bolivia, in the Latin American Andes and Lukla in the Asian Himalayas.

The type of geomorphic and landform change will vary depending on location and the extent of the developments – from levelling land to extending a runway, through to building a completely new artificial island to accommodate a major international airport. Examples of direct landform change associated with air transport include:

- artificial islands
- land reclamation
- land levelling
- coastline alterations
- ridges, embankments and dykes

Impact of air transport on geomorphology and landform change 363

a)

b)

Figure 13.1a Palm Jumeirah, Dubai *Figure 13.1b* Pearl Island, Doha
Source: Google Maps (n.d.)

13.4 Artificial islands

Perhaps the greatest level of geomorphic and landform change occurs when building completely new islands, in some cases by the coastline and in others, significantly offshore. An artificial island is one constructed by humans rather than by natural processes, on water bodies such as the sea, lakes and rivers. They are created by processes such as reclamation of land by dredging material from the sea floor or adjacent areas and building up the island, using materials such as rock and sand, with structures often formed of concrete and steel. These are not new phenomena. Artificial islands known as *crannogs* dot the lakes and waterways of Scotland and Ireland, being built during the Iron Age (800–43 B.C.) and there is evidence that they may be even older (National Geographic, 2019).

Artificial islands now serve a variety of purposes:

- **Residential and commercial** development, such as the Palm Jumeirah in Dubai and Pearl Island in Doha, Qatar (Figure 13.1) and one of the largest artificial islands in the world at Flevopolder, Netherlands.
- **Industrial and logistics** hubs, such as Port Island, Kobe, Japan.
- **Tourism and recreation**, such as the Resort Islands in the Maldives.
- **Transportation infrastructure** – airports, roads and bridges – such as the 55km-long Hong Kong-Zhuhai-Macau bridge.

Artificial islands involve some of the greatest feats of engineering and provide the opportunity to expand habitable land, especially in geographically restricted areas. However, the issue of environmental impacts and sustainability means that these must be carefully designed, constructed and managed.

ATG Case study 13.1

Kansai International Airport (KIX), Osaka Bay, Japan

Perhaps the best example of an airport constructed on an artificial island (actually, two artificial islands) is Kansai, in Japan, with the first island covering around 510 hectares and the second around 545 hectares, constructed 5km off the most populated island of Japan – Honshu (Figure 13.2). It was opened in 1994 to relieve pressure on the nearby Osaka International Airport, providing air services for the nearby cities of Osaka, Kyoto and Kobe. Its location away from built-up areas allowed it to operate 24 hours a day (Mesri and Funk, 2016).

a)

b)

Figure 13.2a Location of Kansai International and Kobe Airports

Figure 13.2b Detailed location of Kansai International Airport

Source: Google Maps (n.d.)

Construction began in 1987 and one of the first considerations for engineers was how to overcome the high earthquake and typhoon risk in the area (with storm surges of up to 3m) in its development. It was also in water more than 18m deep. The seabed in the bay is alluvial clay – loose clay made up of several materials. As weight was added to the seabed – by building the island – water would be squeezed out of the clay. This meant the clay would shrink – and the island would sink as a result.

Engineers installed sand drains measuring nearly 1m in the bay to speed up the shrinking process. A sand drain is a deep hole drilled into the seabed and filled with sand. A heavy weight – often more sand – is placed on top of the drain. This compacts the clay by forcing moisture outward along the sand columns.

The sea wall was finished in 1989 (made of rock and 48,000 tetrahedral concrete blocks). Three mountains were excavated for 21,000,000m³ of landfill. Over three years, 10,000 workers and 10 million work hours, using 80 ships, were needed to complete the 30-metre layer of earth over the sea floor and inside the sea wall. In 1990, a 3km bridge was completed to connect the island to the mainland at Rinku Town (European Space Agency, 2024).

> The airport cost an initial $14 billion to construct but by 2008, this had risen to $20 billion, largely due to the island sinking because of the soft soils in Osaka Bay. At the time, it was the most expensive civil works project in modern history (Institution of Civil Engineers, 2024). Since it opened in 1994, it has sunk an estimated 38 feet and this is a key problem for reclaimed land – often being like a "wet sponge". It is one which will affect Kansai for many years to come.

Airports built on reclaimed land are prominent in Japan. Not just in Kansai, but in 2006, Kobe Airport (Figure 13.2a) was opened, primarily handling domestic flights.

13.5   Land reclamation

Land reclamation has been used for centuries and is the process of creating new land from water sources such as oceans, seas, lakes and riverbeds. It essentially involves transforming these areas into productive land suitable for residential, commercial, industrial or transport uses as well as agriculture and other human activity. It is becoming a more common practice in coastal areas to address the needs of urban areas, both in terms of population growth and economic development.

Sengupta *et al.* (2023) researched 135 cities between 2000 and 2020, with populations more than 1 million, and discovered that 78% had used land reclamation, adding a total of 253,000ha of additional land to the earth's surface – equivalent to the size of Luxembourg.

A common method involves filling the area with large amounts of heavy rock and/or cement, then filling with clay and soil until the desired height is obtained. One of the first major land reclamations occurred in the 1970s when the Port of Rotterdam in the Netherlands was extended (Stauber, Chariton and Apte, 2016).

From an airport perspective, coastal areas can be beneficial due to a lower likelihood of topographical obstacles and often less existing land use, especially in areas where there is a lack of existing land, or noise issues can limit developments. Even in coastal airports next to major cities, utilising take off and/or landings over the sea can reduce the issue of noise for the local population.

One of the first major airports to use land reclamation was Singapore Changi (SIN) in the mid-1970s, which was built with over 40 million cubic metres of sand reclaimed from the seabed. SIN opened in 1981, with a total area of 1,300ha, of which 870ha was reclaimed from the sea – 670ha using seafill and 200ha using landfill (Changi Airport, 2007). The Changi East Project is the airport's attempt to provide for future growth, spanning 1,080ha. The reclamation of land began in 2014, on soft marine clay, which is almost toothpaste-like in texture and will cater for a fifth terminal and a three-runway system, including 40km of new taxiways, amongst other facilities (Changi Airport, 2017).

ATG Case study 13.2

Hong Kong International Airport, Chek Lap Kok (HKG)

Chek Lap Kok Airport, in Hong Kong, was built on partially reclaimed land (see Case Study 1.3) in 1998, to replace the ageing Kai Tak, which could not be expanded in the densely built Kowloon and had serious noise pollution issues. Kai Tak itself was located where part of the land was reclaimed. At 1,248ha, HKG was almost four times the size of the old Kai Tak airport – equal to the size of the Kowloon peninsula (Figure 13.3). The airport increased the land area of Hong Kong by 1%. Nine-hundred hectares of land were reclaimed and the three-runway system (a subsequent phase to build a third runway) will add a further 650ha.

It was formed off Lantau Island, by assimilating the island of Chek Lap Kok and the smaller Lam Chau via land reclamation (75% of the airport's area). Both these were once mountainous islands and over 200 million tonnes of rock were removed and 600 million tonnes of earth replaced to extend the land area combining the two islands. The original 100m peak was reduced to 7m above sea level (Arch2O, 2024); 69 million m³ of silt came from the seabed and 76 million m³ of sand and 121 million m³ of rock were imported (Szabó, Dávid and Lóczy, 2010) making it one of the largest projects ever of its kind.

The project also involved surface transport links and the construction of roads, rail links, bridges and tunnels, including the Tsing Ma bridge, which at 2.2km is one of the longest suspension bridges in the world.

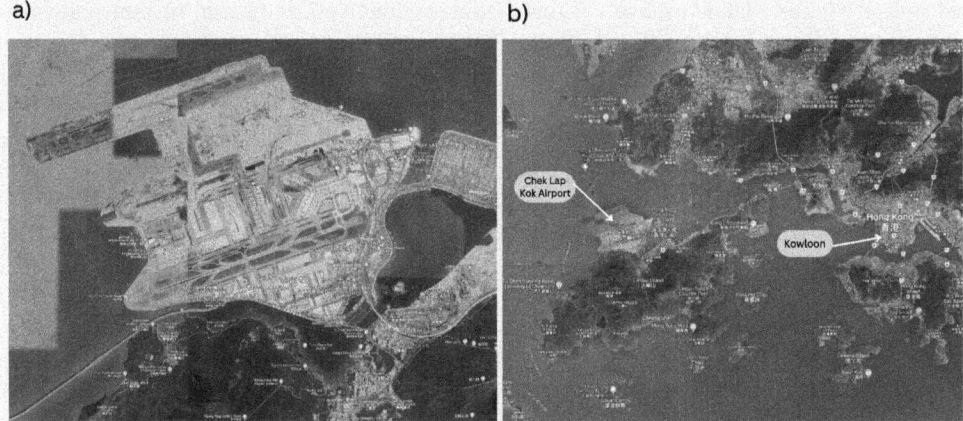

Figure 13.3a Hong Kong International Airport, Chek Lap Kok

Figure 13.3b The location of Chek Lap Kok and Kowloon

Source: adapted from Google Maps (n.d.)

Another significant project involving land reclamation was the construction of a 3,300m runway at Brisbane Airport (BNE). Site preparation took five years and was situated on a reclaimed portion of the Brisbane River delta, extremely soft, up to 3m deep, and as with SIN, comparable to the consistency of toothpaste. The soft, waterlogged soil needed to be loaded with sand to provide stability, and the dredging and reclamation works in Moreton Bay were a monumental challenge – 11 million m³ of sand was dredged. There were 330,000 wick drains installed – these operate in a similar way to a straw, allowing a quicker exit of the water from the soil, to speed up the settling process. The settling process then took three years before the construction was finalised (Brisbane Airport, 2021).

Land reclamation also brings environmental challenges. Reclaiming land can disrupt marine and coastal ecosystems affecting habitats and biodiversity. It can often be vulnerable to erosion, subsidence and flooding, and can be very expensive – economically, socially and ecologically.

Sengupta et al. (2023) also suggested that 70% of recent reclamation had occurred in areas potentially exposed to extreme sea level rise by 2100, creating significant challenges for sustainable coastal development.

Research by Murray et al. (2022) revealed that 4,000sq.km. of tidal wetlands have been lost globally over 20 years, caused by global change and human actions, and around three-quarters of the net global tidal wetland decrease has happened in Asia. However, ecosystem restoration and natural processes have played their part in reducing total losses.

As urban development and population growth continues, the desire for airport development via land reclamation will continue, but this must be done by balancing the benefits with the need to protect ecosystems and communities, if developments are to be sustainable.

13.6   Land levelling

Airports require an extensive amount of flat land, for stable, safe and efficient airport operations. Even though runways are rarely completely flat (Chapter 7), they still require essentially flat land, as does the broader airport infrastructure. In any development, it is unlikely that the entire surface chosen will meet this criteria exactly, and therefore some element of land levelling may be required. Even relatively flat land may require earthworks, drainage systems and providing for soil stability.

Land levelling techniques include cut and fill grading techniques – cutting high areas and filling low areas – to provide a level uniform surface. Soil compaction can compress the soil to increase stability and density.

One of the largest airport projects in recent times requiring land levelling was the new Istanbul Airport, inaugurated in 2018, covering around 7,600ha, of which over 6,000ha was state-owned forest and there were also open-pit coal mines in the area, which had to be filled with soil. An Environmental Baseline and Impact Assessment prepared by Environ (2015) stated that the topography of the site was uneven with a terrain elevation difference of several tens of metres from one portion of the site to another.

ATG Case study 13.3

Toncontín Airport, Tegucigalpa, Honduras (TGU)

The removal of hilly topography is occasionally required to facilitate runway development, extensions or merely to improve safety. TGU often appears in the "most dangerous airports to fly into" articles and tv programmes due to the surrounding mountainous terrain (Figure 13.4), short runway (1,863m) and wind gusts at a reasonably high altitude (1,005m). On approach, aircraft take a 45° turn to the runway inside the valley to make their final descent. In 2007, a large portion of the hillside on the south side of the runway was bulldozed immediately before the threshold to make the approach easier and the runway was also lengthened to 2,163m.

Figure 13.4 Toncontín Airport and surrounding terrain
Source: adapted from Google Earth (n.d.)

In the previous two sections on artificial islands and land reclamation, land levelling was a key component. However, unlike these examples, which are usually extensive, land levelling can also be on a smaller scale, such as for runway extensions, new hangars, etc.

Differing soil conditions and topographical variations can pose problems in land levelling for airport developments and it is important that Environmental Impact Assessments and mitigation strategies are used to minimise any adverse effects.

13.7   Coastline alteration

Coastlines change shape over time because of natural processes, climate change and human activity, from aspects such as land reclamation, sea defence construction, dredging and urban and transport development. Approximately 40% of the world's population lives within 100km of the seashore (Wang *et al.*, 2023) and the increasing demand for land resources has resulted in coastline alterations.

As pointed out by Pijet-Migon and Migon (2018), land reclamation is also coastline alteration, but there are more localised changes associated with alterations such as runway extensions. For example, at Gibraltar Airport, a runway extension to allow the operation of medium-size jet aircraft was built into the bay and approximately half was on human-made land.

Whilst these alterations can result in positive outcomes such as expanding land for infrastructure development and to provide natural hazard protection, they can also have potentially negative effects on shoreline morphology:

- **Habitat destruction.** The loss of marine habitats such as mangroves and coral reefs and a possible biodiversity decline and the alteration of ecosystem dynamics.

- **Water quality.** Dredging can increase sedimentation, which can have impacts on marine life and runoff from construction can also cause water pollution.
- **Subsidence** can be created due to the weight of new constructions increasing flood risks. The example of Kansai in Case Study 13.1 illustrates issues with subsidence.
- **Erosion.** Construction of features such as sea walls can cause sediment build up and erosion, destabilising shorelines.

ATG Case study 13.4

Madeira (Funchal) Airport (FNC)

FNC is the only airport serving the island of Madeira, in the Atlantic Ocean, off the western coast of Morocco. It is in a geographically challenging setting on a largely mountainous island of volcanic origin. There is very little land suitable for airport construction, with its location on a foreland on the east coast, on what is essentially a *tabletop runway* (see Section 7.5).

It was first opened in 1964 with a 1,600m runway. A runway of this length could not accommodate many aircraft types and was also a difficult and technically demanding landing. With tourism developing on the island, the runway was extended to 1,800m in 1986 to enable more aircraft to land. However, there were still serious limits and only smaller narrowbody aircraft could land, some of which were payload restricted. For example, Monarch Airlines operated its A320 aircraft with a 180-seat configuration, however operations to FNC meant a payload restriction to 157 passengers, reducing the likelihood of operating viable commercial services.

In 2002, the runway was extended again, to 2,781m, making operations with a range of aircraft more feasible. As the available land had been used, the extension was built on a bridge (1,020m × 180m), crossing a shallow water bay, 57m above sea level and supported by 180 pillars. It was designed to absorb the impact of one of the largest aeroplanes operating – the Boeing 747.

a)

b)

Figure 13.5a Runway at Madeira Airport with extension
Source: Bingar1234 (2014)

Figure 13.5b Madeira runway extension bridge and pillars
Source: Bartz (2013)

370 *Fundamentals of Global Air Transport Geography*

ATG Case study 13.5

Sangster International Airport, Montego Bay, Jamaica (MBJ)

In 2024, a $70 million runway extension (and associated works) was opened at MBJ, lengthening the runway by 408m to 3,060m (Figure 13.6), as well as providing runway end safety areas (RESAs) at each end, meeting ICAO Annex 14 requirements. The airport is located with the Caribbean Sea to the north and west and residential areas to the south and east.

Figure 13.6a The location of Sangster International Airport on Jamaica

Figure 13.6b Runway extension project area

Source: adapted from Google Maps (n.d.)

The site was located on a coastal platform, approximately 1.2m above sea level. MBJ was originally constructed on land which was a large and extensive mangrove-lined lagoon. The extension will remove close to 8 acres of identified wetlands, over 5 acres classed as in "good" condition. Shoreline protection will consist of the construction of a 3.5m high boulder revetment (MBJ Airport, 2019).

In theory, the runway extension will allow higher payloads to be operated, however as the B747-400 was currently being operated, the benefits in terms of larger aircraft are still to be realised (Echevarne, 2024).

13.7.1 Coastal erosion and deposition

As stated previously, many airports are located on or near the coast. However, the seas and oceans can create significant erosion risk, potentially affecting the airports and surrounding areas.

ATG Did you know?

Coastal erosion and deposition

Coastal erosion is the loss of coastal land and occurs when shoreline materials such as sediments are picked up and moved away by the action of the waves. Coastal erosion takes place with *destructive waves* and cliffs are one of the main pieces of evidence for the impact of destructive waves.

Coastal deposition is the opposite of coastal erosion and is the creation of new landforms along the shoreline, whereby materials such as sediment are deposited by water. *Constructive waves* create the build-up of shoreline material.

Erosion is directly related to deposition and vice versa.

Erosion and deposition are natural processes, however human activities and environmental factors can mean these can have increasingly problematic effects. Erosion occurs through several main processes:

1. **Hydraulic action** – the force of the water against the cliffs. Compression occurs in rocky areas when air enters rock cracks. Air is trapped in cracks by the rising tide; as waves crash against the rock, the air inside the crack is rapidly compressed and decompressed causing cracks to spread and rock to break off. This a key process in cave formation.
2. **Abrasion** – occurs when rocks and other material are thrown against the coastline by strong waves, causing more material to break off and be carried away.
3. **Attrition** – occurs when material such as rocks and stones carried by waves hit against each other wearing them down. This is how sand and round pebbles are formed.
4. **Corrosion** – or solution/chemical weathering, occurs when the sea water dissolves certain types of rocks. In the UK, chalk and limestone are prone to this type of erosion.

ATG Case study 13.6

Stornoway Airport, Isle of Lewis, Scotland (SYY)

The Isle of Lewis is located off the northwest coast of Scotland and is a key access point for the Scottish Western Isles. SYY provides air links to Glasgow, Edinburgh and Inverness on the mainland, as well as inter-island services, and provides lifeline services for tourism, business and medical services.

Being a coastal airport, it is prone to coastal erosion. In 2023, a £5.3 million programme by airport operator Highlands and Islands Airports Ltd (HIAL) was completed to protect SYY from the effects of coastal erosion. Without this work, it was believed that the sand dunes which shield the airport would have been lost, ultimately putting the airport at risk of flooding during severe weather.

Over 36,000 tonnes of rock armour were placed and landscaped on the beach sides of the runway embankment. A mattress of almost 22,000 tonnes of stone was created beneath the beach, with 2,081 baskets filled with stone built on top to reinforce the beach and stop the sea from eroding the land. In addition, a further 14,058 tonnes of rock armour stone was positioned on the headland at the end of the runway (HIAL Ltd, 2023).

13.8   Ridges, embankments and dykes

Ridges, embankments and dykes are constructed to serve multiple purposes around airports, relating to environmental issues, safety and operational functionality. For example:
1. Flood prevention and groundwater protection
2. Noise barriers
3. Habitat management
4. Infrastructure construction

ATG Did you know?

Ridges, embankments and dykes

Ridges are long, narrow raised land formations with sloping sides.

Embankments are artificial slopes made of earth and/or stones, used to hold back water or support infrastructure, such as a road or railway line.

Dykes are barriers used to regulate or hold back water from a river, lake or even the ocean.

1. Flood prevention and groundwater protection

Ridges, embankments and dykes can assist in managing surface water runoff and preventing flooding, by channelling water to designated drainage systems and to protect low-lying airports from the sea.

Amsterdam Schiphol, in the Netherlands, is built below sea level and is enclosed by embankments known as *polders*. Incheon Airport in Seoul, South Korea, has been developed on the seabed on reclaimed land and is protected by dykes. Shannon Airport in Ireland has east and west embankments of 1.3km and 1.6km respectively, serving as flood defence structures from the adjacent Shannon Estuary (MWP, 2020).

2. Noise barriers

Airports can generate significant noise from aircraft taking off and landing and via ground activities. Chapter 15 discusses the issue of air transport noise in detail, however effective land use planning by airports can involve noise abatement measures. These may be related to using noise-compatible buildings as noise barriers and the use of noise proofed terraced housing instead of semi-detached housing (Murphy and King, 2014) as a first line of defence – for example closest to the runway.

From a landform perspective, ridges and embankments can act as sound barriers, to reduce the impact of noise on local communities by deflecting and absorbing sound waves. Embankments and *earth berms* are natural barriers which can be created during the construction phase. Vegetation such as trees could also be used, however their effectiveness may be more related to the landscape vegetation providing noise attenuating effects through people's emotional processing, termed as "psychological noise reduction" (Yang, Bao and Zhu, 2011).

ATG Case study 13.7

Amsterdam Schiphol Airport, Netherlands (AMS)

The Netherlands is an extremely flat country, undisturbed by hills or mountains, which has facilitated its famous cycling culture. It consists of mainly low-lying plains – hence the "Nether-lands". It has faced unique water management challenges, where a third of the land lies below mean sea-level, and without dunes, dikes and pumps, 65% would be under water at high tide (Hoeksema, 2007).

AMS is the largest airport in the Netherlands, one of the busiest in Europe and is the main hub for one of the oldest airlines in the world (established 1919) – KLM. It covers an area of 27.87sq.km – or 14 times the size of Monaco – and has around 30km of fencing around the perimeter. Schiphol lies at the bottom of what was once Haarlemmer Lake, which was drained in 1852, and the first aircraft landed at Schiphol in 1916. Technically, the airport lies on the bottom of the lake, over 4m below sea level (Schiphol, 2024).

Schiphol has six runways – five long runways used for international air traffic and Schiphol-Oostbaan, which is shorter and used mainly for general aviation, private jets and helicopters (Figure 13.7a). The newest runway (**Schiphol-Polderbaan**) opened in 2003 and is also the longest (3,800m) and busiest.

From an air transport perspective, flat land is perfect for airport construction, but when there is very little elevation all around the airport perimeter, it makes it harder to manage aircraft noise issues. As Figure 13.7a illustrates, the Polderbaan is located quite some way from the other runways and the terminal buildings. When it was being planned, the impact of noise pollution was an issue and to try to reduce this problem, it was constructed to the west of the airport, on the other side of the A5 motorway. It was on reclaimed land from water known as a polder (hence, polderbaan).

The residents in Hoofddorp-Noord experienced extra inconvenience from aircraft ground noise when the Polderbaan runway was opened and as a solution, *noise-reflecting ridges* were constructed between Hoofddorp-Noord and Runway 18R-36L (the Polderbaan) as shown in Figure 13.7b. These are like little wedge-shaped hills, laid out like interlocking diamonds, with slopes that are angled to reflect sound waves upwards. The ridges are 3m high – 2m of that are above ground, and 1m below ground.

374 *Fundamentals of Global Air Transport Geography*

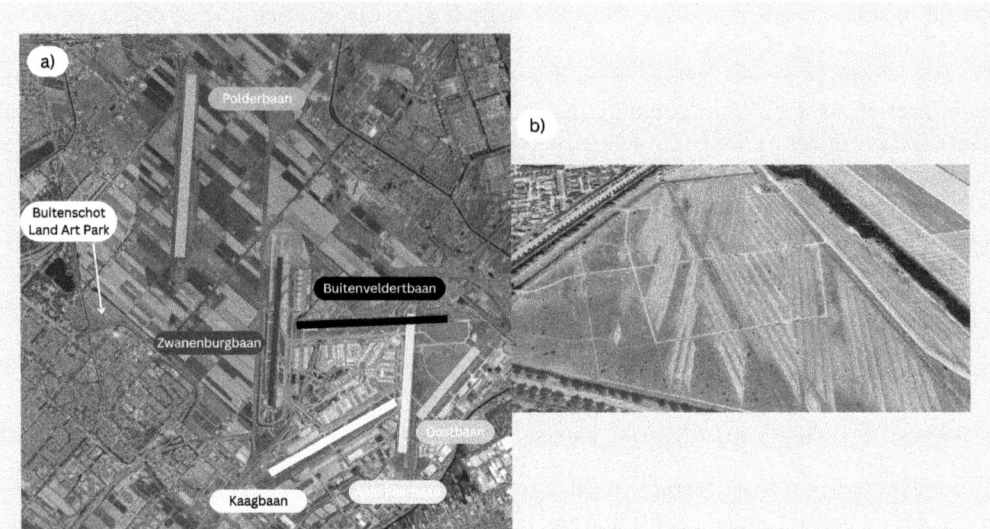

Figure 13.7a The runway layout at AMS
Source: adapted from Google Maps (n.d.)

Figure 13.7b Noise-deflecting ridges at Buitenschot Land Art Park
Source: Google Earth (n.d.)

The idea was the inspiration of the Netherlands Organisation for Applied Scientific Research (TNO). The TNO found that in the autumn, after the surrounding fields had been plowed, noise levels decreased significantly. The farmer's furrows, because they had multiple ridges to absorb the sound waves, deflected the sound and muted the noise (Hansman, 2015). From this, Buitenschot Land Art Park was created around the landscape, for aesthetic and recreational value, providing open spaces for walking and cycling (Schiphol, 2024).

3. Habitat management

Bird strikes are a particular hazard at airports, where birds may be susceptible to ingestion by aircraft engines and indeed other animals such as deer also pose a danger to aircraft taking off and landing, if allowed to roam within the airport grounds (see Section 8.3). Ridges and embankments could be constructed and designed to deter wildlife from nesting or exploring near runways and taxiways. This may only be effective if used in conjunction with other management measures such as bird netting, air cannons and habitat adjustments, to make the airport grounds less desirable.

4. Infrastructure construction

Ridges and embankments can provide support for airport infrastructure such as runways and taxiways. They can raise the ground level, allowing the structures to be built on stable earth. They can also be used to maximise available space, where horizontal expansion is difficult or impossible, and these can also support service roads.

Manchester Airport's second runway was opened in 2001 and during the earthworks phase of construction, a *tunnel* was created at the point where the runway passes over the River Bollin. The river valley was infilled to create an embankment over which the runway crossed the river, where approximately 2.5 million m³ of earth had to be moved from either side of the valley (VHE, 2024).

A Coruña Airport in Spain is located on a hill with the main runway approximately 2,000m in length, running NE to SW, and the SW part required an embankment 25m high. When the runway was extended by 400m from its SW end, as the natural ground surface was sloping down towards the southwest, a new embankment extension 35m high was required (Cañizal et al., 2015).

13.9 Using recycled materials in airport construction

As identified by Douglas and Lawson (2003), the changes to earth surface systems from airport land reclamation are significant and the indirect effects, both on and around airports and from areas where construction materials are obtained, need to be considered. They also emphasise that materials' use and geomorphic change associated with airport construction has resulted in the material flows involved being of geological significance, in terms of mass moved per unit of time. Some of the material flows can be reduced, however, by using recycled materials in airport construction and other airport-related developments.

At Nashville International Airport, USA (BNA), runway 2L/20R was demolished and rebuilt, reopening in 2010. Normally, reconstruction would involve breaking and removing the old concrete and transporting rubble to landfill. At BNA, the Airport Authority opted for a concrete recycling method, crushing and screening concrete rubble at an on-site plant and then placing the aggregate as a base to construct the new runway (Garver, 2024).

As part of their multi-phased modernisation programme, including four new runways, Chicago O'Hare International Airport (ORD) was able to recycle 600,000 tonnes of materials, including concrete and asphalt, bricks, scrap metal, light bulbs and landscaping waste. Over 98% of the debris was recycled and prevented from entering landfill (FAA, 2013).

To minimise the environmental impact of their runway rehabilitation project, Treviso Airport (TSF) near Venice, Italy, completely recycled the materials used from the demolition of their old runway into the new runway pavement. Recycling techniques, such as the cement stabilisation of soil, cement treatment of milled cement concrete and cement-bitumen treatment of reclaimed asphalt, were used (Bocci et al., 2013).

Amsterdam Schiphol (AMS) is testing the operation of a "plastic road". A stretch of road has been constructed on the apron, made from the airport's own plastic waste. The airport believes the benefits include not only recycling of its own plastic, but also being easier to maintain than traditional roads, having a longer design life and being quicker and easier to build, reducing emissions, as well as better and more natural water management (Schiphol, 2024).

Chapter review questions

13.1 What are the different geomorphic and landform changes associated with the air transport industry?

13.2 What are the main types of direct and indirect anthropogenic processes caused by the industry?

13.3 Why have artificial islands been created for airport construction and what are the potential environmental impacts? Do you feel that there will be more of a requirement for these in the future? If so, why?

13.4 Land reclamation is involved in artificial island construction, but also in smaller-scale projects. Research a range of airports which have reclaimed land. What were the purposes? Do you think there were any potential solutions other than reclaiming land?

13.5 Why is land levelling required in almost any airport building project? Research your local airport, either by visiting or remotely. Where is it located from a topographical perspective? Why do you think it is located where it is? Is there any evidence of landform change?

13.6 What are the range of possible impacts of air transport on coastlines? For any ONE of these, choose a case study airport located at the coast. What are the actual or potential negative impacts of the airport on the coastline? Has the airport implemented any processes to reduce its coastline impacts?

13.7 What are the main purposes of airports constructing ridges, embankments or dykes? For any ONE of these, research a range of case study airports and assess their impacts.

13.8 How can using recycled materials in airport construction reduce overall material flows? Research a range of airports in your country. Can you find examples of recycled or reused material in the construction process for any part of the airport?

ATG trivia

The second lives of aircraft

What happens to aircraft when they reach the end of their useful lives? "Useful" can mean for structural and safety reasons or for economic reasons, when newer aircraft are more cost effective to operate. According to EASA (2024), an average of 650 commercial aircraft were retired annually during the last decade and more than 30% of the current fleet in Europe is expected to be retired over the next decade.

With the spotlight firmly on sustainability, the air transport industry must do its best to ensure that reuse and recycling of aircraft and their components occurs where possible. Still serviceable aircraft components may be reused as spares for other aircraft. When aircraft are disassembled, the fuselage can be cut up and parts used in other sectors – for example, the windows are of interest to the interior design industry and the plastic elements are used to create clothing or fleece blankets. Of the total weight of an A320 aircraft, 92% can be recovered (Airbus, 2022).

Aircraft graveyards (or boneyards) are where planes go for long-term storage or to be scrapped. The highest proportion of these are in the southwestern USA, where there is a dry desert climate, which reduces corrosion. The largest boneyard is for military aircraft, at Davis-Monthan Air Force Boneyard in Tucson, Arizona, which is a 2,600-acre open-storage facility, holding around 4,000 aircraft (Figure 13.8). Others storing commercial aircraft include Mojave Air and Space Port, in the Mojave Desert in California; Pinal County Airpark, near Marana; and Arizona and Southern California Logistics Airport, near Victorville, California.

There have also been a few "different" ideas as to how to use these old aircraft. Jumbo Stay is a hotel in a Boeing 747 at Stockholm-Arlanda Airport. In the Costa Rican jungle, you can take to the skies in a Boeing 727 fuselage in a Hotel Costa Verde hotel suite – without leaving the ground. The forward section of an ex-Monarch Airlines DC-10 has been converted into a multi-purpose resource at Manchester Airport's Runway Visitor Park, seating up to 40 passengers and used as a classroom or conference breakout room. Some aircraft may never die!

Figure 13.8 Aerial view of the aircraft boneyard at Davis-Monthan Air Force Base, in Tucson, Arizona

Source: Library of Congress (n.d.)

References

Airbus (2022) *End-of-life reusing, recycling, rethinking | FAST Online | News | Airbus Aircraft*. Available at: https://aircraft.airbus.com/en/newsroom/news/2022-11-end-of-life-reusing-recycling-rethinking.

Anderson, R.S. and Anderson, S.P. (2010) *Geomorphology: the mechanics and chemistry of landscapes*. Cambridge: Cambridge University Press.

Arch2O (2024) *Chek Lap Kok Airport | Foster and Partners*. Available at: www.arch2o.com/chek-lap-kok-airport-foster-partners/.

Bartz, R. (2013) *Madeira Airport runway*. Available at: https://commons.wikimedia.org/wiki/File:Madeira_Airport_Runway_Crop.jpg (Accessed: 5 July 2024).

Bingar1234 (2014) *Madeira*. Available at: https://commons.wikimedia.org/wiki/File:Madeira_(2).jpg.

Bocci, M. *et al.* (2013) 'Runway pavement reconstruction with full material recycling: the case of the airport of Treviso', *Advanced Materials Research*, 723, pp. 1044–1051. Available at: https://doi.org/10.4028/www.scientific.net/AMR.723.1044.

Brisbane Airport (2021) *Brisbane's new runway*. Available at: www.bne.com.au/corporate/projects/future-bne-projects/completed-projects/brisbanes-new-runway.

Cañizal, J. *et al.* (2015) 'High rockfill embankment for the extension of an airport main runway', in. *15th Pan-American Conference on Soil Mechanics and Geotechnical Engineering*, Buenos Aires. Available at: https://doi.org/10.3233/978-1-61499-603-3-188.

Changi Airport (2007) *Some facts on Singapore Changi Airport*. Available at: https://web.archive.org/web/20070101114643/http://www.changiairport.com/changi/en/about_us/fact_sheets/facts_changi_ap.html.

Changi Airport (2017) *Changi East in 2017 and Beyond*. Available at: www.changiairport.com/corporate/media-centre/changijourneys/the-airport-never-sleeps/changi-east-in-2017-and-beyond.html.

Chorley, R.J., Schumm, S.A. and Sugden, D.E. (2019) *Geomorphology*. 1st edn. Abingdon: Routledge.

Coates, D.R. (1976) *Urban geomorphology*. Boulder, Colorado: Geological Society of America. Available at: https://doi.org/10.1130/SPE174.

Crutzen, P.J. (2002) 'Geology of mankind', *Nature*, 415(6867), pp. 23–23. Available at: https://doi.org/10.1038/415023a.

Douglas, I. and Lawson, N. (2003) 'Airport construction: materials use and geomorphic change', *Journal of Air Transport Management*, 9(3), pp. 177–185. Available at: https://doi.org/10.1016/S0969-6997(02)00082-0.

EASA (2024) *Sustainability in the end-of-life phase of aircraft – what happens after airplanes fly their last trip*. Available at: www.easa.europa.eu/en/light/topics/sustainability-end-life-phase-aircraft.

Echevarne, R. (2024) 'Interview. ACI Latin America-Caribbean Director General'.

Environ (2015) *Istanbul new airport ESIA. Environmental baseline and impact assessment. Resource efficiency*. Bath. Available at: www.igairport.aero/media/iomlsl5z/chapter-7_10-resource-efficiency.pdf.

European Environment Agency (2024) *Anthropogenic processes*. Available at: www.eea.europa.eu/themes/water/glossary/anthropogenic-processes.

European Space Agency (2024) *Kansai and Kobe International Airport (Osaka Bay, Japan) – historical views – Earth Watching*. Available at: https://earth.esa.int/web/earth-watching/historical-views/content/-/article/kansai-and-kobe-international-airport-osaka-bay-japan-/.

FAA (2013) *Recycling, reuse and waste reduction at airports. A synthesis document*. Available at: www.faa.gov/sites/faa.gov/files/airports/resources/publications/reports/RecyclingSynthesis2013.pdf.

Garver (2024) *Recycled runway – concrete recycling method saves millions of dollars*. Available at: https://garverusa.com/iq/2010/vol-2-issue-3/recycled-runway/889 (Accessed: 8 July 2024).

Google Earth (n.d.). *Google Earth*. Available at: www.google.co.uk/earth/.

Google Maps (n.d.) *Google Maps*. Available at: www.google.com/maps.

Goudie, A.S. (2019) *Human impact on the natural environment. Past, present and future*. 8th edn. Chichester: John Wiley and Sons.

Goudie, A.S. and Viles, H.A. (2016) *Geomorphology in the anthropocene*. Cambridge: Cambridge University Press.

Hansman, H. (2015) 'This crazy land art deflects noise from Amsterdam's Airport', *Smithsonian Magazine*. Available at: www.smithsonianmag.com/innovation/crazy-land-art-deflects-noise-from-amsterdams-airport-180955398/.

HIAL Ltd (2023) *Airport information for Stornoway Airport, Highlands and Islands Airports Limited*. Available at: www.hial.co.uk/stornoway-airport/airport-information-12/1.

Hoeksema, R.J. (2007) 'Three stages in the history of land reclamation in the Netherlands', *Irrigation and Drainage*, 56(S1), pp. S113–S126. Available at: https://doi.org/10.1002/ird.340.

Institution of Civil Engineers (2024) *Kansai Airport*. Available at: www.ice.org.uk/what-is-civil-engineering/what-do-civil-engineers-do/kansai-airport/.

Library of Congress (n.d.) *Photographs in the Carol M. Highsmith Archive, Library of Congress, Prints and Photographs Division*. Available at: https://loc.getarchive.net/media/aerial-view-of-the-tucson-arizona-area-with-a-focus-on-a-giant-airplane-boneyard-28ad4a.

MBJ Airport (2019) *Runway extension and associated works project at Sangster International Airport, Montego Bay, St. James.* Available at: www.mbjairport.com/content/650/Public%20Notice%20-%20Project%20Document_SIA%20Runway%20Extension%20and%20Associated%20Works%20MBJ.pdf.

Mesri, G. and Funk, J. (2016) 'Closure to "settlement of the Kansai International Airport Islands" by G. Mesri and J. R. Funk', *Journal of Geotechnical and Geoenvironmental Engineering*, 142, p. 07016008. Available at: https://doi.org/10.1061/(ASCE)GT.1943-5606.0001448.

Murphy, E. and King, E.A. (2014) *Environmental noise pollution. Noise mapping, public health and policy.* Burlington, MA: Elsevier.

Murray, N.J. *et al.* (2022) 'High-resolution mapping of losses and gains of earth's tidal wetlands', *Science*, 376(6594), pp. 744–749. Available at: https://doi.org/10.1126/science.abm9583.

MWP (2020) *EIA Screening Report. Embankment refurbishment works, Shannon Airport.* Available at: https://assets.gov.ie/152477/dcc07a4d-05b0-4771-a83a-959ac7f4bb5b.pdf.

National Geographic (2019) *Artificial islands older than Stonehenge stump scientists, Culture.* Available at: www.nationalgeographic.com/culture/article/neolithic-island-older-than-stonehenge-crannog-scotland.

Pijet-Migon, E. and Migon, P. (2018) 'Landform change due to airport building', in M.J. Thornbush and C.D. Allen (eds) *Urban Geomorphology: landforms and processes in cities.* Amsterdam: Elsevier, pp. 101–111.

Schiphol (2024) *Schiphol as a neighbour.* Available at: www.schiphol.nl/en/schiphol-as-a-neighbour/.

Sengupta, D. *et al.* (2023) 'Mapping 21st century global coastal land reclamation', *Earth's Future*, 11(2), p. e2022EF002927. Available at: https://doi.org/10.1029/2022EF002927.

Stauber, J.L., Chariton, A. and Apte, S. (2016) 'Global change', in *Marine ecotoxicology. Current knowledge and future issues.* Oxford: Elsevier.

Steffen, W., Crutzen, P.J. and McNeill, J.R. (2007) 'The anthropocene: are humans now overwhelming the great forces of nature?', *Ambio*, 36(8), pp. 614–621.

Szabó, J., Dávid, L. and Lóczy, D. (2010) *Anthropogenic geomorphology. A guide to man-made landforms.* Dordrecht: Springer. Available at: https://doi.org/10.1007/978-90-481-3058-0.

United Nations (2024) *Population.* Available at: www.un.org/en/global-issues/population.

VHE (2024) *Manchester Airport runway 2, VHE.* Available at: www.vhe.co.uk/project/26/manchester-airport-runway-2/.

Wang, G. *et al.* (2023) 'Analysis of coastline changes under the impact of human activities during 1985–2020 in Tianjin, China', *PLOS ONE*, 18(11), p. e0289969. Available at: https://doi.org/10.1371/journal.pone.0289969.

Yang, F., Bao, Z.Y. and Zhu, Z.J. (2011) 'An assessment of psychological noise reduction by landscape plants', *International Journal of Environmental Research and Public Health*, 8(4), pp. 1032–1048. Available at: https://doi.org/10.3390/ijerph8041032.

14 The impact and mitigation of air transport on climate change

Chapter outcomes

At the end of this chapter, you will be able to:

- Explain the main sources of CO_2 emissions from the air transport industry.
- Understand the importance and range of non-CO_2 air transport emissions.
- Assess the range of areas in which the air transport industry has mitigated its emissions.
- Analyse the importance of aircraft technologies in reducing fuel consumption.
- Understand the importance of reducing weight to reducing aircraft emissions.
- Describe the ways operational improvements can influence aircraft emissions.
- Understand the regulations surrounding aviation net zero 2050 commitments.
- Assess the potential for hydrogen and electric power as future solutions in achieving net-zero.
- Explain the concept of sustainable aviation fuels and understand their potential in reducing emissions.
- Assess the concept of market-based measures to reduce the air transport industry's carbon footprint.

14.1 Introduction

Climate change refers to long-term shifts in temperature and weather patterns. Such shifts can be natural, due to changes in the Sun's activity, volcanic activity and natural cycles such as the El Niño-southern oscillation (ENSO), which is a pattern of changing water temperatures in the Pacific Ocean. In an "El Niño" year, the global temperature warms and in a "La Niña" year, it cools down (Met Office, 2023).

Since the 19th century, however, human activities have been the main driver of climate change, primarily due to the burning of fossil fuels. Burning fossil fuels generates greenhouse gas (GHG) emissions that act like a blanket wrapped around the earth, trapping the Sun's heat and raising temperatures, and the world is now warming faster than at any point in recorded history. Fossil fuels account for over 75% of global GHG emissions (United Nations, 2023). The United Nations Secretary-General António Guterres stated that climate change is a "code red for humanity" (ICAO, 2022a).

The United Nations (2023) has stated that there are seven main human-induced causes of climate change:

1. Generating power
2. Manufacturing goods
3. Cutting down forests
4. Using transportation
5. Producing food
6. Powering buildings
7. Consuming too much

The Intergovernmental Panel on Climate Change (IPCC) is the United Nations body for assessing the science related to climate change. It provides regular assessments of the scientific basis of climate change, its impacts and future risks and options for adaptation and mitigation. The latest assessment cycle is the sixth (AR6) with the synthesis report released in 2023. Some key findings (with *high confidence*) included:

- Human activities, principally through emissions of greenhouse gases, have unequivocally caused global warming, with global surface temperatures reaching 1.1°C above 1850–1900, in 2011–2020.
- In 2019, atmospheric carbon dioxide (CO_2) concentrations (410 parts per million) were higher than at any time in at least 2 million years.
- Concentrations of methane (1,866 parts per billion) and nitrous oxide (332 parts per million) were higher than at any time in at least 800,000 years.
- Global net anthropogenic GHG emissions in 2019 are estimated to be about 12% higher than in 2010 and 54% higher than in 1990.

(IPCC, 2023)

With the earth being a system, changes in one area can influence changes in all others, and therefore the consequences of climate change now include intense droughts, water scarcity, more health risks, rising sea levels, melting polar ice, catastrophic storms and declining biodiversity.

The United Nations (2023) states that thousands of scientists and government reviewers have agreed that limiting global temperature rise to no more than 1.5°C would help avoid the worst climate impacts, yet policies in place currently suggest a 2.8°C temperature rise by 2100. A number of countries have committed to net-zero emissions by 2050; however emissions must be cut in half by 2030 to keep warming below 1.5°.

There are three broad categories of action required:

- cutting emissions
- adapting to climate impacts
- financing required adjustments

Sections 6.7 and 6.8 focused on the impacts climate change has on the air transport industry, however air transport is also a contributor to climate change. This chapter will therefore analyse the sources of air transport emissions, as well as mitigation strategies currently in place to limit air transport impacts and potential future strategies to further reduce these impacts.

14.2 Air transport CO_2 emissions

ICAO serves as a multilateral platform for cooperation on international aviation environmental protection. ICAO member states have agreed to concentrate their aviation environmental collaboration on three core areas:

- climate change and aviation emissions
- local air quality
- aircraft noise

(ICAO, 2023)

According to the IEA (2023), aviation accounted for 2% of global energy-related CO_2 emissions in 2022. This involved emitting around 800 million tonnes of CO_2, approximately 80% of pre-pandemic levels, which according to Lee *et al.* (2020) in Ritchie (2020), was 1.04 billion tonnes in 2018, a figure which has doubled since 1987 (Figure 14.1). Air transport CO_2 emission share stood at 2.5% in 2018.

Whilst CO_2 emissions doubled between 1987 and 2018, passenger numbers more than quadrupled during this time, from less than 1 billion in 1987 to almost 4.3 billion in 2018 (Figure 14.2).

On this broad metric, air transport emissions have become much more efficient on a per-passenger basis, but the issue is the overall increase. Oliver Wyman (2023), in their Global Fleet

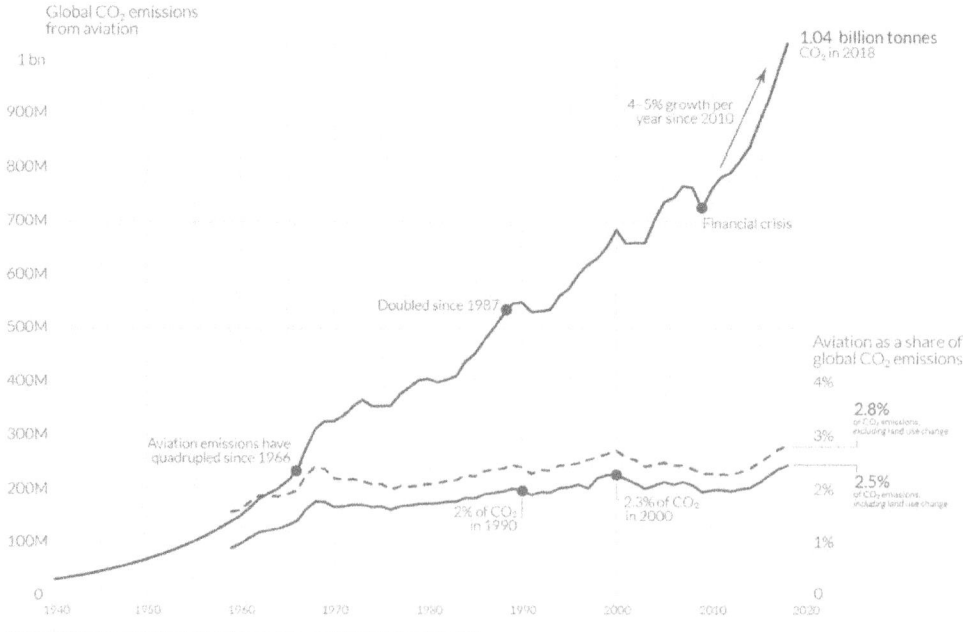

Figure 14.1 Global carbon dioxide emissions from aviation, 1940–2018

Source: Lee *et al.* (2020), in Ritchie (2020)

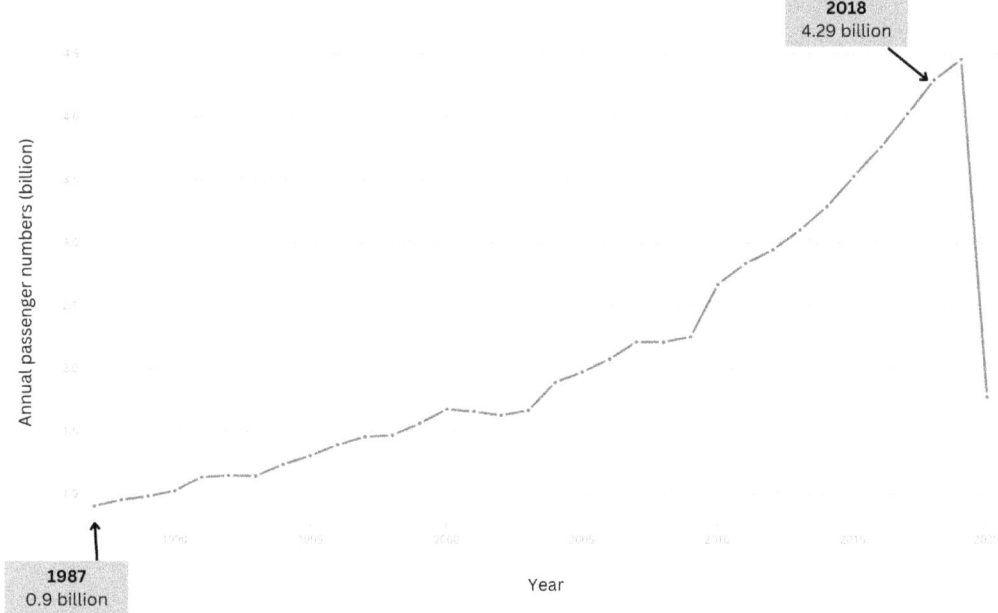

Figure 14.2 World air passenger traffic evolution, 1980–2020
Source: World Bank (2024)

and MRO Market Forecast, expects the worldwide commercial aviation fleet to further expand by 33% between 2023 and 2033, from around 27,400 to over 36,000 aircraft. To attempt to mitigate the overall growing emissions moving forward, it is critical to understand the key sources of air transport emissions.

According to Padhra and Kurnaz (2023), CO_2 emissions can be broadly categorised as:

- **Direct**
  - combustion of jet fuel in aircraft engines
  - production and assembly of aircraft
  - airport operations

- Indirect
  - transport of passengers to and from the airport
  - construction of infrastructure
  - broader supply chain serving the air transport industry

14.2.1 *Direct CO_2 emissions – combustion of jet fuel in aircraft engines*

The main source of direct CO_2 emissions is via aircraft powerplants and the auxiliary power unit (APU), which provides energy for non-propulsive activities, such as aircraft power when on an airport stand. There is a linear correlation between engine thrust and fuel consumption, and between fuel consumption and carbon dioxide emissions (Padhra and Kurnaz, 2023). Jet fuel consumption produces CO_2 at a defined ratio of 3.16kg per 1kg of fuel consumed (EESI, 2022).

384 *Fundamentals of Global Air Transport Geography*

The heavier the aircraft, the more thrust required, the more fuel consumption and more emissions. Therefore, an aircraft flying full will produce higher emissions than the same one with a low passenger or freight load. However, it is preferable to have one full flight than two half-full flights, as per-passenger emissions will be lower on the full flight, therefore higher load factors are not just economically beneficial, but also environmentally.

The highest thrust phase is the take-off cycle, which may utilise between 75–100% of thrust, depending on the aircraft performance and operating parameters such as the runway length, obstacle clearance and temperatures. Airbus A319, A320 and A321 aircraft, for example, have a minimum of 75% maximum rated thrust for take-off. Fuel consumption can vary significantly between aircraft types and Koudis *et al.* (2017) analysed aircraft during their take-off roll at 100% thrust settings and stated that an A321 with V2533-A5 engines consumed 38.8kg of fuel whereas a larger Boeing 747-400 with RB211–524G engines consumed 98.1kg.

Size of aircraft and length of flights have a large impact on CO_2 emissions. Within the EASA area in 2019, only 5.5% of departures were for >4,000km flight distance, but accounted for a 46.1% share of CO_2 and 6.6% of departures were by twin-aisle aircraft, but accounted for 48.1% of CO_2 emissions (EASA, 2022).

ATG Case study 14.1

Fuel burn and % phase of flight

OAG Labs has created a model and database of fuel burn and carbon emissions for flights. They researched a range of flights from London Heathrow (LHR) on 28 January, 2022, to a range of destinations and using a range of aircraft types. The results showed that the longer the flight, the higher the % fuel burn is consumed in the cruise portion. Thus, the cruise accounts for 96.3% of fuel from LHR–Hong Kong and only 62.3% from LHR–Paris CDG (Table 14.1). The higher % of fuel used on the ground in taxiing for shorter flights, such as LHR–Edinburgh and LHR–Paris CDG, demonstrates the importance of taxiing efficiency and reducing airport congestion, and this will be analysed later in the chapter. The findings from Rowland (2022) also demonstrated that larger aircraft (which are heavier) generally burn more fuel, however more modern aircraft such as the A350 burns less than the older B777 and the newer single-aisle A320neo burns less than the classic A320.

Table 14.1 Fuel burn percentage by phase of flight for various routes and aircraft type

Route	A/c type	Taxi-out	Take-off	Climb	Cruise	Approach	Taxi-in
LHR–Dubai	A380	1.8	0.6	1.5	94.5	1.0	0.6
LHR–Hong Kong	A350	1.1	0.4	1.1	96.3	0.7	0.4
LHR–New York JFK	B777	1.7	0.6	1.4	94.6	0.8	0.9
LHR–Madrid	A320	6.0	1.9	4.8	82.5	3.1	1.8
LHR–Lisbon	A320neo	3.6	1.7	4.4	86.4	2.7	1.2
LHR–Edinburgh	A319	11.4	3.5	8.9	68.1	5.8	2.3
LHR–Paris CDG	A319	11.2	4.0	10.3	62.3	6.7	5.5
Average		5.3	1.8	4.7	83.6	3.0	1.6

Source: adapted from Rowland (2022) in Padhra and Kurnaz (2023)

Ultimately, the choice of aircraft, age, engine type, aerodynamic performance and aircraft weight (both structural weights and optional weights such as in-flight entertainment [IFE] and seating types) are all factors which contribute to the level of fuel burn and these will be considered later in the chapter when analysing mitigation strategies.

14.2.2 Direct CO_2 emissions – production and assembly of aircraft

As per the United Nations (2023), one of the main causes of climate change is the manufacturing of goods. According to Boeing (2023), in 2022 there were 24,500 aircraft active in the commercial airline sector. Between 2023 and 2042 they forecast a demand for 42,595 new aircraft to be produced, as a mixture of replacements for older aircraft and growth. Whilst many of these new aircraft may be more fuel efficient than the aircraft they will replace, there will still be emissions created in the manufacturing process, including from supply chain deliveries shipped from different parts of the world, for final assembly.

The life cycle of aircraft emissions should be considered alongside methods such as in-flight fuel burn. From the beginning of the manufacturing process through to the scrapping and/or recycling of aircraft at end of life, there is the potential for harmful emissions at each stage.

Cox, Jemiolo and Mutel (2018) assessed the life cycle of air transport in the context of the commercial air transport fleet in Switzerland, using 72 common aircraft types and flight distances. They summarised the lifecycle inventory into four environmental flows and for a more complete picture of CO_2 emissions, each of these aspects and stages should be considered:

1. **Airport** construction, operation, maintenance and end-of-life
2. **Aircraft** construction, maintenance and end-of-life
3. **Fuel production**, refining and delivery
4. **Aircraft emissions** during operation

Kito (2021) analysed air transport lifecycle CO_2 emissions in Japan between 1965 and 2019 and discovered that for a natural aircraft replacement scenario, cumulative lifecycle CO_2 emissions at the aircraft manufacturing phase for single-aisle aircraft were 7.7Mt and the flying phase were 291.1Mt and 20.8Mt at the manufacturing stage and 719.0Mt at the flying stage for twin-aisle aircraft. The reduction potential of lifecycle CO_2 emissions by introducing new passenger aircraft was 28.2Mt between 1965 and 2019, but the associated cost was US$94 billion and therefore not cost-effective as a CO_2 reduction policy.

14.2.3 Direct CO_2 emissions – airport operations

As shown in Cox, Jemiolo and Mutel (2018), airports are an important source of CO_2 emissions. The main sources of emissions in airport operations are in the turnaround process and during the aircraft taxiing phase, where engine thrust is approximately 7% of maximum (Padhra and Kurnaz, 2023). Anywhere between 2 and 17% of fuel burn can be in taxi-in and taxi-out activities, and proportionately more on the shortest flights (Rowland, 2022). The longer the distance from stand to runway, the higher the fuel burn and the more ramp congestion, the longer the aircraft will be in taxi mode, the higher the fuel burn. As a result, ground efficiency at airports is vital in reducing fuel burn and carbon emissions.

During turnarounds, the use of the auxiliary power unit (APU) can increase emissions versus using the airport ground power unit (GPU). Padhra (2018) surveyed 25,196 turnarounds and concluded that the average turnaround time for a short-haul European airline was 42 minutes, during which the APU fuel burn was on average between 40–67kg, with corresponding CO_2 emissions between 126–212kg per turnaround. The provision of GPUs reduced average emissions by 47.6%.

There are many other sources of airport-related emissions, including airside ground vehicles, supplier delivery vehicles and surface access for passengers and staff. There are many third-party suppliers present within and around airports, such as fuelling, catering, ground handling, maintenance and retail companies – some of which the airport itself may have limited control over regarding emissions.

ATG Case study 14.2

Hong Kong International Airport carbon footprint

As shown in Figure 14.3, 70% of HKIA's carbon footprint is from aircraft-related emissions, including take-off and landing, with 30% being a result of ground emissions. The airport has limited influence over aircraft-related emissions, other than to report them using the Airports Council International (ACI) Airport Carbon Accreditation (ACA) methodology. APU usage has been banned at frontal stands since 2014.

Of the 30% caused by ground emissions, 85% can be significantly influenced by the airport authority, such as electricity and a large proportion of business partner operations – such as ramp handling, airline catering, fuel suppliers – who are signatories to the HKIA net-zero carbon pledge.

Figure 14.3 HKIA Airport carbon footprint breakdown, 2021

Source: adapted from Airport Authority Hong Kong (2022)

14.3 Air transport non-CO_2 emissions

Global air transport operations contribute to anthropogenic climate change via a complex set of processes that lead to a net surface warming, with key emissions including both **CO_2** and **non-CO_2 sources**, including nitrogen oxides (NO_x), water vapour, soot and sulfate aerosols, and increased cloudiness due to contrail formation (Lee *et al.*, 2020). Non-CO_2 GHGs trap more heat within the atmosphere than CO_2 (UNFCCC, 2015)

14.3.1 Radiative forcing

The difference between incoming and outgoing radiation is known as a planet's *radiative forcing* (RF). As with applying a pushing force to a physical object causing it to become unbalanced and move, a climate forcing factor will change the climate system (NOAA, n.d.). A positive RF will result in the planet warming. Including tropospheric and land surface adjustments, this is often called the *effective radiative forcing* (ERF). This quantifies the energy gained or lost by the earth system following an imposed perturbation (such as in GHGs or solar irradiance), therefore is a fundamental driver in the earth's top of atmosphere (TOA) budget.

The change in ERF from 1750 to 2019 by contributing forcing agents is illustrated in Figure 14.4. The total anthropogenic ERF over the industrial era (1750–2019) was 2.72 W m^{-2}. This shows the warming effect of carbon dioxide, methane, nitrous oxide, halogens, ozone, stratospheric water vapour and contrails. The figure has increased from 2.29 W m^{-2} in 2011.

According to Lee *et al.* (2020), global aviation contributes a few percent to anthropogenic RF and non-CO_2 impacts comprise about two-thirds of the net RF. Based on global warming potential – which is the cumulative RD, both direct and indirect effects, over a specified time horizon resulting from the emission of a unit mass of gas related to some reference gas – the indication is that aviation emissions are currently warming the climate at approximately three times the rate of that associated with aviation CO_2 emissions alone.

In terms of the aviation contribution to climate change, and including all factors such as carbon dioxide, nitrogen oxide and effect of contrails, this was calculated to be 3.5% of all human activities that drives climate change (Lee *et al.*, 2020). It is therefore important that in future international agreements, non-CO_2 impacts are given the focus they require.

14.3.2 Aircraft condensation trails (contrails)

The white streaks seen coming off jet airplanes at altitude are called contrails (Figure 14.5). Contrails are clouds that form when water vapour condenses and freezes around small soot

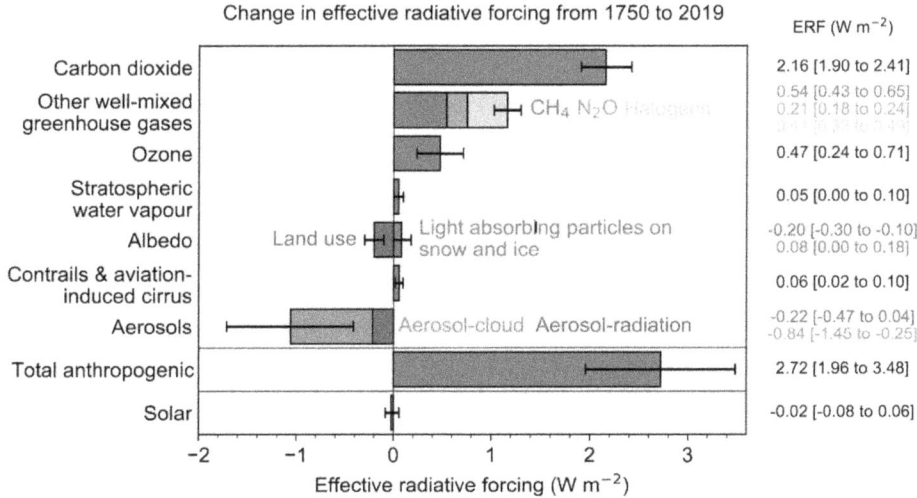

Figure 14.4 Change in effective radiative forcing from 1750 to 2019

Source: Intergovernmental Panel on Climate Change (2023)

388 *Fundamentals of Global Air Transport Geography*

a) Contrails cross the sky as an aircraft is on approach to land at London Heathrow airport

b) Contrails pierce the late evening sky near Ashbourne, Derbyshire, UK

Figure 14.5 Aircraft contrails
Source: author

particles or aerosols in aircraft exhaust. The water vapour comes from the air around the aircraft and the aircraft exhaust fumes. These can then become human-made cirrus clouds (UCAR, 2023), which can trap infrared rays, producing a warming effect up to three times the impact of CO_2 (EESI, 2022). These contrails normally have a relatively short lifespan, usually a matter of hours or even minutes, but their collective influence from thousands of flights can have a serious warming effect.

Although contrails, like clouds, reflect incoming solar energy, they also absorb outgoing surface radiated energy, creating a net warming effect. Some studies suggest that the global warming impact of contrails could be as much as a third to half the total impact from all carbon emissions ever emitted by aircraft (Padhra and Kurnaz, 2023).

The net RF is strongly influenced by the physical characteristics of contrails, which depend on the following:

- Amount of water vapour and particles present in the atmosphere occurring naturally or added by the jet engine exhaust.
- Temperature, humidity and pressure of the ambient atmosphere. The temperature should be below around −40°C.
- Wind and turbulence characteristics of the atmosphere.
- Wingspan and size of the aircraft.
- Propulsive efficiency of the jet engines.
- Type and blend of fuel used to power jet engines.

(Padhra and Kurnaz, 2023)

These will determine the width, depth, evolution and lifespan of the contrails.

In terms of mitigation, a possible approach is to focus on flying in areas of the atmosphere where contrails are less likely, based on horizontal and vertical flight planning. Weather conditions are dynamic and this will mean variations in areas of contrail formation. Research by Avila and Sherry (2016) on ice supersaturated regions in US airspace indicated that contrails were most likely to form between FL320 and FL350 in the months of June, July, August and September. Avila, Sherry and Thompson (2019) stated that on average, 15% of flights generate contrails in the USA and that for most flights which cruise between FL340 and FL360, these could avoid generating contrails with an increase of 2,000- or 4,000-feet increment in cruise flight levels (FLs).

ATG Did you know?

Contrails

Did you know that aircraft contrails can be visible from space?
The enhanced infrared image in Figure 14.6 was taken from the Moderate Resolution Imaging Spectroradiometer (MODIS) aboard NASA's Terra satellite and shows widespread aircraft contrails over the southeastern USA, on 29 January, 2004. The use of satellite data is critical for studying the effects of contrails.

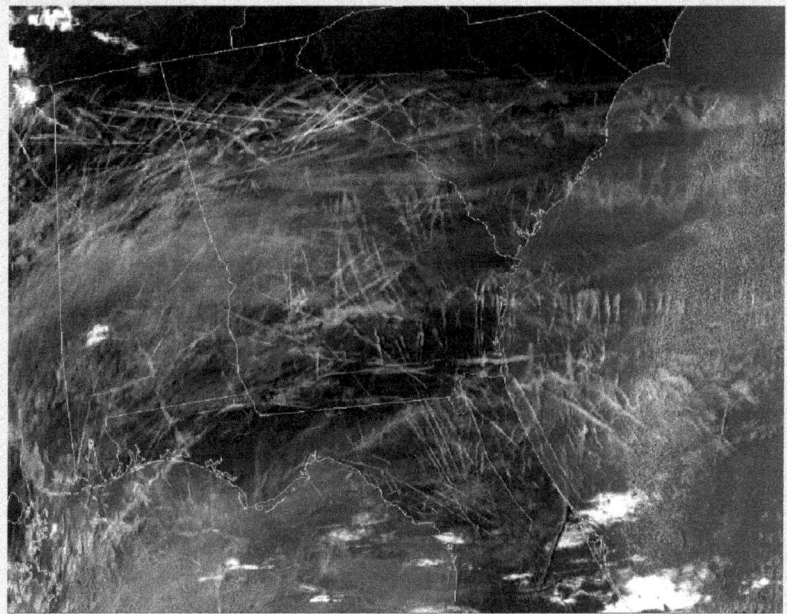

Figure 14.6 Aircraft contrails from space
Source: NASA Earth Observatory (2004)

14.4 Mitigation of air transport emissions

According to Bisignani (2009), aviation had improved its fuel efficiency by 70% over the preceding 40 years. The IPCC (1999) stated that 40% of improvements were from engine efficiency and 30% from airframe efficiency improvements versus the early jet aircraft. Zheng and Rutherford (2020) stated that the average block fuel intensity of new aircraft decreased by 41% from 1970 and 2019, with a compound annual reduction rate of 1.0%. By 2023 (IATA, 2023b) states that the overall fuel efficiency of the commercial aircraft fleet was around 80% better than 50 years prior.

There are four main areas where the air transport industry has been successful in reducing its emissions:

1. Improvements in aircraft and engine technology
2. Service weight reductions
3. Operational improvements
4. Greener on-the-ground infrastructure

14.4.1 Improvements in aircraft and engine technology

Improvements are centred around burning less fuel, so mechanical efficiency, reducing weight and aerodynamic improvements all have a role to play.

Improvements in engine technology

The 40% improvement in fuel efficiency, as a result of more efficient engines, was the highest single contributor to improvements (IPCC, 1999). According to Padhra and Kurnaz (2023), the engines on the commercial jet aircraft produced in 1958 – the Boeing 707 (Pratt and Whitney JT3D) – consumed 53kg of fuel per hour per kilonewton of thrust at maximum thrust setting, however the General Electric GE90, powering the Boeing 777 in 1994, consumed only 33kg per hour per kilonewton.

One of the main improvements has been in the engine *bypass ratio* (BPR).

ATG Did you know?

Turbofan engine bypass ratio

Turbojets are the original jet-powered aircraft and are generally found in military fighter jet aircraft. A *turbofan* engine is a development of the turbojet engine and has many advantages over the turbojet. In a turbofan (bypass) engine (Figure 14.7), the BPR is a comparison between the mass flow rate of air drawn into the engine through the fan disk that goes around the engine core with the mass flow rate of the air that goes through the engine core (SKYbrary, n.d.).

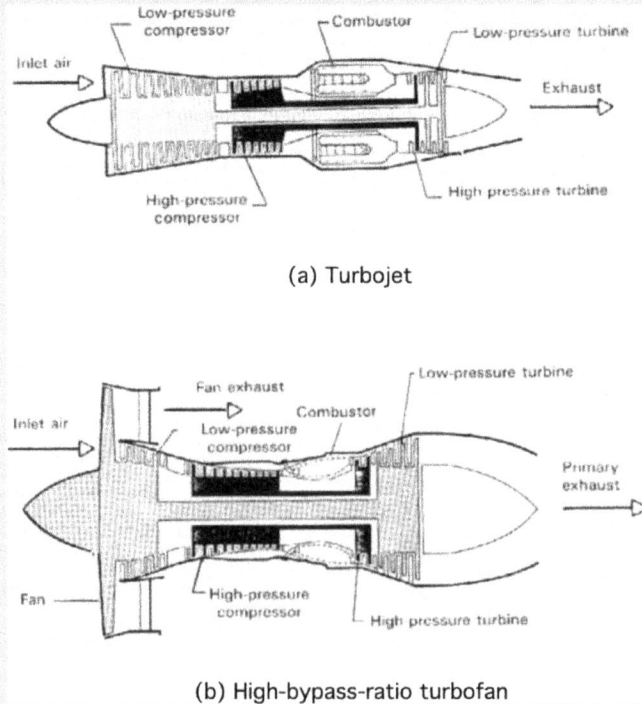

Figure 14.7 High bypass-ratio turbofan engine
Source: NASA (n.d.)

The impact and mitigation of air transport on climate change 391

Engines with higher BPRs tend to be more efficient and have lower fuel consumption (hence emissions). They can produce a greater amount of thrust while consuming the same amount of fuel as an engine with a lower BPR.

In an engine with a BPR of 12:1, for every 13 units of air drawn into the engine, 12 will by-pass the core and 1 will go through it. One of the first production turbofan engines was the Rolls-Royce Conway, which entered service on the Boeing 707-420 in the early 1960s and had a BPR of 0.3:1. The Pratt and Whitney JT8D in the mid-1960s improved the BPR to around 0.96 on the B727 and to around 1.74:1 on the McDonnell Douglas MD-80 series in the 1970s.

Further improvements continued with the CFM56 engine in the 1970s and early 1980s (with a BPR of 5.0–6.8:1) powering aircraft such as the DC-8 and B737-300. Two of the most efficient engines now are the CFM LEAP-1A, which has a BPR around 11.3:1, and Pratt & Whitney PW1100G, with its Geared Turbofan (GTF) architecture and a BPR of 11.4–12.7:1. Both engines power the Airbus A320neo (Kuropatwa, Wegrzyn and Kozuba, 2022).

Improvements in airframe technology

Advances in aerospace materials have led to lighter and stronger airframe structures, with more use of composite materials instead of heavier aluminium. Modern aircraft such as the Boeing 787 and Airbus A350 have a significant proportion of their structure made from composite materials, meaning a reduction of thousands of kilograms in weight. This increases efficiency and reduces fuel consumption and emissions.

Composites, including resins and adhesives, accounted for 9% of the 1990s Boeing 777 structural weight (Boeing, 2014). Aluminium was the largest component at around 50%. With the introduction of the Boeing 787, 50% of the structural weight was from composite materials (Boeing, 2022). Most of the primary structure is made from composite materials, mainly the fuselage. Earlier aircraft from the late 1960s, 1970s into the early 1980s, such as the B747, A300

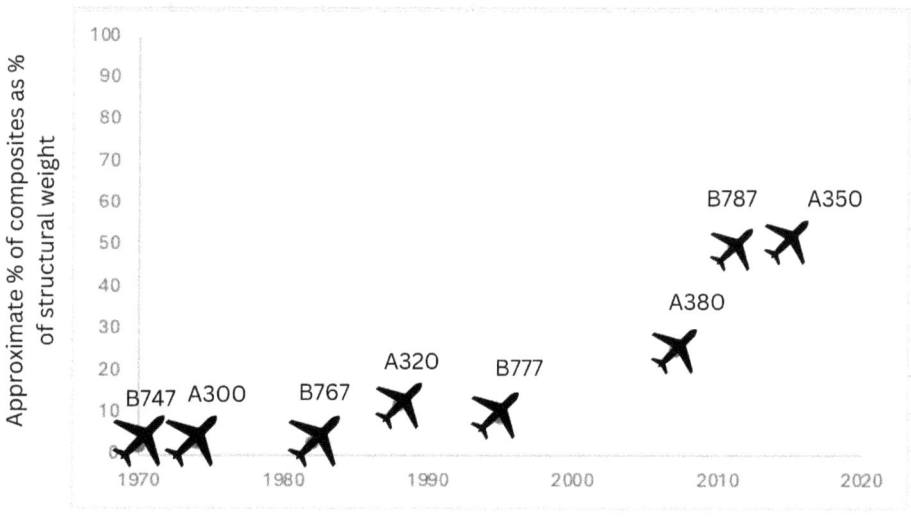

Figure 14.8 Composites as a % of aircraft structural weight
Source: author from aircraft manufacturer data

and B767, had a fraction of the composite materials of today's aircraft (Figure 14.8). Boeing and Airbus claim that this next generation in airframes can reduce fuel use and CO_2 by around 20–25%.

Aerodynamic improvements

Lift generation from the wings causes wingtip vortices, which reduces their ability to generate lift. More lift can be generated by flying at a higher angle of attack which will also induce more drag. Induced drag can be reduced by installing wingtip devices such as winglets and sharklets (Figure 14.9), reducing the powerful vortices from the wingtips, thus the aircraft needs less thrust to maintain the same speed. These make the aircraft heavier overall, hence the longer the flight, the more effective they can be.

Blended winglets lowered block fuel and CO_2 emissions by up to 4% on the Boeing 737 and up to 5% on the Boeing 757 and 767 (Boeing, 2009). JetBlue believed that retrofitting its in-service fleet of A320 aircraft would improve the aerodynamics of the aircraft and reduce emissions by up to 4%. The retrofit installation of split scimitar winglets will improve the efficiency of Ryanair Boeing 737–800 aircraft by up to 1.5% (Ryanair, 2023). Guerrero, Sanguineti and Wittkowski (2020) believed that winglets could achieve fuel burn reductions of 4–6%. Cirium (2022) calculated that the Boeing 737–800 benefits the most from winglets, averaging a 6.69% increase in efficiency but this varies between 4.6–10.5% depending on route.

Figure 14.9 Aircraft wingtip devices
Source: author from Canva.com

ATG Did you know?

Winglets

Winglets were described by an early aerodynamic pioneer – Frederick W. Lanchester – in 1897. He noticed how birds such as eagles flew with their wingtips tilted up and presented a

> paper, "The soaring of birds and the possibilities of mechanical flight". This was six years before the Wright Brothers first powered flight! According to Anderson (1999, in Jarrett, 2014, p.55), "the coming of age of theoretical aerodynamics began with Frederick Lanchester".

In addition, there are other areas which can result in weight savings: keeping the aircraft clean (and thus more aerodynamic), reducing the weight of exterior paint (and perhaps less elaborate designs!) and fuel management awareness training for staff.

14.4.2 Service (operating) weight reductions

If heavier aircraft = more fuel = more emissions, then if the structural weight of an aircraft can be reduced, then the interior operating weights should also be considered.

De Moor (2020) analysed the "cost of weight" on an aircraft and stated that extra weight on aircraft will cost extra fuel. The cost of weight was calculated at 3.02%, therefore for every 100kg of weight reduction, an aircraft consumes approximately 3kg less fuel per hour. As an example, an extra 1kg of weight on a single A320 was calculated as costing $108 per year in extra fuel. There are many service areas which can be considered for weight reductions, however there may be what could be termed a "passenger experience trade-off". In other words, a consideration could be given to the impact any changes could have on the passenger's flight experience and a business decision taken versus saving cost in fuel and emissions.

ATG Case study 14.3

American Airlines, Dallas, USA: the legend of the olive

One of the earliest tests of removing in-flight products at the expense of the passenger experience came in the 1980s, when Robert Crandall, the Head of American Airlines, decided to remove one olive from every salad served to passengers, with a predicted saving of $100,000 a year. The actual savings may differ from this figure; however, the theory was that no one would notice one olive being removed and this principle can extend into many cost (and indeed fuel) saving strategies. Why waste money (and weight) on products if the passengers do not notice or care? This can be a fine line though!

In-flight meal equipment

Replacing metal cutlery with plastic (and now increasingly wood or bamboo, to reduce single use plastic), china plates and bowls with lighter materials and drinks in glass bottles with PET – for airlines with 400-seat aircraft and two meal services, this could result in hundreds of kilograms being saved per sector. Virgin Atlantic made its glassware thinner, redesigned its meal trays and altered its beverage offering for night flights, when fewer people drink. Qantas, on its long London–Perth route, introduced lighter crockery, glassware, cutlery and linen and weight was reduced by 11%

Potable (drinking) water

Airlines will have tanks containing drinking water for passenger drinks. The weight in these tanks can be substantial, especially on long-haul widebody flights. In 2008, Monarch Airlines researched water usage and after a one-month period, concluded that the water tanks, which were being filled to full, never exceeded 50% usage, therefore the amount of water carried could be reduced, saving tens of kilos per flight.

In-flight literature

In-flight magazines and newspapers can weigh a significant amount and many airlines are considering their value and alternatives. Some are now included electronically if the aircraft has an IFE system, for example BA removed printed copies of its High Life magazine in 2020. In 2018, United Airlines used lighter paper for its in-flight magazine, saving up to 170,000 gallons of fuel.

In-flight entertainment

On longer flights, in-flight entertainment (IFE) is often viewed as a necessity from a passenger experience perspective. The weight penalties for this equipment – seat-back screens, servers, cabling, etc. is substantial. According to Erdemir *et al.* (2017) this could account for approximately 4kg per seat. For an aircraft with 500 seats, that is 2,000kg of weight.

A number of carriers removed IFE screens from some of their cabins, such as Etihad, American Airlines and Qantas (Bailey, 2020). The possibilities of streaming onto personal devices means future requirements for seatback IFE may reduce, although bandwidth on aircraft is an issue. Thinner screens may help by still considering the passenger experience, whilst reducing weight.

Aircraft seats

Using lighter weight seats and thinner padding has contributed to weight and fuel savings for many airlines. In 2014, Southwest re-upholstered its seats with lighter material and each aircraft was made 272kg lighter. In 2023, Jazeera Airways stated that the new seats on its A320/A321 aircraft weighed only 6.8kg, using composite and titanium materials, with a saving of 1.2 metric tonnes per aircraft in weight (Dron, 2023).

In-flight trolleys and product

Qantas reduced the weight of trolley carts for international flights from 25 to 18kg. In 2014, Thomas Cook invested in lightweight trolleys, with 1% fuel cost saving. Optimising meals, drinks and in-flight sales items can also result in a reduction in weight, by considering packaging and the number of drinks and retail products sold, to minimise unnecessary weight.

In 2017, United Airlines removed on-board sales of duty-free items such as perfumes, cutting 1.4 million gallons of fuel a year. Airlines considering this may have to also consider possible revenue losses, with airline profit margins being historically wafer thin.

ATG Case study 14.4

All Nippon Airlines (ANA), Japan: reducing CO_2 emissions by reducing the weight of aeroplanes

ANA is focusing on several areas to attempt to achieve net-zero CO_2 emissions:

- Digitalisation of in-flight magazines and newspapers.
- Optimisation of the amount of goods and drinking water to be loaded on board.
- Digitalisation of manuals used by flight crew and cabin attendants.
- Weight reduction of carts used by cabin attendants for in-flight services.

ANA introduced new lightweight service carts between 2015 and 2022, saving up to 10kg per cart, with a weight reduction of around 580kg per Boeing 777-300ER aircraft. Across the fleet, fuel consumption was reduced by approximately 5,700 tonnes annually, contributing to CO_2 emissions equivalent to about 17,500 swimming pools (ANA, 2021).

14.4.3 Operational improvements

For many years, airlines have been seeking ways to reduce their cost base. As fuel is either the number one or number two cost item for airlines every year, reducing fuel burn will reduce costs and emissions and a number of operational improvements have been undertaken to this end.

Continual descent/climb operation

ICAO (2022b) defines these as:
- **Continual climb operation (CCO)**: An operation, enabled by airspace design, procedure design and air traffic control (ATC), in which a departing aircraft climbs without interruption, to the greatest possible extent, by employing optimum climb engine thrust, at climb speeds until reaching the cruise FL.
- **Continual descent operation (CDO)**: An operation, enabled by airspace design, procedure design and ATC, in which an arriving aircraft descends continuously, to the greatest possible extent, by employing minimum engine thrust, ideally in a low drag configuration, prior to the final approach fix/final approach point (Figure 14.10).

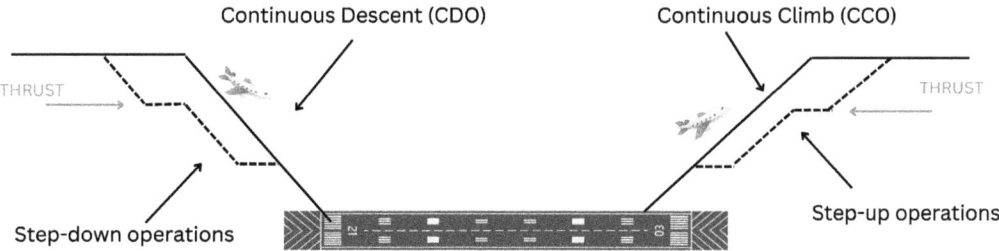

Figure 14.10 CDO and CCO

Source: author

These are designed to deliver environmental and economic benefits – reduced fuel burn, gaseous emissions, noise and fuel costs – without adversely affecting safety. Employing these techniques reduces intermediate level-offs (step-down and step-up) and reduces the need for extra thrust inputs. Eurocontrol (2020) estimates that in Europe, potential savings from optimising CCO and CDO are up to 340,000 tonnes of fuel per year and 1.1 million tonnes of CO_2. Eurocontrol noted, however, that achieving 100% CCO and CDO across the European network may not be possible due to factors such as aircraft separation, weather and ATC workload.

Single-engine taxiing

This procedure involves using only one power unit for twin-engine commercial air transport aircraft, whilst taxiing on the ground at airports. Single-engine taxiing (SET) can reduce fuel burn, CO_2 and NO_x at airports. Stettler *et al.* (2018), using London Heathrow as a case study, believe that without using SET during taxi-in, fuel consumption and pollutant emissions would increase by up to 50% and Heathrow estimated that taxi-related fuel consumption could be reduced by up to 40% using SET. Guo, Zhang and Wang (2014) calculated reductions of up to 50% in fuel consumption due to SET at US airports.

The use of SET is not always possible or advised due to factors such as the additional responsibilities and workload to pilots, on uphill slopes, slippery surfaces or when de-icing operations are required and operators also need to consider the specific airport conditions. Time is a key factor also – time needed for other engines to be started for take-off – especially important at congested airports.

Air traffic management

Air traffic management (ATM) offers some of the best opportunities to improve flight efficiency. Aircraft fly between waypoints in the sky but depending on atmospheric and traffic conditions on the day, air traffic controllers (ATCOs) may be able to facilitate more efficient flight routes. EASA (n.d.) believes that progress achieved in airspace design and deployment of interoperable technologies has reduced CO_2 emissions.

The UK CAA (2023) states that if planes can fly more direct routes and are not held in stacks or on taxi-ways, their environmental footprint will decrease. When aircraft fly internationally, they fly into and through other countries' airspace, and working together can reduce emissions by reducing the need for unnecessary FL or directional changes, hence in Europe the Single European Sky (SES) project has been established.

There are many other areas where operational improvements have resulted in reduced emissions, such as better load planning management, aircraft handling management and flight planning.

14.4.4 Greener on-the-ground infrastructure

According to ICAO (2019), airport-related emissions are estimated to represent between 2–5% of global aviation emissions. Many of these emission sources are within the direct control of airports and others, via third party operators, can be less so. Emissions are from sources such as aircraft on the ground, airside vehicles, power to airport infrastructure and surface access to and from the airport itself. In 2021, Hong Kong International Airport (HKIA) stated that 70% of its carbon footprint was aircraft-related and 30% related to ground emissions.

Improvements have been achieved in a wide variety of airports, in terms of greener energy usage, electric airside vehicles and policies to encourage public transport usage.

Airport Carbon Accreditation scheme

In 2009, Airports Council International (ACI), which represents the collective interests of airports around the world and works with international organisations such as ICAO and national governments, established Airport Carbon Accreditation (ACA) and is the institutionally endorsed, global carbon management certification programme for airports. It independently assesses and recognises the efforts of airports to manage and reduce their carbon emissions through six levels of certification:

1. Mapping
2. Reduction
3. Optimisation
4. Neutrality
5. Transformation
6. Transition

The scheme recognises that airports are at different stages in their carbon management journeys and utilises internationally recognised methodologies, with a common framework for active carbon management with measurable outcomes. From July 2017 to June 2018, airports achieved a reduction of 347,026 tonnes of CO_2, which was equal to the CO_2 sequestered by 8,158,834 seedlings grown for ten years! (Airport Carbon Accreditation, 2023). In 2023, over 500 airports were certified under the accreditation process.

ATG Case study 14.5

Cochin International Airport, India – world's first solar-powered airport

In 2015, Cochin International Airport, in the Indian state of Kerala, opened a 12MWp solar power plant, becoming the world's first airport to operate completely on solar power, comprising 46,150 solar panels across 45 acres of land and achieving power neutrality.

14.5 The future

The air transport industry has been successful in reducing its carbon footprint in many areas, especially at the unit level, however if the industry is to achieve its goals and continue to grow, much more needs to be done. For example, the IEA (2023) has stated that although there have been major improvements in energy intensity, these have not been sufficient to counterbalance energy growth demand in recent years.

14.5.1 *IATA strategy towards net zero 2050*

At the 77th IATA Annual General Meeting in October 2021, a resolution was passed by IATA member airlines committing them to achieving net-zero carbon emissions from their operations by 2050. This pledge brings air transport in line with the objectives of the Paris Agreement to limit global warming to well below 2°C. To succeed, it will require coordinated efforts from the entire air transport industry as well as significant government support (IATA, 2021).

IATA considers that the mitigation will come in four major areas and related percentage contributions (Figure 14.11).

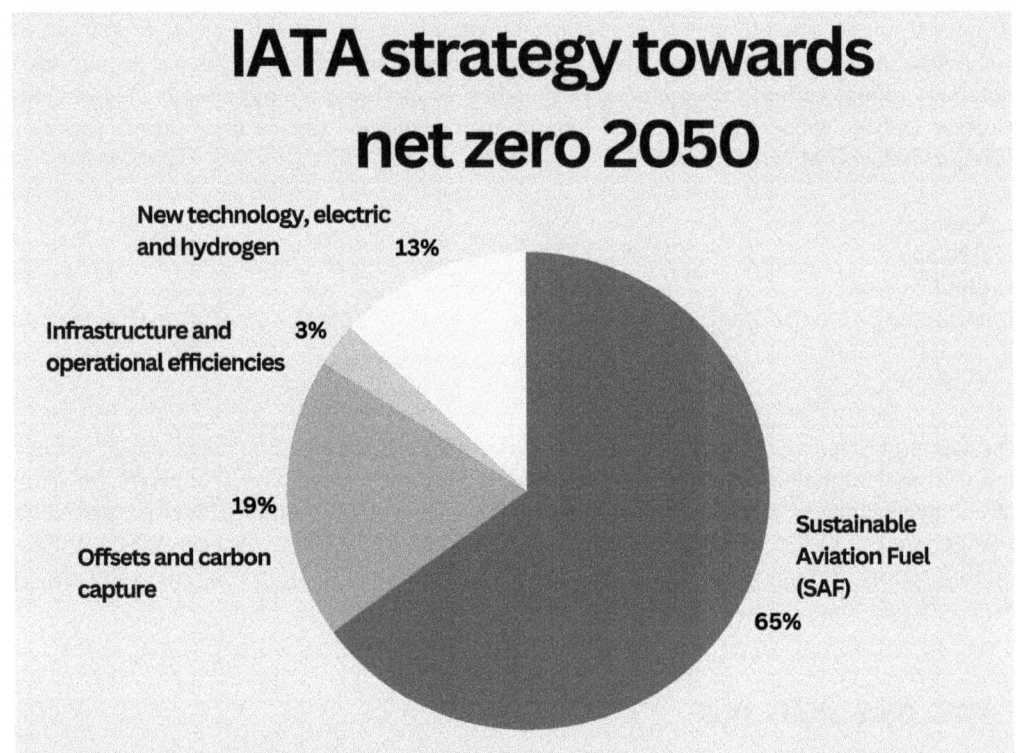

Figure 14.11 IATA strategy towards net zero 2050
Source: adapted from IATA (2021)

14.5.2 *ICAO long-term global aspirational goal*

At the 41st ICAO assembly in October 2022, ICAO member states adopted a collective long-term global aspirational goal (LTAG) of net-zero carbon emissions by 2050. The LTAG does not attribute specific obligations to individual states but recognises each state's specific circumstances and capabilities and each state will contribute within their own national time frame.

The achievements will rely on the combined efforts of multiple CO_2 reduction measures and acknowledges the difficulties in decarbonising the air transport industry.

ATAG (2021) in their Waypoint 2050 report identified different scenarios for how the air transport industry could reach net zero 2050 and in what proportions:

1. Sustainable aviation fuels (SAF) – between 53% and 71%
2. New technologies – between 12% and 34%
3. Operational and infrastructure efficiency – between 7% and 10%
4. Offsets or other carbon mitigation options – between 6% and 8%

Figure 14.12 Futuristic blended wing body aircraft
Source: author from Canva.com

14.5.3 Aircraft technologies

New aircraft technologies are focused on increased efficiency and reductions in carbon emissions over the long term. As per ICAO, these may only have an impact from the late 2030s. They may involve a range of aerodynamic improvements, propulsion, aircraft systems and structures and potential energies. One solution may be the blended wing body (BWB) aircraft, an example illustrated in Figure 14.12.

Two main technologies which are being researched today and which may contribute towards the net zero 2050 targets, are hydrogen and electric power.

Hydrogen

It is proposed that hydrogen (H_2) as an energy source will play a key role in the path towards net zero 2050 ICAO (2022c). According to Clean-Aviation.eu (2020), novel and disruptive aircraft, aero-engine and systems innovations, in combination with H_2 technologies, can help to reduce global warming effects by 50–90%. This could be H_2 combustion in turbine engines or by fuel cells, which generate electricity to ultimately provide thrust.

H_2 can be produced and consumed without creating CO_2 and is widely available in water, which makes it attractive as a solution, as does its low mass. When produced from clean electricity, H_2 has no CO_2 emissions over its life cycle and can be an excellent option as raw material to produce SAF or for direct use on board aircraft. Figure 14.13 illustrates the various roles – both liquid and gaseous – H_2 can play in reaching net zero.

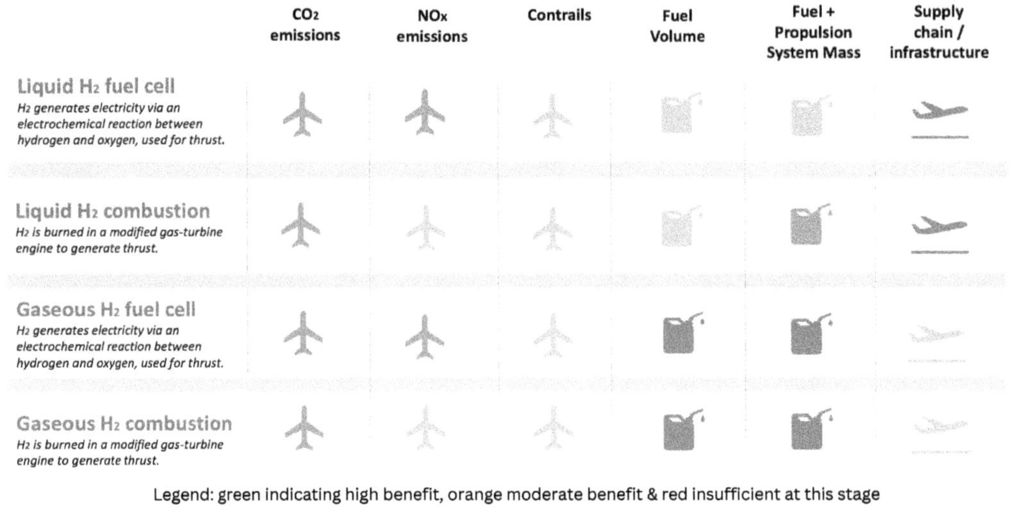

	CO₂ emissions	NOx emissions	Contrails	Fuel Volume	Fuel + Propulsion System Mass	Supply chain / infrastructure
Liquid H₂ fuel cell — H₂ generates electricity via an electrochemical reaction between hydrogen and oxygen, used for thrust.						
Liquid H₂ combustion — H₂ is burned in a modified gas-turbine engine to generate thrust.						
Gaseous H₂ fuel cell — H₂ generates electricity via an electrochemical reaction between hydrogen and oxygen, used for thrust.						
Gaseous H₂ combustion — H₂ is burned in a modified gas-turbine engine to generate thrust.						

Legend: green indicating high benefit, orange moderate benefit & red insufficient at this stage

Figure 14.13 The role of hydrogen in reaching net zero in aviation
Source: EASA (2023)

According to research by Clean Sky 2 and Fuel Cells and Hydrogen 2 Joint Undertakings (2020), H_2 combustion could reduce climate impact in flight by 50–75% and fuel-cell propulsion by 75–90%. This could be achieved by:

- the overall aircraft efficiency with lighter fuel tanks and fuel cell systems
- liquid hydrogen distribution within the aircraft
- turbines capable of burning hydrogen with low NO_x emissions
- development of efficient refuelling technologies enabling flow rates comparable to kerosene

However, H_2 has lower energy density meaning more on-board storage is needed to fly and cover the same distance – this would reduce payload and revenue-generating opportunities for airlines. Existing fuel tanks would only allow for very short routes and storage issues of H_2 in gas or liquid forms would exist, due to the current location of fuel tanks within aircraft wings. In addition EASA (2023) believes that as a result, commuter, regional and shorter-range routes could be the first to be targeted, but in larger planes and for longer routes, significant changes in aircraft design due to the volume storage needs would be required. The potential future use of H_2 in aviation will require significant research, development and investment, as well as regulations to ensure safety and economic viability, and can only be realised through the coordinated and global efforts of all stakeholders (ICAO, 2022c).

Electric aircraft

In recent years, electric technological developments have created all-electric transportation in areas such as the automotive and rail sectors, and hence much research is ongoing into the feasibility of using these technologies in the air transport sector. Within aircraft, instead of internal combustion engines, electric motors would drive conventional propellers or sets of multiple small fans. Electric energy is stored in batteries or potentially in fuel cells and CO_2 emissions during operations would be zero. Non-CO_2 effects such as NO_x emissions and contrails could also be eradicated.

According to IATA (2023b), small electric test aircraft up to nine seats are already flying, aircraft up to 19 seats are planned for the later 2020s and regional aircraft in the 2030s.

The size and weight of batteries is a large impediment to usage in all but small aircraft and for short distances. Battery technology is still not advanced enough to cater for the high-energy requirements of long-range and high payload flying. For example, small propeller aircraft, such as a Cessna 172 (four seats), are a similar size to cars, and it is this size which is witnessing prototypes being developed. Large propeller aircraft, such as the Bombardier Q400, which can seat 80–90 passengers, have 30-times greater power requirements at take-off (Padhra and Kurnaz, 2023), which current batteries cannot provide in the space available on aircraft.

Lifecycle emissions would also depend on the primary energy mix for electricity generation, requiring renewable sources, which considering the size of the global fleet, would be a huge challenge.

ATG Case study 14.6

Airbus ZEROe aircraft concepts

Airbus is planning to bring to market the world's first H_2-powered commercial aircraft by 2035. This will involve exploring both H_2 combustion and H_2 fuel cell technologies. There are four ZEROe concepts:

1. **Turbofan.** Two hybrid-hydrogen engines to provide thrust.
2. **Turboprop.** Two hybrid-hydrogen turboprop engines, which drive eight-bladed propellors, to provide thrust.
3. **Blended wing body (BWB).** Two hybrid-hydrogen turbofan engines providing thrust but in a new aircraft design concept, assisting in hydrogen storage and distribution.
4. **Fully electrical concept.** Based on a fully electrical propulsion system powered by fuel cells, having a much shorter range than the turbofan or BWB designs.

These are exploratory conceptual aircraft but provide examples of research into H_2-powered commercial aircraft. There is an acknowledgement that deployment of H_2 infrastructure at airports is a prerequisite to support the widescale scale-up and adoption of H_2 aircraft.

Hybrid-electric

Higher potential exists in the short–medium term with more hybrid solutions. It may be a necessary intermediate step towards full-electric propulsion for larger aircraft. For example, the Boeing 787 houses six electrical generators on board, compared to three in older aircraft. These are used to provide power for aircraft systems and lighting and also recharge the two sets of lithium-ion batteries on board the aircraft (Padhra and Kurnaz, 2023).

Hybrid technology involves a combination of conventional fuel and electrical power. The combustion and electric propulsion systems could potentially be used in combination during take-off to provide maximum thrust and the combustion engine could be throttled back when the aircraft is in its cruise phase or descending. This could reduce fuel consumption by up to 5% (Airbus, 2021).

IATA (2023b) believes that hybrid-electric aircraft on a new airframe body, such as a blended wing, could contribute to achieving CO_2 reductions of up to 40%.

14.5.4 Sustainable aviation fuel

Aviation has historically been powered by petroleum-based fossil fuels. Sustainable aviation fuel (SAF) is a term for non-petroleum-based fuels which, according to the UK CAA, can reduce lifecycle CO_2 emissions by over 70% compared to conventional fossil jet fuel.

> **ATG Did you know?**
>
> **Sustainable aviation fuel**
>
> SAFs are commonly referred to as biofuels and these are called "drop-in" fuels, since very little, if any, modifications are required to existing aircraft and engines. It can be produced from a number of sources, including waste oil and fats, green and municipal waste, and non-food crops. It can also be produced synthetically via a process that captures carbon directly from the air.

SAF is "sustainable" because the raw feedstock may not compete with food crops or water supplies. Whereas fossil fuels add to the overall level of CO_2 by emitting carbon that had been previously locked away, SAF recycles the CO_2 which has been absorbed by the biomass used in the feedstock during its life (IATA, 2021).

ICAO has categorised SAF feedstocks into the following categories:

- **Primary and co-products** are the main products of a production process. These products have significant economic value and elastic supply (i.e., there is evidence that there is a causal link between feedstock prices and the quantity of feedstock being produced). Examples include *jatropha and palm oil*.
- **By-products** are secondary products with inelastic supply and economic value. Examples include *palm fatty acid distillate*.
- **Wastes** are materials with inelastic supply and no economic value. A waste is any substance or object which the holder discards or intends or is required to discard. Examples include *municipal solid wastes* and *used cooking oil*.
- **Residues** are secondary materials with inelastic supply and little economic value. Examples include *agricultural residues such as manure and straw and forestry residues such as bark and branches*.

Both IATA (2021) – at 65% – and ATAG (2021) – at 53–71% – consider that SAF has the highest potential to contribute towards net zero 2050. SAF production tripled from 100 million litres in 2021 to 300 million litres in 2022, however IATA predicts an expected SAF requirement of 449 billion litres by 2050, to achieve net zero. The IEA predicts that biofuels will reach around 10% of fuel demand by 2030.

Historic regulations have meant that biofuels must be blended with traditional kerosene-based fuels up to a maximum of 50%. There are also a few other challenges in the use of SAF:

- Limited infrastructure for production and distribution of biofuels.
- Costs much higher than jet fuel.

The impact and mitigation of air transport on climate change 403

- Limited airport infrastructure for aircraft supply.
- Large quantity of biomass required for smaller quantity of biofuel.
- Pressures on land use and possible competition with food crops, depending on the type of biofuel production – the "food versus fuel" debate.
- How sustainable is the lifecycle process?

In future, governments are expected to be crucial to develop policies which efficiently accelerate the commercial production and deployment of SAF. In November 2023, during the Third ICAO Conference on Aviation and Alternative Fuels (CAAF/3) held in Dubai, member states agreed to strive to achieve a collective global aspirational vision to reduce CO_2 emissions in international aviation by 5% by 2030, compared to zero cleaner energy use.

ATG Case study 14.7

Virgin Atlantic and SAF

In 2008, Virgin Atlantic operated the first flight by a commercial airline to be powered in part with biofuel, with one of its Boeing 747-400 carrying a 20% biofuel blend in one of its fuel tanks, between London and Amsterdam.

In November 2023, Virgin operated the world's first transatlantic flight using only SAF, from London Heathrow to New York JFK, using an SAF blend of 88% hydroprocessed esters and fatty acids (HEFA) and 12% synthetic aromatic kerosene (SAK), using a Boeing 787 powered with Rolls-Royce Trent 1000 engines. The flight saved the lifecycle equivalent of 95 tonnes of CO_2 or 64% of the emissions of a standard LHR–JFK flight (Virgin Atlantic, 2024).

14.5.5   *Operations and infrastructure*

ATAG (2021) and IATA (2021) believe that improvements in these areas could account for 7% and 10% respectively, towards net zero 2050 goals. As shown in sections 14.4.3 and 14.4.4, the industry has already made significant operational, air traffic management (ATM) and infrastructural improvements and it is hoped that these areas will continue to yield further reductions in emissions.

The ICAO Committee on Aviation Environmental Protection (CAEP)'s Working Group 2 – Airports and Operations (WG2) addresses environmental issues relating to airport, aircraft operations near airports and aircraft operations in general. The aim is to develop global best practices in operational and infrastructure areas, leading to improvements in environmental management policies. Operations and ATM offers the highest potential for reducing emissions in the short-term.

Initiatives in ATM include:

- The **Single European Sky (SES)** is an initiative to improve the way Europe's airspace is managed. It is currently fragmented and divided along national borders, which can lead to inefficiencies and airspace users flying further than they need to, increasing emissions. The **Single European Sky ATM Research (SESAR)** project is an ATC modernisation programme aimed at developing a new generation ATM system.

- **Next Generation Air Transportation System (NextGen)** is the US FAA's programme to modernise the US National Airspace System. Improving efficiencies should assist in reducing emissions.
- The use of **artificial intelligence (AI)** – or airspace intelligence – has the ability to reduce the environmental impact of air traffic (Eurocontrol, 2023)

Single-engine taxiing of aircraft on the ground could be taken a step further with the TaxiBot concept, whereby the TaxiBot ("pushback tug") remains attached to the aircraft and the pilots are in control during the taxiing process after pushback. The tug is disconnected at a pre-designated point on the taxiway, where the aircraft can commence engine start-up and get ready for take-off. This is being used at Delhi International Airport and has led to approximate savings of 532kg CO_2 per aircraft for the average taxiing time of 14 minutes (ICAO, 2022a).

The use of **solar power** has been utilised in a unique way at Mombasa International Airport from 2019, consisting of solar power generation facilities and mobile airport gate electric equipment. The facility provides pre-conditioned air and compatible electricity that runs on solar energy to service aircraft during ground operations, eliminating CO_2 emissions from aircraft parked at the gate. The system has reduced on average over 704 tonnes of CO_2 annually.

14.5.6 Market-based measures: carbon offsetting and reduction scheme for international aviation (CORSIA)

A further important method for improving the industry's carbon footprint involves market-based measures (MBM). These are what can be called "carrot and stick" economic measures, whereby airports and airlines can be penalised for their emissions and/or incentivised to reduce emissions. There are three types of MBM:

1. Taxes and levies
2. Carbon offsetting
3. Emissions trading

Taxes, which could be in many different areas, such as fuel, airport or navigation, are often seen as unfair and especially so if countries implement different structures (or none), which result in individual air transport organisations being taxed more on a like-for-like basis than others.

Carbon offsets are tradeable "rights" linked to activities that lower the amount of CO_2 in the atmosphere. By buying these certificates, a person or group can fund projects, instead of taking actions to lower their own emissions. In other words, "offset" their own CO_2 emissions with an equal amount elsewhere (MIT, 2022), such as reforestation and peat restoration.

Emissions trading schemes, often referred to as "cap and trade". First, a "cap" is set, which is a limit on the number of emissions and is determined by countries and organisations who set targets. Trading refers to trade in emissions capacity, thus those who exceed their limits buy allowances from those who have spare capacity. These allowances are limited, meaning they have financial value, with supply and demand the key factors in price. In 2005, the EU set up the world's first international emissions trading system (EU-ETS) and is now in its fourth phase (2021–2030); however, many non-EU countries operating in EU airspace, especially developing countries, have deemed this to be unfair.

In 2016, ICAO adopted CORSIA as a global MBM solution to address CO_2 emissions from international aviation. This became applicable in January 2019 and requirements are compiled within the SARPs of Annex 16 (Environmental Protection). This was the first time a single industry sector agreed to a global market-based measure in the field of climate change (IATA, 2023a). CORSIA offsets aviation emissions by facilitating a reduction in emissions from other sectors, via emissions allowances.

CORSIA is being implemented in phases, with voluntary participation from 2021–2023, the first phase in 2024–2026 and then requiring the participation of all member states from 2027–2035, exempting those from least developed countries (LDCs), small island developing states (SIDS), landlocked developing countries (LLDCs) and states with less than 0.5% of global international revenue tonne kilometre (RTK) in 2018 (IATA, 2023a). These would include many countries within Africa and the western part of Latin America.

From 1 January, 2019, any aeroplane operator has had to report its annual CO_2 emissions to the state it is attributed. From 1 January, 2021, the state has had to calculate annual offsetting requirements for each operator. Figures during the pilot phase related to 2019 emissions levels, which were to provide the baseline for carbon neutral growth from 2020 onwards, with the aim of stabilising aviation's net CO_2 emissions. It is anticipated that CORSIA will mitigate around 2.5 billion tonnes of CO_2 between 2021 and 2035 (ATAG, 2020).

ATG Case study 14.8

Galapagos Ecological Airport – one of the world's "greenest" airports

Galapagos Ecological Airport achieved carbon neutrality status as part of the ACI Airport Carbon Accreditation framework in 2018, becoming the first airport in the Latin America–Caribbean region to achieve this landmark. The airport runs on 100% renewable energy, with 35% coming from photovoltaic panels installed on the terminal walkways and 65% by windmills strategically located in the airport area.

It was also the first airport to be awarded the US Green Building Council LEED Gold certification for the airport's design and construction, optimising the use of energy, lighting and water consumption and the use of recycled and ecologically manufactured materials.

Some of its carbon offsetting projects involve distributing efficient cookstoves into poverty-stricken regions of Peru, conserving threatened tropical rainforest in the Peruvian Amazon and delivering solar water heaters to Indian communities and businesses in Bangalore, India.

ATG Case study 14.9

British Airways – "Perfect Flight"

In September 2021, British Airways, Heathrow Airport, NATS, bp, Airbus and Glasgow Airport collaborated in a project called the "Perfect Flight", on a flight from London Heathrow to Glasgow, using SAF, offsetting remaining emissions and using the most fuel-efficient journey on the fuel-efficient A320neo. The flight achieved a 62% CO_2 emissions reduction compared to a similar flight from 2010 (34% from more efficient aircraft and operations and 28% from SAF usage) with verified carbon offsets (British Airways, 2024).

Chapter review questions

14.1 Explain the types of air transport industry CO_2 emissions.
14.2 Explain the sources of non-CO_2 air transport emissions. Why are contrails potentially so harmful? What could be done to reduce their formation?
14.3 How have aircraft technology improvements helped the air transport industry to reduce its emissions? For any ONE aircraft, analyse these improvements and how it has reduced emissions versus a comparable aircraft from a previous era.
14.4 Which ways can airlines reduce service weight on their aircraft? How do you think these may influence the passenger experience? For any ONE airline, research and analyse what changes they have made in their aircraft to reduce its service weight.
14.5 What are the various ways in which operational improvements can reduce aircraft emissions? Research your local major airport – have any of these operational improvements been used?
14.6 Assess the potential for both hydrogen and electric-powered aircraft to be utilised in the air transport industry. What are the major issues to be overcome and to what extent do you believe that these will be able to significantly contribute towards long-term emissions reductions?
14.7 Describe THREE different sources of SAF. What are the potential issues in the global implementation of SAF on a wider scale?
14.8 Explain the principles of market-based measures in the context of the ICAO CORSIA scheme. Research your favourite airline – do they offset their emissions and if so, which projects do they support? Do they offer passengers the opportunity to voluntarily offset their emissions? Conduct a poll of your classmates or family – what percentage of people would be prepared to pay these fees? How important do you believe these schemes are in helping airlines to achieve net-zero goals?

ATG trivia

Plant a tree? Restore peat swamps? Install "friendly" stoves? The world of carbon offsetting

Whilst air transport organisations are undertaking offsetting programmes to reduce their net carbon emissions, there is also the opportunity at many airlines for passengers to voluntarily offset their own emissions, by paying to offset the emissions created by their own air travel journey. Projects include:

- **Cathay Pacific** – offering access to clean cooking in Bangladesh via the Bondhu Chula (friendly stove), reducing fuel consumption by approximately 50%.
- **American Airlines** – protecting and restoring peat in Indonesia and providing clean cookstoves in Honduras.
- **Wizz Air** – reforestation in Uganda and converting methane emissions from landfill waste into electricity in Turkey.
- **Lufthansa** – construction of biogas plants in Nepal and providing the opportunity to buy SAF.

References

Airbus (2021) *Hybrid and electric flight*. Available at: www.airbus.com/en/innovation/low-carbon-aviation/hybrid-and-electric-flight.

Airport Authority Hong Kong (2022) *Sustainability Report 2022*. Available at: www.hongkongairport.com/iwov-resources/file/sustainability/sustainability-report/AA_Sustainability_Report_202122_ENG.pdf.

Airport Carbon Accreditation (2023) *Airport Carbon Accreditation – Home*. Available at: www.airportcarbonaccreditation.org/.

ANA (2021) *Airlines' initiatives for the SDGs*. Available at: www.ana.co.jp/en/gb/offers-and-announcements/ana-future-promise/co2-reduction-2021-08-04-02/.

Anderson, J.D. (1999) *A history of aerodynamics and its impact on flying machines*. Cambridge: Cambridge University Press.

ATAG (2020) *CORSIA explained*. Available at: https://aviationbenefits.org/environmental-efficiency/climate-action/offsetting-emissions-corsia/corsia/corsia-explained/ (Accessed: 9 November 2023).

ATAG (2021) *Waypoint 2050*. Available at: https://aviationbenefits.org/environmental-efficiency/climate-action/waypoint-2050/.

Avila, D. and Sherry, L. (2016) 'Method for analysis of Ice Super Saturated Regions (ISSR) in the U.S. airspace', in *2016 Integrated Communications Navigation and Surveillance (ICNS). 2016 Integrated Communications Navigation and Surveillance (ICNS)*, pp. 10B2-1–10B2-9. Available at: https://doi.org/10.1109/ICNSURV.2016.7486319.

Avila, D., Sherry, L. and Thompson, T. (2019) 'Reducing global warming by airline contrail avoidance: a case study of annual benefits for the contiguous United States', *Transportation Research Interdisciplinary Perspectives*, 2, p. 100033. Available at: https://doi.org/10.1016/j.trip.2019.100033.

Bailey, J. (2020) *Low-cost carriers could undergo an entertainment revolution, simple flying*. Available at: https://simpleflying.com/rise-of-seatback-entertainment/.

Bisignani, G. (2009) *The IATA technology roadmap report*, 3rd edn, p. 50. Available at: www.escholar.manchester.ac.uk/api/datastream?publicationPid=uk-ac-man-scw:106699&datastreamId=FULL-TEXT.PDF.

Boeing (2009) *AERO – blended winglets improve performance*. Available at: www.boeing.com/commercial/aeromagazine/articles/qtr_03_09/article_03_1.html.

Boeing (2014) *The Boeing 777 family*. Available at: www.boeing.com/farnborough2014/pdf/BCA/bck-777%20Family%20Backgrounder.pdf.

Boeing (2022) *787 by design: by design: advanced composite use*. Available at: www.boeing.com/commercial/787/by-design/#/advanced-composite-use.

Boeing (2023) *Commercial market outlook*. Available at: www.boeing.com/commercial/market/commercial-market-outlook/index.page.

British Airways (2024) *Sustainability at British Airways 2024*. Available at: https://mediacentre.britishairways.com/factsheet/details/217.

CAA (2023) *Air traffic management*. Available at: www.caa.co.uk/consumers/environment/environmental-stakeholders/air-traffic-management/.

Cirium (2022) *The impact of winglets on fuel consumption and aircraft emissions*. Available at: www.cirium.com/thoughtcloud/impact-winglets-on-fuel-consumption-and-aircraft-emissions/.

Clean Sky 2 and Fuel Cells and Hydrogen 2 Joint Undertakings (2020) *Hydrogen-powered aviation*. Available at: https://cleansky.paddlecms.net/sites/default/files/2021-10/20200507_Hydrogen-Powered-Aviation-report.pdf.

Clean-Aviation.eu (2020) *Hydrogen-powered aviation*. Available at: www.clean-aviation.eu/hydrogen-powered-aviation.

Cox, B., Jemiolo, W. and Mutel, C. (2018) 'Life cycle assessment of air transportation and the Swiss commercial air transport fleet', *Transportation Research Part D: Transport and Environment*, 58, pp. 1–13. Available at: https://doi.org/10.1016/j.trd.2017.10.017.

Dron, A. (2023) *Seat retrofit for Jazeera Airways designed to save fuel, space | Aviation Week Network*. Available at: https://aviationweek.com/air-transport/interiors-connectivity/seat-retrofit-jazeera-airways-designed-save-fuel-space.

EASA (2022) *European Aviation Environmental Report 2022*. Available at: www.easa.europa.eu/en/light/topics/european-aviation-environmental-report-2022.

EASA (2023) *Hydrogen and its potential in aviation*. Available at: www.easa.europa.eu/en/light/topics/hydrogen-and-its-potential-aviation.

EASA (n.d.) *Air traffic management and operations*, EASA Eco. Available at: www.easa.europa.eu/eco/eaer/topics/air-traffic-management-and-operations.

EESI (2022) *Issue brief | The growth in greenhouse gas emissions from Commercial Aviation (2019, revised 2022) | White Papers*. Available at: www.eesi.org/papers/view/fact-sheet-the-growth-in-greenhouse-gas-emissions-from-commercial-aviation.

Erdemir, G. et al. (2017) *Project PISCES: developing an in-flight entertainment system for smart devices*. Available at: https://ceur-ws.org/Vol-1853/p09.pdf.

Eurocontrol (2020) *Continuous climb and descent operations (CCO / CDO)*. Available at: www.eurocontrol.int/concept/continuous-climb-and-descent-operations.

Eurocontrol (2023) *Artificial intelligence*. Available at: www.eurocontrol.int/artificial-intelligence.

Guerrero, J.E., Sanguineti, M. and Wittkowski, K. (2020) 'Variable cant angle winglets for improvement of aircraft flight performance', *Meccanica*, 55(10), pp. 1917–1947. Available at: https://doi.org/10.1007/s11012-020-01230-1.

Guo, R., Zhang, Y. and Wang, Q. (2014) 'Comparison of emerging ground propulsion systems for electrified aircraft taxi operations', *Transportation Research Part C: Emerging Technologies*, 44, pp. 98–109. Available at: https://doi.org/10.1016/j.trc.2014.03.006.

IATA (2021) *Our commitment to fly net zero by 2050*. Available at: www.iata.org/en/programs/environment/flynetzero/.

IATA (2023a) *Fact sheet: CORSIA*. Available at: www.iata.org/en/iata-repository/pressroom/fact-sheets/fact-sheet---corsia/.

IATA (2023b) *Net zero 2050: new aircraft technology*. Available at: www.iata.org/en/iata-repository/pressroom/fact-sheets/fact-sheet-new-aircraft-technology/.

ICAO (2019) *2019 environmental peport. Aviation and environment*. Available at: www.icao.int/environmental-protection/Documents/ICAO-ENV-Report2019-F1-WEB%20(1).pdf.

ICAO (2022a) *2022 environmental report*. Available at: www.icao.int/environmental-protection/Pages/envrep2022.aspx.

ICAO (2022b) *CCO-CDO Workshop*. Abu Dhabi.

ICAO (2022c) *Hydrogen, a key solution to decarbonize aviation*. Available at: www.icao.int/Meetings/a41/Documents/WP/wp_514_en.pdf.

ICAO (2023) *Environmental protection*. Available at: www.icao.int/environmental-protection/Pages/default.aspx.

IEA (2023) *Aviation*. Available at: www.iea.org/energy-system/transport/aviation.

Intergovernmental Panel on Climate Change (2023) *Climate Change 2021 – The Physical Science Basis: Working Group I Contribution to the Sixth Assessment Report of the Intergovernmental Panel on Climate Change*. 1st edn. Cambridge: Cambridge University Press. Available at: https://doi.org/10.1017/9781009157896.

IPCC (1999) *Aviation and the global atmosphere*. Available at: https://archive.ipcc.ch/ipccreports/sres/aviation/index.php?idp=133.

IPCC (2023) *Climate Change 2023 – synthesis report: summary for policymakers*. Available at: www.ipcc.ch/report/ar6/syr/downloads/report/IPCC_AR6_SYR_SPM.pdf.

Jarrett, P. (2014) 'FW Lanchester and the Great Divide', *The Journal of Aeronautical History*, 14(2), pp. 54–105.

Kito, M. (2021) 'Impact of aircraft lifetime change on lifecycle CO_2 emissions and costs in Japan', *Ecological Economics*, 188, p. 107104. Available at: https://doi.org/10.1016/j.ecolecon.2021.107104.

Koudis, G.S. et al. (2017) 'Airport emissions reductions from reduced thrust takeoff operations', *Transportation Research Part D*, 52, pp. 15–28. Available at: http://dx.doi.org/10.1016/j.trd.2017.02.004.

Kuropatwa, M., Wegrzyn, N. and Kozuba, J. (2022) 'Turbofan engines efficiency, historical trends, and future prediction':, *Safety & Defense*, 8(2), pp. 82–90. Available at: https://doi.org/10.37105/sd.186.

Lee, D.S. et al. (2020) 'The contribution of global aviation to anthropogenic climate forcing for 2000 to 2018', *Atmospheric Environment*, 244, p. 117834. Available at: https://doi.org/10.1016/j.atmosenv.2020.117834.

Met Office (2023) *Causes of climate change*. Available at: www.metoffice.gov.uk/weather/climate-change/causes-of-climate-change.

MIT (2022) *Carbon offsets*, MIT Climate Portal. Available at: https://climate.mit.edu/explainers/carbon-offsets.

de Moor, S. (2020) 'A fuel efficiency masterclass – part two of three', *Aircraft Analytics*. Available at: https://aircraft-analytics.com/insights/a-fuel-efficiency-masterclass-part-two-of-three/.

NASA (n.d.) *Chapter 10: Technology of the jet airplane*. Available at: https://history.nasa.gov/SP-468/ch10-3.htm.

NASA Earth Observatory (2004) *Aircraft contrails*. Available at: https://earthobservatory.nasa.gov/images/4435/aircraft-contrails.

NOAA (n.d.) *Climate forcing*. Available at: www.climate.gov/maps-data/climate-data-primer/predicting-climate/climate-forcing.

Oliver Wyman (2023) *Global fleet and MRO market forecast 2023–2033*. Available at: www.oliverwyman.com/our-expertise/insights/2023/feb/global-fleet-and-mro-market-forecast-2023-2033.html.

Padhra, A. (2018) 'Emissions from auxiliary power units and ground power units during intraday aircraft turnarounds at European airports', *Transportation Research Part D: Transport and Environment*, 63, pp. 433–444. Available at: https://doi.org/10.1016/j.trd.2018.06.015.

Padhra, A. and Kurnaz, S. (2023) 'Aviation and climate change: becoming a climate-neutral industry (pp. 84–108)', in *Challenges and opportunities for aviation stakeholders in a post-pandemic world*. IGI Global.

Ritchie, H. (2020) 'Climate change and flying: what share of global CO_2 emissions come from aviation?', *Our World in Data* [Preprint]. Available at: https://ourworldindata.org/co$_2$-emissions-from-aviation.

Rowland, B. (2022) *How much fuel does a plane use during flight? | OAG*. Available at: www.oag.com/blog/which-part-flight-uses-most-fuel.

Ryanair (2023) 'Ryanair cuts carbon emissions by 165,000 tones with winglet retrofit', *Ryanair's Corporate Website*. Available at: https://corporate.ryanair.com/news/ryanair-cuts-carbon-emissions-by-165000-tonnes-with-winglet-retrofit/.

SKYbrary (n.d.) *Bypass ratio | SKYbrary Aviation Safety*. Available at: https://skybrary.aero/articles/bypass-ratio.

Stettler, M.E.J. et al. (2018) 'The impact of single engine taxiing on aircraft fuel consumption and pollutant emissions', *The Aeronautical Journal*, 122(1258), pp. 1967–1984. Available at: https://doi.org/10.1017/aer.2018.117.

UCAR (2023) *Contrails | Center for Science Education*. Available at: https://scied.ucar.edu/image/multiple-contrails.

UNFCCC (2015) *Non CO_2 GHGs*. Available at: https://unfccc.int/resource/climateaction2020/tep/thematic-areas/non-co2-ghgs/index.html.

United Nations (2023) *What is climate change?* Available at: www.un.org/en/climatechange/what-is-climate-change.

Virgin Atlantic (2024) *Virgin Atlantic's Flight100 saved 95 tonnes of CO_2 and demonstrated environmental benefits of sustainable aviation fuel*. Available at: https://corporate.virginatlantic.com/gb/en/media/press-releases/virgin-atlantic-flight100-saved-95-tonnes-of-co2.html.

World Bank (2024) *World Bank Open Data*. Available at: https://data.worldbank.org.

Zheng, S. and Rutherford, D. (2020) 'Fuel burn of new commercial jet aircraft: 1960 to 2019'. Available at: https://theicct.org/sites/default/files/publications/Aircraft-fuel-burn-trends-sept2020.pdf.

15 The impact and mitigation of air transport on noise

Chapter outcomes

At the end of this chapter, you will be able to:

- Explain the reasons why aircraft noise is considered a problem.
- Describe and explain the various sources of aircraft noise.
- Identify and explain the different ways the air transport industry measures noise.
- List and analyse the global ICAO regulations as they relate to air transport noise.
- Explain the principles of the Balanced Approach to Aircraft Noise Management.
- Understand why supersonic air transport poses a particular noise problem.

15.1   Introduction

According to ICAO (n.d.a), aircraft noise is the most significant cause of adverse community reaction related to the operation and expansion of airports. The noise footprint from each commercial aircraft movement has reduced massively since the introduction of the jet age, however aircraft still create noise.

Politically, community opposition to air transport because of noise can have a large impact on policies and regulations. Despite the reduction in aircraft noise over the years, one of the key issues is industry growth, and although individual noise events may be much quieter, cumulatively there are now many more noise events. Due to this increase in movements and with the planned growth in air transport movements over the next 15 to 20 years, if the industry is to successfully undertake its planned growth, solutions will have to be found in many parts of the world to reduce the issue of aircraft and airport noise.

15.2   Why is noise a problem?

The UK government, in its Aviation Policy Framework (gov.uk, 2013), recognised that noise was the primary concern of local communities near airports, but also that a balance needs to be achieved between the negative impacts of noise and the positive economic impact of the air transport industry.

ATG Did you know?

The difference between sound and noise

Imagine you are studying in your front room at home. Your friend and your dog are also there, and the dog starts to playfully bark. Your friend is enjoying the barking and messing around with the dog, whereas you are not, as you are trying to study. Your friend is likely experiencing *sound*, whilst you are experiencing *noise*.

Scientifically, sound and noise are technically the same, as they are vibrations that we pick up with our ears, and the bigger the waves, the stronger the vibrations are and the louder the sound is. *Sounds are something we hear in general; however, noise is sound that is unwanted.* In the case of the dog, your friend is experiencing a pleasant sound, whereas you are experiencing an unpleasant noise. However, if the dog keeps on barking over a period, or perhaps the barking gets louder, then it may also become noise to your friend. Noise, therefore, has an objective, physical component as well as a subjective component, which considers an individual's perception or reaction to a sound (FAA, 2022b).

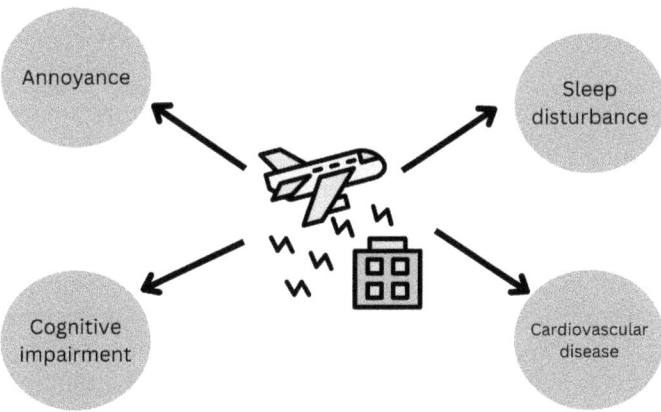

Figure 15.1 Most common health effects of aircraft noise exposure
Source: author

The extent to which noise will be an issue between airports and local communities will vary based on a variety of factors, such as the individuals themselves, the airport location and the population centres, the volume of aircraft movements, the types of aircraft and relations between the airport and community. Understanding the perception of noise (or sound) is critical for the air transport industry, to consider its impacts and possible mitigations.

There are a number of different possible health outcomes associated with aircraft noise exposure and the UK CAA (2024a) lists four of the most common adverse effects (Figure 15.1).

15.2.1 Annoyance

This is the most widespread subjective response to noise and can be identified as dissatisfaction, displeasure or feeling of resentment when noise interferes with thoughts, feelings or activities. This can also be accompanied by stress-related symptoms, leading to changes in heart rate and blood pressure (Clark, 2015). The sound itself plays a role as do social and psychological factors. Factors such as attitudes to the source and whether noise could be reduced by the source also influence annoyance responses (WHO, 1999). Janssen *et al.* (2011), in a study of 34 airports, found a significant increase over the years in noise annoyance at a given level of aircraft noise exposure. A study by the UK CAA (2018) found that people were more highly annoyed by aircraft noise than 30 years previously.

15.2.2 Cognitive impairment

Negative effects on reading comprehension and memory in school children have been suggested, as well as on attention, perception, mood and learning. A study by Stansfeld and Matheson (2003) found that aircraft noise exposure was associated with psychological symptoms and in children, it impaired reading comprehension and long-term memory. Stansfeld and Clark (2015) found that for environmental noise exposure, such as aircraft noise, there was robust evidence for a negative effect on children's cognitive skills such as reading and memory and on standardised academic test scores. In a study on school children around Frankfurt/Main airport in Germany, Klatte *et al.* (2017) found that increasing noise exposure was linearly associated with less positive ratings of quality of life and decreasing reading performance.

15.2.3 Sleep disturbance

Aircraft noise is intermittent in its nature and exposure during the night may result in sleep disturbance, such as awakenings, sleep structure changes such as changes to sleep stages, arousals in heart rate and body movements. People may be aware of these, or it may result in next-day fatigue (CAA, 2024a). In 2011, the WHO estimated sleep disturbance to be the most adverse non-auditory effect of environmental noise exposure. Research by Hume, Brink and Basner (2012) concluded that there was evidence for the effect of night-time aircraft noise exposure on sleep disturbance. There is evidence that aircraft noise influences the time spent in different sleep stages, with aircraft noise reducing slow-wave sleep and rapid eye movement (REM) sleep, with possible implications for early morning (0400–0630) flight hours at airports (Clark, 2015).

The consequences of interrupted sleep from aviation noise can be classified in three timescales (Benz *et al.*, 2022):

1. **Immediate reactions to night noise**, such as an increase in blood pressure and heart rate, causing changes from deeper to lighter sleep, awakenings and general sleep loss.
2. **Short-term reactions to night noise.** Due to the reduction in sleep, cognitive performance may be lower and there may be negative impacts on mood and wellbeing.
3. **Long-term reactions.** Chronic sleep loss and recurring interruptions of sleep are a major risk factor for cardiovascular and metabolic diseases, as is the risk of developing hypertension.

15.2.4 Cardiovascular disease

Aircraft noise at high levels can be considered a stressor on the body and research has found an association with an increased risk of developing cardiovascular disease (CVD) which includes all the diseases of the heart and circulation, such as heart attack and stroke, explained in Benz *et al.* (2022). Saucy *et al.* (2020) conducted research in their TraNQuiL study on the acute triggering effects of aircraft noise at night on cardiovascular mortality near Zurich Airport, Switzerland. The study findings suggested that night-time aircraft noise events may trigger cardiovascular deaths and may be of particular importance in relation to ischaemic heart disease (IHD) and heart failure (CAA, 2022).

ATG Case study 15.1

The discussion on the health effects of aircraft noise (DEBATS) study, France

The DEBATS study was conducted by the Gustave Eiffel University and was the first large-scale research programme in France to evaluate the possible effects of aircraft noise exposure on the health of residents living near airports. Participants were first interviewed in 2013 at the start of the study and then in 2015 and 2017. Evrard *et al.* (2021) presented the results at the ICBEN Conference on Noise as a Public Health Problem in 2021. A total of 1,244 residents near Paris-Charles de Gaulle (CDG), Lyon-Saint Exupèry (LYS) and Toulouse-Blagnac (TLS) airports were questioned and the results suggested that exposure to aircraft noise, in France as elsewhere, creates a range of self-reported health issues.

Given these possible negative effects and the planned future growth of the air transport industry, there needs to be a way of reducing the noise footprint where possible.

15.2.5 The impacts of aircraft noise on biodiversity

The effects of air transport noise are not limited to humans. As ICAO stated in its 2022 Environment Report, "the aviation sector can have adverse effects on biodiversity in a number of ways, including … the effects of light and noise pollution on particular species" (ICAO, 2022). This relates to fish, mammals, reptiles, amphibians and invertebrates.

Wolfenden *et al.* (2019) researched the behaviour of wild chiffchaff birds at both Manchester and Amsterdam Schiphol Airports (and nearby control sites) and concluded that they demonstrated a negative relationship between noise exposure and song frequency, responded more aggressively to song playback and were five times more likely to attack a speaker emitting bird song than their counterparts who lived away from airport noise.

Alquezar and Macedo (2019) examined the existence of Natural Protected Areas of high priority conservation located within noise-impacted areas of Brazilian airports. They highlighted the main effects of noise on wildlife behavioural changes, such as increased alertness, vocal behaviour modifications and lower reproductive success. As birds depend upon communication for reproductive purposes, continuous noise exposure has been reported to cause decreases in nest success and nestling growth rates. This can in turn reduce population sizes and decreased species richness and diversity (CAA, 2023b).

15.3 Sources of air transport noise

Most of the noise created by the air transport sector is from the aircraft themselves – via the engines, the airframe and the aircraft systems. Engine noise tends to dominate on take-off whilst airframe noise may be most noticeable on landing. Noise will also be created on the ground within the airport environment via vehicular traffic and the surface traffic travelling to and from an airport can create annoyance to the local populations.

> **ATG Did you know?**
>
> **Aircraft noise**
>
> Aircraft noise has decreased dramatically since the early days of the jet age, by approximately 90% since jet aircraft entered service in the 1960s (Gély and Márki, 2022; NATS, n.d.). This is primarily due to the reduction of noise at source by improved engine technologies, airframe designs, aerodynamics and lighter-weight materials.

15.3.1 Jet engine noise

Noise from jet engines can occur from several sources and some of the key components are shown in Figure 15.2. *Propellor-driven aircraft* produce noise from the rotation of the propellors and the interaction of the propellor blades with the surrounding air.

Jet noise

- **Sheer layer noise** is generated at the boundary between the high velocity exhaust (nozzle) jet and the surrounding air. The difference in velocity between the fast-moving jet and the slower ambient air creates turbulence and vortices leading to noise.
- **Shock-associated noise** is generated when the high-speed exhaust jet interacts with the surrounding air, causing shockwaves.

Aeroacoustic noise

- **Turbomachinery noise** is caused by rotating blades and other components of the engine, especially in the compressor and turbine sections.
- The **combustion** process in the engine can create noise, especially in afterburning engines where fuel is injected into the exhaust stream for additional thrust.
- Airflow through the **engine inlets and exhaust nozzles** can contribute to noise generation.

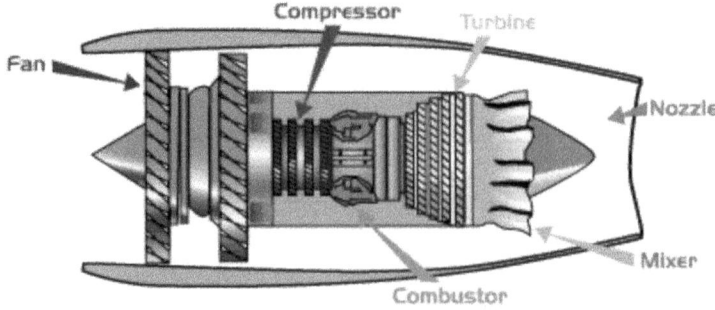

Figure 15.2a Jet engine

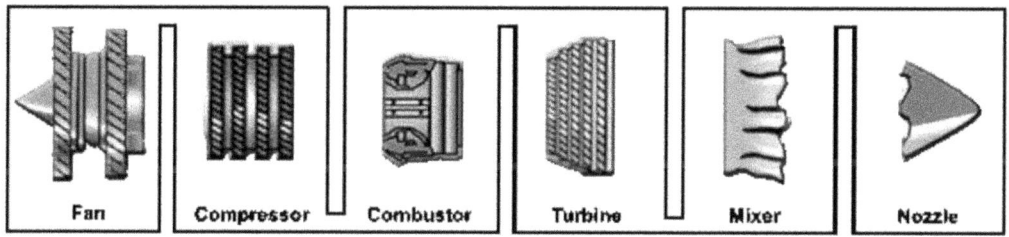

Figure 15.2b Jet engine components
Source: NASA GRC (2021)

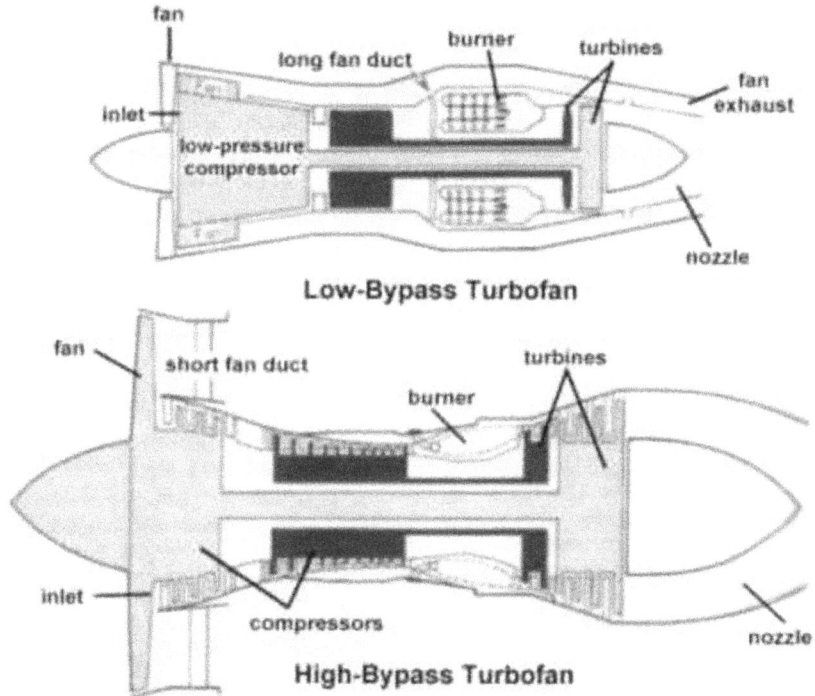

Figure 15.3 Low-bypass and high-bypass ratio engines
Source: NASA GRC (2021)

Fan and compressor noise

- In aircraft engines with turbofan configurations, the large fan at the front of the engine can also cause noise, both from the rotation of the fan blades and the interaction of the fan with the surrounding air.

Engine components and structural vibrations

- Vibrations and mechanical movements within the engine and its components can contribute to overall noise production.

416 *Fundamentals of Global Air Transport Geography*

To reduce noise in the early engines, exhaust modification occurred, to be fitted with internal **mixer** units (Figure 15.2). This increased the frequency of the noise and it also became absorbed by the atmosphere quicker, as well as reducing flow velocity. However, these were not very efficient due to the added drag and weight.

Modern engines are generally **high bypass ratio (BPR)** engines and this concept was explained in Section 14.4. Most of the air which enters the engine bypasses the core, which reduces the difference in velocity between the still air and expelled air, reducing noise. Figure 15.3 illustrates the difference between older low-BPR engines versus the newer high-BPR engines.

15.3.2 Airframe noise

Noise is produced by the airflow moving around the aircraft, with the principle being that the smoother the aircraft, the less noise it will make. There are a few airframe components which are responsible for creating noise (Figure 15.4). Many improvements have been made over the years to streamline aircraft design and develop new technologies to reduce noise, however when the aircraft is in the approach phase, many of the control surfaces required will result in the aircraft moving from a "clean" to "dirty" status, when the extendable surfaces are in their extended state, disturbing airflow, creating noise.

Aerodynamic noise

- During flight, air flows over the surfaces of the aircraft, including fuselage and wings and the associated turbulence can create noise.

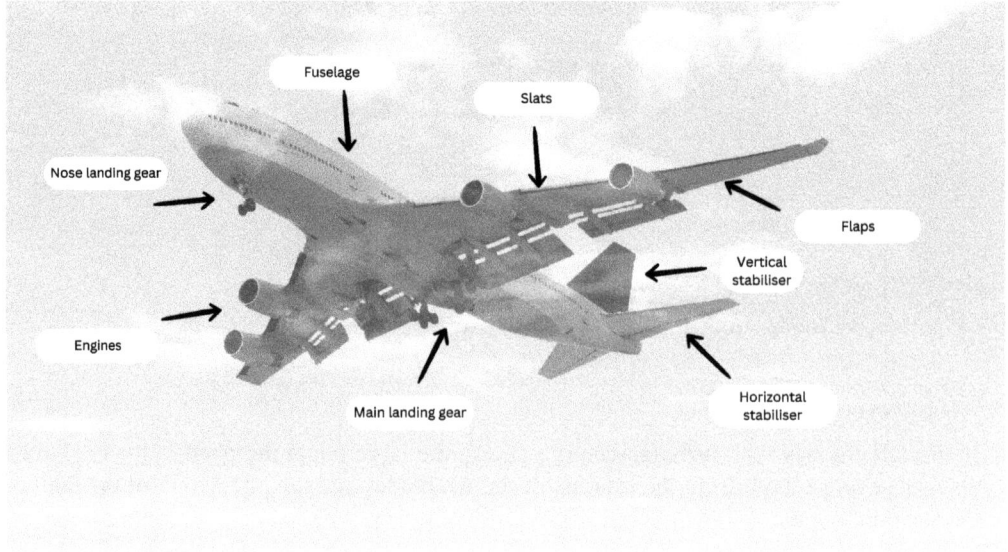

Figure 15.4 Sources of aircraft noise
Source: author

High-lift devices

- Aircraft use high-lift devices such as flaps and slats during take-off and landing and the disruption of airflow, as well as their movement, can create noise.

Landing gear

- On take-off and landing, the landing gear is retracted and extended and this movement and the disrupted airflow can create noise.

15.3.3 Air-conditioning and the auxiliary power unit

The aircraft air conditioning system will also generate noise. The functioning of these systems involves an air cycle machine with a turbine and compressor assembly, which can create significant noise, especially inside the cabin and for personnel around the aircraft.

The auxiliary power unit (APU) is a small turboshaft engine used to power aircraft systems and provides air for the main engine start-up, hence generates noise (Maaz, 2022). A way to reduce APU noise is to use a ground power unit (GPU) supplied at the airport, which is connected to the electrical system of an aircraft, with an added advantage of reduced fuel burn and emissions.

15.3.4 Ground vehicles

Noise pollution from ground vehicles is a major environmental challenge for airport management and is generally from two major sources.

Airport vehicles

The functioning of an airport requires many vehicles, both from the airport themselves and the many third-party operators which may be present. These will include airside coaches, baggage loading equipment, catering trucks, fuel trucks and cleaning vehicles. At a major international airport, there can be hundreds of vehicles driving at any one time, potentially creating noise pollution.

Surface access

Non-transiting departing and arriving passengers must access the airport using some form of surface transport, such as cars, taxis and buses, which may all generate noise, both around the airport and surrounding neighbourhoods, which may be used to access the airport. Airports which have cargo facilities will also require trucks and lorries for cargo deliveries. In addition, airport shops and catering units require resupplies and these vehicles may create environmental noise around the airport.

15.4 Measuring noise

To establish the extent of any noise generated by the air transport industry, it is important to measure the noise and there are various ways this can be achieved. The way people experience noise from all sources can differ and individual perception/annoyance can also differ for a multitude of reasons, and this is one of the difficulties in setting noise levels.

418 *Fundamentals of Global Air Transport Geography*

The noise level of aircraft can vary depending on a few factors:
- Aircraft height.
- Is it directly overhead or laterally displaced?
- Is it arriving or departing (level of thrust and air resistance around the aircraft footprint)?
- Weather conditions, especially wind direction and strength.

15.4.1 Decibels

The human ear can contend with a broad range of sound levels and for this, the decibel (dB) scale is used. This encapsulates the energy of sound with reference to the smallest audible sound of 0dB, thus a sound ten times more powerful is 10dB and a sound 100 times more powerful is 20dB. According to the US CDC (2022), normal conversation is about 60dB and a motorcycle engine is about 95dB. The loudest sounds we hear without pain are around 120dB, and noise above 70dB, over a prolonged period, may start to damage hearing.

> **ATG Did you know?**
>
> **The loudest aircraft**
>
> According to Guinness World Records (2024), the loudest aircraft likely ever flown was the Republic XF-84H Thunderscreech military aircraft. No scientific measurements were ever made, however during engine run-up tests at Edwards Air Force Base, California, in July 1955, the base received noise complaints from as much as 40km (25 miles) away. The screeching whine of the aircraft's propeller required ground crew communication to be conducted using signal flags and lights.

15.4.2 "A-weighted" decibels (dBA)

The human ear also responds to different pitches or frequencies of sound differently. Frequency is an important characteristic of sound and is the rate of repetition of the sound pressure oscillations that reach our ears, expressed in units called Hertz (Hz). The normal frequency range is from 20Hz to a high of 10–20,000Hz, but people are most sensitive to sounds between approximately 500–2,000Hz – the voice range (FAA, 2023). To correlate sound with its perception, the sound energy spectrum is "weighted", and the dBA scale is used. This is most used to correlate with human response to environmental noise (Figure 15.5).

15.4.3 Maximum A-weighted noise level (L_{max})

A-weighted sound levels vary with time and how close the aircraft is as it approaches and travels into the distance. This variation makes it convenient to calculate a measure for the loudest part of the flight – the L_{max}. This figure does not explain cumulative noise exposure, but merely one dimension of an event.

15.4.4 Sound exposure level

Sound exposure level (SEL) is the sound exposure level of an aircraft event, measured in dBA of a one-second burst of steady noise that contains the same total A-weighted sound energy as

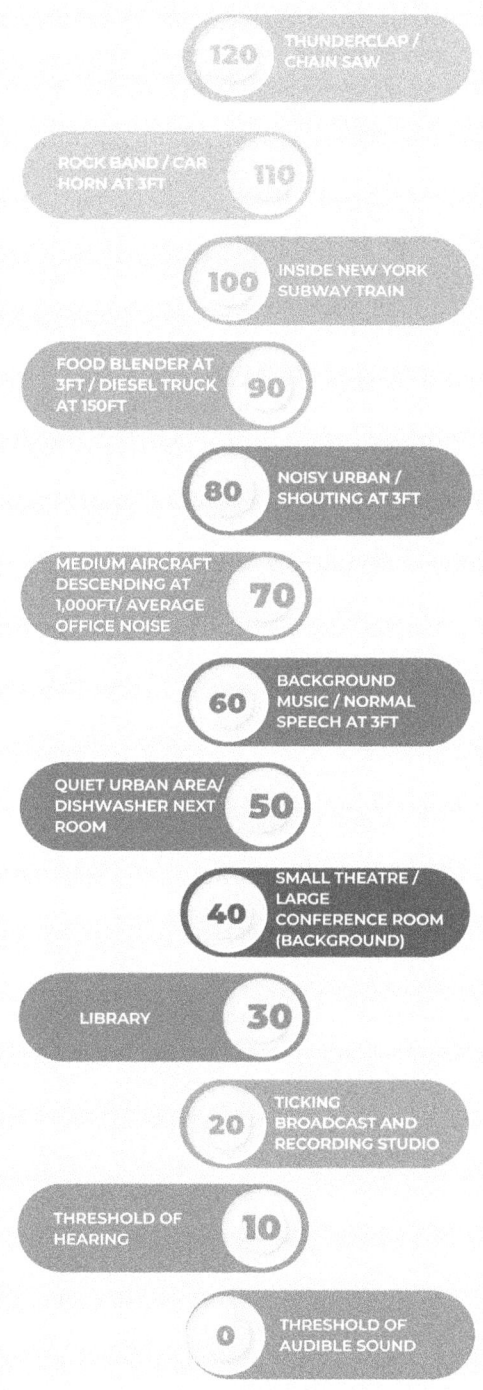

Figure 15.5 Comparative dBA noise sources
Source: author

the whole event. SEL is often used to characterise the likelihood of sleep disturbance relating to aircraft noise (FAA, 2023). Studies have found that SEL above 90dBA generally leads to sleep disturbance. SEL **footprints** can be used to work out the areas where take-off creates an SEL over 90dBA, to inform decisions about whether a particular type of aircraft should be permitted to operate at night, or to influence airport construction or extension in populated areas (CAA, 2024a).

15.4.5   Equivalent sound level (L_{eq}/L_{Aeq})

The L_{eq} measures the average acoustic energy over a period, to take account of the cumulative effect of multiple noise events – for example, the aggregate sound at a location which has aircraft flyovers throughout the day.

According to NATS (n.d.), a common use is the L_{eq} 16h metric. This describes the cumulative noise exposure from aircraft noise events over a 16-hour period. In the UK, *noise contours* are then created connecting areas with the same noise exposure between 0700 and 2300 hours – the UK description of daytime. Average summer L_{eq} noise contour maps are typically produced annually from measurements taken between 16 June and 15 September. Contours are normally plotted from 51dBA to 72dBA L_{eq} at 3dB intervals (CAA, 2024b). These make it possible to identify how many people live in areas where there is significant annoyance from noise.

Global research has found that annoyance due to aircraft noise is correlated with this cumulative metric. The UK government considers an L_{eq} of over 57dBA to represent the noise level for the onset of significant community annoyance (CAA, 2024a).

ATG Case study 15.2

Noise monitoring at London Heathrow, UK (LHR)

The UK government's overall policy on aviation noise is "to limit and, where possible, reduce the number of people in the UK significantly affected by aircraft noise as part of a policy of sharing benefits of noise reduction with industry in support of sustainable development" (Heathrow, 2024).

As part of this policy, LHR has an array of fixed and mobile noise monitors located around the airport. These are used to monitor compliance with the UK Department for Transport noise limits, to assist with annual noise contour mapping and air traffic improvement initiatives and to measure noise in specific community locations. In 2022, departure measurements were taken between 6.0 and 26.1km from aircraft start-of-roll (SOR) and between 2.8 and 23.0km to touchdown for arrivals, and there were 12 fixed and 23 mobile noise monitor sites (CAA, 2023a).

15.4.6   Day–night average sound level (DNL/L_{den})

L_{den} is a variant of L_{eq} which includes a 10dB weighting for noise events at night and a 5dB weighting for events during evening periods (1900–2300 hrs), reflecting the potential for increased sensitivity to noisy events during those time periods. In the UK, analysts identify how many homes and residents are located where the L_{eq} is over 57dBA or the L_{den} is over 55dBA.

15.4.7   Effective perceived noise in decibels (EPNdB)/effective perceived noise level (EPNL)

This is the measure of the relative noisiness of an individual aircraft pass-by event. The value applies to individual aircraft and their specific engines and is used for aircraft noise certification for commercial aeroplanes, as it accounts for the subjective effects of aircraft noise on human beings. It consists of an integration over the noise duration of the perceived noise level (PNL) adjusted for spectral irregularities (PNLT), normalised to a reference duration of 10 seconds (ICAO, 2017). ICAO SARPs in Annex 16, Vol I, state that separate ratings are given for approach, lateral and flyover reference measurement points.

Reference noise measurement points

1. **Approach.** This point is located on the extended centre line of the runway, 2,000m from the threshold. This corresponds to a position of 120 metres vertically, on level ground, below the 3° aircraft descent path.
2. **Lateral.** The two points are located on a line parallel to and 450 metres from the runway centre line, where the take-off noise level is at maximum during take-off. The certified noise level corresponds to the average of these two levels.
3. **Flyover.** The reference point is located on the extended centre line of the runway, at 6,500 metres from the start of aircraft roll.

Cumulative levels are defined as the arithmetic sum of the certification levels at each of the three points.

15.5   ICAO noise regulations

In the early 1960s, with the introduction of the first generation of jet aircraft such as the de Havilland Comet and the Boeing 707 and the acceleration of their use in international flights, the negative side of commercial aviation began to be recognised, in the form of noise. In November 1966, the international conference on the reduction of noise and disturbance caused by civil aircraft (also known as the London Noise Conference) was held with the objective of reaching an international solution through ICAO. The following year, the Fifth Air Navigation Conference of ICAO held at Montréal in November 1967 made certain recommendations on the subject. Based upon these recommendations, the 16th Session of the ICAO Assembly, held in September 1968 in Buenos Aires, adopted *Resolution A16-3* calling for an international conference to consider the problem of aircraft noise in the vicinity of airports and seeking to establish international specifications and guidance material relating to aircraft noise (ICAO, 2024).

In 1969, a "Special Meeting on Aircraft Noise in the Vicinity of Aerodromes" was held in Montréal, with a view to preparing new standards. A draft **Annex 16** on aircraft noise, arising from the work done by this Special Meeting, was processed. The Council formed the Committee on Aircraft Noise (CAN), to examine aircraft noise certification problems; this Committee held its first meeting from 28 September to 2 October, 1970.

- **Chapter 2**: The first SARPs of Annex 16 – Aircraft Noise (which became Volume I) were adopted by the ICAO Council on 2 April, 1971, became effective on 2 August, 1971, and applicable on 6 January, 1972. These became the **Chapter 2 Noise Standard**. This Standard defined the three noise reference measurement points for certification and set noise limits as a direct function of maximum take-off mass (MTOM), to recognise that heavier aeroplanes

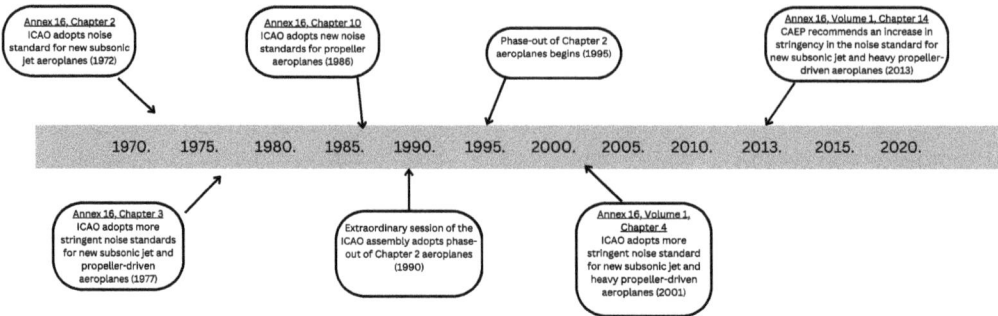

Figure 15.6 ICAO Annex 16 timeline
Source: adapted from ICAO (n.d.b)

produced more noise than lighter aeroplane types. Figure 15.6 illustrates the timeline of Annex 16 developments over the years.
- **Chapter 3**: After the Chapter 2 standards were introduced, engine technologies further developed and much higher bypass ratio jet engines were introduced into service, which resulted in a reduction in noise. As a result, in 1977, the ICAO Annex 16 noise standards became more stringent and this became the **Chapter 3 Noise Standard** for aircraft certification.

In 1983, the **Committee on Aviation Environmental Protection (CAEP)** was established and this replaced the CAN as well as replacing the Committee on Aircraft Engine Emissions (CAEE). This recognised the increasing importance of environmental protection for international aviation. The goal of CAEP was adding or revising standards and recommended practices (SARPs) associated with environmental issues, specially related to noise and engine emissions. It is a technical committee, reporting to the ICAO Council, and includes members from a variety of government and non-governmental organisations, and it seeks to protect the environment without sacrificing the viability of the international aviation system (Kearns, 2021).

- **Chapter 4**: In the next couple of decades following 1977, further noise reduction technologies were incorporated into engine and airframe designs, leading to incremental improvements in aircraft noise performance. This resulted in a further stringency increase in the Annex 16 Noise Standard which became the **Chapter 4 Standard**, which was approved in 2001, to be applicable for aeroplane type designs on or after 1 January, 2006. This meant a cumulative noise margin of 10 EPNdB below the Chapter 3 Standard.

ATG Did you know?

Cumulative noise margin

Cumulative noise margin is the sum of the individual margins (the difference between certified noise level and noise limit) at each of the three Chapter 3 noise measurement points expressed in EPNdB.

- **Chapter 14**: In February, 2013, the CAEP/9 meeting recommended an amendment to Annex 16, Volume I, including an increase in stringency of 7 EPNdB (cumulative) relative to the Chapter 4 levels. In 2014, this recommendation was adopted by the ICAO Council as the new Annex 16, Vol I, **Chapter 14 noise standard** for jet and propeller-driven aeroplanes. It was applicable to new aeroplane types submitted for certification on or after 31 December, 2017, and on or after 31 December, 2020, for aircraft less than 55 tonnes in mass.

The progression of ICAO noise standards for aeroplanes has resulted in a cumulative reduction of 33 EPNdB below the original Chapter 2 Noise Standard (ICAO, 2015).

As a result of the new Chapter 14 noise standard, ICAO (n.d.d) expects the number of people affected by significant aircraft noise will be reduced and that more than 1 million people could be removed from the DNL of 55dB affected areas between 2020 and 2036.

What this has ultimately meant for people on the ground near airports is that aircraft operating now are much quieter than before.

The average margin to the Chapter 3 limit for both single-aisle (due to aircraft such as the Airbus A320neo and Boeing 737MAX) and twin-aisle (due to aircraft such as the Airbus A350 and Boeing 787) has improved by around 10 EPNdB since 2002, as well as there being a similar reduction in the regional jet category since 2014, due to aircraft such as the Airbus A220 and Embraer E2-series.

Aircraft certified during the last ten years, such as the A350, A320neo, B787 and B737MAX, have a cumulative margin of 5 to 15 EPNdB below the latest Chapter 14 Standard (EASA, 2022). From the introduction of one of the earliest commercial jet aircraft (Boeing 707) in the late 1950s, through to the A320neo 60 years later, the FAA estimates an 83% EPNdB noise reduction (FAA, 2024).

ATG Case study 15.3

Boeing 737 series, Washington, USA

The Boeing 737 is a narrow-body aircraft produced by Boeing at its Everett and Renton factories in Washington State, USA. Across its variants, it has been flying since 1967 and alongside the Airbus A320 series, is one of the best-selling aircraft of all time. In relation to noise standards, the original -100 and -200 variants had to meet the levels imposed in 1972 with the Chapter 2 standards. When the "Classic" -300, -400 and -500 were certified, these had to meet the Chapter 3 standards as did most of the Next Generation aircraft, except for the 737–900ER, which was certified in 2007, after the Chapter 4 noise standards were introduced. The most recent MAX variant -8 was also introduced in the Chapter 4 category, whilst the -9 and -8200 were covered by the Chapter 14 requirements.

Chapter 2 aircraft, such as the original Boeing 737-200, were phased out in a number of countries due to their noise profile. For example, a ban on Chapter 2 aircraft took effect in the EU from April 2002. In the USA, all certified aircraft must be at least Chapter 3 (called Stage 3 in the USA), excepting flights such as for humanitarian aid or flying to the USA for maintenance (FAA, 2022a). According to EASA, most Next Generation aircraft are Chapter 4 compliant.

Table 15.1 Boeing 737 variants

Aircraft variant	Series	FAA type certification date
B737-100	Original	15 December, 1967
B737-200		21 December, 1967
B737-200C		29 October, 1968
B737-300	Classic	14 November, 1984
B737-400		2 September, 1988
B737-500		12 February, 1990
B737-700	Next Generation	7 November, 1997
B737–800		13 March, 1998
B737-600		12 August, 1998
B737–900		17 April, 2001
B737–900ER		20 April, 2007
B737–8	MAX	8 March, 2017
B737–9		15 February, 2018
B737–8200		31 March, 2021

Source: adapted from EASA (2023)

15.6 ICAO Balanced Approach to Aircraft Noise Management

Limiting or reducing the number of people affected by significant aircraft noise is one of ICAO's key environmental goals. The main overarching policy on aircraft noise is the Balanced Approach to Aircraft Noise Management, which was adopted by the ICAO Assembly in 2001. Guidance is published in ICAO Doc 9829, 2nd Edition, updated in 2010, and is intended to apply to any airport being served by international air traffic, which has a perceived noise problem. The Balanced Approach consists of identifying the noise problem at an airport and then analysing the various measures available to reduce noise through the exploration of four principal elements (Figure 15.7) with the goal of addressing the noise problem in the most cost-effective manner (ICAO, 2010).

This provides ICAO contracting states with an internationally agreed approach to address aircraft noise problems where they occur – at individual airports – in an environmentally responsive and economically responsible way. The Balanced Approach gives ICAO contracting states a flexible way to identify a specific noise problem and remedies that are targeted and tailored to the individual airport situation in a transparent process. Individual states have ultimate responsibility to develop appropriate noise solutions at their airports, with due regard to ICAO rules and policies.

15.6.1 Reduction of noise at source

The first element is to reduce the noise at source. This has been controlled since the 1970s by setting noise limits for aircraft, as explained in Section 15.5, and has been responsible for most of the noise reductions from this period until the present day. Aircraft noise is generated by turbulent flows of air over and around surfaces, such as air going into and out of the engine, and air flowing around the airframe – fuselage, wings and other aerodynamic surfaces such as flaps, slats and landing gear.

The impact and mitigation of air transport on noise 425

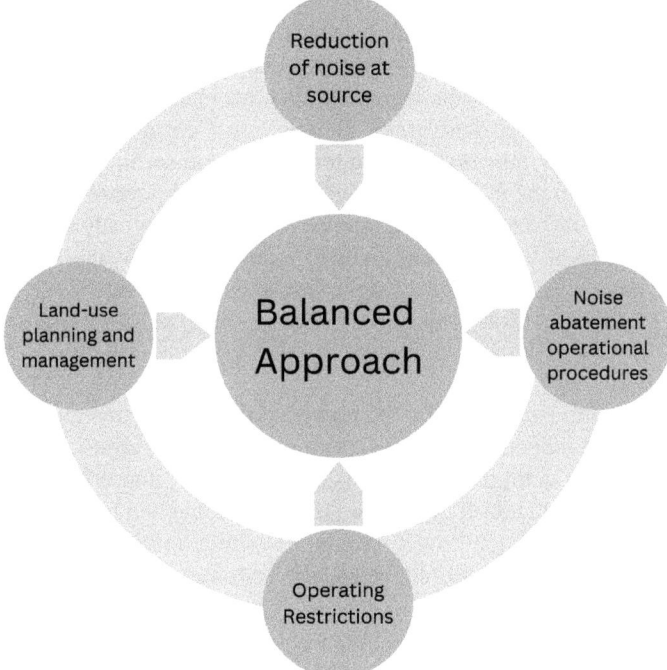

Figure 15.7 ICAO Balanced Approach to Aircraft Noise Management
Source: author

Engine technologies

BYPASS RATIOS

The largest factor driving down aircraft noise has been higher BPRs. High BPR engines are a type of jet engine where most of the air bypasses the engine core, contributing to thrust via the large fan at the front. This design improves fuel efficiency and reduces noise, making it ideal for commercial air transport. The large fan and slower exhaust velocity also lower operational costs and emissions. Modern high bypass engines power most commercial aircraft, such as the Boeing 777 and Airbus A350, due to these advantages and the ratio has risen significantly over time since the early turbofan designs in the early 1960s.

One of the first jet engines used commercially – the Pratt and Whitney JT3D – which powered the Boeing 707 – had a BPR of 1.42:1. In comparison, the Pratt and Whitney Geared turbofan (GTF) engine family has one of the highest BPRs ever built. The PW1100G variant, which powers the A320neo, has a BPR of 12.5:1 (MTU, 2024).

One possible future noise reduction technology will be to take this process further, with an ultra-high bypass ratio engine, although technological challenges exist.

Fan noise

Aircraft engine fan noise is a significant component of the overall noise generated by jet engines, particularly in high BPR engines. The noise is primarily produced by the interaction of the fan blades with the surrounding air, including the turbulence and vortices created as the blades move. As the fan blades rotate at high speeds, they generate broadband noise, including low-frequency sounds that can travel long distances. Advances in blade design, such as shaping and spacing, and the use of noise-reducing materials have helped reduce fan noise.

Research at the Rolls-Royce University Technology Centre (UTC) at the University of Southampton (2024) on turbofan noise has suggested sound absorbing "liners" placed on the inner surfaces of an engine in the intake and the exhaust are an important method for reducing fan noise. UTC also states that there is growing interest in fuel efficient, advanced open rotor (AOR) powered aircraft. Bleed valves used in engines to manage air flow, mainly in the approach to landing condition, have been identified as important sources of noise.

Airframe noise

Airframe noise is the sound generated by the non-propulsive components of an aircraft, such as the wings, landing gear and control surfaces, during flight. This noise is caused by the interaction of airflow with these components, leading to turbulence, and other aerodynamic effects. Airframe noise becomes particularly significant during landing and take-off when the aircraft is closer to the ground and moving at lower speeds. Reducing airframe noise is crucial for minimising the overall noise impact of aircraft, especially in communities near airports, and involves optimising design and materials.

According to Astley (2015), reducing airframe noise is challenging. The use of flaps and slats and the deploying of landing gear at approach are required to maintain lift and slow the aircraft, however noise is subsequently created.

The scope for noise technology reductions of conventional aircraft seems to be limited. Newer designs and configurations may be required but these will bring their own challenges (ICAO, n.d.c).

ATG Case study 15.4

Noise chevrons

Noise chevrons are serrated edges incorporated into the exhaust nozzles of aircraft engines to reduce noise (Figure 15.8) They are found at the rear of the engine where the hot jet exhaust meets the cooler bypass air or ambient air. Chevrons work by smoothing the mixing of these airflows, reducing turbulence and the associated noise. This design helps lower the intensity of jet noise, particularly during take-off and landing, which are critical phases for noise abatement. Chevrons are commonly seen on modern high BPR engines, such as on the Boeing 787.

Boeing defined this technology as part of their second Quiet Technology Demonstrator (QTD2), working with General Electric and NASA, and adopted it for many of its newest models, such as the 787, 747–8 and 737 MAX.

The impact and mitigation of air transport on noise 427

Figure 15.8a An early chevron test article with symmetrical notches
Source: NASA (n.d.)

Figure 15.8b An early chevron test article inside an engine nozzle
Source: NASA (n.d.)

Figure 15.8c Jet engine nozzle
Source: Canva.com

It is important to note that not all noise annoyance generated around airports is from aircraft. For example, generators, plant, construction, airport road traffic and maintenance can all play a role and indeed London City Airport categorises noise complaints into these aspects of ground noise in its Noise Action Plan (London City Airport, 2018).

15.6.2 Land-use planning and management

The ICAO assembly has stated that:

> the number of people affected by aircraft noise is dependent on the way in which the use of land surrounding an airport is planned and managed and in particular the extent to which residential development and other noise-sensitive activities are controlled.
>
> (ICAO, 2010)

The objective is to direct incompatible land use (such as houses and schools) away from the airport environs and to encourage compatible land use (such as industrial and commercial use) to locate around the airport facilities. This is not always possible at existing airports.

Local or regional authorities are often responsible for land-use planning and management functions necessary to implement urbanisation control measures around an airport. There are various ways land-use planning can be achieved (Table 15.2). The use of noise contours, as explained in Section 15.4, is crucial in this process, to ascertain affected areas more accurately.

Table 15.2 Categorisation of types of land-use planning and management measures

Instruments	Examples
Planning instruments	Comprehensive planning, noise zoning, subdivision regulations, transfer of development rights, easement acquisition
Mitigating instruments	Building codes, noise insulation programmes, land acquisition and relocation, transaction assistance, real estate disclosure, noise barriers
Financial instruments	Capital improvements, tax incentives, noise-related airport charges to assist in funding noise mitigation efforts

Source: ICAO (2010)

ATG Case study 15.5

Noise insulation scheme, Glasgow Airport, UK (GLA)

GLA is one of a number of UK and European airports which provides financial assistance to local residents experiencing noise disturbances from airport operations. The aim of the airport noise insulation scheme (NIS) is to work collaboratively with the community to ensure that the indoor living spaces of eligible households and noise-sensitive buildings are not unduly adversely affected by noise from aircraft. To achieve this, from 1 November, 2023, eligible properties were able to apply for funding towards proportionate insulation works to ensure that habitable rooms have satisfactory internal protection from noise that may be generated by aircraft taking off and landing at GLA. These had to be situated within the current 63dB 16-hour summer day L_{eq} noise contour area (Glasgow Airport, 2024).

ATG Case study 15.6

Noise-related airport charges: Zurich Airport, Switzerland (ZRH)

ZRH attempts to incentivise airlines to use the quietest aircraft, via its pricing structure. The noise charge model stipulates that every aircraft be classified into one of five noise categories based on noise measurements. The noise charge model for jet aircraft is based on average peak noise values during take-off, as measured at specific monitoring stations at Zurich Airport.

Noise class I comprises the noisiest and noise class V the least noisy aircraft types. The noise charge increases with each noise class and ranges from CHF 0 (noise class V) to CHF 2,000 (noise class I) as shown in Table 15.3.

During the night between 2200 and 0600, additional night-time noise charges are levied, which increase in half-hourly increments. Since 2013, flights between 0600 and 0700 and between 2100 and 2200 have also had to pay a noise charge supplement.

Table 15.3 Zurich Airport noise charges

Class	I	II	III	IV	V
Charge in Swiss Francs (CHF)	2,000	400	40	10	–
Example aircraft types	Boeing B747-400	Airbus A380–800	Boeing B777-200	Airbus A350–900	Airbus A220-100

Source: Flughafen Zurich (2024)

15.6.3 Noise abatement operational procedures

The way aircraft are operated in day-to-day operations may also have an impact on ground noise. Implementation of noise abatement operational procedures can minimise the number of people affected by noise, considering that safety should always remain the number one priority.

ICAO Doc 8168: Procedures for Air Navigation Services – Aircraft Operations – provides guidance to air transport operators. ICAO (2010) also provides examples for abatement procedures, not all of which will be possible at all airports:

Use of flight departure and approach routings

- **Noise preferential routes.** These can be established to ensure that departing and arriving aircraft avoid overflying noise-sensitive areas.
- **SID/STAR procedures.** Aircraft should operate to and from airports using standard instrument departure (SID) and if appropriate standard terminal arrival (STAR) procedures. These procedures provide obstacle clearance protection to the aircraft and enable the development of noise abatement flight tracks for the aircraft.
- **Dispersed flight tracks.** Successive departing aircraft may be dispersed on different flight tracks over wide-ranging areas.
- **Automated arrival and departure procedures** – for example based on area navigation (RNAV) procedures and systems using on-board flight management systems (FMSs) to provide accuracy and control when operating SIDs and STARs to minimise the noise exposure area.

Use of runways

- **Noise preferential runways.** Using the specific runway which will create least noise on initial departure and final approach (bearing in mind flight safety factors, such as wind direction, etc.).
- **Displaced thresholds.** Displacing the commencement of the take-off and landing threshold.

Use of approach procedures

- **Descent profiles.** Maintaining higher than normal approach altitudes/angles and instrument landing system (ILS) glide slope interception from a higher altitude.
- **Reduced power/drag techniques.** Delaying as much as possible wing flap extension and landing gear deployment, whilst ensuring a safe operation, involving engine power changes.
- **Continuous descent operations (CDO).** This usually relates to the approach phase – continual descent approach (CDA) – between 6,000ft and the glide slope interception and will allow an uninterrupted descent from cruising altitude (Figure 14.11). It will reduce ground noise, as it reduces the overall engine thrust required during descent and keeping the aircraft higher for longer, as opposed to the traditional step-down approach (SDA). Eurocontrol (2020) estimates benefits of between 1–5dB.

A study by Eurocontrol (2022) on the proportion of flights that achieve CDO from the cruising altitude at the top 25 European airports (Table 15.4) showed that Oslo Gardermoen Airport was top at 59.1%

Use of reverse thrust

- Reverse thrust can assist an aircraft braking, especially on contaminated runways. However, this could be limited to reverse idle and perhaps at specific times, such as during the night, again allowing for safety.

Ground-based operational procedures

- **Limiting aircraft engine ground running.**
- **High power engine runs only in designated areas.** These can be especially effective if using noise insulation such as ground run-up enclosures, which are in use at Dubai International Airport.
- **Use of ground power unit (GPU)** instead of the aircraft auxiliary power unit (APU), to provide power to the aircraft whilst on stand.

Table 15.4 Proportion of flights achieving CDO from the cruising altitude at the top 25 European airports

Airport	% CDO from top of descent
Oslo Gardermoen	59.1
Copenhagen	51.4
Stockholm Arlanda	49.3
Manchester	44.1
Dublin	37.9
Athens	36.9

Source: adapted from Eurocontrol (2022)

15.6.4   *Operating restrictions*

Noise concerns have led to some states phasing out older, noisier aircraft, for example the EU in 2002 with Chapter 2 aircraft. However, these can have significant economic implications for the airlines concerned and it is important that a balance be struck between the needs of airlines, airports and environmental interests. ICAO has urged member states to only adopt operating restrictions where such action is supported by a prior assessment of anticipated benefits and possible adverse impacts. Chapter 7.1.7 in ICAO (2010) provides examples of possible operating restrictions:

Cap rules

Defining a maximum number of operations not to be exceeded at an airport for any given period of the year. For example, in 2023, the Mexican government announced a reduction in flight capacity from 52 to 43 take-offs and landings per hour at Mexico City International Airport (CAPA, 2023).

Noise quotas

Cap the total noise level from aircraft operations over a given period. This can be based on a historic noise level or a future goal.

Night-time restrictions

Flights at night are of particular concern, mainly due to sleep requirements and many airports have some aspect of night-time flight restrictions.

Curfews

These may be global or aircraft-specific partial operating restrictions that prohibit take-off and/or landing during an identified time period. Geneva Airport (2024) does not permit take-offs or landings between 0030 and 0600, unless for exceptional reasons.

ATG Case study 15.7

Noise reduction initiatives at London City Airport, UK (LCY)

LCY is located approximately six miles east of the City of London and is surrounded by urban and residential areas. The Environmental Noise (England) Regulations 2006 requires operators of civil airports in England to produce Noise Action Plans to manage airport noise. Key aspects of LCY noise initiatives include:

- Aircraft movement limits. 592 per weekday, 100 on Saturdays and 200 on Sundays.
- Reduced operating hours. 0630–2230 on weekdays, 0630–1300 on Saturdays and 1230–2230 on Sundays.
- Use of SIDs and STARs.

- Aircraft approaching LCY follow a steep approach angle of 5.5° (instead of around 3° at most other airports), which helps keep aircraft higher for longer, reducing the noise impact on local communities.
- Use of RNAV1 on flightpaths.
- Noise and flight track monitoring system.
- Quiet operating procedures, such as minimum use of reverse thrust and GPU usage.
- Incentives and penalties scheme based on departure noise levels.
- Annual noise contours.
- Sound insulation schemes, offering sound insulation treatment to eligible residential properties within the 57, 63 and 66dB $L_{eq,16h}$ noise contours (London City Airport, 2018).

Despite these initiatives, there is still much opposition to any expansion, both from local communities and in government (London Assembly, 2023), illustrating the challenges many airports face in balancing their operational and growth requirements with the problem of airport noise.

In the longer term, completely new aircraft designs, such as blended wings and perhaps even morphing aircraft, could potentially lead to major reductions in airframe noise and improved environmental impact (Astley, 2015). Noise reductions at scale are unlikely to come from existing designs and are more likely from disruptive designs and new propulsive systems. Pieren *et al.* (2024) suggested that the blended wing body concept could substantially reduce noise annoyance, with the reduced sound levels likely coming from the acoustic shielding by the body of the extended fuselage, with reductions of between 10–20dB.

Chapter review questions

15.1 Analyse the various health implications associated with aircraft noise. How does aircraft noise also affect biodiversity?

15.2 What are the main sources of aircraft noise? What are the main sources of ground noise at airports? For any ONE of these sources, analyse solutions which have been proposed to reduce the noise impact. What is the likelihood of success and are there any associated benefits and/or risks?

15.3 Explain the various ways in which the air transport industry measures noise. Using aircraft manufacturer data, can you find either the dB or EPNdB levels for any ONE aircraft type? How have the noise levels improved from previous variants of the aircraft or from previous aircraft in the same category?

15.4 Explain the development of ICAO regulations on noise chapters. For any ONE airport, can you find any discriminatory pricing based on noise category and to what extent?

15.5 The first element of the Balanced Approach to Aircraft Noise Management is the reduction of noise at source. Looking forward to proposed new engine and/or aircraft technologies, analyse the impact any of these new technologies may have on noise.

15.6 What are some of the key noise reduction strategies available to airports in terms of land-use planning and management? For your local airport, can you establish what their approach is to managing noise?

15.7 Explain the noise abatement operational procedures which have been introduced by the air transport industry. For your local airport, can you establish which measures they have introduced? Do the flight paths avoid the main population centres and if so, how?

15.8 Why should operating restrictions at airports only be introduced after other possible solutions have been attempted? At the main hub airport in your country, are there any operating restrictions in place? To what extent do you feel that these have constrained air transport growth and are they proportionate versus the local noise problem?

ATG trivia

Supersonic flight and the sonic boom

The iconic supersonic commercial aircraft, Concorde, was operated by Air France and British Airways (seven airframes each) from 1976 until the type was retired in 2003. Routes were almost exclusively operated on transoceanic flights (largely between Paris/London and across the Atlantic to destinations such as New York, Washington and Barbados). In addition to being very expensive to operate and maintain, environmental factors played a role in Concorde's retirement. The aircraft had large issues with noise, both at take-off – a take-off at Washington Dulles Airport in 1977 measured 119.4dB (Thibault, 2019) – and the "sonic boom", which essentially meant the aircraft could not be operated supersonically over land in many countries, limiting its sales potential.

ATG Did you know?

Sonic boom

A sonic boom is the thunder-like noise a person on the ground hears when an aircraft or other type of aerospace vehicle flies overhead faster than the speed of sound/supersonic. Air reacts like a fluid to supersonic objects. As objects travel through the air, the air molecules are pushed aside with great force and this forms a shock wave much like a boat creates a bow wave. The bigger and heavier the aircraft, the more air it displaces. The shock wave forms a cone of pressurised air molecules which move outward and rearward in all directions and extend to the ground. As the cone spreads across the landscape along the flight path, they create a continuous sonic boom along the full width of the cone's base. The sharp release of pressure, after the build-up by the shock wave, is heard as the sonic boom (NASA, 2003).

One of the keys to making future commercial supersonic air transport feasible is reducing the sonic boom. NASA has developed an experimental aircraft – X-59 – designed to reduce the sonic boom to a "sonic thump". The X-59 is planned to undergo test flying and provide the data to regulators, to hopefully review the rules prohibiting commercial supersonic flight over land.

References

Alquezar, R. and Macedo, R. (2019) 'Airport noise and wildlife conservation: what are we missing?', *Perspectives in Ecology and Conservation*, 17. Available at: https://doi.org/10.1016/j.pecon.2019.08.003.

Astley, J. (2015) *Whisper it – jet engines are getting quieter*, The Conversation. Available at: http://theconversation.com/whisper-it-jet-engines-are-getting-quieter-44331.

Benz, S. et al. (2022) 'Impact of aircraft noise on health', in L. Leylekian, A. Covrig and A. Maximova (eds) *Aviation noise impact management: technologies, regulations, and societal well-being in Europe*. Cham: Springer International Publishing, pp. 173–195. Available at: https://doi.org/10.1007/978-3-030-91194-2_7.

CAA (2018) *Aircraft noise and annoyance: recent findings. CAP 1588*. London. Available at: https://publicapps.caa.co.uk/docs/33/CAP1588_FEB18.pdf.

CAA (2022) *Aircraft noise and sleep disturbance: an update (2014–2022)*. Available at: https://publicapps.caa.co.uk/docs/33/CAP2370.pdf.

CAA (2023a) *CAP1149. Noise monitor positions at Heathrow, Gatwick and Stansted Airports*. Available at: https://publicapps.caa.co.uk/docs/33/CAP1149_Ed9.pdf.

CAA (2023b) 'The effects of aircraft noise on biodiversity. CAP2517'.

CAA (2024a) *Aviation noise and health*. Available at: www.caa.co.uk/consumers/environment/noise/aviation-noise-and-health/.

CAA (2024b) *Measuring and modelling noise*. Available at: www.caa.co.uk/consumers/environment/noise/measuring-and-modelling-noise/.

CAPA (2023) *Mexican standoff! Caps at Mexico City International Airport could have far-reaching consequences*. Available at: https://centreforaviation.com/analysis/reports/mexican-standoff-caps-at-mexico-city-international-airport-could-have-far-reaching-consequences-659524.

CDC (2022) *What noises cause hearing loss? | NCEH*. Available at: www.cdc.gov/nceh/hearing_loss/what_noises_cause_hearing_loss.html.

Clark, C. (2015) *Aircraft noise effects on health*. Prepared for the Airports Commission. Available at: https://assets.publishing.service.gov.uk/media/5a819b09e5274a2e87dbe879/noise-aircraft-noise-effects-on-health.pdf.

EASA (2022) *Aircraft noise*, EASA Eco. Available at: www.easa.europa.eu/eco/eaer/topics/technology-and-design/aircraft-noise.

EASA (2023) *Type-certificate data sheet. No. EASA.IM.A.120 for Boeing 737*. Available at: www.easa.europa.eu/en/downloads/7297/en.

Eurocontrol (2020) *Continuous climb and descent operations (CCO / CDO)*. Available at: www.eurocontrol.int/concept/continuous-climb-and-descent-operations.

Eurocontrol (2022) *EUROCONTROL Data Snapshot #32*. Available at: www.eurocontrol.int/publication/eurocontrol-data-snapshot-32-focusing-continuous-descent-operations-top-25-airports.

Evrard, A.-S. et al. (2021) 'Health effects of aircraft noise: overview of the cross-sectional DEBATS study's results'. Available at: https://icben.ethz.ch/2021/ICBEN%202021%20Papers/full_paper_28700.pdf.

FAA (2022a) *Aircraft noise levels & stages*. Available at: www.faa.gov/noise/levels.

FAA (2022b) *Fundamentals of noise and sound*. Available at: www.faa.gov/regulations_policies/policy_guidance/noise/basics (Accessed: 3 January 2024).

FAA (2023) *Appendix E: basics of noise*. Available at: www.faa.gov/media/28146.

FAA (2024) *Advisory circulars (ACs) – search results*. Available at: www.faa.gov/regulations_policies/advisory_circulars/index.cfm/go/document.list/.

Flughafen Zurich (2024) *Noise charges – Flughafen Zuerich*. Available at: www.flughafen-zuerich.ch/en/company/responsibility/noise-and-sound-insulation/noise-charges.

Gély, D. and Márki, F. (2022) 'Understanding the basics of aviation noise', in L. Leylekian, A. Covrig, and A. Maximova (eds) *Aviation noise impact management: technologies, regulations, and societal well-being in Europe*. Cham: Springer International Publishing, pp. 1–9. Available at: https://doi.org/10.1007/978-3-030-91194-2_1.

Geneva Airport (2024) *Measures and actions for the local residents*. Available at: www.gva.ch/en/Site/Geneve-Aeroport/Developpement-durable/Gestion-du-bruit-(1)/Gestion-du-bruit (Accessed: 11 January 2024).

Glasgow Airport (2024) *Noise Insulation Scheme (NIS)*. Available at: www.glasgowairport.com/about-us/noise/noise-insulation-scheme-nis/.

gov.uk (2013) *Aviation Policy Framework*. Available at: www.caa.co.uk/media/lygj4eso/aviation-policy-framework.pdf.

Guinness World Records (2024) *Loudest aircraft*. Available at: www.guinnessworldrecords.com/world-records/633410-loudest-aircraft.

Heathrow (2024) *Our noise strategy*. Available at: www.heathrow.com/company/local-community/noise/making-heathrow-quieter/our-noise-strategy.

Hume, K.I., Brink, M. and Basner, M. (2012) 'Effects of environmental noise on sleep', *Noise & Health*, 14(61), pp. 297–302. Available at: https://doi.org/10.4103/1463-1741.104897.

ICAO (2010) *Doc9829 AN/451 Guidance on the Balanced Approach to Aircraft Noise Management*. 2nd Ed.

ICAO (2015) *Aircraft noise technology and international noise standards. Dr. Neil Dickson, Environment Officer*. Available at: www.icao.int/Meetings/EnvironmentalWorkshops/Documents/2015-Warsaw/3_2_Aircraft-Noise-Technology-and-International-Noise-Standards.pdf.

ICAO (2017) *Annex 16 — Environmental Protection – Vol. I – Aircraft Noise*. 8th edn. Available at: https://elibrary.icao.int/reader/289652/&returnUrl%3DaHR0cHM6Ly9lbGlicmFyeS5pY2FvLmludC9teS1saWJyYXJJ5?productType=eBook&themeName=Blue-Theme.

ICAO (2022) *2022 Environmental Report*. Available at: www.icao.int/environmental-protection/Pages/envrep2022.aspx.

ICAO (2024) *The postal history of ICAO*. Available at: https://applications.icao.int/postalhistory/annex_16_environmental_protection.htm.

ICAO (n.d.a) *Aircraft noise*. Available at: www.icao.int/environmental-protection/pages/noise.aspx.

ICAO (n.d.b) *CAEP working group 1*. Available at: www.icao.int/environmental-protection/Pages/CAEP-WG1.aspx.

ICAO (n.d.c) *Noise reduction technology*. Available at: www.icao.int/environmental-protection/Pages/Noise-Reduction-Technology.aspx.

ICAO (n.d.d) *Reduction of noise at source*. Available at: www.icao.int/environmental-protection/pages/Reduction-of-Noise-at-Source.aspx.

Janssen, S.A. et al. (2011) 'Trends in aircraft noise annoyance: the role of study and sample characteristics', *The Journal of the Acoustical Society of America*, 129(4), pp. 1953–1962. Available at: https://doi.org/10.1121/1.3533739.

Kearns, S.K. (2021) *Fundamentals of international aviation*. 2nd edn. Abingdon: Routledge (Aviation Fundamentals).

Klatte, M. et al. (2017) 'Effects of aircraft noise on reading and quality of life in primary school children in Germany: results from the NORAH study', *Environment and Behavior*, 49(4), pp. 390–424. Available at: https://doi.org/10.1177/0013916516642580.

London Assembly (2023) *Airport's expansion threatens even more noise for suffering residents | London City Hall*. Available at: www.london.gov.uk/who-we-are/what-london-assembly-does/london-assembly-press-releases/airports-expansion-threatens-even-more-noise-suffering-residents.

London City Airport (2018) *London City Airport noise action plan*. Available at: www.londoncityairport.com/corporate/environment/noise-management-and-monitoring/noise-action-plan.

Maaz, M.A. (2022) *The main sources of aircraft noise & the steps taken to reduce them*, Simple Flying. Available at: https://simpleflying.com/aircraft-noise-sources-reduction-measures/.

MTU (2024) *Pratt & Whitney GTF™ engine – MTU Aero Engines*. Available at: www.mtu.de/engines/commercial-aircraft-engines/narrowbody-and-regional-jets/gtf-engine-family/.

NASA (2003) *Sonic booms*. Available at: www.nasa.gov/wp-content/uploads/2021/09/120274main_fs-016-dfrc.pdf.

NASA (n.d.) *STEM LEARNING: Noise–Chevron Design Educator Guide*. Available at: www.nasa.gov/wp-content/uploads/2018/07/chevrons-educator-guide-v6.pdf.

NASA GRC (2021) *Engines*. Available at: www.grc.nasa.gov/www/k-12/UEET/StudentSite/engines.html.

NATS (n.d.) *Measuring noise*. Available at: www.nats.aero/environment/noise-and-emissions/measuring-noise/.

Pieren, R. *et al.* (2024) 'Perception-based noise assessment of a future blended wing body aircraft concept using synthesized flyovers in an acoustic VR environment—the ARTEM study', *Aerospace Science and Technology*, 144, p. 108767. Available at: https://doi.org/10.1016/j.ast.2023.108767.

Saucy, A. *et al.* (2020) 'Individual aircraft noise exposure assessment for a case-crossover study in Switzerland', *International Journal of Environmental Research and Public Health*, 17(9), p. 3011. Available at: https://doi.org/10.3390/ijerph17093011.

Stansfeld, S. and Clark, C. (2015) 'Health effects of noise exposure in children', *Current Environmental Health Reports*, 2(2), pp. 171–178. Available at: https://doi.org/10.1007/s40572-015-0044-1.

Stansfeld, S.A. and Matheson, M.P. (2003) 'Noise pollution: non-auditory effects on health', *British Medical Bulletin*, 68, pp. 243–257. Available at: https://doi.org/10.1093/bmb/ldg033.

Thibault, O. (2019) *Concorde: technical feat, financial fiasco*. Available at: https://phys.org/news/2019-03-concorde-technical-feat-financial-fiasco.html.

University of Southampton (2024) *Leading the way in aircraft noise reduction*. Available at: www.southampton.ac.uk/engineering/research/impact/leading_the_way_in_aircraft_noise_reduction.page#publications.

WHO (1999) *Guidelines for community noise*. Available at: www.who.int/publications/i/item/a68672.

Wolfenden, A.D. *et al.* (2019) 'Aircraft sound exposure leads to song frequency decline and elevated aggression in wild chiffchaffs', *Journal of Animal Ecology*, edited by E. Derryberry, 88(11), pp. 1720–1731. Available at: https://doi.org/10.1111/1365-2656.13059.

Index

Note: **Bold** page numbers refer to tables; *italic* page numbers refer to figures.

a la carte ancillary revenues **289**, 290
Abate, M. 306
abrasion 215, 371
absolute humidity 104–105
absolute instability 109
absolute stability 109
ACI Europe 230, 301
Addis Ababa Bole International Airport (case study) 62, 305
adiabatic lapse rate 108–109; dry adiabatic lapse rate (DALR) 108–109; saturated adiabatic lapse rate (SALR) 109
advection 90, 91, *121*
advertising sold by the airline **289**
AENA 345
aeroacoustic noise 414
aerodrome reference codes 192–193, **192**
aerodynamic improvements 390, 392–393, 399
aerodynamic noise 416–417
aeroplane ear 93
aeroplane flight controls 80–81; axes of rotation 80–81; flight control surfaces 80–81
aeroplane service ceilings **87**
aerospace 31, 55, 303, 391, 434
aerotropolis 16–17, 313, 321–323
AFRAA 270
Africa 41, 62, 152, 161, 163, 164, *169*, 232, 235, 267, 269–271, 279, **298**, 299, 305–307, 315, 328, 338, 340, 405; airline economic performance 299–300
African Development Bank 306
ailerons 80–81
Air Accidents Investigation Branch (AAIB) UK 125, 126
air connectivity 301–307, 350; in Africa (case study) 305–306, 307; benefits 302–303; circular and symbiotic relationship between economic growth and *301*; global and regional 303–307, *303*, *304*, *305*
air density 92–95, 104, 177; effects on pressure 94; effects of temperature and water vapor 94

Air France 39, 40, **51**, 56, 259, 264, 299, 434
Air Inuit, Northern Canada 197
air masses 10, 159, 163–164, *163*; Continental Polar 163; Continental Tropical 163; Maritime Arctic 163; Maritime Polar 163; Maritime Tropical 163; returning Maritime Polar 164
air navigation regions 66
air quality 276, 320, 359, 382
air service liberalisation in Nigeria (case study) 271
air traffic management (ATM) 396, 403
air transport industry 30–69; aircraft 48–59; airline industry 36–48; airports 59–65: airspace 65–69; aviation value chain 36; origins of 32–36
air transport noise *see* noise
air-conditioning and the auxiliary power unit, noise 417
Airbus 52, 54–56, **56**, 57, *58*, 59, 127, 138, 199, 300, 308, *308*, 344, 362, 376, 392, 405
Airbus A220 53, **56**, **192**, 300, **430**
Airbus A320 series 41, 49, **51**, 53, 56, **56**, **87**, 138, 178, **192**, 196, 226, 284, 290, 362, 369, 376, **384**, **384**, 391, 392, 393, 394, 423
Airbus A350 52, 53, 54, 56, **56**, 62, 63, 82, **87**, 127, 299, **384**, 391, 423, 425, **430**
Airbus A380 24, 32, 49, **52**, 53, 54, 56, **87**, 149, **150**, 191, **192**, 195, 201, 237, 299, 362, **384**, **430**
Airbus aircraft currently in production **56**
Airbus and Boeing aircraft on passengers versus range *58*
Airbus ZEROe aircraft concepts (case study) 401
aircraft 48–59; aircraft and engine historical developments 49–53; extended-range twin-engine operations performance standards (ETOPS) 52–53; future 58–59; key developments in commercial aircraft development 1783–1947 **50**; key developments in commercial aircraft development 1949–2023 **51–52**; manufacturers 54–57; performance and payload range 57–58; types of airliner 53–54

440 *Index*

aircraft accident involving Dornier DO 228–202 at Bodo Airport Norway 4 December 2003 (case study) 153
aircraft and the future 58–59; demand 58, *59*; design 58–59
aircraft boneyards 376–377
aircraft demand 58, *59*; new to 2042 *308*
aircraft design 58–59
aircraft "drywashing": Emirates Airline/Dubai International Airport (case study) 237
aircraft manufacturers 54–57; Airbus 54–56, **56**; ATR 57; Boeing 54, **55**; Bombardier 56–57; De Havilland Canada 57; Embraer 56–57
aircraft performance and payload range 57–58; Airbus and Boeing aircraft on passengers versus range *58*
aircraft seats 394
aircraft servicing 236
aircraft take-off performance 104, 177–178
aircraft taxiing 6, 77, 283, 384–385, 396, 404
aircraft technologies 64, 302, 389–393, 399–402; electric 400–401; hybrid-electric 401–402; hydrogen 399–400
aircraft, types of airliner 53–54; narrowbody aircraft 53; regional aircraft 53–54; widebody aircraft 53
aircraft weight reductions 393–395; aircraft seats 394; All Nippon Airlines, Japan: reducing CO_2 emissions by reducing the weight of aeroplanes 395; in-flight entertainment (IFE) 394; in-flight literature 394; in-flight meal equipment 393; in-flight trolleys and product 394; potable water 394
aircraft wing and lift 79–80, *79*
airframe icing 122, 123, 152
airframe noise 416–417; aerodynamic 416–417; high-lift devices 417; landing gear 417
airframe technology 391–392
airline business models 39–48; airline differences from the "traditional" FSNC or LCC business models **44**; business aviation 47; cargo carriers 48; carrier-within-a-carrier (CWC) 45–46; charter carriers 43–45; FSNCs 40; FSNC and LCC characteristics **41**; hybrid 43; LCCs 41–43; regional carriers 45; specialist/niche carriers 46
airline consolidation 63
airline deregulation: Africa 269–271; air service liberalisation in Nigeria (case study) 271; Single African Air Transport Market (SAATM) 270; Yamoussoukro Decision (YD) 269–270
airline deregulation: air transport liberalisation arrangements within ASEAN countries **269**; Asia Pacific 268–269
airline deregulation: Europe 262–264; the Three Packages 262–263

airline deregulation: Latin America and the Caribbean 264–266; liberalisation in Argentina (case study) 266
airline deregulation: Middle East 266–267
airline deregulation: USA 260–262
Airline Deregulation Act of 1978 260–261, 330
airline industry 36–48; airline business models 39–48; metric terminology 38; performance 38–39
airline ownership restrictions 14, 265
airport(s) 59–65; business model 61; in China (case study) 64–65; constraints 63–64; Ethiopian Airlines and Addis Ababa Bole International Airport (case study) 62; an evolving industry 62–63; top ten for aircraft movements, 2023 **61**; top ten for cargo, 2023 **62**; top ten for passenger numbers, 2023 **61**; types of 59–63
airport(s), biodiversity 228–230
airport carbon accreditation (ACA) scheme 386, 397, 405
airport charges, noise related Zurich Airport 430, **430**
airport city 16–17, 239, 313, 321–323; Eastern Airport City Project and U-Tapao Rayong-Pattaya International Airport, Thailand (case study) 323, *324*; Incheon Airport, Seoul, South Korea (case study) 322
airport(s), climate change impact on 177–180, *177*
airport codes 26–27; IATA 27; ICAO 26, *26*
airport construction, using recycled materials in 375
airport drainage systems 237–238
airport(s), earthquake hazards 209–210
airport earthquake mitigation 210–211
airport(s), example earliest **190**
airport land use 317–318
airport master plans 191–192
airport operations, direct CO_2 emissions 385–386
airport regulations, international 191–193; aerodrome reference codes 192–193, **193**; ICAO Annex 14 aerodromes 191
airport(s), runways 193–208; grass 195–197; gravel 197; ice 198–199; mountain 201–202; paved surfaces 199–201; sand 197–198; slopes 207–208; tabletop 201–203; unpaved surfaces 195–199; waterways 203–206, *204*, *205*, *206*
airport slots 63–64
airport(s), top five in the world by land area **193**
airport(s), tsunami hazards 211–213
airport tsunami mitigation 213–214
airports, types of 59–63; airline consolidation 63; alliance hubs 61; business airport 61; business model 61; cargo hub 61; global 59; hub-and-spoke networks 62; local 60; longer-range aircraft 63; low-cost 61; purpose 60; regional/national 59; route network 60; secondary airports 63; traffic characteristics 60; transit 61

airport vehicles, noise 417
airport(s), water disposal 241–243
airport(s), water handling capacity 237–241; airport drainage systems 237–238; hazards–coastal flooding 238–239; hazards–river flooding 239–240, *240*; landslides 240–241
airport(s), water supply 236–237; aircraft "drywashing": Emirates Airline/Dubai International Airport (case study) 237; Rajiv Gandhi International Airport, Hyderabad, India: water sustainability through efficient devices, recycling and replenishment (case study) 236–237
airport(s) within the top 100 exposed to high take-off weight-restriction risk 181–182
airspace 14, 32, 65–69, 252–253, 255, 256, 257, 266, 308, 388, 395, 396, 403–404; flight information regions (FIRs) 66, *67*; horizontal boundary 65; Kármán line 66; types of 67–69; vertical boundary 65
airspace classification 68–69
airspace closure 215, 217, 253
Alaska Airlines 45, 197
Albania 263, **352**
Alberts, H.C. 338
All Nippon Airlines, Japan: reducing CO_2 emissions by reducing the weight of aeroplanes 395
alliance hubs 61
Alquezar, R. 413
altimetry 95–97; errors and corrections 97; QFE 97; QNE 97; QNH 96–97; types of altitude 95–96
altitude 65, 66, 84, 86, 88, 93, 95–97, 109, 110, 113, 123, 144, **146**, 148, 149, 154, 181, 183, 218, 319, 367, 387, 431
altocumulus clouds *111*, 112, 114
altostratus clouds *111*, 112, 114, 115, 165
American Airlines 16, 38, **39**, 43, 179, 267, 316, 343, 394, 406; the legend of the olive (case study) 393
American Samoa 353
Amores, A. 238
Amsterdam Schiphol Airport, Netherlands 17, 40, 61, 179, **190**, 238, 239, 256, 261, 284, 305, 320–321, 323, 372–375, 403, 413; Buitenschot Land Art Park 374; flood protection at (case study) 238–239
anabatic winds *see* valley winds
Anchorage Airport, Alaska 61, **62**, 71, *71*
ancillary revenue 38, **41**, 43, 275, *285*, *288*, 288–290, 291; a la carte features **289**; advertising sold by the airline **289**; commission-based products **289**; fare or product bundles **289**; frequent flyer activities **289**; Wizz Air, Hungary (case study) 290
Anderson, J.D. 393

Anderson, R.S. 361
Anderson, S.P 361
anemometer 130, *132*
Annex 16 **34**, 404, 421, *422*, 423
annoyance, noise 22, 412, 414, 417, 420, 429, 433
Antarctica, ice runways (case study) 198–199, *199*
anthropocene 173, 360–361
anthropogenic geomorphology 361–362
anthropogenic processes, direct **361**; indirect **361**
Apte, S. 365
Arab Air Carriers Organisation (AACO) 267
Arctic front 164
areas of human impact in geomorphology 361–362, *363*
Argentina 145, 198, 260, 264, **265**, 266
argon **82**, **83**
Aristotle 6
artificial islands 19, **19**, 207, 318, 361–364, 368
ASEAN Multilateral Agreement on Air Services (MAAS) **269**
ASEAN Multilateral Agreement on the Full Liberalisation of Air Freight Services **269**
ASEAN Multilateral Agreement on the Full Liberalisation of Passenger Air Services (MAFLPAS) **269**
ash clouds 215, 217
Asia-Pacific region 232, **298**, *305*; airline economic performance 298
asphalt runways 190, 195, 200, 201
Associate Members of United Nations regional commissions 329, *330*
Association of Caribbean States (ACS) 266
Association of Southeast Asian Nations (ASEAN) **269**
Astley, J. 426, 433
ATAG 187, 300, 302, 303, 306, 333, 398, 402, 403, 405
ATM *see* air traffic management
the atmosphere and air transport 77–98; altimetry 95–97; components of 82–83; International Standard Atmosphere (ISA) 86–87; layers 83–86; pressure and density 92–95; principles of flight 78–82; temperature 88–92
atmospheric circulation 10, 78, 158–162; effects of three-cell circulation model 160–162; three-cell circulation model 159–160
atmospheric pressure 92–95; high and low pressure 93–94; measurement and units 93
atmospheric stability 6, 108–110, **109**; adiabatic lapse rate 108–109; types of 109–110
atomisation 24
ATR 54, 57, **87**, 421
attrition 371
automated arrival and departure procedures 430
Auxiliary Power Unit (APU) 383, 385–386, 417, 431

available seat kilometres (ASKs) 38, 42, **285**, 294, 343; global share of LCC ASKs **42**
available tonne kilometres (ATKs) 38, **285**
Aviation Environment Foundation 229
aviation value chain 32, 36, *37*, 232, 275, 280–284; airlines 282; aviation fuel production 282–283; fuel hedging 283; manufacturers 283; revenue by subsector, 2022 *281*
Avila, D. 388
A-weighted decibels (dBA) 418
axes of rotation 80–81
axial tilt 159

Badcock, M. 352
Baker, D. 321
Baker Islands 353, *354*
banks, landforms 238, 361
Bao, Z.Y. 373
Barcus, H.R. 338
barometer 93, 95
Barra Airport, Outer Hebrides, Scotland (case study) 46, 197, *198*, 328
Barry, M.A. 232
Barry, R. 137, 139, 163
Basner, M. 412
Baumeister, S. 176, 177
Beaufort wind force scale 131, **132**
Bednarek, J. 323
Beijing Daxing Airport, China 317, 362
Bell, M. 231
Benoist Model XIV 48, 49, *50*, *204*
Benson, M. 351
Benz, S. 412, 413
Bermuda I Agreement 259
Bicker, R. 230
bilateral air service agreements (ASAs) 14, 258–260, 264, 270; Bermuda I Agreement 259
bilateral and national regulations 253
biodiversity at airports 228–230; London Gatwick Airport, UK: Biodiversity Action Plan (case study) 230; Maun International Airport, Botswana: restoration and regeneration (case study) 229; Salvador Bahia Airport, Brazil: fauna management (case study) 230; Vancouver International Airport, Canada: a salmon-safe airport (case study) 229
biodiversity, changes to 176
biodiversity, effects of aircraft noise on 413
biogeography 5, 11–12, 221–228; habitat modification and fragmentation 222; wildlife management at airports 223–227
biological hazards 230–234; air transport and pandemics 231; COVID-19 232–234; disease transmission 230–231; high-efficiency particulate air (HEPA) filters 231–232; severe acute respiratory syndrome (SARS) 231–232
biosphere 175, 222

bird strikes 224–227, 374; first and first fatality 225; highest 227; locations *225*; "Miracle on the Hudson River", New York City, US Airways 1549 (case study) 226, *226*
Bisignani, G. 389
"black swans" 280
blended wing body (BWB) *399*, 401, 433
blue ice runways *see* ice runways
Bocci, M. 375
Boehm, S. 175
Boeing 21, 49, **51**, 53, 54, *58*, *59*, 127, 197, 283, 308, 343, 362, 385, 390, 391, 392, 421; aircraft currently in production **55**
Boeing 737 Series 42, 43, 46, 49, **55**, **87**, 178, **192**, 197, 283, 299, 392, (case study) 423, **424**, 426
Boeing 747 Series 40, 49, *55*, 122, 149, **192**, 196, 218, 260, 369, 377, 384, 403, 426, **430**
Boeing 767 Series 53, 54, **55**, 62, **192**, 392
Boeing 777 Series **51**, 53, 125, 178, **192**, 275, 390, 391, 395, 425, **430**
Boeing 777-8 and 777-9 (B777X) folding wingtips (case study) 192–193
Boeing 787 **52**, **55**, 63, **87**, 167, **192**, 199, 391, 401, 403, 423, 426
Bombardier 45, 47, 53, 56–57, **192**, 223, 401
Boniface, B. 22
Boston Logan Airport, USA 43, 53, 228
Boucsein, B. 316, 318, 323
Bowen, J. 30, 59, 78, 232
Boyle, M. 4, 12, 13, 347
BPR *see* engine bypass ratio
branches of geography 4–5, *5*
Bretton Woods Conference 277
Brink, M. 412
Brisbane Airport, Australia 366; "soil like toothpaste" (case study) 201
British Airways (BA) 16, 40, 46, 48, **51**, 125, 167, 218, 259, 264, 289–290, 316, 343–344, 394, 405, 434; "Perfect Flight" (case study) 405
Brown, T. 231
Bryant, E. 168, **171**, 211, 214
Budd, L. 65, 231, 261
building design 210
Buis, A. 159
Buitenschot Land Art Park, Netherlands 374
Bulgaria 18, 350, **352**
Burbidge, R. 176
Burghouwt, G. 62
Burrell, K. 352
business airport 61
business aviation 47; NetJets (case study) 47
Butler, G. 352
Button, K. 268, 349
Buys Ballot's Law 141

CAA, South Africa 223
CAA, UK 340, 396, 402, 411, 412, 413, 420

CAAF *see* ICAO Conference on Aviation and Alternative Fuels
cabin humidity 126–127, 231
CAEP *see* ICAO Committee on Aviation Environmental Protection (CAEP)
Caldecott, B. 181, **182**
California Bearing ratio (CBR) 195
Canada geese 226
canals 361
Cañizal, J. 375
Canterbury Nor'wester winds *see* foehn winds
CAPA 41, 42, 46, 256, *291*, *297*, 299, 343, 432
carbon dioxide *see* CO_2
carbon footprint 386, 396–397, 404
carbon offsets 404, 405; (case study) 406
Card, D. 261
cardiovascular disease, noise 22, 412–413
cargo carriers 48
cargo hub airport 61, 71
cargo tonne kilometres (CTKs) 38, **39**
Caribbean 42, 66, 181, 264–266, *305*, 329, 347, 405
Caribbean Community (CARICOM) 265, **265**
carrier-within-a-carrier 45–46; Go (case study) 46
CASK *see* cost per available seat kilometre
catalytic tourism impacts 300
Cathay Pacific 19, 48, 290, 406; Cathay Cargo (case study) 48
Caves, R.E. 134, 192, 196, 200, 207
celsius (°C) 88
centrality 59
Centralny Port Komunikacyjny (CPK) Airport, Poland 320
centre of gravity (CG) 80, 82
centripetal acceleration 137, 139
Cessna 47, **87**, 401
Chapter 2 Noise Standard 421, *422*, 423
Chapter 3 Noise Standard 422, *422*, 423
Chapter 4 Noise Standard 422, *422*
Chapter 14 Noise Standard *422*, 423
characteristics of airline operations 284–285; derived demand 284; homogenous product 284; perishable product 285
Chariton, A. 365
charter carriers 43–45; TUI Group (case study) 45
Chek Lap Kok Airport, Hong Kong (case study) 19, *20*, 318, 321, 362; carbon footprint (case study) 386; Hong Kong International Airport (case study) 365–366
Chen, M. 322
Chertock, M. 235
Chicago Conference *see* Chicago Convention
Chicago Convention 1944 14, 32, 33, **34**, 36, 65, 251, 254–258
Chicago O'Hare Airport 10, 60, *61*, 316, 362, 375
China 48, 129, 177, **194**, 231–232, 267–269, 277, 279, 297, 298, 300, 303, 304, 308, 314, 315, 317, 342, **346**, **348**; airports (case study) 64–65, **64**, **65**; growth in (case study) 343
Chinook winds *see* foehn winds
Chorley, R. 137, 139, 163, 187, 361
CIA 187
cirrocumulus clouds *111*, 112, 113
cirrostratus clouds *111*, 112, 114, 165
cirrus clouds *111*, 112, 113, 114, 165, 388
cities, top three for connectivity per world region 2019 305; world's largest 315
Civil Aviation Authority (CAA) 36, 253
Clague, J.J. 209, 214, 239, 240
Clark, C. 412
clear air turbulence (CAT) 84, 146, 149, 167, **171**, 176, 180
clear ice 123
Clermont-Ferrand, France 199
climate *see* climatology
climate change 5, 10, 19, 20–21, 78, 92, 172–175, *175*, 235, 276, 330, 345, 359, 368, 380–405; contemporary 173–175; disruption in the air 180–181; disruption to passenger demand 181–182; greenhouse gas and 172; history 172–173; impacts on air transport industry 175–182; impacts on Greek airports (case study) 178
climate change, future air transport strategies 397–405; aircraft technologies 399–402; IATA strategy towards Net Zero 2050 397–398; ICAO long-term aspirational goal (LTAG) 398; market-based measures (MBM): carbon offsetting and reduction scheme for international aviation (CORSIA) 404–405; operations and infrastructure 403–404; sustainable aviation fuels (SAF) 398, 402–403
climate change, impact and mitigation of air transport on 380–406; airport carbon accreditation scheme 397; CO_2 emissions 382–386; the future 397–405; mitigation of emissions 389–397; non-CO_2 emissions 386–389
climatology 5, 7–10, 77–78, 157–186, 188; air masses 163–164; air transport and 157–186; climate change 172–175; fronts 164–166; global atmospheric circulation 158–162; impacts of climate change on air transport industry 175–182; jet streams 166–168; tropical cyclones 168–172
clouds 110–118; classification 110–113; formation 116–118; types–high *111*, 112, 113–114; types–low *111*, 112, 114–116; types–middle *111*, 112, 114
CO_2 21, 82–83, 173, *174*
CO_2 emissions, air transport 382–386; airport operations 385–386; combustion of jet fuel in aircraft engines 383–385; Hong Kong International Airport carbon footprint (case study) 386; production and assembly of aircraft 385

coalescence 103, 118
coastal erosion **171**, 370–371
coastal flooding 12, 238–239; flood protection at Amsterdam Schiphol Airport, Netherlands (case study) 238–239
coastline alteration 368
Coates, D.R. 362
Cochin International Airport, India (case study) 397
Coffel, E. 178
cognitive impairment, noise 22, 412
Cohen, S. 351
cold front 114, 164–165, *165*
Coleman, J. 77
College Park Airport, Maryland 190, **190**
combustion of jet fuel in aircraft engines 383–385
commission-based ancillary products **289**
comparative dBA noise sources *419*
compass points 130, 135
competition 15, 40, 56, 62, 259, 260, 261, 262, 264, 267, 270, 275, 279, 284, 287, 293, 298, 303, 339, 403
complementarity 22–23
composite materials **52**, 127, 154, 294, 391–392, *391*, 394
Concorde 23, 49, **51**, 52, 66, 86, **87**, 110, 154, 272, 434
concrete runways 190, 195, 199, 200, 201
condensation 103, 104, 106, 107, 109, 116, 121, 144, 161
condensation trails *see* contrails
conditional instability 109–110
conduction 90, 91
Continental Polar air mass 163
Continental Tropical air mass 163
continual climb operations (CCO) 395–396, *395*
continuous descent operations (CDO) 395–396, 431, **431**
contrails 21, 113, 387–388, *388*, *389*, 400
controlled airspace 67, 68, 293
convection 85, 90, 91, 108, 116, 117–118, 142, 148, 150
convective turbulence 121, 146, *147*
Convention on International Civil Aviation *see* Chicago Convention
Cooper, C. 22
Cooper, R. 22
Coordinated Universal Time (UTC) 272, 353
Coriolis force/effect 137, 138–139, *139*, 140, 141, 159, 168
Cornell Law School 262
corrosion 127, 212, 215, 371, 376
CORSIA *see* market-based measures
cost per available seat kilometre (CASK) **285**, 294, *294*, 295, 296
costs, airline **41**, 42, 43, 49, **51**, **52**, 56, 63, 69, 133, 177, 180, 193, 200, 232, 253, 259, **285**, 286, 287, 289, 292–296

Coulter, L. 176, 177
Courchevel Altiport, France (case study) 207, *208*
COVID-19 12, 31, 36, 221, 232, 278, 280–282, 286, 287, 291, 295–296, 302, 333
Cox, B. 385
Crandall, R. 393
crannogs 363
Creedy, S. 218
Cristea, A.D. 267
crosswinds 134, 135, 136
Crouch, T. 129
Crutzen, P.J. 360
cultural diaspora 347
cumulative noise margin 422
cumuliform clouds 112
cumulonimbus clouds 110, *111*, 112, *115*, 116, 118, 119, 150, 161, 164
cumulus clouds *111*, 112, 115, 116, 117, 119, 150, 164, 165
cyclones *see* tropical cyclones
cyclostrophic wind 140, 141

Dallas Fort Worth International Airport 27, **61**, **193**, *194*, 362
Darién National Park, Panama (case study) 327, *328*
Dastrup, R.A. 13
datum 96, 97
Dávid, L. 361, 366
Davie, T. 234
Davies, T.R. 209, 214, 239, 240
Davis-Monthan Air Force Boneyard, Tucson, Arizona 376–377, *377*
Dawson, R.J. 179, 238
day-night average sound level (DNL/L_{den}) 420
DC-3 49, **50**, 195, 240
de Havilland 32, 49, **51**, 57, **192**, 203, 223, 421
de Havilland Canada Dash 8 54, 57, 223
de Havilland Canada DHC-2 203
de Havilland Canada DHC-6 Twin Otter 32, **87**, 197, 203
de Moor, S. 393
De Wit, J. 62
Debbage, K.G. 14
decibels (dB) 418
Decroly, J.-M. 345
de-icing 124–125, 241–242, *243*; operations at Oslo Gardermoen Airport, Norway (case study) 242, *243*
demand generation 339
demographics 339, 342
Dempsey, P.S. 259
density altitude 95–97
Denver International Airport, USA 10, **44**, **61**, **182**, **193**, *194*, 313, 362
deposition 19, 104, 118, 370, 371
deregulation *see* airline deregulation
derived demand 23, 38, 284

Index   445

Derudder, B. 339
descent profiles 431
desertification 10, 158, 176
designation 259, 263
dewpoint 102, 106, 107, 117, 120, 121, 144
diaspora 347
direct economic impacts 300
disaster response 333
dispersed flight tracks 430
displaced thresholds 431
Dobruszkes, F. 345
Doganis, R. 15, 44, 252, 254, 259, 261, 284, 286, 296
Don Mueang International Airport, Thailand 63, 239, *240*, 323, *324*
Douglas, I. 19, 375
drag 79, *79*, 80, 81, 133, 392, 395, 416, 431
dredging **19**, 201, **361**, 363, 366, 368, 369
Dresner, M. 48
drizzle 115, 119, 120, 165
Dubai International Airport 23, 59, 60, **61**, 71, **182**, 237, 267, 298, 299, 303, 346, **347**, **384**, 431
Duncan, T. 351
Dunn, G. 299
Duram, L.A. 18
dust storm 120
dykes 238, 239, 359, 361–362, 372

Earhart, A. 354
earthquakes 209–211; airport mitigation 210–211; hazards and airports 209–210; in Nepal, 2015 (case study) 210; Sabiha Gökçen Airport, Istanbul: seismic engineering (case study) 211
EASA 86, 123, 335, 376, 384, 396, *400*, 423, **424**
Eastern Airport City Project and U-Tapao Rayong-Pattaya International Airport, Thailand (case study) 323, *324*
Eastern Economic Corridor (EEC), Thailand 323
ECAC region and flood risk (case study) 179, 181
Echevarne, R. 265, 266, 327, 370
ecology 222
economic development and air transport 300–307; benefits of air connectivity 302–303; economic impacts 300; global and regional air connectivity 303–307; importance of air connectivity 301
economic geography 14–15, 275–309; air transport and economic development 300–308; aviation value chain 280–284; economic outlook for the air transport industry 308–309; economic performance of the airline industry 284–300; factors influencing air transport growth 276–280
economic outlook for air transport industry 308–309, *308*
economic performance of the airline industry 284–300; characteristics of airline operations 284–285; costs 292–296; declining yields 286–287; growth in revenue and ancillaries 287–292; high growth industry 286; profitability 296–297; regional economic performance 297–300
economic terminology 285
Edmunds, J. 40, 56
EESI 383, 388
effective perceived noise in decibels (EPNdB) 421
effective perceived noise level (EPNL) 421
effective radiative forcing (ERF) 387, *387*
El Niño 7–9, *8*, 380
electric aircraft 335, 400–401; Airbus ZEROe aircraft concepts (case study) 401
elevators 80–81
ELR 108–110
embankments **19**, 361, 362, 372–374
Embraer 43, 45, 47, 53, 56–57, 423; Unidade Gaviao Peixoto Airport 194
emergency response 210, 212
emissions trading schemes (ETS) 404–405
emissions, CO_2 382–386; airport operations 385–386; combustion of jet fuel in aircraft engines 383–385; Hong Kong International Airport carbon footprint (case study) 386; production and assembly of aircraft 385
emissions, mitigation of 389–397; greener on-the-ground infrastructure 389, 396–397; improvements in aircraft and engine technology 390–393; operational improvements 395–396; service (operating) weight reductions 393–395
emissions, non-CO_2 386–389; radiative forcing (RF) 387; contrails 387–388, *389*
engine bypass ratio (BPR) 390–391, 416, 422, 425
engine components and structural vibrations, noise 415–416
engine icing 122, 123–124
engine technology 21, 389–391
enhancing trade and market access in remote areas 327–328
environmental geography 4, 5 18–22, 357
environmental lapse rate *see* ELR
Equatorial trough 161
equivalent sound level (L_{eq}/L_{Aeq}) 420
Eratosthenes 4
Erdemir, G. 394
Eric, T.N. 306
eruptions of Eyjafjallajökull Volcano, Iceland, April 2010 (case study) 217
essential air service (EAS) 45, 330–331, *332*
Ethiopian Airlines 299, 306; Addis Ababa Bole International Airport and (case study) 62
Etihad Airways 48, 267, 298, 394
ETOPS **51**, 52–53
EU enlargement in 2004 (case study) 350, 352
EU/non-EU Open Skies 263–264
Eurocontrol 176, 178, 179, 180, 181, 396, 404, 431, **431**
Europe, airline economic performance 299

European Commission 256, 263–264, 314
European Common Aviation Area (ECAA) 263
European Parliament 14, 35, 262–264
European Union (EU) 14, 18, 86, 256, 262–264, 266, 268, 269, 270, 299, 306, 329, 335, 349–350, 352, 404, 423, 432
Eurostar 24
EU-US Open Skies (case study) 264
Evans, W.N. 261
evaporation 102, 103, 105, 107, 118, 234
Evrard, A.-S. 413
exosphere *84*, 86
extended-range twin-engine operations performance standards *see* ETOPS
extreme time zones *see* time zones

FAA 53, 57, 66, 83, 86–88, 91, 93–95, 97, 101–103, 105, 109, 110, **112**, 114, 115, 118, 134, 135, 138, 140, 141, 145, 146, 148–150, 152, 166, 168, 179, 195, 223, 333, 375, 411, 418, 420, 423, **424**
factors influencing air transport growth 276–280; economic growth 276–278; globalisation 279; gross domestic product (GDP) 276–278; income growth 278–279; liberalisation 279; political stability and security 280; reduction in airfares 279, *280*
Fageda, X. 330
fahrenheit (°F) 88
fan and compressor noise 415
fare or product bundles **289**
Faulks, R.W. 23
fauna management, Salvador Bahia Airport, Brazil (case study) 230
Federal Aviation Administration *see* FAA
Ferrel cell 159, *160*
financial performance *see* economic performance
Finger, M. 268
Finlay, M. 62
firefighting 221, 236, 241
Five Freedoms Agreement 33, *34*, 257
flight control surfaces 80–81
flight information regions (FIRs) 66–67, *67*
Flightradar24 71, *254*, *257*, 272
floatplanes 203
flood prevention and groundwater protection 372
flood protection at Amsterdam Schiphol Airport, Netherlands (case study) 238–239
flooding: at Don Mueang International Airport, Bangkok, Thailand (case study) 239, *240*; at La Vanguardia Airport, Colombia (case study) 240
flying boats 203, 204, *204*
flying vehicles (ATG trivia) 334–335
foehn winds 144–145, *145*
fog 120–122; advection 121; frontal *121*, 122; The hazard of fog: crash of KLM4805 and Pan Am 1736 at Tenerife Los Rodeos Airport 27 March 1977 122; orographic 122; radiation 120, *121*; steam 121
folding wingtips 192–193
Ford Field, Michigan, USA 190, 199
Fouberg, E.H. 12, 13
four forces of flight 79–80, *79*; drag 80; hazards due to icing 124; lift 79; thrust 80; weight 80
Frame, T.H.A. 180
France 25, 32, **39**, **50**, **51**, 61, 129, 164, 181, **190**, 199, 207, *208*, 258, 299, 340, 350, 413
Freedoms of the Air 14, 33, *34*, 254, 255–258, *255*, 257, 258
Freestone, R. 321
frequent flyer activities, ancillary revenues **289**
frequent flyer programmes (FFP) 40, **41**, 43
Frey, R.P. 114, 140, 141, 142, 143, 150, 157
friction force 139–140
frontal depressions 159, 160, 166
frontal lifting 108, 116, *117*
fronts, global 164; Arctic 164; ITCZ 164; Mediterranean 164; Polar 164
fronts, weather 164–166; cold 164–165; occluded 165–166, stationary 165; warm 165
FSNC 15, 40–41, **41**, 43, **44**, 264, 284, 285, 295; KLM Royal Dutch Airlines (case study) 40
FSNC and LCC characteristics **41**, **44**
fuel burn 80, 130, 133, 167, 180, 256, 384–386, 392, 395–396, 417; and % phase of flight **384**
fuel hedging 283
fuel production, aviation 282–283, 385
full-service network carriers *see* FSNC
Funk, J. 364
Furnace Creek, California 98

Galapagos Ecological Airport (case study) 405
The Galunggung Glider 218
Garver 375
Gasco, A. 316, 318, 323
GDP *see* gross domestic product
Gély, D. 414
General Electric **51**, 390, 426
geomorphic change *see* landform change
geomorphology 10–12, 187–188, 200, 234, 359–379; Amsterdam Schiphol Airport, Netherlands (case study) 373–374; anthropogenic 361–362; areas of human impact in 361–362, *363*; artificial islands 363–365; coastline alteration 368–370; Kansai International Airport, Osaka Bay, Japan (case study) 364–365; land levelling 367–368; Madeira (Funchal) Airport (case study) 369; ridges, embankments and dykes 372–374; Sangster International Airport, Montego Bay, Jamaica (case study) 370; Stornoway Airport, Isle of Lewis, Scotland (case study) 371–372; Toncontin Airport, Tegucigalpa, Honduras (case study) 367, *368*; urban 362; using recycled materials in airport construction 375

geophysical hazards 11, 187–188, 209–215; *see also* earthquakes; tsunamis; volcanoes
geostrophic wind 140, *140*, 141
Gibraltar Airport 146, 368
Glasgow Airport 405; noise insulation scheme (case study) 429
global atmospheric circulation *see* atmospheric circulation
global cities 17, 305, 315, 318, 339, 343, 346; Index **340**
globalisation 13, 49, 232, 252, 268, 279, 318, 338, 339, 360
Go (case study) 46
Goetz, A.R. 62, 318, 323
GOL Linhas Aéreas Inteligentes S.A, **44**, 290; (case study) 42
Goodwin, S. 157
Goudie, A.S. 360, **361**
gradient wind 141
Graham, A. 318, 320
Graham, B. 62
grass runways 190, 195–197
Gratton, G. 175, 176, 178, 180
gravel kit example *196*
gravel runways 195, 197; Air Inuit, Northern Canada (case study) 197
great circle route 70–71, *71*; between New York and Madrid *71*; between Tokyo and Louisville via Anchorage *71*
Greek airports (case study) 178
greener on-the-ground infrastructure 21, 389, 396–397
greenhouse gas (GHG) 78, 82, 98, 102, 172–173, 175, 181, 308, 380–381
Greenwich Mean Time (GMT) 272
Griggs, G. 179, 238, 239
Gross Domestic Product (GDP) 16, 31, 64, 181, 270, 276–279, *276*, *277*, *278*, 282, 300, 306, 308, 315, 326, 329, 330, 340, 341, 344, 350; annual GDP growth per country 2022 *278*; annual world % GDP growth 1982–2022 *277*; India (case study) 279; 'real' 277; twenty countries with the largest GDP in 2022 *276*
ground power unit (GPU) 385, 417, 431, 433
ground speed 167, 168
ground vehicles, noise 417; airport vehicles 417; surface access 417
groundwater flow 103
growth in revenue and ancillaries 287–292
Guerrero, J.E. 392
Guinness World Records 195, 207, 218, 418
Gulf airspace blockade: Qatar and Saudi Arabia 253, *254*
Gulf Cooperation Council (GCC) 267
Guo, R. 396

habitat: destruction 368; management 12, 221, 227, 372, 374; modification and fragmentation 222
Hadley cell 159, *160*

hail 103, 110, 119, 151
Halfacree, K. 338
Hannam, K. 18, 349, 350, 352
Hansen, J. 173
Hardiman, J. 168, 200
Hart, J. 105
Harvey, A. 188
Havana Convention, 1928 33, 254
Hayward, J. 40, 212
hazards *see* coastal flooding; river flooding; thunderstorms; turbulence
Hazen, H.D. 338
headwinds 130, 133, *133*, 299
healthcare access 333–334; Royal Flying Doctor Service, Australia (case study) 333–334
Heinz, S. 306
Helm winds *see* foehn winds
Herbertson, A.J. 78
high power engine runs 431
high pressure 79, 93–94, 97, 138, 141, 160–161
high-efficiency particulate air (HEPA) filters 126–127, 230–231
high-lift devices, noise 417
Highland Council 331
Highsmith, Carol M. *see* Library of Congress
Hillberry, R. 267
hoar frost 123
Hoeksema, R.J. 373
Holden, J. 4, 222, 228, 238
Holloway, S. 16, 314, 315, 316
homogenous product 284
Hong Kong International Airport *see* Chek Lap Kok Airport, Hong Kong
Horton, R. 178
Howland Islands 353, *354*
hub-and-spoke networks 23, *24*, **41**, 60, 62, 71, 260, 298
Hubbard, N. 271, 306
Huffman Prairie Field, Dayton, Ohio, USA 190
human geography 4,5, 12–18, 25, 249, 314; subdisciplines of 13
Hume, K.I. 412
humidity 6, 77, 97, 102, 203, 104–108, 161, 163, 164, 181, 388; absolute 104; dewpoint 106; latent heat 107, *108*; measuring 105; relative 105–106
Humphreys, B. 255, 258, 260, 268, 270
Hungary, Wizz Air **352**
hurricane hunters 183
Hurricane Maria (24 September 2017) *170*
hurricanes *see* tropical cyclones
hybrid carriers 43; JetBlue (case study) 43
hybrid-electric aircraft 401–402
hydraulic action 371
hydrogen-powered aircraft 21, 59, 64, 309, 399–401
hydrology 5, 12, 19, 25, 188, 221, 222, 234–243, *243*; air transport and 236; hazards–coastal

flooding 238–239; hazards–river flooding 239–240; landslides 240–241; regulations 235; water disposal at airports 241–243; water handling capacity at airports 237–238; water supply at airports 236–237
hygrometer 105
Hyndman, D. 208, 210, 215

IATA 21, 26–27, 36, *37*, 38, *39*, 48, 52, 58, 133, 231, *233*, 236, 251, 259, 268, 270, 278, 282, *286*, *287*, *288*, *292*, *296*, **298**, 299, 301–303, *303*, *304*, *305*, 307, 308, 326, 342, 344, 345, 389, 398, 401, 402, 403, 405
IATA codes *see* airport codes
IATA strategy towards Net Zero 2050 58, 309, 397–400, *398*, *400*, 402–403
ICAO 14, 22, 26, *27*, 31–36, 47, 48, 58, 66–68, 83, 87–88, 93, 106, 122, 130, 135, 146, 149, **150**, 176, 190–192, 195, 200, 207, 208, **212**, 212–214, 225–231, 236–237, 239, 241–242, 251–256, 258, 265, *269*, 270, 301, 320, 370, 380, 382, 395–400, 402–404, 410, 413, 421–433 ; Balanced Approach to Aircraft Noise Management 424–433; Standards and Recommended Practices (SARPs) 14, 33, 35–36, 191, 252, 255, 270, 422
ICAO Annex 14 aerodromes 191
ICAO Annexes 1–19 of the Chicago Convention **34**
ICAO Balanced Approach to Aircraft Noise Management 424–433, *425*; land-use planning and management 429–430, **429**, **430**; noise abatement operational procedures 430–431, **431**; operating restrictions 432–433; reduction of noise at source 424–429, *427*, *428*
ICAO codes *see* airport codes
ICAO Committee on Aviation Environmental Protection (CAEP) 403, 422–423
ICAO Conference on Aviation and Alternative Fuels (CAAF) 403
ICAO Doc 9137 Airport Services Manual Part 3 227
ICAO long term aspirational goal (LTAG) 398
ICAO noise regulations *see* noise regulations, ICAO
ice at the airport hazard: serious incident involving ATR72-212A on departure from Manchester Airport, UK on 4 March 2016 125
ice in-flight hazard: accident of British Airways Boeing 777-236ER at London Heathrow Airport (LHR) on 17 January 2008 125–126
ice runways 198–199; Antarctica (case study) 198–199, *199*
icing 107, 113, 114, 115, 122–126, 152, 161, 165, 176, 183; aircraft ground de/anti-icing 124–125; airframe 123; aviation hazards associated with 124; engine 123–124
IdeaWorks Company *288*, 289, 290

IEA 21, 382, 397, 402
impact of air transport on urban geography 320–324; aerotropolis and airport city 321–323; Eastern Airport City Project and U-Tapao Rayong-Pattaya International Airport, Thailand (case study) 323, *324*
impact of urban geography on air transport 316–320; airport land use 317–318; categorisation 318; environmental 320; site selection and spatial planning 318; surface access 319–320
imperial diaspora 347
Incheon Airport, Seoul, South Korea 17, **62**, (case study) 322, 372
income growth 278–279
India **50**, **61**, 267, 268, 277, (case study) 279, 298, 303, 304, 308, 314, 315, 341, 342, **346**, 347, **348**, 405
indicated airspeed 167
indirect economic impacts 300
Indonesia 213, 406, air transport in (case study) 341, *342*
induced economic impacts 300
infiltration 103
in-flight entertainment (IFE) 40, 49, 70, 283, 295, 385, 394
in-flight literature 394
in-flight meal equipment 393
in-flight trolleys 151, 283, 293, 394
infrastructure construction 372, 374–375
instrument flight rules (IFR) 68–69
integrators 32, 48
Intergovernmental Panel on Climate Change *see* IPCC
The Interim Agreement on International Civil Aviation 33
intermediacy 59
*inter*metropolitan approach 16, 315–316
The International Air Services Transit Agreement 33, 256
International Air Transport Agreement *see* Five Freedoms Agreement
International Business Aviation Council (IBAC) 47
International Civil Aviation Organisation *see* ICAO
International Energy Agency *see* IEA
International Organisation for Migration 347, **348**
International Standard Atmosphere *see* ISA
inter-tropical convergence zone (ITCZ) 161–162, 164
intervening opportunities 22–23
InterVistas 271, 306, *307*
*intra*metropolitan approach 16, 316
Iowa State University *136*, *137*
IPCC 20, 173, 175, 180, 381, 389–390
IPCC Sixth Assessment Report (AR6) on climate change 175, 381
ISA 6, 86–87, **87**, 92, 93, 96, 154; aeroplane service ceilings **87**; values at a range of altitudes **87**; values at sea level *87*

Ismaila, D.A. 271, 306
Ison, S. 261
Istanbul Airport, Turkey 23, 60, **61**, 211, 258, 317, 367
Italy **352**

Jackman, A. 13
James, D. 105
Jannus, T 48, **50**
Janssen, S.A. 412
Jarrett, P. 393
Jazeera Airways 267, 394
Jeffrey, R. 39
Jemiolo, W. 385
jet engine noise 414–416; aeroacoustic 414; engine components and structural vibrations 415–416; fan and compressor 415; jet 414; low-bypass and high-bypass ratio engines *415*
jet streams 84, 133, 140, 148, 158, 166–168, *167*, 180; impacts on air transport 167–168
JetBlue **42**, (case study) 43, 343, 392
JetStar **44**, 46
JNCC 228
Joshi, G. 168
Joshi, M. 180
Juancho E. Yrausquin Airport, Island of Saba, Caribbean Netherlands Antilles: the shortest commercial runway in the world 203

Kai Tak Airport, Hong Kong 19, *148*, 362, 365, 366, (case study) 147–148
Kano *162*
Kansai International Airport, Osaka Bay, Japan 171, 180, 188, 318, **347**, 364, *364*, 365, 369
Kaplan, D.H. 16, 314, 315, 316
Kármán line 66
Kasarda, J. 16, 17, 321–322
Kashe, K. 229
katabatic winds *see* mountain winds
Kazda, A. 134, 192, 196, 200, 207
Kearns, S. 31, 32, 60, 79, 85, 86, 122, 157, 172, 187, 422
Kelvin (K) scale 88
Kessides, I. 261
key developments in commercial aircraft development 1783–1947 50
key developments in commercial aircraft development 1949–2023 **51–52**
Kim, S.-H. 181
King Fahd International Airport, Saudi Arabia 10, 193, **193**, 316, 360
King, E.A. 373
Kiribati 353, *354*
Kito, M. 385
Kitty Hawk **50**, 129
Klatte, M. 412
KLM 4805 and Pan Am 1736 crash at Tenerife Los Rodeos Airport 27 March 1977 122

KLM Royal Dutch Airlines 32, **37**, 40, 46, 122, 264, 373, (case study) 40
Knox, D. 352
Kobe Airport 364–365
Korup, O. 209, 214, 239, 240
Koudis, G.S. 384
Kozuba, J. 391
Kraemer, M.U.G. 349
Kundu, A. 78
Kurnaz, S. 21, 172, 176, 177, 178, 180, 383, **384**, 385, 388, 390, 401
Kuropatwa, M. 391
Kutbi, N. 226
Kuzma, S. 235
Kuznets, S. 277

La Niña 7–9, *8*, 380
labour costs *293*, 294
labour diaspora 347
labour mobilities 350–351
Lagos 161, *162*
Lanchester F.W. 392–393
land levelling 362, 367–368
land reclamation **19**, **361**, 362, 365–368, 375
land-use planning and management 429–430, **429**, **430**; noise insulation scheme, Glasgow Airport (case study) 429
landform change 19–20, 359–379; Amsterdam Schiphol Airport, Netherlands (case study) 373–374; areas of human impact 361–362, *363*; artificial islands 363–365; coastline alteration 368–370; Kansai International Airport, Osaka Bay, Japan (case study) 364–365; land levelling 367–368; Madeira (Funchal) Airport (case study) 369; ridges, embankments and dykes 372–374; Sangster International Airport, Montego Bay, Jamaica (case study) 370; Stornoway Airport, Isle of Lewis, Scotland (case study) 371–372; Toncontin Airport, Tegucigalpa, Honduras (case study) 367, *368*; using recycled materials in airport construction 375
landforms 5, 10–12, 19, 25, 187–189, *189*, 193, 200, 209
landforms, spatial scales 188, *189*; classification 189; physical geography of airports 189–190
landing gear, noise 417
landscaping 201, 221, 229, 236, 375
Lankford, T.T. 92, 108, 109, *109*, 110, 146, 149, 166
Laroe, C. 232
Larsson, A. 351
Las Vegas, USA **61**, **182**, 344
latent heat 107, *108*, 123
Latin America 264–266, 279, **298**, 299, *305*, 326, 327, 405; airline economic performance 299
Latin American Aviation Commission (LACAC) 265, **265**

Law, K. 77
Lawson, N. 19, 375
LCC 15, 18, 40–43, **41**, **44**, 46, 61, 267, 284, 285, 295, 299, 306, 344, 352; GOL Linhas Aéreas Inteligentes S.A. (case study) 42; ULCC–Ryanair (case study) 43
Lean, G.L. 352
LeDuc, J.W. 232
Lee, D.S. 21, *382*, 386, 387
Lee, S.H. 180
leisure mobilities 351–352
liberalisation *see* airline deregulation
liberalisation in Argentina 266
lifestyle mobilities 351
lift 7, 66, 78, 79, *79*, 80, 81, 133, 148, 149, 177, 392, 426
lightning 116, 150, 151, 152, 153, 161
Lindsey, R. 82
load factor **285**, 291, *292*
location of tropical cyclones 168, **169**
Lockwood, J.G. 158, 159, 161, 166
Lóczy, D. 361, 366
Loh, C. 53
Lohmann, G. 46
London Gatwick Airport, UK 43, 344; Biodiversity Action Plan (case study) 230
London Heathrow Airport, UK 16, 23, 46, **51–52**, 60, **61**, 63, 106, *107*, 125, 179, 267, 272, 299, 302, 305, 316, 317, 320, 343, 344, 346, **347**, 384, 396, 403, 405, 420
London Luton Airport, UK 196, 316, 317, 352
London Stansted Airport, UK 16, 46, 63, 316, 319
longitudinal runway slopes 207–208
the loudest aircraft 418
low pressure 79, 93–94, 97, 138, 142, 144, 159–161, 164
low-cost airports 61
low-cost carriers *see* LCC
low-level wind shear alert system (LLWAS) 148
LTAG *see* ICAO long-term aspirational goal
Lv, Y. 180
Lyon.Saint Exupèry Airport 25, 413

Maaz, M.A. 417
MacDonald, D. 105
Macedo, R. 413
Mach speed **51**, 154
Madeira (Funchal Airport) case study 369, *369*
Madhav, N. 231
Madrid Convention, 1926 33, 254
"The mail plane"–remote air services subsidy (RASS) scheme (case study) 46
Maksel, R. 166
Maldives 238, 244, **329**, 362, 363; seaplanes (case study) 205, *205*, *206*
Manchester Airport 125, 375, 377, 413, **431**
Mara, F. 211

Mariscal Sucre Quito International Airport Ecuador (case study) 135, *136*, 319, *319*
Maritime Arctic air mass 163
Maritime Polar air mass 163
Maritime Tropical air mass 163
market-based measures (MBM): carbon offsetting and reduction scheme for international aviation (CORSIA) 404–405
Markhvida, K. 36, **37**
Márki, F. 414
Marshall, T. 5, 13
Martello, W.E. 261
massification 24
Matheson, M.P. 412
Matthews, T. 93
Mattoo, A. 267
Maun International Airport, Botswana: restoration and regeneration (case study) 229
maximum A-weighted noise level (L_{max}) 418
McCabe, K. 130, 166
McClatchey, J. 163
McDonnell Douglas 391
McKinsey 234
McLean, A.R. 232
McNeill, J.R. 360
McSweeney, K. 4, 18
measuring noise 417–421; A-weighted decibels (dBA) 418; day-night average sound level (DNL/L_{den}) 420; decibels (dB) 418; effective perceived noise in decibels (EPNdB)/effective perceived noise level (EPNL) 421; equivalent sound level (L_{eq}/L_{Aeq}) 420; maximum A-weighted noise level (L_{max}) 418; noise monitoring at London Heathrow (case study) 420; reference noise measurement points 421; sound exposure level 418, 420
mechanical turbulence 146, *147*
Mediterranean front 164
Meijer, G. 254
Mendes De Leon, P. 62
Mercator projection 70–71
Merkert, R. 325, 328, 330
mesosphere *84*, 86
Mesri, G. 364
Met Office 6, 88, 106, 110, *111*, 116, 130, 144, 145, 150, 159–161, 166, 168, 174, 380
METAR 106, *107*
meteor showers 86
meteorology 5, 6–7, 157; air transport and the atmosphere 77–100; air transport and water vapour 101–128; air transport and wind 129–156
Middle East 62, 66, 256, 266–267, 290, **298**, 303, *305*, 308, 346, 352; airline economic performance 298
Migon, P. **19**, 368
migration 331, 338–339, 347–352; Poland and LCCs–post-2004 (case study) 352

"Miracle on the Hudson River", New York City, US Airways 1549 (case study) 226, *226*
The missing minute 310
MIT 173, 404
mixed ice 123
mobilities 5, 18, 49, 338, 340, 347–352; EU enlargement in 2004 (case study) 350; labour 350–351; leisure 351–352; lifestyle 351
Moderate Resolution Imaging Spectroradiometer (MODIS) 389
Mojave Air and Space Port, Mojave Desert, California 377
Monarch Airlines 44, 369, 394
Monioudi, I.N. 181
Moody, Captain E. 218
mountain runways 201 202, *202*
mountain winds 142–143
MTU 425
multilateral regulations 32–33, 252, 253, 254–258, 265, 269, **269**, 382
multimodal airports 17, 319–320, 335
Murphy, A.B. 12, 13
Murphy, E. 373
Murray, N.J. 367
Mutel, C. 385
Myroniuk, V. 322

narrowbody aircraft 49, 53, 54, 56, 58, 62, 308, 309, 362, 369
NASA 21, 66, 78, 79, 80, 86, 98, 102, 104, 138, 157, 159, 166, *170*, 173, 174, *175*, *217*, 235, *240*, *314*, *335*, 389, *390*, 415, 426, *427*, 434, 435
Nashville International Airport 375
National Air Traffic Services (NATS) 23, 66, 405, 414, 420
National Transportation Safety Board (NTSB) USA 134, 149
nationality and sovereignty 252–254
natural hazards 208–209
net profit 36, 285, 296, 297, **298**, 299
net profit margin 38, **285**
Net Zero 2050 *see* IATA strategy towards Net Zero
NetJets (case study) 47
neutral stability 109, 110
New York airport system, USA (case study) 340, **341**
New York JFK Airport, USA **52**, **61**, 179, 200, 238, 272, 316, **341**, 343, 344, 346, **347**, **384**, 403
New York LaGuardia Airport, USA **341**
Newark Liberty International Airport New Jersey, USA (case study) 136, *137*, 238, **341**
Newbold, K.B 18
Newton's laws of motion 78, **78**
Next Generation Air Transportation System (NextGen) 404
NextGen *see* Next Generation Air Transportation System (NextGen)

Nigeria 315; air service liberalisation in (case study) 271, 306; and the ITCZ (case study) 161, *162*
night-time restrictions 432
nimbostratus clouds *111*, 112, 113, 115, 165
nine Freedoms of the Air 33, **255**, 258
the 1991 Mount Pinatubo (Philippines) eruptions and their effects on aircraft operations (case study) 215
nitrogen **82**, **83**, 116, 242
nitrogen oxides (NOx) 21, 386, 387
Niue 353
Njoya, E.T. 327
NOAA 7, 9, 66, **82**, **83**, *84*, *85*, 101, *102*, 103, 104, 135, 150, *151*, *162*, 163, 166, *167*, *169*, 171, 183, 211, 213, 387
noise abatement operational procedures 430–431, **431**; automated arrival and departure procedures 430; continuous descent operations (CDO) 431, **431**; descent profiles 431; dispersed flight tracks 430; displaced thresholds 431; ground power unit (GPU) 431; high power engine runs only in designated areas 431; limiting aircraft engine ground running 431; noise preferential runways 431; preferential routes 430; reduced power/drag techniques 431; reverse thrust 431; SID/STAR procedures 430
noise barriers 372, 373, *374*, **429**
noise cap rules 432
noise chevrons (case study) 426–428, **427**, **428**
noise contours 420, 429, 433
noise curfews 432
noise, impact and mitigation of air transport on 410–433; ICAO Balanced Approach to Aircraft Noise Management 424–433; ICAO noise regulations 421–424; measuring 417–421; sources of 413–417; why is noise a problem 410–413
Noise insulation scheme, Glasgow Airport (case study) 429
noise monitoring at London Heathrow, UK (case study) 420
noise operating restrictions 432–433; cap rules 432; curfews 432; night-time restrictions 432; quotas 432; reduction initiatives at London City Airport (case study) 432–433
noise preferential routes 430
noise preferential runways 431
noise quotas 432
noise reduction at source 424–429, *427*, *428*; airframe 426; engine technologies 425; noise chevrons (case study) 426–428, **427**, **428**
noise reduction initiatives at London City Airport (case study) 432–433
noise regulations, ICAO 421–424; Annex 16 421; Boeing 737 series (case study) 423, **424**; Chapter 2 Standard 421, *422*; Chapter 3

Standard 422, *422*; Chapter 4 Standard 422, *422*; Chapter 14 Standard *422*, 423; cumulative noise margin 422

noise-related airport charges, Zurich Airport, Switzerland (case study) 430, **430**

noise, sources of 413–417, *416*; air-conditioning and the auxiliary power unit 417; airframe 416–417; ground vehicles 417; jet engine 414–416

noise, why a problem 410–413; annoyance 412; cardiovascular disease 412–413; cognitive impairment 412; DEBATS study France (case study) 413; impact on biodiversity 413; most common health effects of aircraft noise exposure *411*; sleep disturbance 412

non-CO_2 emissions 386–389; contrails 387–388, *389*; radiative forcing (RF) 387

non-urban areas 325

nonaeronautical revenues 17, 318, 321

Noonan, F. 354

Nordhaus, W.D. 173

Norse Atlantic Airways (case study) 199, *199*

North America, airline economic performance 297

O'Connell, J.F. 267, 306
O'Reilly, K. 349, 351
OAG 24, **42**, 266, 284, 322, **346**, **347**, 384
occluded front *165*, 165–166
oldest airlines still operating, examples **37**
Olipra, L. 352
Oliver Wyman 48, 382
Ooishi, W. 166
open skies 252, 258, 261–271
operating profit **285**, 296, *296*
operating profit margin **285**, 296, *296*, 297, **298**
operational improvements 395–396; continual climb operation (CCO) 395–396; continual descent operation (CDO) 395–396
operations and infrastructure 403–404; Single European Sky (SES) 403; Single European Sky ATM Research (SESAR) 403
origins of air transport 32–36; Annexes of the Chicago Convention **34**; Chicago Convention, 1944 33; Five Freedoms Agreement 33, *34*; Havana Convention, 1928 33; ICAO 34–36; The Interim Agreement on International Civil Aviation 33; The International Air Services Transit Agreement 33; Madrid Convention, 1926, 33; Paris Convention, 1919 32; Provisional International Civil Aviation Organisation (PICAO) 33; sovereignty 32, 33; Two Freedoms Agreement 33
orographic lifting 108, 116, *117*, *121*
Oslo Gardermoen Airport 242, *243*, 431, **431**
Oster Jr, C.V. 261
outside air temperature (OAT) 154
ownership and control 259, 269, 306

ownership restrictions *see* airline ownership restrictions

oxygen 82–83, **82**, **83**, 84, 93, 95, 116

Pacione, M. 18, 315
Padhra, A. 21, 172, 176, 177, 178, 180, 383, **384**, 385, 388, 390, 401
Palm Jumeriah, Dubai *363*
Pancer-Cybulska, E. 352
Pande, P. 53
Paris Charles de Gaulle Airport, France 64, 384, **384**, 413
Paris Climate Agreement 173
Paris Convention, 1919 14, 32, 252, 254
Paris, C.M. 352
paved runways 199–201
pavements *see* paved runways
Pearce, B. 300
Pearl Island, Doha *363*
Pearson, J. 343
Pek, S. 181, **182**
Pentelow, L. 181
perishable product 48, 285, 325
Peters, L. 25, 203
Petley, D. 241
Pheil, A. 30, **50**
Phoenix Airfield, Antarctica 199
Phoenix Sky Harbor Airport, USA 97, 106, *107*, 178, **182**
physical geography 4–12, 75, 314; airports 189–190; subdisciplines of 5
Pieren, R. 433
Pijet-Migon, E. **19**, 368
Pinal County Airpark, USA 377
pitch 80–81, 125
plains 158, 361, 373
plant uptake 103
point-to-point networks 15, 16, 23, *24*, **41**, 303, 316
Poland 18, 320, 350; and LCCs–post-2004 (case study) 352
Polar cell 159, *160*
Polar front 160, 164
polar jet stream 166, *167*, 168, 180
political geography 13–14, 251–271; airline deregulation–Africa 269–271; airline deregulation–Asia Pacific 268–269; airline deregulation–Europe 262–264; airline deregulation–Latin America and the Caribbean 264–266; airline deregulation–Middle East 266–267; airline deregulation USA 260–262; bilateral air service agreements (ASA) 258–260, 264; bilateral and national regulations 253; Chicago Convention 254–258; multilateral regulations 253, 254–258; nationality and sovereignty 252–254; open skies 261–264, 266, 267, 268, 271; Russian airspace (case study) 256, *257*

Index   453

political stability and security 280
population density 339; distribution 339, 341, 342; growth 279, 315, 338, 339, 342, 345, 365, 367
population geography 5, 13, 18, 338–354; demand generation 339–344; migration 347–352; mobilities 347–352; seasonality 344–345; tourism and route networks 344–347
potable water 236, 394
Prather, C.D. 134
Pratt and Whitney 283, 390, 391, 425
precipitation 6, 7, 10, 77, 83, 103, 107, 108, **109**, 110, 114–116, 118–120, 143, 144, 150, 151, 158, 161, 164–166, 176, 177, 181; growth 118; types 118–119
pressure gradient force 137–138, 140, 141, 143
Price, M. 78
principles of flight 78–82; aeroplane flight controls 80–81; aircraft wing and lift 79–80, *79*; centre of gravity (CG) 80, 82; four forces of flight *79*; Newton's laws of motion 78; winglets 80
production and assembly of aircraft 383, 385
Prosser, M.C. 180
Provisional International Civil Aviation Organisation (PICAO) 33
public service obligation (PSO) 46, 302, 330–331

Qamdo Bamda Airport **194**, 195
Qantas 37, **44**, 46, 54–55, *55*, 63, 290, 393–394; Australia–London–the "Kangaroo Route" (case study) 54; route development between London and Sydney from 1947 *55*
QFE 96, 97, *96*
QNE 97
QNH 96–97, *96*
Quantick, H.R. 161
Quiet Technology Demonstrator 426
Quinn, N.W. 234

radiative forcing (RF) 387, *387*
rain 6, 7, 77, 94, 101, 103, 110, 115, 116, 118, 119, 120, 123, 133, 150, 151, 159, 161–165, **171**, 208, 244
Rajiv Gandhi International Airport, Hyderabad, India: water sustainability through efficient devices, recycling and replenishment (case study) 236–237
RASK *see* revenue per available seat kilometre
Recchi, E. 349
reduced power/drag techniques 431
reduction in airfares 279, *280*
reference noise measurement points 421
regional aircraft 53–54, 56, 57, 401; regional jets 45, 53, 56, 423; turboprop 54
regional airline economic performance 297–300
regional carriers 45; SkyWest Airlines (case study) 45
regional jets 45, 53, 56, 60, 423
regional/national airport 59

Reid, K. 210
relative humidity (RH) 105–106, *106*, 126–127
remote geography *see* rural geography
reservoirs 211, 361
return on invested capital (ROIC) 275, 281, 282, 283
returning Maritime Polar air mass 164
revenue passenger kilometres (RPKs) 38, **39**, **285**, *286*, **298**
revenue per available seat kilometre (RASK) **285**, 290, *291*, 292, 294, *294*, 295
revenue, airline 14, 36, 39, 45, 182, 232, **285**, 287–290, *288*, *291*, 294, 295, 296, *297*, 343, 394
reverse thrust 198, 431, 433
ridges, embankments and dykes 362, 372–374; flood prevention and groundwater protection 372; habitat management 374; infrastructure construction 374–375; noise barriers 372, 373, *374*
rime ice 123
Riordan, D. 78
Ritchie, H. 325, 382
river flooding 12, 239–240; at Don Mueang International Airport, Bangkok, Thailand (case study) 239, *240*; at La Vanguardia Airport, Colombia (case study) 240
Rodgers, C. P. 225
Rodrigue, J.-P. 22, 23, 30, 59, 78
roll 80–81
Rolls-Royce **52**, 391, 403, 426
Romania 18, 256, 350, **352**
Roser, M. 325
Ross, D. 139
route networks 23, 339, 344; world's busiest 346, **346**, **347**
Royal Flying Doctor Service, Australia (case study) 333–334
Royal Meteorological Society 105, **132**, 152, 160
RPKs *see* revenue passenger kilometres
rudder 80–81
runoff 103, 228, 229, 237, 369, 372
runway at Brisbane Airport "soil like toothpaste" (case study) 201
runway orientation 134, 135; Mariscal Sucre International Airport Quito Ecuador 135–*136*; Newark Liberty International Airport New Jersey USA 136–*137*
runway slopes 207–208; Courchevel Altiport, France (case study) 207, *208*; longitudinal 207–208; transverse 208
runways 193–208; example longest commercial **194**, 195; grass 195–197; gravel 197; ice 198–199; mountain 201–202; paved surfaces 199–201; sand 197–198; slopes 207–208; tabletop 201–203; unpaved surfaces 195–199; waterways 203–206
rural geography 17–18, 313, 314, air transport and 324–334; Colombia–Andean geography (case

study) 326; Darién National Park, Panama (case study) 327, *328*; disaster response and healthcare access 333–334; economic development and air transport in remote areas 327–330; enhancing trade and market access 327–328; essential air service (EAS) and public service obligation (PSO) 330–331, *332;* rural air service in East Malaysia (case study) 332; tourism 328–330; Wick John O'Groats Airport to Aberdeen International Airport, Scotland–PSO (case study) 331

rural population *17*, 324
Russia 39, **194**, 256–257, 278, 328
Russian airspace (case study) 256, *257*
Rutherford, D. 389
Ryanair 38, **39**, 41, **42**, 46, 53, 63, 262, 279, 283, 289–291, *292*, 295, 297, 299, 350, 352, 392; ULCC (case study) 43
Ryley, T. 176, 177

Sabiha Gökçen Airport, Istanbul: seismic engineering (case study) 211
Saccoccia, L. 235
Sadraey, M.H. 78, 82, 83, 84, 85, 104
SAF *see* sustainable aviation fuels
SAF feedstocks 402
Salamooska, J. 349
Salewski, C. 316, 318, 323
Salvador Bahia Airport, Brazil: fauna management (case study) 230
Samborska, V. *325*
Samoa 353
sand runways 197, *198*; Barra Airport, Outer Hebrides, Scotland (case study) 197, *198*
sandstorm 120
Sangster International Airport, Montego Bay, Jamaica 179, 181, 370, *370*
Sanguineti, M. 392
Saucy, A. 413
Saunders, L. 261
Schumer, C. 175
Schumm, S.A 187, 361
Scott, D.J 181
SDGs 235, **235**
sea level rise 176, 178, 238, 367
seaplanes in the Maldives 205, *206*
seasonality 344–345
the second lives of aircraft 376–377
secondary airports **41**, **44**, 63, 309
Sedona Airport, Arizona, USA *202*
SEL footprints 420
Semeyutin, A. 306
Sengupta, D. 365, 367
severe acute respiratory syndrome (SARS) 231–232
Sforza, L. 351
Shanghai Pudong Airport, China **62**, 178, 179, 362

Sheller, M. 18, 349, 350, 352
Sherry, L. 388
SID/STAR procedures 430
Sindhamani, V. 239
Singapore Changi Airport 61, 228, 238, 322, **347**, 365; Jewel Changi Airport–the "Rain Vortex" 244; wildlife management (case study) 228
Single African Air Transport Market (SAATM) 270, 299
Single European Sky (SES) 396, 403
Single European Sky ATM Research (SESAR) 403
single-engine taxiing 283, 396, 404
site selection 134, 190, 318
Skiles, J. 226
SKYbrary 35, **87**, 96, 97, 106, 119, 130, 133, 143, 149, 166, 223, 242, 390
SkyWest Airlines (case study) 45
sleep disturbance, noise 22, 412, 420
sleet 7, 103, 110, 119
Small Island Developing States (SIDS) 181, 325, 405, (case study) 329, *330*
snow 6, 77, 92, 104, 119, 120, 163, 198, 234, 241
social and cultural exchange 339
solar power 397, 404, 405
solar radiation 86, 88, 90, *90*, 92, 158, 159
sonic boom 434–435
sonic thump 435
soot 21, 386–387
Sorensen, J. 289
sound and noise 411
sound exposure level (SEL) 418, 420
South Carolina Geological Survey 189
Southern California Logistics Airport, USA 377
Southwest Airlines 38, **39**, 41, **42**, 43, **44**, 260, 279, 290, 295, 394
sovereignty 13, 32, 33, 35, 65, 252–255
specialist niche carriers 46; "The mail plane"–remote air services subsidy (RASS) scheme (case study) 46
Squire, R. 13
St. Petersburg-Tampa Airboat Line 30, **50**, 204
stability *see* atmospheric stability
Standard Arrival procedures (STAR) *see* SID/STAR
Standard Instrument Departure (SID) *see* SID/STAR
Standards and Recommended Practices, ICAO (SARPs) 14, 33, 35–36, 191, 252, 255, 270, 422
Stansfeld, S. 412
stationary front 165, *165*
Stauber, J.L. 365
Steffen, W. 360
Stettler, M.E.J. 396
Stevenson screen 88, *89*
Stockholm-Arlanda Airport 322, 377, **431**
Storer, L. 180

Stornoway Airport, Isle of Lewis, Scotland (case study) 371–372
Stratiform clouds 112
stratocumulus clouds *111*, 112, 116, 165
stratosphere 83, *84*, 85–86, 114, 166, 215
stratus clouds *111*, 112, 115, 119, 165
Suau-Sanchez, P. 345
subdisciplines of human geography *13*
subdisciplines of physical geography *5*
sublimation 104, 234
subsidence *19*, 176, 209, **361**, 367, 369
subtropical jet stream 166, *167*
Sugden, D.E. 187, 361
sulfate 386
Sullenberger, Chesley 226
supersonic flight and the sonic boom (case study) 434–435
surface access 59, 319–320, 339, 345, 386, 396; noise 417
surface temperature variations 92
Sushama, L. 178
sustainable aviation fuels (SAF) **52**, 309, 398–399, 402–403, 405–406; Virgin Atlantic and SAF (case study) 403
Sustainable Development Goals (SDGs) *see* SDGs
Suvarnabhumi Airport, Thailand 179, 239, 323, *324*, 362
Syed, M. 252
Szabó, J. 361, 366
Szostak, E. 352

tabletop runways 201–203, *202*, *203*, 369
tailwinds 130, 133, *133*; crash of Raytheon Hawker 800XP at Aspen Colorado 21 February 2022 134
take-off roll 95, 134, 224, 384, 421
Talcha Airport, Nepal *202*
tariffs 254, 259, 264
Tavares, R. 230
TaxiBot 404
temperature 6, 7, 9, 20, 77, 78, 82, 83, *85*, 86, **87**, 88–92, 98, 101, 103, 104–109, 110, 118–125, 142, 144, 145, 163, 164, 174, *175*, 176, 178, 181, 231, 238, 380, 381, 388; advection 91; conduction 91; convection 91; defining 88; effect of cloud and wind on 92; measurement and units 88; nature of the surface; solar and terrestrial radiation 88–90; surface variations 92
Tenzing-Hillary Airport, Lukla, Nepal (case study) 144, 207, 362
terracing 361
terrestrial radiation 88, 90, 92
thermal turbulence *see* convective turbulence 146, *147*
thermosphere *84*, 86
Thibault, O. 434
Thomas Cook 394

Thomas, H.S.C. 222
Thompson, T. 178, 388
Thomson, Sir Adam 284
three-cell circulation model 159–162; effects of 160–162; equatorial trough and inter-tropical convergence zone (ITCZ) 161–162; Ferrel cell 159–160; Hadley cell 159; high and low pressure 160–161; Nigeria and the ITCZ (case study) 161–162; Polar cell 160
the Three Packages–European liberalisation 262–263
thrust **51**, 78, 79, 80, 104, 126, 177, 198, 383–385, 390–392, 395, 396, 399, 401, 414, 418, 425, 431, 433
Thulemark, M. 351
thunder 116, 150, 151, 152
thunderstorm life cycle 150–151; decaying stage 151; mature stage 151; towering cumulus stage 150
thunderstorms 150–153; aircraft accident involving Dornier DO 228–202 at Bodo Airport; hazards to aviation 151–152
time travel: GMT, UTC and local times 272
time zones 272, 353, *354*
Tokyo 53, 60, **61**, 71, 212, 256, *257*, 315, **340, 346, 347**
Tolcha, T. 306
Toncontin Airport, Tegucigalpa, Honduras (case study) 367, *368*
top ten airports for aircraft movements, 2023 **61**
top ten airports for cargo, 2023 **62**
top ten airports for passenger numbers, 2023 **61**
topography 11, 88, 134, 188, 318, 325, 328, 360, 367
Torricelli, E. 93
Toulouse.Blagnac Airport 413
tourism 16, 18, 31, 45, 49, 175, 181, 204, 267, 270, 276, 280, 300, 302, 306, 316, 320, 322, 325, 326, 328–331, 339–341, 344–346, 349, 351–352, 359, 363, 369, 371; environmental and social challenges in 345–346; overtourism in Venice (case study) 345–346; in remote areas 328–330
tourism dependant locations 181, **329**, 330; Small Island Developing States (SIDS) (case study) 329, *330*
trade diaspora 347
trade winds 7, 161
traffic rights 14, 33, 254–259, 263–265, 270
TraNQuiL noise study 413
transferability 23
transit airports 61
transpiration 103, 234
transport geography 22–25; concept of space 22–23; core principles 23–24; elements of 23; modes of 22
transport mode choice: France (case study) 25; London to Paris (case study) 24

transverse runway slopes 208
Tretheway, M.W 36, **37**
Treviso Airport, Venice, Italy 375
Triple Tree Aerodrome, South Carolina 196
Troll Airfield, Antarctica 199
tropical cyclones 101, 168–172; classification 170, **171**; conditions required for formation 168; hazards 170, **171**; locations 168, **169**
troposphere 83–84, *85*, 90, 93, 159, 161, 163, 166, 168
tsunami 211–214; airport mitigation 213–214; deadliest–Indian Ocean 26 December 2004 (case study) 213; hazards and airports 211–213; near Sendai, Japan 2011 (case study) 213, 214; vulnerable airports *212*
TUI Group (case study) 45
turbofan engines 390, *390*, 391, 401, 415, 425–426
turboprop aircraft 45, **51**, 54, 57, **87**, 183, 334, 401
turbulence 7, 84, 85, 92, 112–116, 121, 142, 144, 145–150, 180–181; classifications **146**; clear air turbulence (CAT) 149; convective turbulence 146, *147*; mechanical turbulence 146, *147*; wake vortex turbulence 149, **150**; wind shear 148
turf runways *see* grass runways
Turmann, A. 351
Two Freedoms Agreement 33, 256, 257
Typhoon Jebi, Kansai International Airport, Japan, September 2018 171
Typhoon Odette, Mactan-Cebu International Airport, Philippines, December 2021 171
typhoons *see* tropical cyclones

UCAR 82, 91, 92, 164, 388
UK Aviation Policy Framework 410
Ullman, E. 22
ultra-high bypass ratio engine (UHBR) 425
unbundling 279, 289
uncontrolled airspace 68
UNDRR 11, 188
UNFCCC 173, 386
Union Glacier blue-ice runway, Antarctica 198–199
unit cost *see* cost per available seat kilometre
unit revenue *see* revenue per available seat kilometre
United Airlines **39**, 60, 136, 149, 272, 344, 394
United Nations (UN) 13, 14, 20, 32, 34, 36, 65, 173, 235, 252, 255, 305, 315, 325, 329, *330*, 333, 338, 341, 344, 347, *348*, 362, 380, 381, 385; Humanitarian Air Service (UNHAS) 333
United States Antarctic Program (USAP) 199
University Corporation for Atmospheric Research *see* UCAR
University of Illinois 141
UNOOSA 66
UNWTO 328, 330

urban air mobility 334, *335*
urban geography 13, 15–17, and air transport 313–324; Eastern Airport City Project and U-Tapao Rayong-Pattaya International Airport, Thailand (case study) 323, *324*; impact of air transport on 320–324; impact on air transport 316–320; Incheon Airport, Seoul (case study) 322; Mariscal Sucre Quito International Airport Ecuador (case study) 319; urbanisation and 314–316
urban geomorphology 362
urban population 16, 315, 343–344
urbanisation 16, 235, 314–316, 318, 323–324, 338–339, 342, 344, 360, **361**, 429
Urry, J. 18, 349, 350, 352
US Department of State 262, 271
US Department of Transportation 261, 331
U.S. Geological Survey *216*
U.S. national Weather Service (NWS) StormReady Programme, Tampa, USA 171–172
US open skies 261–262
US-EU open skies (case study) 264
utilisation **41**, 44, **44**, 268

valley winds 142–143
Vancouver International Airport, Canada: a salmon-safe airport (case study) 229
Vasigh, B. 300
Vega, H. 349
Venice, overtourism in (case study) 345–346
venturi effect 143
victim diaspora 347
Vidi, L. 93
Viles, H.A. 360, 361
Virgin Atlantic **52**, 167, 343, 393; SAF and (case study) 403
Virgin Galactic 66
visibility 119–122; dust storm and sandstorm 120; fog 120–122; obstructions 120; precipitation 120
visual flight rules (VFR) 68–69, 120
volcanic ash clouds *see* ash clouds
volcanoes 214–217; ash advisory centres (VAACs) 217; ash detection systems 217; eruptions of Eyjafjallajökull Volcano, Iceland, April 2010 (case study) 217; hazards and air transport 215, *216*; mitigation 217; the 1991 Mount Pinatubo (Philippines) eruptions and their effects on aircraft operations (case study) 215
Vorage, P. 239
Vostok, Antarctica 98

wake vortex turbulence 149, *149*, **150**
Walmsley, S. 83, **87**, 105, 108, 109, 123, 142, 144
Wang, G. 368
Wang, Q. 396
warm front 122, 165, *165*

Warnock-Smith, D. 271, 306
water disposal management 242
water handling capacity at airports *see* airports, water handling capacity
water quality 222, 228, 229, 239, 241, 369
water supply at airports *see* airports, water supply
water, types of 241
water vapour 6, 21, 78, 82–83, **82**, **83**, 94; and air transport 101–126; atmospheric stability 108–110; clouds 110–118; hazards–icing 122–126; hazards–visibility 119–122; humidity 104–108; hydrologic (water) cycle 102–104; precipitation 103, 118–119
waterways 22, 195, 203–206; floatplanes 203; flying boats *204*; seaplanes in the Maldives (case study) 205, *206*
Wegrzyn, N. 391
weighted average cost of capital (WACC) 281, 282, 283
Weiss, R.A. 232
WHO 126, 231
Whyte, R. 46
Wick John O'Groats Airport to Aberdeen International Airport, Scotland–PSO (case study) 331
Wickson, M. **87**, 90, 109, 161
widebody aircraft 49, **51**, 53, 54, 56, 58, 62, 308, 309, 320, 362, 394
wildlife hazard mitigation 227–228; ICAO Doc 9137 Airport Services Manual Part 3 227
wildlife management at airports 223–228; bird strikes 224–227, 228; hazard mitigation 227–228; non-avian wildlife hazards 223–224
Williams Field, Antarctica 199
Williams, G. 267
Williams, P.D. 176, 180
wind and air transport 129–156; direction and speed 130–137; forces affecting 137–140; hazards–thunderstorms 150–153; hazards–turbulence 145–150; local 141–145; types 140–141
wind direction 7, 130–137, 140, 146, 165, 176, 190, 418, 431; crosswinds 134; defining and measuring 130; hazard of tailwinds–crash of Raytheon Hawker 800XP at Aspen Colorado 21 February 2022 134; headwinds 133; multiple/crosswind runway orientation Newark Liberty International Airport New Jersey USA 136, *137*; runway orientation and 134–137; single runway orientation Mariscal Sucre International Airport Quito Ecuador 135, *136*; tailwinds 133; wind vane and windsock 130–131, *132*
wind, forces affecting 137–140; centripetal acceleration 139; Coriolis force/effect 138–139; friction force 139–140; pressure gradient force 137–138

wind, local 141–145; foehn 144–145; land breeze 141–142; mountain 142–143; sea breeze 141–142; valley 142–143
wind rose 135–137; Mariscal Sucre International Airport Quito Ecuador *136*; Newark Liberty International Airport New Jersey USA *137*
wind shear 146, 148, 151, 161, 167, 180
wind speed 130–135; anemometer 130, *132*, Beaufort wind force scale 131, **132**; defining and measuring 130; windsock 131, *132*
wind types 140–141
windsock 131, *132*, 141
winglets 80, 392–393
Wittkowski, K. 392
Wizz Air, Hungary 42, **42**, 267, 294, 299, 350, 352, 406; ancillary revenues (case study) 290
WMO 90, 98, 173
Wolfe, S.A. 144
Wolf's Fang runway, Antarctica 199
Wolfenden, A.D. 413
Woods, M. 324
working age population 343–344, 347
World Bank 16, *17*, *64*, 234, 266, 269, 276, *277*, 314, 315, 324, 341, 343, *383*
World Economic Forum 329, **329**
World Food Programme (WFP) 333
World Health Organization *see* WHO
World Meteorological Organisation *see* WMO
World Population Review 315, 326, 342
world's largest cities 315
Wright Brothers 48, **50**, 129, 189, 190, 225, 393

XF-84H Thunderscreech 418

Yamoussoukro Decision (YD) 269–270
Yan, Y.Y. 142
Yang, F. 373
yaw 80–81
Yesudian, A.N. 179, 238
yield **285**, 286–287, *287*, 291, *292*; declining 286–287; % annual change in, 2005–2023 *287*

Zaniewski, K.J. 338
Zhan, Y. 279
Zhang, Y. 396
Zhao, J. 178
Zheng, S. 389
Zhou, T. 178
Zhou, Y. 178
Zhu, Z.J. 373
Zimmerer, K.S. 4, 18
Zonda winds *see* foehn winds
Zou, L. 48
Zurich Airport, TraNQuiL noise study 413

For Product Safety Concerns and Information please contact our EU representative GPSR@taylorandfrancis.com
Taylor & Francis Verlag GmbH, Kaufingerstraße 24, 80331 München, Germany

www.ingramcontent.com/pod-product-compliance
Ingram Content Group UK Ltd.
Pitfield, Milton Keynes, MK11 3LW, UK
UKHW052217271125
465515UK00012B/232